CLASSICAL MECHANICS
Systems of Particles and Hamiltonian Dynamics

Springer
New York
Berlin
Heidelberg
Hong Kong
London
Milan
Paris
Tokyo

```
SFL
QA805 .G675 2000
c.2
Greiner, Walter, 1935-
Classical mechanics : syste
of particles and Hamiltonia
dynamics
```

THEORETICAL PHYSICS

Greiner
Quantum Mechanics
An Introduction 3rd Edition

Greiner
Quantum Mechanics
Special Chapters

Greiner · Müller
Quantum Mechanics
Symmetries 2nd Edition

Greiner
Relativistic Quantum Mechanics
Wave Equations 2nd Edition

Greiner · Reinhardt
Field Quantization

Greiner · Reinhardt
Quantum Electrodynamics
2nd Edition

Greiner · Schramm · Stein
Quantum Chromodynamics

Greiner · Maruhn
Nuclear Models

Greiner · Müller
Gauge Theory of Weak Interactions
2nd Edition

Greiner
Classical Mechanics
Point Particles and Relativity
(in preparation)

Greiner
Classical Mechanics
Systems of Particles and Hamiltonian Dynamics

Greiner
Classical Electrodynamics

Greiner · Neise · Stöcker
Thermodynamics and Statistical Mechanics

Walter Greiner

CLASSICAL MECHANICS

Systems of Particles and Hamiltonian Dynamics

Foreword by D. Allan Bromley

With 266 Figures

Springer

Walter Greiner
Institut für Theoretische Physik
Johann Wolfgang Goethe-Universität
Robert Mayer Strasse 10
Postfach 11 19 32
D-60054 Frankfurt am Main
Germany
greiner@th.physik.uni-frankfurt.de

Library of Congress Cataloging-in-Publication Data
Greiner, Walter, 1935–
  Classical mechanics: systems of particles and Hamiltonian dynamics/Walter Greiner.
      p. cm.—(Classical theoretical physics)
    Includes bibliographical references and index.
    ISBN 0-387-95128-8 (softcover: alk. paper)
      1. Mechanics, Analytic.   I. Title   II. Series.
  QA805 .G675 2000
  531—dc21   00-059584

ISBN 0-387-95128-8   Printed on acid-free paper.

Translated from the German *Mechanik: Teil 2*, by Walter Greiner, published by Verlag Harri Deutsch, Thun, Frankfurt am Main, Germany, © 1989.

© 2003 Springer-Verlag New York, Inc.
All rights reserved. This work may not be translated or copied in whole or in part without the written permission of the publisher (Springer-Verlag New York, Inc., 175 Fifth Avenue, New York, NY 10010, USA), except for brief excerpts in connection with reviews or scholarly analysis. Use in connection with any form of information storage and retrieval, electronic adaptation, computer software, or by similar or dissimilar methodology now known or hereafter developed is forbidden.
The use in this publication of trade names, trademarks, service marks, and similar terms, even if they are not identified as such, is not to be taken as an expression of opinion as to whether or not they are subject to proprietary rights.

Printed in the United States of America.

9 8 7 6 5 4 3 2 1   SPIN 10778257

www.springer-ny.com

Springer-Verlag   New York Berlin Heidelberg
A member of BertelsmannSpringer Science+Business Media GmbH

Foreword

More than a generation of German-speaking students around the world have worked their way to an understanding and appreciation of the power and beauty of modern theoretical physics—with mathematics, the most fundamental of sciences—using Walter Greiner's textbooks as their guide.

The idea of developing a coherent, complete presentation of an entire field of science in a series of closely related textbooks is not a new one. Many older physicians remember with real pleasure their sense of adventure and discovery as they worked their ways through the classic series by Sommerfeld, by Planck, and by Landau and Lifshitz. From the students' viewpoint, there are a great many obvious advantages to be gained through the use of consistent notation, logical ordering of topics, and coherence of presentation; beyond this, the complete coverage of the science provides a unique opportunity for the author to convey his personal enthusiasm and love for his subject.

These volumes on classical physics, finally available in English, complement Greiner's texts on quantum physics, most of which have been available to English-speaking audiences for some time. The complete set of books will thus provide a coherent view of physics that includes, in classical physics, thermodynamics and statistical mechanics, classical dynamics, electromagnetism, and general relativity; and in quantum physics, quantum mechanics, symmetries, relativistic quantum mechanics, quantum electro- and chromodynamics, and the gauge theory of weak interactions.

What makes Greiner's volumes of particular value to the student and professor alike is their completeness. Greiner avoids the all too common "it follows that...," which conceals several pages of mathematical manipulation and confounds the student. He does not hesitate to include experimental data to illuminate or illustrate a theoretical point, and these data, like the theoretical content, have been kept up to date and topical through frequent revision and expansion of the lecture notes upon which these volumes are based.

Moreover, Greiner greatly increases the value of his presentation by including something like one hundred completely worked examples in each volume. Nothing is of greater importance to the student than seeing, in detail, how the theoretical concepts and tools under study are applied to actual problems of interest to working physicists. And, finally, Greiner adds brief biographical sketches to each chapter covering the people responsible for the development of the theoretical ideas and/or the experimental data presented. It was Auguste Comte (1789–1857) in his *Positive Philosophy* who noted, "To understand a science it is necessary to know its history." This is all too often forgotten in modern physics teaching, and the bridges that Greiner builds to the pioneering figures of our science upon whose work we build are welcome ones.

Greiner's lectures, which underlie these volumes, are internationally noted for their clarity, for their completeness, and for the effort that he has devoted to making physics an integral whole. His enthusiasm for his sciences is contagious and shines through almost every page.

These volumes represent only a part of a unique and Herculean effort to make all of theoretical physics accessible to the interested student. Beyond that, they are of enormous value to the professional physicist and to all others working with quantum phenomena. Again and again, the reader will find that, after dipping into a particular volume to review a specific topic, he or she will end up browsing, caught up by often fascinating new insights and developments with which he or she had not previously been familiar.

Having used a number of Greiner's volumes in their original German in my teaching and research at Yale, I welcome these new and revised English translations and would recommend them enthusiastically to anyone searching for a coherent overview of physics.

D. Allan Bromley
Henry Ford II Professor of Physics
Yale University
New Haven, Connecticut, USA

Preface

Theoretical physics has become a many faceted science. For the young student, it is difficult enough to cope with the overwhelming amount of new material that has to be learned, let alone obtain an overview of the entire field, which ranges from mechanics through electrodynamics, quantum mechanics, field theory, nuclear and heavy-ion science, statistical mechanics, thermodynamics, and solid-state theory to elementary-particle physics; and this knowledge should be acquired in just eight to ten semesters, during which, in addition, a diploma or master's thesis has to be worked on or examinations prepared for. All this can be achieved only if the university teachers help to introduce the student to the new disciplines as early on as possible, in order to create interest and excitement that in turn set free essential new energy.

At the Johann Wolfgang Goethe University in Frankfurt am Main, we therefore confront the student with theoretical physics immediately, in the first semester. Theoretical Mechanics I and II, Electrodynamics, and Quantum Mechanics I—An Introduction are the courses during the first two years. These lectures are supplemented with many mathematical explanations and much support material. After the fourth semester of studies, graduate work begins, and Quantum Mechanics II—Symmetries, Statistical Mechanics and Thermodynamics, Relativistic Quantum Mechanics, Quantum Electrodynamics, Gauge Theory of Weak Interactions, and Quantum Chromodynamics are obligatory. Apart from these, a number of supplementary courses on special topics are offered, such as Hydrodynamics, Classical Field Theory, Special and General Relativity, Many-Body Theories, Nuclear Models, Models of Elementary Particles, and Solid-State Theory.

This volume of lectures, *Classical Mechanics: Systems of Particles and Hamiltonian Dynamics*, deals with the second and more advanced part of the important field of classical mechanics. We have tried to present the subject in a manner that is both interesting to the student and easily accessible. The main text is therefore accompanied by many exercises and examples that have been worked out in great detail. This should make the book useful also for students wishing to study the subject on their own.

Beginning the education in theoretical physics at the first university semester, and not as dictated by tradition after the first one and a half years in the third or fourth semester, has brought along quite a few changes as compared to the traditional courses in that discipline. Especially necessary is a greater amalgamation between the actual physical problems and the necessary mathematics. Therefore, we treat in the first semester vector algebra and analysis, the solution of ordinary, linear differential equations, Newton's mechanics of a mass point, and the mathematically simple mechanics of special relativity.

Many explicitly worked-out examples and exercises illustrate the new concepts and methods and deepen the interrelationship between physics and mathematics. As a matter of fact, the first-semester course in theoretical mechanics is a precursor to theoretical physics. This changes significantly the content of the lectures of the second semester addressed here. Theoretical mechanics is extended to systems of mass points, vibrating strings and membranes, rigid bodies, the spinning top, and the discussion of formal (analytical) aspects of mechanics, that is, Langrange's, Hamilton's formalism, and Hamilton–Jacobi formulation of mechanics. Considered from the mathematical point of view, the new features are partial differential equations, Fourier expansion, and eigenvalue problems. These new tools are explained and exercised in many physical examples. In the lecturing praxis, the deepening of the exhibited material is carried out in a three-hour-per-week *theoretica*, that is, group exercises where eight to ten students solve the given exercises under the guidance of a tutor.

We have added some chapters on modern developments of nonlinear mechanics (dynamical systems, stability of time-dependent orbits, bifurcations, Lyapunov exponents and chaos, systems with chaotic dynamics), being well aware that all this material cannot be taught in a one-semester course. It is meant to stimulate interest in that field and to encourage the students' further (private) studies.

The last chapter is devoted to the history of mechanics. It also contains remarks on the lives and work of outstanding philosophers and scientists who contributed importantly to the development of science in general and mechanics in particular.

Biographical and historical footnotes anchor the scientific development within the general context of scientific progress and evolution. In this context, I thank the publishers Harri Deutsch and F.A. Brockhaus (*Brockhaus Enzyklopädie*, F.A. Brockhaus, Wiesbaden—marked by [BR]) for giving permission to extract the biographical data of physicists and mathematicians from their publications.

We should also mention that in preparing some early sections and exercises of our lectures we relied on the book *Theory and Problems of Theoretical Mechanics*, by Murray R. Spiegel, McGraw-Hill, New York, 1967.

Over the years, we enjoyed the help of several former students and collaborators, in particular, H. Angermüller, P. Bergmann, H. Betz, W. Betz, G. Binnig, J. Briechle, M. Bundschuh, W. Caspar, C. v. Charewski, J. v. Czarnecki, R. Fickler, R. Fiedler, B. Fricke, C. Greiner, M. Greiner, W. Grosch, R. Heuer, E. Hoffmann, L. Kohaupt, N. Krug, P. Kurowski, H. Leber, H.J. Lustig, A. Mahn, B. Moreth, R. Mörschel, B. Müller, H. Müller, H. Peitz, G. Plunien, J. Rafelski, J. Reinhardt, M. Rufa, H. Schaller, D. Schebesta, H.J. Scheefer, H. Schwerin, M. Seiwert, G. Soff, M. Soffel, E. Stein, K.E. Stiebing, E. Stämmler, H. Stock, J. Wagner, and R. Zimmermann. They all made their way in science and society, and meanwhile work as professors at universities, as leaders in industry, and in other places. We particularly acknowledge the recent help of Dr. Sven Soff during the preparation of the English manuscript. The figures were drawn by Mrs. A. Steidl.

The English manuscript was copy-edited by Heather Jones, and the production of the book was supervised by Francine McNeill of Springer-Verlag New York, Inc.

Walter Greiner
Johann Wolfgang Goethe-Universität
Frankfurt am Main, Germany

Contents

Foreword v

Preface vii

Examples xviii

I NEWTONIAN MECHANICS IN MOVING COORDINATE SYSTEMS 1

1 Newton's Equations in a Rotating Coordinate System 3

Introduction of the operator $\widehat{D}$ 7
Formulation of Newton's equation in the rotating coordinate system 7
Newton's equations in systems with arbitrary relative motion 8

2 Free Fall on the Rotating Earth 10

Perturbation calculation 12
Method of successive approximation 14
Exact solution 15

3 Foucault's Pendulum 23

Solution of the differential equations 26
Discussion of the solution 28

II MECHANICS OF PARTICLE SYSTEMS   39

4 Degrees of Freedom   41

Degrees of freedom of a rigid body   41

5 Center of Gravity   43

6 Mechanical Fundamental Quantities of Systems of Mass Points   66

Linear momentum of the many-body system   66
Angular momentum of the many-body system   67
Energy law of the many-body system   69
Transformation to center-of-mass coordinates   72
Transformation of the kinetic energy   73

III VIBRATING SYSTEMS   81

7 Vibrations of Coupled Mass Points   83

The vibrating chain   90

8 The Vibrating String   105

Solution of the wave equation   107
Normal vibrations   109

9 Fourier Series   125

10 The Vibrating Membrane   136

Derivation of the differential equation   136
Solution of the differential equation: Rectangular membrane   138
Inclusion of the boundary conditions   140
Eigenfrequencies   141
Degeneracy   141
Nodal lines   142
General solution (inclusion of the initial conditions)   143
Superposition of node line figures   145
The circular membrane   146
Solution of Bessel's differential equation   149

IV MECHANICS OF RIGID BODIES 163

11 Rotation About a Fixed Axis 165

Moment of inertia (elementary consideration) 166
The physical pendulum 171

12 Rotation About a Point 190

Tensor of inertia 190
Kinetic energy of a rotating rigid body 193
The principal axes of inertia 194
Existence and orthogonality of the principal axes 196
Transformation of the tensor of inertia 200
Tensor of inertia in the system of principal axes 202
Ellipsoid of inertia 203

13 Theory of the Top 216

The free top 216
Geometrical theory of the top 217
Analytical theory of the free top 220
The heavy symmetric top: Elementary considerations 233
Further applications of the top 239
The Euler angles 249
Motion of the heavy symmetric top 253

V LAGRANGE EQUATIONS 269

14 Generalized Coordinates 271

Quantities of mechanics in generalized coordinates 276

15 D'Alembert Principle and Derivation of the Lagrange Equations 279

Virtual displacements 279

16	**Lagrange Equation for Nonholonomic Constraints**	314
17	**Special Problems**	324

Velocity-dependent potentials 324
Nonconservative forces and dissipation function (friction function) 328
Nonholonomic systems and Lagrange multipliers 330

VI HAMILTONIAN THEORY 339

18	**Hamilton's Equations**	341

The Hamilton principle 351
General discussion of variational principles 354
Phase space and Liouville's theorem 364
The principle of stochastic cooling 369

19	**Canonical Transformations**	380
20	**Hamilton–Jacobi Theory**	386

Visual interpretation of the action function S 400
Transition to quantum mechanics 410

VII NONLINEAR DYNAMICS 417

21	**Dynamical Systems**	419

Dissipative systems: Contraction of the phase-space volume 421
Attractors 423
Equilibrium solutions 425
Limit cycles 432

22	**Stability of Time-Dependent Paths**	442

Periodic solutions 443
Discretization and Poincaré cuts 444

| 23 | **Bifurcations** | 452 |

Static bifurcations 452
Bifurcations of time-dependent solutions 457

| 24 | **Lyapunov Exponents and Chaos** | 460 |

One-dimensional systems 460
Multidimensional systems 462
Stretching and folding in phase space 466
Fractal geometry 467

| 25 | **Systems with Chaotic Dynamics** | 475 |

Dynamics of discrete systems 475
One-dimensional mappings 476

VIII ON THE HISTORY OF MECHANICS 513

| 26 | **Emergence of Occidental Physics in the Seventeenth Century** | 515 |

Notes 522
Recommendations for further reading on theoretical mechanics 535

Index 537

Examples

1.1	Angular velocity vector ω	7
1.2	Position vector **r**	7
2.1	Eastward deflection of a falling body	18
2.2	Eastward deflection of a thrown body	18
2.3	Superelevation of a river bank	19
2.4	Difference of sea depth at the pole and equator	20
3.1	Chain fixed to a rotating bar	29
3.2	Pendulum in a moving train	30
3.3	Formation of cyclones	34
3.4	Movable mass in a rotating tube	35
5.1	Center of gravity for a system of three mass points	45
5.2	Center of gravity of a pyramid	45
5.3	Center of gravity of a semicircle	46
5.4	Center of gravity of a circular cone	47
5.5	Momentary center and pole path	49
5.6	Scattering in a central field	51
5.7	Rutherford scattering cross section	56
5.8	Scattering of a particle by a spherical square well potential	60
5.9	Scattering of two atoms	64
6.1	Conservation of the total angular momentum of a many-body system: Flattening of a galaxy	68
6.2	Conservation of angular momentum of a many-body problem: The pirouette	69
6.3	Reduced mass	74
6.4	Movement of two bodies under the action of mutual gravitation	75
6.5	Atwoods fall machine	77
6.6	Our solar system in the Milky Way	78

7.1	Two equal masses coupled by two equal springs	86
7.2	Coupled pendulums	87
7.3	Eigenfrequencies of the vibrating chain	97
7.4	Vibration of two coupled mass points, two dimensional	99
7.5	Three masses on a string	100
7.6	Eigenvibrations of a three-atom molecule	103
8.1	Kinetic and potential energy of a vibrating string	112
8.2	Three different masses equidistantly fixed on a string	114
8.3	Complicated coupled vibrational system	116
8.4	Mathematical supplement: The Cardano formula	118
9.1	Inclusion of the initial conditions for the vibrating string by means of the Fourier expansion	128
9.2	Fourier series of the sawtooth function	129
9.3	Vibrating string with a given velocity distribution	130
9.4	Fourier series for a step function	132
9.5	On the unambiguousness of the tautochrone problem	133
10.1	The longitudinal chain: Poincaré recurrence time	155
10.2	Orthogonality of the eigenmodes	160
11.1	Moment of inertia of a homogeneous circular cylinder	168
11.2	Moment of inertia of a thin rectangular disk	170
11.3	Moment of inertia of a sphere	172
11.4	Moment of inertia of a cube	173
11.5	Vibrations of a suspended cube	173
11.6	Roll off of a cylinder: Rolling pendulum	175
11.7	Moments of inertia of several rigid bodies about selected axes	179
11.8	Cube tilts over the edge of a table	181
11.9	Hockey puck hits a bar	182
11.10	Cue pushes a billiard ball	184
11.11	Motion with constraints	186
11.12	Bar vibrates on springs	187
12.1	Tensor of inertia of a square covered with mass	198
12.2	Transformation of the tensor of inertia of a square covered with mass	206
12.3	Rolling circular top	207
12.4	Ellipsoid of inertia of a quadratic disk	210
12.5	Symmetry axis as a principal axis	211
12.6	Tensor of inertia and ellipsoid of inertia of a system of three masses	212
12.7	Friction forces and acceleration of a car	214
13.1	Nutation of the earth	225
13.2	Ellipsoid of inertia of a regular polyhedron	226
13.3	Rotating ellipsoid	227
13.4	Torque of a rotating plate	228

13.5	Rotation of a vibrating neutron star	229
13.6	Pivot forces of a rotating circular disk	230
13.7	Torque on an elliptic disk	232
13.8	Gyrocompass	240
13.9	Tidal forces, and lunar and solar eclipses: The Saros cycle	241
13.10	The sleeping top	260
13.11	The heavy symmetric top	261
13.12	Stable and unstable rotations of the asymmetric top	267
14.1	Small sphere rolls on a large sphere	272
14.2	Body glides on an inclined plane	273
14.3	Wheel rolls on a plane	273
14.4	Generalized coordinates	275
14.5	Cylinder rolls on an inclined plane	275
14.6	Classification of constraints	276
15.1	Two masses on concentric rollers	282
15.2	Two masses connected by a rope on an inclined plane	283
15.3	Equilibrium condition of a bascule bridge	284
15.4	Two blocks connected by a bar	289
15.5	Ignorable coordinate	291
15.6	Sphere in a rotating tube	293
15.7	Upright pendulum	294
15.8	Stable equilibrium position of an upright pendulum	295
15.9	Vibration frequencies of a three-atom symmetric molecule	297
15.10	Normal frequencies of a triangular molecule	300
15.11	Normal frequencies of an asymmetric linear molecule	302
15.12	Double pendulum	303
15.13	Mass point on a cycloid trajectory	306
15.14	String pendulum	308
15.15	Coupled mass points on a circle	309
15.16	Lagrangian of the asymmetric top	311
16.1	Cylinder rolls down an inclined plane	317
16.2	Particle moves in a paraboloid	319
16.3	Three masses coupled by rods glide in a circular tire	321
17.1	Charged particle in an electromagnetic field	327
17.2	Motion of a projectile in air	330
17.3	Circular disk rolls on a plane	333
17.4	Centrifugal force governor	335
18.1	Central motion	346
18.2	The pendulum in the Newtonian, Lagrangian, and Hamiltonian theories	346
18.3	Hamiltonian and canonical equations of motion	348
18.4	A variational problem	351
18.5	Catenary	355

18.6	Brachistochrone: Construction of an emergency chute	357
18.7	Derivation of the Hamiltonian equations	363
18.8	Phase diagram of a plane pendulum	364
18.9	Phase-space density for particles in the gravitational field	368
18.10	Cooling of a particle beam	374
19.1	Example of a canonical transformation	384
19.2	The harmonic oscillator	384
20.1	The Hamilton–Jacobi differential equation	388
20.2	Angle variable	392
20.3	Solution of the Kepler problem by the Hamilton–Jacobi method	393
20.4	Formulation of the Hamilton–Jacobi differential equation for particle motion in a potential with azimuthal symmetry	395
20.5	Solution of the Hamilton–Jacobi differential equation of Example 20.4	396
20.6	Formulation of the Hamilton–Jacobi differential equation for the slant throw	398
20.7	Illustration of the action waves	401
20.8	Periodic and multiply periodic motions	404
20.9	The Bohr–Sommerfeld hydrogen atom	411
20.10	On Poisson brackets	413
20.11	Total time derivative of an arbitrary function depending on q, p, and t	415
21.1	Linear stability in two dimensions	427
21.2	The nonlinear oscillator with friction	430
21.3	The van der Pol oscillator with weak nonlinearity	438
21.4	Relaxation vibrations	439
22.1	Floquet's theory of stability	447
22.2	Stability of a limit cycle	449
24.1	The baker transformation	473
25.1	The logistic mapping	477
25.2	Logistic mapping and the Bernoulli shift	486
25.3	The periodically kicked rotator	489
25.4	The periodically driven pendulum	496
25.5	Chaos in celestial mechanics: The staggering of Hyperion	503

PART I

NEWTONIAN MECHANICS IN MOVING COORDINATE SYSTEMS

1 Newton's Equations in a Rotating Coordinate System

In classical mechanics, Newton's laws hold in all systems moving uniformly relative to each other (i.e., inertial systems) if they hold in one system. However, this is no longer valid if a system undergoes accelerations. The new relations are obtained by establishing the equations of motion in a fixed system and transforming them into the accelerated system.

We first consider the *rotation* of a coordinate system (x', y', z') about the origin of the inertial system (x, y, z) where the two coordinate origins coincide. The inertial system is denoted by L ("laboratory system") and the rotating system by M ("moving system").

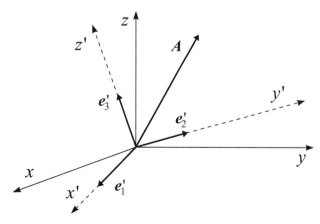

Figure 1.1. Relative position of the coordinate systems x, y, z and x', y', z'.

In the primed system the vector $\mathbf{A}(t) = A'_1 \mathbf{e}'_1 + A'_2 \mathbf{e}'_2 + A'_3 \mathbf{e}'_3$ changes with time. For an observer resting in this system this can be represented as follows:

$$\left.\frac{d\mathbf{A}}{dt}\right|_M = \frac{dA'_1}{dt}\mathbf{e}'_1 + \frac{dA'_2}{dt}\mathbf{e}'_2 + \frac{dA'_3}{dt}\mathbf{e}'_3.$$

The index M means that the derivative is being calculated from the moving system. In the inertial system (x, y, z) $\mathbf{A}$ is also time dependent. Because of the rotation of the primed system the unit vectors $\mathbf{e}'_1, \mathbf{e}'_2, \mathbf{e}'_3$ also vary with time; i.e., when differentiating the vector $\mathbf{A}$ from the inertial system, the unit vectors must be differentiated too:

$$\left.\frac{d\mathbf{A}}{dt}\right|_L = \frac{dA'_1}{dt}\mathbf{e}'_1 + \frac{dA'_2}{dt}\mathbf{e}'_2 + \frac{dA'_3}{dt}\mathbf{e}'_3 + A'_1 \dot{\mathbf{e}}'_1 + A'_2 \dot{\mathbf{e}}'_2 + A'_3 \dot{\mathbf{e}}'_3$$

$$= \left.\frac{d\mathbf{A}}{dt}\right|_M + A'_1 \dot{\mathbf{e}}'_1 + A'_2 \dot{\mathbf{e}}'_2 + A'_3 \dot{\mathbf{e}}'_3.$$

Generally the following holds: $(d/dt)(\mathbf{e}'_\gamma \cdot \mathbf{e}'_\gamma) = \mathbf{e}'_\gamma \cdot \dot{\mathbf{e}}'_\gamma + \dot{\mathbf{e}}'_\gamma \cdot \mathbf{e}'_\gamma = (d/dt)(1) = 0$. Hence, $\mathbf{e}'_\gamma \cdot \dot{\mathbf{e}}'_\gamma = 0$. The derivative of a unit vector $\dot{\mathbf{e}}_\gamma$ is always orthogonal to the vector itself. Therefore the derivative of a unit vector can be written as a linear combination of the two other unit vectors:

$$\dot{\mathbf{e}}'_1 = a_1 \mathbf{e}'_2 + a_2 \mathbf{e}'_3,$$
$$\dot{\mathbf{e}}'_2 = a_3 \mathbf{e}'_1 + a_4 \mathbf{e}'_3,$$
$$\dot{\mathbf{e}}'_3 = a_5 \mathbf{e}'_1 + a_6 \mathbf{e}'_2.$$

Only 3 of these 6 coefficients are independent. To show this, we first differentiate $\mathbf{e}'_1 \cdot \mathbf{e}'_2 = 0$, and obtain

$$\dot{\mathbf{e}}'_1 \cdot \mathbf{e}'_2 = -\dot{\mathbf{e}}'_2 \cdot \mathbf{e}'_1.$$

Multiplying $\dot{\mathbf{e}}'_1 = a_1 \mathbf{e}'_2 + a_2 \mathbf{e}'_3$ by $\mathbf{e}'_2$ and correspondingly $\dot{\mathbf{e}}'_2 = a_3 \mathbf{e}'_1 + a_4 \mathbf{e}'_3$ by $\mathbf{e}'_1$, one obtains

$$\mathbf{e}'_2 \cdot \dot{\mathbf{e}}'_1 = a_1 \quad \text{and} \quad \mathbf{e}'_1 \cdot \dot{\mathbf{e}}'_2 = a_3,$$

and hence $a_3 = -a_1$. Analogously one finds $a_6 = -a_4$ and $a_5 = -a_2$.

The derivative of the vector $\mathbf{A}$ in the inertial system can now be written as follows:

$$\left.\frac{d\mathbf{A}}{dt}\right|_L = \left.\frac{d\mathbf{A}}{dt}\right|_M + A'_1(a_1 \mathbf{e}'_2 + a_2 \mathbf{e}'_3) + A'_2(-a_1 \mathbf{e}'_1 + a_4 \mathbf{e}'_3)$$
$$+ A'_3(-a_2 \mathbf{e}'_1 - a_4 \mathbf{e}'_2)$$
$$= \left.\frac{d\mathbf{A}}{dt}\right|_M + \mathbf{e}'_1(-a_1 A'_2 - a_2 A'_3) + \mathbf{e}'_2(a_1 A'_1 - a_4 A'_3)$$
$$+ \mathbf{e}'_3(a_2 A'_1 + a_4 A'_2).$$

From the evaluation rule for the vector product,

$$\mathbf{C} \times \mathbf{A} = \begin{vmatrix} \mathbf{e}_1' & \mathbf{e}_2' & \mathbf{e}_3' \\ C_1 & C_2 & C_3 \\ A_1' & A_2' & A_3' \end{vmatrix}$$
$$= \mathbf{e}_1'(C_2 A_3' - C_3 A_2') - \mathbf{e}_2'(C_1 A_3' - C_3 A_1') + \mathbf{e}_3'(C_1 A_2' - C_2 A_1'),$$

it follows by setting $\mathbf{C} = (a_4, -a_2, a_1)$ that

$$\left.\frac{d\mathbf{A}}{dt}\right|_L = \left.\frac{d\mathbf{A}}{dt}\right|_M + \mathbf{C} \times \mathbf{A}.$$

We still have to show the physical meaning of the vector $\mathbf{C}$. For this purpose we consider the special case $d\mathbf{A}/dt|_M = 0$; i.e., the derivative of the vector $\mathbf{A}$ in the moving system vanishes. $\mathbf{A}$ moves (rotates) with the moving system; it is tightly "mounted" in the system. Let φ be the angle between the axis of rotation (in our special case the z-axis) and $\mathbf{A}$. The component parallel to the angular velocity $\boldsymbol{\omega}$ is not changed by the rotation.

The change of $\mathbf{A}$ in the laboratory system is then given by

$$dA = \omega\, dt\, A \sin\varphi \qquad \text{or} \qquad \left.\frac{dA}{dt}\right|_L = \omega A \sin\varphi.$$

This can also be written as

$$\left.\frac{d\mathbf{A}}{dt}\right|_L = \boldsymbol{\omega} \times \mathbf{A}.$$

The orientation of $(\boldsymbol{\omega} \times \mathbf{A})dt$ also coincides with $d\mathbf{A}$ (see Figure 1.2). Since the (fixed)

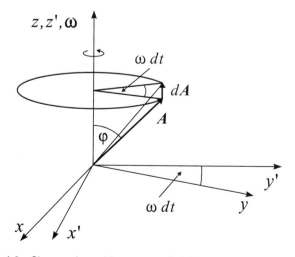

Figure 1.2. Change of an arbitrary vector **A** tightly fixed to a rotating system.

vector **A** can be chosen arbitrarily, the vector **C** must be identical with the angular velocity $\boldsymbol{\omega}$ of the rotating system M. By insertion we obtain

$$\left.\frac{d\mathbf{A}}{dt}\right|_L = \left.\frac{d\mathbf{A}}{dt}\right|_M + \boldsymbol{\omega} \times \mathbf{A}. \tag{1.1}$$

This can also be seen as follows (see Figure 1.3): If the rotational axis of the primed system coincides during a time interval dt with one of the coordinate axes of the nonprimed system, e.g., $\boldsymbol{\omega} = \dot{\varphi}\mathbf{e}_3$, then

$$\dot{\mathbf{e}}_1' = \dot{\varphi}\mathbf{e}_2' \quad \text{and} \quad \dot{\mathbf{e}}_2' = -\dot{\varphi}\mathbf{e}_1',$$

i.e.,

$$a_1 = \dot{\varphi}, \quad a_2 = a_4 = 0, \quad \text{and hence} \quad \mathbf{C} = \dot{\varphi}\mathbf{e}_3' = \boldsymbol{\omega}.$$

In the general case $\boldsymbol{\omega} = \omega_1\mathbf{e}_1 + \omega_2\mathbf{e}_2 + \omega_3\mathbf{e}_3$, one decomposes $\boldsymbol{\omega} = \sum \boldsymbol{\omega}_i$ with $\boldsymbol{\omega}_i = \omega_i\mathbf{e}_i$, and by the preceding consideration one finds

$$\mathbf{C}_i = \boldsymbol{\omega}_i; \quad \text{i.e.,} \quad \mathbf{C} = \sum_i \mathbf{C}_i = \sum_i \boldsymbol{\omega}_i = \boldsymbol{\omega}.$$

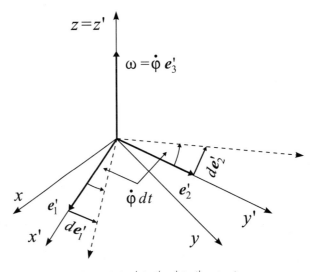

Figure 1.3. $|d\mathbf{e}_1'| = |d\mathbf{e}_2'| = \dot{\varphi} \cdot dt$.

Introduction of the operator $\widehat{D}$

To shorten the expression $\partial F(x,\ldots,t)/\partial t = \partial F/\partial t$, we introduce the operator $\widehat{D} = \partial/\partial t$. The inertial system and the accelerated system will be distinguished by the indices L and M, so that

$$\widehat{D}_L = \left.\frac{\partial}{\partial t}\right|_L \quad \text{and} \quad \widehat{D}_M = \left.\frac{\partial}{\partial t}\right|_M.$$

The equation $\left.\dfrac{d\mathbf{A}}{dt}\right|_L = \left.\dfrac{d\mathbf{A}}{dt}\right|_M + \boldsymbol{\omega} \times \mathbf{A}$ then simplifies to

$$\widehat{D}_L \mathbf{A} = \widehat{D}_M \mathbf{A} + \boldsymbol{\omega} \times \mathbf{A}.$$

If the vector $\mathbf{A}$ is omitted, the equation is called an operator equation

$$\widehat{D}_L = \widehat{D}_M + \boldsymbol{\omega} \times,$$

which can operate on arbitrary vectors.

Example 1.1: Angular velocity vector ω

$$\left.\frac{d\boldsymbol{\omega}}{dt}\right|_L = \left.\frac{d\boldsymbol{\omega}}{dt}\right|_M + \boldsymbol{\omega} \times \boldsymbol{\omega}.$$

Since $\boldsymbol{\omega} \times \boldsymbol{\omega} = 0$, it follows that

$$\left.\frac{d\boldsymbol{\omega}}{dt}\right|_L = \left.\frac{d\boldsymbol{\omega}}{dt}\right|_M.$$

These two derivatives are evidently identical for all vectors that are parallel to the rotational plane, since then the vector product vanishes.

Example 1.2: Position vector r

$$\left.\frac{d\mathbf{r}}{dt}\right|_L = \left.\frac{d\mathbf{r}}{dt}\right|_M + \boldsymbol{\omega} \times \mathbf{r},$$

in operator notation this becomes

$$\widehat{D}_L \mathbf{r} = \widehat{D}_M \mathbf{r} + \boldsymbol{\omega} \times \mathbf{r},$$

where $(d\mathbf{r}/dt)|_M$ is called the *virtual* velocity and $(d\mathbf{r}/dt)|_M + \boldsymbol{\omega} \times \mathbf{r}$ the *true* velocity. The term $\boldsymbol{\omega} \times \mathbf{r}$ is called the *rotational velocity*.

Formulation of Newton's equation in the rotating coordinate system

Newton's law $m\ddot{\mathbf{r}} = \mathbf{F}$ holds only in the inertial system. In accelerated systems, there appear additional terms. First we consider again a pure rotation.

For the acceleration we have

$$\ddot{\mathbf{r}}_L = \frac{d}{dt}(\dot{\mathbf{r}})_L = \widehat{D}_L(\widehat{D}_L \mathbf{r}) = (\widehat{D}_M + \boldsymbol{\omega}\times)(\widehat{D}_M \mathbf{r} + \boldsymbol{\omega} \times \mathbf{r})$$
$$= \widehat{D}_M^2 \mathbf{r} + \widehat{D}_M(\boldsymbol{\omega} \times \mathbf{r}) + \boldsymbol{\omega} \times \widehat{D}_M \mathbf{r} + \boldsymbol{\omega} \times (\boldsymbol{\omega} \times \mathbf{r})$$
$$= \widehat{D}_M^2 \mathbf{r} + (\widehat{D}_M \boldsymbol{\omega}) \times \mathbf{r} + 2\boldsymbol{\omega} \times \widehat{D}_M \mathbf{r} + \boldsymbol{\omega} \times (\boldsymbol{\omega} \times \mathbf{r}).$$

We replace the operator by the differential quotient:

$$\left.\frac{d^2\mathbf{r}}{dt^2}\right|_L = \left.\frac{d^2\mathbf{r}}{dt^2}\right|_M + \left.\frac{d\boldsymbol{\omega}}{dt}\right|_M \times \mathbf{r} + 2\boldsymbol{\omega} \times \left.\frac{d\mathbf{r}}{dt}\right|_M + \boldsymbol{\omega} \times (\boldsymbol{\omega} \times \mathbf{r}). \tag{1.2}$$

The expression $(d\boldsymbol{\omega}/dt)|_M \times \mathbf{r}$ is called the *linear acceleration*, $2\boldsymbol{\omega} \times (d\mathbf{r}/dt)|_M$ the *Coriolis acceleration*, and $\boldsymbol{\omega} \times (\boldsymbol{\omega} \times \mathbf{r})$ the *centripetal acceleration*.

Multiplication by the mass m yields the force $\mathbf{F}$:

$$m\left.\frac{d^2\mathbf{r}}{dt^2}\right|_M + m\left.\frac{d\boldsymbol{\omega}}{dt}\right|_M \times \mathbf{r} + 2m\boldsymbol{\omega} \times \left.\frac{d\mathbf{r}}{dt}\right|_M + m\boldsymbol{\omega} \times (\boldsymbol{\omega} \times \mathbf{r}) = \mathbf{F}.$$

The basic equation of mechanics in the rotating coordinate system therefore reads (with the index M being omitted):

$$m\frac{d^2\mathbf{r}}{dt^2} = \mathbf{F} - m\frac{d\boldsymbol{\omega}}{dt} \times \mathbf{r} - 2m\boldsymbol{\omega} \times \mathbf{v} - m\boldsymbol{\omega} \times (\boldsymbol{\omega} \times \mathbf{r}). \tag{1.3}$$

The additional terms on the right-hand side of equation (1.3) are *virtual forces* of a dynamical nature, but actually they are due to the acceleration term. For experiments on the earth the additional terms can often be neglected, since the angular velocity of the earth $\omega = 2\pi/T$ ($T = 24$ h) is only $7.27 \cdot 10^{-5}$ s^{-1}.

Newton's equations in systems with arbitrary relative motion

We now drop the condition that the origins of the two coordinate systems coincide. The general motion of a coordinate system is composed of a rotation of the system and a translation of the origin. If $\mathbf{R}$ points to the origin of the primed system, then the position vector in the nonprimed system is $\mathbf{r} = \mathbf{R} + \mathbf{r}'$.

For the velocity we have $\dot{\mathbf{r}} = \dot{\mathbf{R}} + \dot{\mathbf{r}}'$, and in the inertial system we have as before

$$m\left.\frac{d^2\mathbf{r}}{dt^2}\right|_L = \mathbf{F}|_L = \mathbf{F}.$$

By inserting $\mathbf{r}$ and differentiating, we obtain

$$m\left.\frac{d^2\mathbf{r}'}{dt^2}\right|_L + m\left.\frac{d^2\mathbf{R}}{dt^2}\right|_L = \mathbf{F}.$$

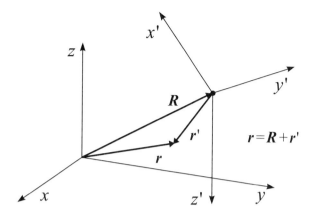

Figure 1.4. Relative position of the coordinate systems x, y, z and x', y', z'.

The transition to the accelerated system is performed as above (equation (1.3)), but here we still have the additional term $m\ddot{\mathbf{R}}$:

$$m\frac{d^2\mathbf{r}'}{dt^2}\bigg|_M = \mathbf{F} - m\frac{d^2\mathbf{R}}{dt^2}\bigg|_L - m\frac{d\boldsymbol{\omega}}{dt}\bigg|_M \times \mathbf{r}' - 2m\boldsymbol{\omega}\times\mathbf{v}_M - m\boldsymbol{\omega}\times(\boldsymbol{\omega}\times\mathbf{r}'). \tag{1.4}$$

2 Free Fall on the Rotating Earth

On the earth, the previously derived form of the basic equation of mechanics holds if we neglect the rotation about the sun and therefore consider a coordinate system at the earth center as an inertial system.

$$m\ddot{\mathbf{r}}'|_M = \mathbf{F} - m\ddot{\mathbf{R}}|_L - m\dot{\boldsymbol{\omega}} \times \mathbf{r}'|_M - 2m\boldsymbol{\omega} \times \dot{\mathbf{r}}'|_M - m\boldsymbol{\omega} \times (\boldsymbol{\omega} \times \mathbf{r}'). \tag{2.1}$$

The rotational velocity $\boldsymbol{\omega}$ of the earth about its axis can be considered constant in time; therefore, $m\dot{\boldsymbol{\omega}} \times \mathbf{r}' = 0$.

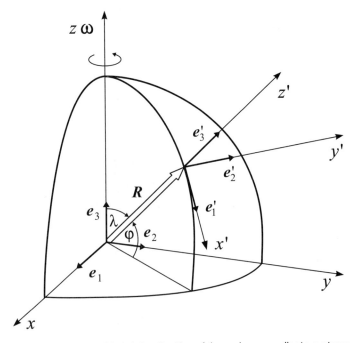

Figure 2.1. Octant of the globe: Position of the various coordinate systems.

The motion of the point **R**, i.e., the motion of the coordinate origin of the system (x', y', z'), still has to be recalculated in the moving system. According to (2.1), we have

$$\ddot{\mathbf{R}}\big|_L = \ddot{\mathbf{R}}\big|_M + \dot{\boldsymbol{\omega}}\big|_M \times \mathbf{R} + 2\boldsymbol{\omega} \times \dot{\mathbf{R}}\big|_M + \boldsymbol{\omega} \times (\boldsymbol{\omega} \times \mathbf{R}).$$

Since **R** as seen from the moving system is a time-independent quantity and since $\boldsymbol{\omega}$ is constant, this equation finally reads

$$\ddot{\mathbf{R}}\big|_L = \boldsymbol{\omega} \times (\boldsymbol{\omega} \times \mathbf{R}).$$

This is the centripetal acceleration due to the earth's rotation that acts on a body moving on the earth's surface. For the force equation (2.1) one gets

$$m\ddot{\mathbf{r}}' = \mathbf{F} - m\boldsymbol{\omega} \times (\boldsymbol{\omega} \times \mathbf{R}) - 2m\boldsymbol{\omega} \times \dot{\mathbf{r}}' - m\boldsymbol{\omega} \times (\boldsymbol{\omega} \times \mathbf{r}').$$

Hence, in free fall on the earth—contrary to the inertial system—there appear virtual forces that deflect the body in the x'- and y'-directions.

If only gravity acts, the force **F** in the inertial system is $\mathbf{F} = -\gamma M m \mathbf{r}/r^3$. By insertion we obtain

$$m\ddot{\mathbf{r}}' = -\gamma \frac{Mm}{r^3}\mathbf{r} - m\boldsymbol{\omega} \times (\boldsymbol{\omega} \times \mathbf{R}) - 2m\boldsymbol{\omega} \times \dot{\mathbf{r}}' - m\boldsymbol{\omega} \times (\boldsymbol{\omega} \times \mathbf{r}').$$

We now introduce the experimentally determined value for the gravitational acceleration **g**:

$$\mathbf{g} = -\gamma \frac{M}{R^3}\mathbf{R} - \boldsymbol{\omega} \times (\boldsymbol{\omega} \times \mathbf{R}).$$

Here we have inserted in the gravitational acceleration $-\gamma M\mathbf{r}/r^3$ the radius $\mathbf{r} = \mathbf{R} + \mathbf{r}'$ and kept the approximation $\mathbf{r} \approx \mathbf{R}$, which is reasonable near the earth's surface. The second term is the centripetal acceleration due to the earth's rotation, which leads to a decrease of the gravitational acceleration (as a function of the geographical latitude). The reduction is included in the experimental value for **g**. We thus obtain

$$m\ddot{\mathbf{r}}' = m\mathbf{g} - 2m\boldsymbol{\omega} \times \dot{\mathbf{r}}' - m\boldsymbol{\omega} \times (\boldsymbol{\omega} \times \mathbf{r}').$$

In the vicinity of the earth's surface ($r' \ll R$) the last term can be neglected, since ω^2 enters and $|\omega|$ is small compared to $1/s$. Thus the equation simplifies to

$$\ddot{\mathbf{r}}' = \mathbf{g} - 2(\boldsymbol{\omega} \times \dot{\mathbf{r}}') \quad \text{or} \quad \ddot{\mathbf{r}}' = -g\mathbf{e}_3' - 2(\boldsymbol{\omega} \times \dot{\mathbf{r}}'). \tag{2.2}$$

The vector equation is solved by decomposing it into its components. First one suitably evaluates the vector product. From Figure 2.1 one obtains, with $\mathbf{e}_1, \mathbf{e}_2, \mathbf{e}_3$ the unit vectors of the inertial system and $\mathbf{e}_1', \mathbf{e}_2', \mathbf{e}_3'$ the unit vectors of the moving system, the following relation:

$$\begin{aligned}\mathbf{e}_3 &= (\mathbf{e}_3 \cdot \mathbf{e}_1')\mathbf{e}_1' + (\mathbf{e}_3 \cdot \mathbf{e}_2')\mathbf{e}_2' + (\mathbf{e}_3 \cdot \mathbf{e}_3')\mathbf{e}_3' \\ &= (-\sin\lambda)\mathbf{e}_1' + 0\,\mathbf{e}_2' + (\cos\lambda)\mathbf{e}_3'.\end{aligned}$$

Because $\boldsymbol{\omega} = \omega\mathbf{e}_3$, one gets the component representation of $\boldsymbol{\omega}$ in the moving system:

$$\boldsymbol{\omega} = -\omega\sin\lambda\,\mathbf{e}_1' + \omega\cos\lambda\,\mathbf{e}_3'.$$

Then for the vector product we get

$$\boldsymbol{\omega} \times \dot{\mathbf{r}}' = (-\omega \dot{y}' \cos \lambda) \mathbf{e}_1' + (\dot{z}' \omega \sin \lambda + \dot{x}' \omega \cos \lambda) \mathbf{e}_2' - (\omega \dot{y}' \sin \lambda) \mathbf{e}_3'.$$

The vector equation (2.2) can now be decomposed into the following three component equations:

$$\begin{aligned} \ddot{x}' &= 2\dot{y}' \omega \cos \lambda, \\ \ddot{y}' &= -2\omega (\dot{z}' \sin \lambda + \dot{x}' \cos \lambda), \\ \ddot{z}' &= -g + 2\omega \dot{y}' \sin \lambda. \end{aligned} \quad (2.3)$$

This is a system of three coupled differential equations with ω as the coupling parameter. For $\omega = 0$, we get the free fall in an inertial system. The solution of such a system can also be obtained in an analytical way. It is, however, useful to learn various approximation methods from this example. We will first outline these methods and then work out the exact analytical solution and compare it with the approximations.

In the present case, the *perturbation calculation* and the method of *successive approximation* offer themselves as approximations. Both of these methods will be presented here. The primes on the coordinates will be omitted below.

Perturbation calculation

Here one starts from a system that is mathematically more tractable, and then one accounts for the forces due to the perturbation which are small compared to the remaining forces of the system.

We first integrate the equations (2.3):

$$\begin{aligned} \dot{x} &= 2\omega y \cos \lambda + c_1, \\ \dot{y} &= -2\omega (x \cos \lambda + z \sin \lambda) + c_2, \\ \dot{z} &= -gt + 2\omega y \sin \lambda + c_3. \end{aligned} \quad (2.4)$$

In free fall on the earth the body is released from the height h at time $t = 0$; i.e., for our problem, the initial conditions are

$$\begin{aligned} z(0) &= h, & \dot{z}(0) &= 0, \\ y(0) &= 0, & \dot{y}(0) &= 0, \\ x(0) &= 0, & \dot{x}(0) &= 0. \end{aligned}$$

From this we get the integration constants

$$c_1 = 0, \quad c_2 = 2\omega h \sin \lambda, \quad c_3 = 0,$$

and obtain

$$\begin{aligned} \dot{x} &= 2\omega y \cos \lambda, \\ \dot{y} &= -2\omega (x \cos \lambda + (z - h) \sin \lambda), \end{aligned} \quad (2.5)$$

PERTURBATION CALCULATION

$$\dot{z} = -gt + 2\omega y \sin \lambda.$$

The terms proportional to ω are small compared to the term gt. They represent the perturbation. The deviation y from the origin of the moving system is a function of ω and t; i.e., in the first approximation the term $y_1(\omega, t) \sim \omega$ appears. Inserting this into the first differential equation, we find an expression involving ω^2. Because of the consistency in ω we can neglect all terms with ω^2, i.e., we obtain to first order in ω

$$\dot{x}(t) = 0, \qquad \dot{z}(t) = -gt,$$

and after integration with the initial conditions we get

$$x(t) = 0, \qquad z(t) = -\frac{g}{2}t^2 + h.$$

Because $x(t) = 0$, in this approximation the term $2\omega x \cos \lambda$ drops out from the second differential equation (2.5); there remains

$$\dot{y} = -2\omega(z - h) \sin \lambda.$$

Inserting z leads to

$$\dot{y} = -2\omega \left(h - \frac{1}{2}gt^2 - h \right) \sin \lambda$$
$$= \omega g t^2 \sin \lambda.$$

Integration with the initial condition yields

$$y = \frac{\omega g \sin \lambda}{3} t^3.$$

The solutions of the system of differential equations in the approximation $\omega^n = 0$ with $n \geq 2$ (i.e., consistent up to linear terms in ω) thus read

$$x(t) = 0,$$
$$y(t) = \frac{\omega g \sin \lambda}{3} t^3,$$
$$z(t) = h - \frac{g}{2} t^2.$$

The fall time T is obtained from $z(t = T) = 0$:

$$T^2 = \frac{2h}{g}.$$

From this one finds the *eastward deflection* ($\mathbf{e}_2'$ points east) as a function of the fall height:

$$y(t = T) = y(h) = \frac{\omega g \sin \lambda 2h}{3g} \sqrt{\frac{2h}{g}}$$
$$= \frac{2\omega h \sin \lambda}{3} \sqrt{\frac{2h}{g}}.$$

Method of successive approximation

If one starts from the known system (2.5) of coupled differential equations, these equations can be transformed by integration to integral equations:

$$x(t) = 2\omega \cos \lambda \int_0^t y(u)\, du + c_1,$$

$$y(t) = 2\omega h t \sin \lambda - 2\omega \cos \lambda \int_0^t x(u)\, du - 2\omega \sin \lambda \int_0^t z(u)\, du + c_2,$$

$$z(t) = -\frac{1}{2} g t^2 + 2\omega \sin \lambda \int_0^t y(u)\, du + c_3.$$

Taking into account the initial conditions

$x(0) = 0,$ $\quad \dot{x}(0) = 0,$
$y(0) = 0,$ $\quad \dot{y}(0) = 0,$
$z(0) = h,$ $\quad \dot{z}(0) = 0,$

the integration constants are

$c_1 = 0, \quad c_2 = 0, \quad c_3 = h.$

The iteration method is based on replacing the functions $x(u), y(u), z(u)$ under the integral sign by appropriate initial functions. In the first approximation, one determines the functions $x(t), y(t), z(t)$ and then inserts them as $x(u), y(u), z(u)$ on the right-hand side to get the second approximation. In general there results a *successive approximation to the exact solution* if $\omega \cdot t = 2\pi t/T$ ($T = 24$ hours) is sufficiently small.

By setting $x(u), y(u), z(u)$ to zero in the above example in the zero-order approximation, one obtains in the first approximation

$x^{(1)}(t) = 0,$
$y^{(1)}(t) = 2\omega h t \sin \lambda,$
$z^{(1)}(t) = h - \frac{g}{2} t^2.$

To check the consistency of these solutions up to terms linear in ω, we have to check only the second approximation. If there is consistency, there must *not* appear terms that involve ω linearly:

$$x^{(2)}(t) = 2\omega \cos \lambda \int_0^t y^{(1)}(u)\, du = 2\omega \cos \lambda \int_0^t 2\omega h (\sin \lambda) u\, du$$

$$= 4\omega^2 h \cos \lambda \sin \lambda \frac{t^2}{2} = f(\omega^2) \approx 0.$$

Like $x^{(1)}(t)$, $z^{(1)}(t)$ is consistent to first order in ω:

$$z^{(2)}(t) = h - \frac{1}{2}gt^2 + 2\omega \sin \lambda \int_0^t y^{(1)}(u)\, du$$

$$= h - \frac{g}{2}t^2 + 2\omega \sin \lambda \int_0^t 2\omega h (\sin \lambda) u\, du$$

$$= h - \frac{g}{2}t^2 + i(\omega^2).$$

On the contrary, $y^{(1)}(t)$ is not consistent in ω, since

$$y^{(2)}(t) = 2\omega h t \sin \lambda - 2\omega \cos \lambda \int_0^t x^{(1)}(u)\, du - 2\omega \sin \lambda \int_0^t z^{(1)}(u)\, du$$

$$= 2\omega h \sin \lambda \cdot t - 2\omega h \sin \lambda \cdot t + g\omega (\sin \lambda) \frac{t^3}{3}$$

$$= g\omega (\sin \lambda) \frac{t^3}{3} \neq 2\omega h (\sin \lambda) t + k(\omega^2).$$

We see that in this second step the terms linear in ω once again changed greatly. The term $2\omega h t \sin \lambda$ obtained in the first iteration step cancels completely and is finally replaced by $g\omega (\sin \lambda) t^3 / 3$. A check of $y^{(3)}(t)$ shows that $y^{(2)}(t)$ is consistent up to first order in ω.

Just as in the perturbation method discussed above, we get up to first order in ω the solution

$$x(t) = 0,$$
$$y(t) = \frac{g\omega \sin \lambda}{3} t^3,$$
$$z(t) = h - \frac{g}{2}t^2.$$

We have of course noted long ago that the method of successive approximation (iteration) is equivalent to the perturbation calculation and basically represents its conceptually clean formulation.

Exact solution

The equations of motion (2.3) can also be solved exactly. For that purpose, we start again from

$$\ddot{x} = 2\omega \cos \lambda \dot{y}, \tag{2.3a}$$

$$\ddot{y} = -2\omega (\sin \lambda \dot{z} + \cos \lambda \dot{x}), \tag{2.3b}$$

$$\ddot{z} = -g + 2\omega \sin \lambda \dot{y}. \tag{2.3c}$$

By integrating (2.3a) to (2.3c) with the above initial conditions, one gets

$$\dot{x} = 2\omega \cos \lambda y, \tag{2.5a}$$

$$\dot{y} = -2\omega(\sin \lambda z + \cos \lambda x) + 2\omega \sin \lambda h, \tag{2.5b}$$

$$\dot{z} = -gt + 2\omega \sin \lambda y. \tag{2.5c}$$

Insertion of (2.5a) and (2.5c) into (2.3b) yields

$$\ddot{y} + 4\omega^2 y = 2\omega g \sin \lambda t \equiv ct. \tag{2.6}$$

The general solution of (2.6) is the general solution of the homogeneous equation and *one* particular solution of the inhomogeneous equation, i.e.,

$$y = \frac{c}{4\omega^2}t + A \sin 2\omega t + B \cos 2\omega t.$$

The initial conditions at the time $t = 0$ are $x = y = 0$, $z = h$, and $\dot{x} = \dot{y} = \dot{z} = 0$. It follows that $B = 0$ and $2\omega A = -c/4\omega^2$, i.e., $A = -c/8\omega^3$ and therefore

$$y = \frac{c}{4\omega^2}t - \frac{c}{8\omega^3}\sin 2\omega t = \frac{c}{4\omega^2}\left(t - \frac{\sin 2\omega t}{2\omega}\right),$$

i.e.,

$$y = \frac{g \sin \lambda}{2\omega}\left(t - \frac{\sin 2\omega t}{2\omega}\right). \tag{2.7}$$

Insertion of (2.7) into (2.5a) yields

$$\dot{x} = g \sin \lambda \cos \lambda \left(t - \frac{\sin 2\omega t}{2\omega}\right).$$

From the initial conditions, it follows that

$$x = g \sin \lambda \cos \lambda \left(\frac{t^2}{2} - \frac{1 - \cos 2\omega t}{4\omega^2}\right). \tag{2.8}$$

Equation (2.7) inserted into (2.5c) yields

$$\dot{z} = -gt + 2\omega \sin \lambda \left[\frac{g \sin \lambda}{2\omega}\left(t - \frac{\sin 2\omega t}{2\omega}\right)\right],$$

$$\dot{z} = -gt + g \sin^2 \lambda \left(t - \frac{\sin 2\omega t}{2\omega}\right),$$

and integration with the initial conditions yields

$$z = -\frac{g}{2}t^2 + g \sin^2 \lambda \left(\frac{t^2}{2} - \frac{1 - \cos 2\omega t}{4\omega^2}\right) + h. \tag{2.9}$$

Summarizing, one finally has

$$x = g \sin \lambda \cos \lambda \left(\frac{t^2}{2} - \frac{1 - \cos 2\omega t}{4\omega^2}\right),$$

$$y = \frac{g \sin \lambda}{2\omega} \left(t - \frac{\sin 2\omega t}{2\omega} \right), \tag{2.10}$$

$$z = h - \frac{g}{2}t^2 + g \sin^2 \lambda \left(\frac{t^2}{2} - \frac{1 - \cos 2\omega t}{4\omega^2} \right).$$

Since $\omega t = 2\pi$ fall time/1 day, i.e., very small ($\omega t \ll 1$), one can expand (2.10):

$$x = \frac{gt^2}{6} \sin \lambda \cos \lambda (\omega t)^2,$$

$$y = \frac{gt^2}{3} \sin \lambda (\omega t), \tag{2.11}$$

$$z = h - \frac{gt^2}{2} \left(1 - \frac{\sin^2 \lambda}{3} (\omega t^2) \right).$$

If one considers only terms of first order in ωt, then $(\omega t)^2 \approx 0$, and (2.11) becomes

$$x(t) = 0,$$

$$y(t) = \frac{g\omega t^3 \sin \lambda}{3}, \tag{2.12}$$

$$z(t) = h - \frac{g}{2}t^2.$$

This is identical with the results obtained by means of perturbation theory. However, (2.10) is *exact*!

The eastward deflection of a falling mass seems at first paradoxical, since the earth rotates toward the east too. However, it becomes transparent if one considers that the mass in the height h at the time $t = 0$ in the *inertial system* has a larger velocity component toward the east (due to the earth rotation) than an observer on the earth's surface. It is just this "excessive" velocity toward the east which for an observer on the earth lets the stone fall toward the east, but not $\perp$ downward. For the throw upward the situation is the opposite (see problem 2.2).

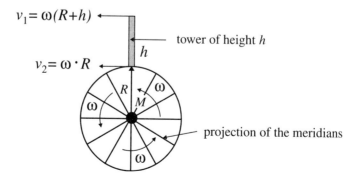

Figure 2.2. Cut through the earth in the equatorial plane viewed from the North Pole: *M* is the earth center, and ω the angular velocity.

Example 2.1: Eastward deflection of a falling body

As an example, we calculate the eastward deflection of a body that falls at the equator from a height of 400 m.

The eastward deflection of a body falling from the height h is given by

$$y(h) = \frac{2\omega \sin \lambda h}{3} \sqrt{\frac{2h}{g}}.$$

The height $h = 400$ m, the angular velocity of the earth $\omega = 7.27 \cdot 10^{-5}$ rad s^{-1}, and the gravitational acceleration is known.

Inserting the values in $y(h)$ yields

$$y(h) = \frac{2 \cdot 7.27 \cdot 400 \text{ rad m}}{3 \cdot 10^5 \text{ s}} \sqrt{\frac{2 \cdot 400 \text{ s}^2}{9.81}},$$

where rad is a dimensionless quantity. The result is

$$y(h) = 17.6 \text{ cm}.$$

Thus, the body will be deflected toward the east by 17.6 cm.

Example 2.2: Exercise: Eastward deflection of a thrown body

An object will be thrown upward with the initial velocity v_0. Find the eastward deflection.

Solution If we put the coordinate system at the starting point of the motion, the initial conditions read

$$z(t = 0) = 0, \quad \dot{z}(t = 0) = v_0,$$
$$y(t = 0) = 0, \quad \dot{y}(t = 0) = 0,$$
$$x(t = 0) = 0, \quad \dot{x}(t = 0) = 0.$$

The deflection to the east is given by y, the deflection to the south by x; $z \neq 0$ denotes the height h above the earth's surface.

For the motion in y-direction we have, as has been shown (see equation (2.4)),

$$\frac{dy}{dt} = -2\omega(x \cos \lambda + z \sin \lambda) + C_2.$$

The motion of the body in x-direction can be neglected; $x \approx 0$. If one further neglects the influence of the eastward deflection on z, one immediately arrives at the equation

$$z = -\frac{g}{2}t^2 + v_0 t,$$

which is already known from the treatment of the free fall without accounting for the earth's rotation. Insertion into the above differential equation yields

$$\frac{dy}{dt} = 2\omega \left(\frac{g}{2}t^2 - v_0 t\right) \sin \lambda,$$

$$y(t) = 2\omega \left(\frac{g}{6}t^3 - \frac{v_0}{2}t^2\right) \sin \lambda.$$

EXACT SOLUTION

At the turning point (after the time of ascent $T = v_0/g$), the deflection is

$$y(T) = -\frac{2}{3}\omega \sin\lambda \frac{v_0^3}{g^2}.$$

It points toward the west, as expected.

Example 2.3: Exercise: Superelevation of a river bank

A river of width D flows on the northern hemisphere at the geographical latitude φ toward the north with a flow velocity v_0. By which amount is the right bank higher than the left one?

Evaluate the numerical example $D = 2\,\text{km}$, $v_0 = 5\,\text{km/h}$, and $\varphi = 45°$.

Solution For the earth, we have $m\dfrac{d^2\mathbf{r}}{dt^2} = -mg\mathbf{e}_3' - 2m\boldsymbol{\omega} \times \mathbf{v}$ with

$$\boldsymbol{\omega} = -\omega \sin\lambda\,\mathbf{e}_1' + \omega \cos\lambda\,\mathbf{e}_3'.$$

The flow velocity is $\mathbf{v} = -v_0\mathbf{e}_1'$, and hence,

$$\boldsymbol{\omega} \times \mathbf{v} = -\omega v_0 \sin\varphi\,\mathbf{e}_2'.$$

Then the force is

$$m\ddot{\mathbf{r}} = \mathbf{F} = -mg\mathbf{e}_3' + 2m\omega v_0 \sin\varphi\,\mathbf{e}_2' = F_3\mathbf{e}_3' + F_2\mathbf{e}_2'.$$

$\mathbf{F}$ must be perpendicular to the water surface (see Figure 2.3). With the magnitude of the force

$$F = \sqrt{4m^2\omega^2 v_0^2 \sin^2\varphi + m^2g^2}$$

one can, from Figure 2.3, determine $H = D\sin\alpha$ and $\sin\alpha = F_2/F$. For the desired height H one obtains

$$H = D\frac{2\omega v_0 \sin\varphi}{\sqrt{4\omega^2 v_0^2 \sin^2\varphi + g^2}} \approx \frac{2D\omega v_0 \sin\varphi}{g}.$$

For the numerical example one gets a bank superelevation of $H \approx 2.9\,\text{cm}$.

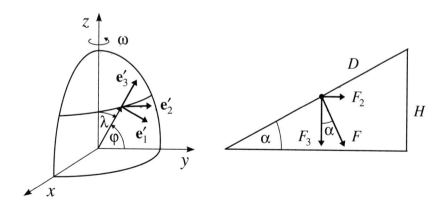

Figure 2.3.

Example 2.4: Exercise: Difference of sea depth at the pole and equator

Let a uniform spherical earth be covered by water. The sea surface takes the shape of an oblate spheroid if the earth rotates with the angular velocity ω.

Find an expression that approximately describes the difference of the sea depth at the pole and equator, respectively. Assume that the sea surface is a surface of constant potential energy. Neglect the corrections to the gravitational potential due to the deformation.

Solution

$$\mathbf{F}_{\text{eff}}(r) = -\frac{\gamma m M}{r^2}\mathbf{e}_r + m\omega^2 r \sin\vartheta \, \mathbf{e}_x, \quad r' = r \cdot \sin\vartheta,$$

$$V\Big|_{r_1}^{r_2} = -\int_{r_1}^{r_2} \mathbf{F}_{\text{eff}}(r) \cdot d\mathbf{r}$$

$$= -\int_{r_1}^{r_2} \left(-\frac{\gamma m M}{r^2}\mathbf{e}_r + m\omega^2 r \sin\vartheta \, \mathbf{e}_x\right) dr \cdot \mathbf{e}_r$$

$$= -\frac{\gamma m M}{r}\Big|_{r_1}^{r_2} - \frac{m\omega^2 r^2 \sin^2\vartheta}{2}\Big|_{r_1}^{r_2}.$$

We therefore define

$$V_{\text{eff}}(r) = -\frac{\gamma m M}{r} - \frac{m\omega^2 r^2}{2}\sin^2\vartheta. \tag{2.13}$$

Let

$$r = R + \Delta r(\vartheta); \qquad \Delta r(\vartheta) \ll R.$$

The potential at the surface of the rotating sphere is constant by definition:

$$V(r) = -\frac{\gamma m M}{R} + V_0.$$

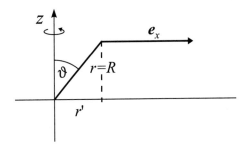

Figure 2.4.

EXACT SOLUTION

According to the formulation of the problem, the earth's surface is an equipotential surface. From this it follows that the attractive force acts normal to this surface. Because of the constancy of the potential along the surface, no tangential force can arise.

$$V(r) = -\frac{\gamma m M}{R}\left(1 - \frac{\Delta r}{R}\right) - \frac{m}{2}\omega^2 R^2 \left(1 + 2\frac{\Delta r}{R}\right)\sin^2\vartheta$$
$$\stackrel{!}{=} -\frac{\gamma m M}{R} + V_0.$$

From this it follows that

$$V_0 = \frac{\gamma m M}{R^2}\Delta r - \frac{m}{2}\omega^2 R^2 \sin^2\vartheta - m\omega^2 R\, \Delta r \sin^2\vartheta.$$

As can be seen by inserting the given values, the last term can be neglected:

$$\frac{\gamma m M}{R^2} \gg m\omega^2 R \sin^2\vartheta.$$

From this it follows that

$$\frac{\gamma m M}{R^2}\Delta r = V_0 + \frac{m}{2}\omega^2 R^2 \sin^2\vartheta,$$

or explicitly for the difference $\Delta r(\vartheta)$,

$$\Delta r(\vartheta) = \frac{R^2}{\gamma m M}\left(V_0 + \frac{m}{2}\omega^2 R^2 \sin^2\vartheta\right). \tag{2.14}$$

The second requirement for the evaluation of the deformation is the volume conservation. Since one can assume $\Delta r \ll R$, we can write this requirement as a simple surface integral

$$\int_{\vartheta=0}^{\pi/2}\int_{\varphi=0}^{2\pi} da \cdot \Delta r(\vartheta) = 0, \tag{2.15}$$

and hence, because of the rotational symmetry in φ,

$$\int_0^{\pi/2}\left(V_0 + \frac{m\omega^2 R^2 \sin^2\vartheta}{2}\right) 2\pi R \cdot R \sin\vartheta\, d\vartheta = 0,$$

from which follows

$$\int_0^{\pi/2}\left(V_0 \sin\vartheta + \frac{m\omega^2 R^2 \sin^3\vartheta}{2}\right) d\vartheta = 0.$$

With $\int_0^{\pi/2}\sin\vartheta\, d\vartheta = 1$ and $\int_0^{\pi/2}\sin^3\vartheta\, d\vartheta = \frac{2}{3}$, one gets

$$V_0 + \frac{m}{3}\omega^2 R^2 = 0,$$
$$V_0 = -\frac{m}{3}\omega^2 R^2.$$

By inserting this result into (2.14), one obtains

$$\Delta r(\vartheta) = \frac{R^4}{\gamma M} \frac{\omega^2}{2} \left(\sin^2 \vartheta - \frac{2}{3} \right).$$

In the last step $\gamma M / R^2$ is to be replaced by g; thus we have found an approximate expression for the difference of the sea depth:

$$\Delta r(\vartheta) = \frac{\omega^2 R^2}{2g} \left(\sin^2 \vartheta - \frac{2}{3} \right). \tag{2.16}$$

By inserting the given values

$$R = 6370\,\text{km}, \quad g = 9.81\,\frac{\text{m}}{\text{s}^2}, \quad \omega = \frac{2\pi}{T} = 7.2722 \cdot 10^{-5}\,\frac{1}{\text{s}},$$

we get

$$d = \Delta r \left(\frac{\pi}{2} \right) - \Delta r(0) \approx 10.94\,\text{km}.$$

If one wants to include the influence of the deformation on the gravitational potential, one needs the so-called spherical surface harmonics. They will be outlined in detail in part 3 of the lectures (electrodynamics).

3 Foucault's Pendulum

In 1851, Foucault[1] found a simple and convincing proof of the earth rotation: A pendulum tends to maintain its plane of motion, independent of any rotation of the suspension point. If such a rotation is nevertheless observed in a laboratory, one can only conclude that the laboratory (i.e., the earth) rotates.

The drawing shows the arrangement of the pendulum and fixes the axes of the coordinate system.

We first derive the equation of motion of the Foucault pendulum. For the mass point we have

$$\mathbf{F} = \mathbf{T} + m\mathbf{g}, \tag{3.1}$$

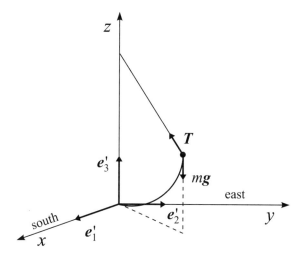

Figure 3.1. Principle of the Foucault pendulum.

---

[1] *Jean Bernard Léon Foucault* [fuk'o], French physicist, b. Sept. 18, 1819 Paris–d. Feb. 11, 1868. In 1851, Foucault performed his famous pendulum experiment in the Panthéon in Paris as a proof of the earth's rotation. In the same year he proved by means of a rotating mirror that light propagates in water more slowly than in air, which was important for confirming the wave theory of light. He investigated the eddy currents in metals detected by D.F. Arago (Foucault–currents), and also studied light and heat radiation together with A.H.L. Fizeau.

where **T** is a still unknown tension force along the pendulum string. In the basic equation that holds for moving reference frames,

$$m\ddot{\mathbf{r}} = \mathbf{F} - m\frac{d\boldsymbol{\omega}}{dt} \times \mathbf{r} - 2m\boldsymbol{\omega} \times \mathbf{v} - m\boldsymbol{\omega} \times (\boldsymbol{\omega} \times \mathbf{r}), \tag{3.2}$$

the linear forces and the centripetal forces can be neglected, because for the earth's rotation $d\omega/dt = 0$ and $t \cdot |\omega| \ll 1$, $t^2\omega^2 \approx 0$ ($t \approx$ pendulum period). By inserting equation (3.1) into the simplified equation (3.2), we get

$$m\ddot{\mathbf{r}} = \mathbf{T} + m\mathbf{g} - 2m\boldsymbol{\omega} \times \mathbf{v}. \tag{3.3}$$

As is obvious from this equation, the earth's rotation is expressed for the moving observer by the appearance of a virtual force, the Coriolis force. The Coriolis force causes a rotation of the vibrational plane of the pendulum. The string tension T can be determined from (3.3) by noting that

$$\mathbf{T} = (\mathbf{T} \cdot \mathbf{e}_1')\mathbf{e}_1' + (\mathbf{T} \cdot \mathbf{e}_2')\mathbf{e}_2' + (\mathbf{T} \cdot \mathbf{e}_3')\mathbf{e}_3' = T\frac{\mathbf{T}}{T}$$
$$= T\frac{(-x, -y, l-z)}{\sqrt{x^2 + y^2 + (l-z)^2}} \approx T\frac{(-x, -y, l-z)}{l}. \tag{3.4}$$

In the last step, we presupposed a very large pendulum length l, so that $x/l \ll 1$, $y/l \ll 1$, and $z/l \lll 1$. Evaluation of the scalar product therefore yields

$$\mathbf{T} = T\left(-\frac{x}{l}\mathbf{e}_1' - \frac{y}{l}\mathbf{e}_2' + \frac{z-l}{l}\mathbf{e}_3'\right). \tag{3.5}$$

Before inserting (3.5) into (3.3), it is practical to decompose (3.3) into individual components. For this purpose one has to evaluate the vector product $\boldsymbol{\omega} \times \mathbf{v}$:

$$\boldsymbol{\omega} \times \mathbf{v} = \begin{vmatrix} \mathbf{e}_1' & \mathbf{e}_2' & \mathbf{e}_3' \\ -\omega\sin\lambda & 0 & \omega\cos\lambda \\ \dot{x} & \dot{y} & \dot{z} \end{vmatrix}$$
$$= -\omega\cos\lambda\,\dot{y}\mathbf{e}_1' + \omega(\cos\lambda\,\dot{x} + \sin\lambda\,\dot{z})\mathbf{e}_2' - \omega\sin\lambda\,\dot{y}\mathbf{e}_3'. \tag{3.6}$$

By inserting (3.5) and (3.6) into equation (3.3) with $m\mathbf{g} = -mg\mathbf{e}_3'$, we obtain a coupled system of differential equations:

$$m\ddot{x} = -\frac{x}{l}T + 2m\omega\cos\lambda\,\dot{y},$$
$$m\ddot{y} = -\frac{y}{l}T - 2m\omega(\cos\lambda\,\dot{x} + \sin\lambda\,\dot{z}), \tag{3.7}$$
$$m\ddot{z} = \frac{l-z}{l}T - mg + 2m\omega\sin\lambda\,\dot{y}.$$

To eliminate the unknown string tension T from the system (3.7), we adopt the already mentioned approximations:

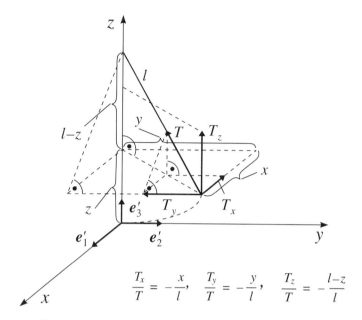

Figure 3.2. Projection of the string tension **T** onto the axes e'_i.

The pendulum string shall be very long, but the pendulum shall oscillate with small amplitudes only. From this it follows that $x/l \ll 1$, $y/l \ll 1$, and $z/l \lll 1$, since the mass point moves almost in the x, y–plane. Hence, for calculating the string tension we use the approximation

$$\frac{l-z}{l} = 1, \qquad m\ddot{z} = 0, \tag{3.8}$$

and obtain from the third equation (3.7)

$$T = mg - 2m\omega \sin\lambda \, \dot{y}. \tag{3.9}$$

Insertion of (3.9) and (3.7) after division by the mass m yields

$$\ddot{x} = -\frac{g}{l}x + \frac{2\omega \sin\lambda}{l}x\dot{y} + 2\omega \cos\lambda \, \dot{y},$$
$$\ddot{y} = -\frac{g}{l}y + \frac{2\omega \sin\lambda}{l}y\dot{y} - 2\omega \cos\lambda \, \dot{x}. \tag{3.10}$$

Equation (3.10) represents a system of nonlinear, coupled differential equations; nonlinear since the mixed terms $x\dot{y}$ and $y\dot{y}$ appear. Since the products of the small numbers ω, x, and $\dot{y}$ (or ω, y, and $\dot{y}$) are negligible compared to the other terms, (3.10) can be considered equivalent to the following equations (3.11):

$$\ddot{x} = -\frac{g}{l}x + 2\omega \cos\lambda \, \dot{y}, \qquad \ddot{y} = -\frac{g}{l}y - 2\omega \cos\lambda \, \dot{x}. \tag{3.11}$$

These two linear (but coupled) differential equations describe the vibrations of a pendulum under the influence of the Coriolis force to a good approximation. In the following we will describe a method of solving (3.11).

Solution of the differential equations

For solving (3.11), we introduce the abbreviations $g/l = k^2$ and $\omega \cos \lambda = \alpha$, multiply $\ddot{y}$ by the imaginary unit $i = \sqrt{-1}$, and obtain

$$\ddot{x} = -k^2 x - 2\alpha i^2 \dot{y}$$
$$i\ddot{y} = -k^2 i y - 2\alpha i \dot{x} \tag{3.12}$$
$$\overline{\ddot{x} + i\ddot{y} = -k^2(x + iy) - 2\alpha i (\dot{x} + i\dot{y})}$$

The abbreviation $u = x + iy$ is obvious:

$$\ddot{u} = -k^2 u - 2\alpha i \dot{u} \quad \text{or} \quad 0 = \ddot{u} + 2\alpha i \dot{u} + k^2 u. \tag{3.13}$$

The equation (3.13) is solved by the *ansatz* useful for all vibration processes,

$$u = C \cdot e^{\gamma t}, \tag{3.14}$$

where γ is to be determined by inserting the derivatives into (3.13):

$$C\gamma^2 e^{\gamma t} + 2\alpha i C \gamma e^{\gamma t} + k^2 C e^{\gamma t} = 0 \quad \text{or} \quad \gamma^2 + 2i\alpha\gamma + k^2 = 0. \tag{3.15}$$

The two solutions of (3.15) are

$$\gamma_{1/2} = -i\alpha \pm ik\sqrt{1 + \alpha^2/k^2}. \tag{3.16}$$

Since $\alpha^2 = \omega^2 \cos^2 \lambda$ because ω^2/k^2 is small compared to 1 ($\omega^2/k^2 = T^2_{\text{pend}}/T^2_{\text{earth}} \ll 1$, where T_{pend} is the pendulum period and $T_{\text{earth}} = 1$ day), it further follows that

$$\gamma_{1/2} = -i\alpha \pm ik. \tag{3.17}$$

The most general solution of the differential equation (3.13) is the linear combination of the linearly independent solutions

$$u = A \cdot e^{\gamma_1 t} + B \cdot e^{\gamma_2 t}, \tag{3.18}$$

where A and B must be fixed by the initial conditions and are of course complex, i.e., can be decomposed into a real and an imaginary part:

$$u = (A_1 + iA_2)e^{-i(\alpha-k)t} + (B_1 + iB_2)e^{-i(\alpha+k)t}. \tag{3.19}$$

The Euler relation $e^{-i\varphi} = \cos\varphi - i\sin\varphi$ allows one to split (3.19) into $u = x + iy$:

$$x + iy = (A_1 + iA_2)[\cos(\alpha - k)t - i\sin(\alpha - k)t]$$
$$+ (B_1 + iB_2)[\cos(\alpha + k)t - i\sin(\alpha + k)t], \tag{3.20}$$

SOLUTION OF THE DIFFERENTIAL EQUATIONS

from which it follows after separating the real and the imaginary parts

$$x = A_1 \cos(\alpha - k)t + A_2 \sin(\alpha - k)t + B_1 \cos(\alpha + k)t + B_2 \sin(\alpha + k)t,$$
$$y = -A_1 \sin(\alpha - k)t + A_2 \cos(\alpha - k)t - B_1 \sin(\alpha + k)t + B_2 \cos(\alpha + k)t. \quad (3.21)$$

Let the initial conditions be

$$x_0 = 0, \qquad \dot{x}_0 = 0,$$
$$y_0 = L, \qquad \dot{y}_0 = 0,$$

i.e., the pendulum is displaced by the distance L toward the east and released at the time $t = 0$ without initial velocity. Inserting $x_0 = 0$ in (3.21), one gets

$$B_1 = -A_1.$$

Differentiating (3.21) and setting $\dot{x}_0 = 0$ yields

$$B_2 = A_2 \frac{k - \alpha}{k + \alpha}.$$

As already noted in (3.16), $\alpha \ll k$ and thus $B_2 \approx A_2$. From (3.21) one now obtains

$$x = A_1 \cos(\alpha - k)t + A_2 \sin(\alpha - k)t - A_1 \cos(\alpha + k)t + A_2 \sin(\alpha + k)t,$$
$$y = -A_1 \sin(\alpha - k)t + A_2 \cos(\alpha - k)t + A_1 \sin(\alpha + k)t + A_2 \cos(\alpha + k)t. \quad (3.22)$$

We still have to include the initial conditions for y_0 and $\dot{y}_0$. From $\dot{y}_0 = 0$ and (3.22) we get

$$-A_1(\alpha - k) + A_1(\alpha + k) = 0 \Rightarrow A_1 = 0.$$

From $y_0 = L$ and equation (3.22) we get

$$2A_2 = L \Rightarrow A_2 = \frac{L}{2}.$$

By inserting these values one obtains

$$x = \frac{L}{2} \sin(\alpha - k)t + \frac{L}{2} \sin(\alpha + k)t,$$
$$y = \frac{L}{2} \cos(\alpha - k)t + \frac{L}{2} \cos(\alpha + k)t.$$

Using the trigonometric formulae

$$\sin(\alpha \pm k) = \sin\alpha \cos k \pm \cos\alpha \sin k, \quad \cos(\alpha \pm k) = \cos\alpha \cos k \mp \sin\alpha \sin k,$$

it follows that

$$x = L \sin\alpha t \cos kt, \qquad y = L \cos\alpha t \cos kt.$$

The two equations can be combined into a vector equation:

$$\mathbf{r} = L \cos kt [\sin(\alpha t)\mathbf{e}_1 + \cos(\alpha t)\mathbf{e}_2]. \quad (3.23)$$

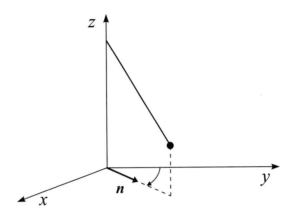

Figure 3.3. The unit vector $\mathbf{n}(t)$ rotates in the x,y–plane.

Discussion of the solution

The first factor in (3.23) describes the motion of a pendulum that vibrates with the amplitude L and the frequency $k = \sqrt{g/l}$. The second term is a unit vector $\mathbf{n}$ that rotates with the frequency $\alpha = \omega \cos \lambda$ and describes the rotation of the vibration plane:

$$\mathbf{r} = L \cos kt \, \mathbf{n}(t),$$
$$\mathbf{n}(t) = \sin \alpha t \, \mathbf{e}_1 + \cos \alpha t \, \mathbf{e}_2.$$

(3.23) also tells us in what direction the vibrational plane rotates. For the northern hemisphere $\cos \lambda > 0$, and after a short time $\sin \alpha t > 0$ and $\cos \alpha t > 0$, i.e., the vibrational plane rotates clockwise. An observer in the southern hemisphere will see his pendulum rotate counter-clockwise, since $\cos \lambda < 0$.

At the equator the experiment fails, since $\cos \lambda = 0$. Although the component $\omega_x = -\omega \sin \lambda$ takes its maximum value there, it cannot be demonstrated by means of the Foucault pendulum.

Following the path of the mass point of a Foucault pendulum, one finds rosette trajectories. Note that the shape of the trajectories essentially depends on the initial conditions (see Figure 3.4). The left side shows a rosette path for a pendulum released at the maximum displacement; the pendulum shown on the right side was pushed out of the rest position.

Because of the assumption $\alpha \ll k$ in (3.16), equation (3.23) does not describe either of the two rosettes exactly. According to (3.23), the pendulum always passes the rest position, although the initial conditions were adopted as in the left figure.

DISCUSSION OF THE SOLUTION

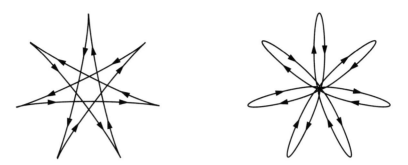

Figure 3.4. Rosette paths of the Foucault pendulum.

Example 3.1: Exercise: Chain fixed to a rotating bar

A vertical bar AB rotates with constant angular velocity ω. A light nonstretchable chain of length l is fixed at one end to the point O of the bar, while the mass m is fixed at its other end. Find the chain tension and the angle between chain and bar in the state of equilibrium.

Solution Three forces act on the body, viz.

(1) the gravitation (weight): $\mathbf{F}_g = -mg\mathbf{e}_3$;

(2) the centrifugal force: $\mathbf{F}_z = -m\boldsymbol{\omega} \times (\boldsymbol{\omega} \times \mathbf{r})$;

(3) the chain tension force: $\mathbf{T} = -T\sin\varphi\, \mathbf{e}_1 + T\cos\varphi\, \mathbf{e}_3$.

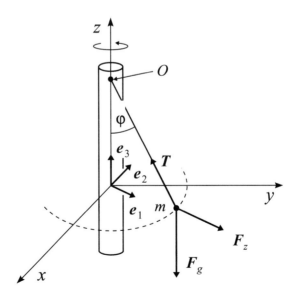

Figure 3.5. $\mathbf{e}_1, \mathbf{e}_2, \mathbf{e}_3$ are the unit vectors of a rectangular coordinate system rotating with the bar; $\mathbf{T}$ is the chain tension force, $\mathbf{F}_g$ is the weight of the mass m; $\mathbf{F}_z$ is the centrifugal force.

Since the angular velocity has only one component in the $\mathbf{e}_3$-direction, $\boldsymbol{\omega} = \omega \mathbf{e}_3$, and

$$\mathbf{r} = l(\sin\varphi\, \mathbf{e}_1 + (1 - \cos\varphi)\mathbf{e}_3),$$

we find for the centrifugal force

$$\mathbf{F}_z = -m(\boldsymbol{\omega} \times (\boldsymbol{\omega} \times \mathbf{r}))$$

the expression

$$\mathbf{F}_z = +m\omega^2 l \sin\varphi\, \mathbf{e}_1.$$

If the body is in equilibrium, the sum of the three forces equals zero:

$$0 = -mg\mathbf{e}_3 + m\omega^2 l \sin\varphi\, \mathbf{e}_1 - T\sin\varphi\, \mathbf{e}_1 + T\cos\varphi\, \mathbf{e}_3.$$

When ordering by components, we obtain

$$0 = (m\omega^2 l \sin\varphi - T\sin\varphi)\mathbf{e}_1 + (T\cos\varphi - mg)\mathbf{e}_3.$$

Since a vector vanishes only if every component equals zero, we can set up the following component equations:

$$m\omega^2 l \sin\varphi - T\sin\varphi = 0, \tag{3.24}$$

$$T\cos\varphi - mg = 0. \tag{3.25}$$

One solution of equation (3.24) is $\sin\varphi = 0$. It represents a state of unstable equilibrium that happens if the body rotates on the axis AB. In this case the centrifugal force component vanishes. A second solution of the system is found by assuming $\sin\varphi \neq 0$. We can then divide equation (3.24) by $\sin\varphi$ and get the tension force T:

$$T = m\omega^2 l \tag{3.26}$$

and after elimination of T from (3.25) we get the angle φ between the chain and the bar:

$$\cos\varphi = \frac{g}{\omega^2 l}.$$

Since the chain OP with the mass m in P moves on the surface of a cone, this arrangement is called the cone pendulum.

Example 3.2: Exercise: Pendulum in a moving train

The period of a pendulum of length l is given by T. How will the period change if the pendulum is suspended at the ceiling of a train that moves with the velocity v along a curve with radius R?

(a) Neglect the Coriolis force. Why can you do that?

(b) Solve the equations of motion (with Coriolis force!) nearly exactly (analogous to Foucault's pendulum).

Solution (a) The backdriving force is

$$F_R = -mg\sin\varphi + \frac{mv^2}{R'}\cos\varphi.$$

One has $v(x) = \omega(R+x) = \omega(R + l\sin\varphi)$ and $R' = R + x$. Hence, it follows that
$$F_R = -mg\sin\varphi + m\omega^2(R + l\sin\varphi)\cos\varphi.$$

The differential equation for the motion therefore reads
$$m\ddot{s} = -mg\sin\varphi + m\omega^2(R + l\sin\varphi)\cos\varphi.$$

Since $s = l\varphi$, $\ddot{s} = l\ddot{\varphi}$, it follows that $l\ddot{\varphi} = -g\sin\varphi + \omega^2(R + l\sin\varphi)\cos\varphi$, or

$$\ddot{\varphi} + \frac{g}{l}\sin\varphi - \omega^2\left(\frac{R}{l} + \sin\varphi\right)\cos\varphi = 0. \tag{3.27}$$

For small amplitudes, $\cos\varphi \approx 1$ and $\sin\varphi \approx \varphi$, i.e.,

$$\ddot{\varphi} + \frac{g}{l}\varphi - \omega^2\left(\frac{R}{l} + \varphi\right) = 0$$

or

$$\ddot{\varphi} + \left(\frac{g}{l} - \omega^2\right)\varphi - \omega^2\frac{R}{l} = 0. \tag{3.28}$$

Here, the Coriolis force was neglected, since the angular velocity $\dot{\varphi}$ and hence $\dot{x}$ is small compared to the rotational velocity $v = \omega(R + x)$, i.e., $\boldsymbol{\omega} \times \dot{\mathbf{x}} \cong 0$. The solution of the homogeneous differential equation is

$$\varphi_h = \sin\left(\sqrt{\frac{g}{l} - \omega^2}\, t\right).$$

The particular solution of the inhomogeneous differential equation is

$$\varphi_i = \frac{\omega^2(R/l)}{(g/l) - \omega^2}.$$

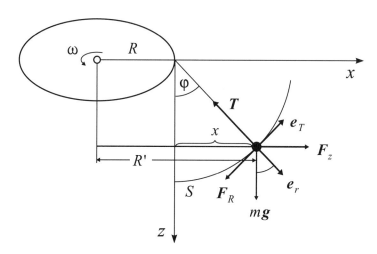

Figure 3.6.

The general solution of (3.27) is therefore

$$\varphi = \varphi_h + \varphi_i = \sin\left(\sqrt{\frac{g}{l} - \omega^2}\, t\right) + \frac{\omega^2 (R/l)}{(g/l) - \omega^2}.$$

Hence, the vibration period is

$$T = \frac{2\pi}{\sqrt{(g/l) - \omega^2}}.$$

For $\omega = \sqrt{g/l}$ the period becomes infinite, since the centrifugal force exceeds the gravitational force. This interpretation gets to the core of the matter, although the formula (3.28) holds only for small angular velocities: For large angular velocities the approximation of small vibration amplitudes x in equation (3.27) is no longer allowed, because the pendulum mass is being pressed outward due to the centrifugal force, i.e., to large values of x.

(b) The equations of motion read

$$m\ddot{\mathbf{r}} = \mathbf{F} - m\frac{d\boldsymbol{\omega}}{dt} \times \mathbf{r} - 2m\boldsymbol{\omega} \times \mathbf{v} - m\boldsymbol{\omega} \times (\boldsymbol{\omega} \times (\mathbf{r} + \mathbf{R})).$$

With

$$\boldsymbol{\omega} = -\omega \mathbf{e}_z, \quad \mathbf{R} = R\mathbf{e}_x, \quad -2m\boldsymbol{\omega} \times \mathbf{v} = 2m\omega(-\dot{y}, \dot{x}, 0),$$

and

$$-m\boldsymbol{\omega} \times (\boldsymbol{\omega} \times (\mathbf{r} + \mathbf{R})) = m\omega^2 (R + x, y, 0),$$

one finds

$$\begin{aligned}
m\ddot{x} &= -T_x - 2m\omega\dot{y} + m\omega^2(R+x) = -\frac{x}{l}T - 2m\omega\dot{y} + m\omega^2(R+x), \\
m\ddot{y} &= -T_y + 2m\omega\dot{x} + m\omega^2 y = -\frac{y}{l}T + 2m\omega\dot{x} + m\omega^2 y, \\
m\ddot{z} &= -T_z + mg \phantom{(R+x) - 2m\omega\dot{y} + m\omega^2} = -\frac{z}{l}T + mg.
\end{aligned} \qquad (3.29)$$

In the following, we will assume that the pendulum length is large, i.e., for small amplitudes $z \approx l$ ($\dot{z} = \ddot{z} = 0$). The string tension is then given by $T = mg$:

$$\ddot{x} = \left(\omega^2 - \frac{g}{l}\right)x - 2\omega\dot{y} + \omega^2 R,$$
$$\ddot{y} = \left(\omega^2 - \frac{g}{l}\right)y + 2\omega\dot{x}.$$

With the substitution $u = x + iy$, it follows further that

$$\ddot{u} \left(\omega^2 - \frac{g}{l}\right) u + 2i\omega\dot{u} + \omega^2 R. \qquad (3.30)$$

For the homogeneous solution of the differential equation (3.30) one gets with the *ansatz* $u_{\text{hom}} = c\, e^{\gamma t}$ the characteristic polynomial

$$\gamma^2 - \left(\omega^2 - \frac{g}{l}\right) - 2i\omega\gamma = 0.$$

DISCUSSION OF THE SOLUTION

The homogeneous solution then takes the form

$$u_{\text{hom}} = c_1 \exp\left(i(\omega + \sqrt{g/l})t\right) + c_2 \exp\left(i(\omega - \sqrt{g/l})t\right).$$

The particular solution is simply obtained as

$$u_{\text{part}} = \frac{\omega^2 R}{(g/l) - \omega^2}.$$

From this it follows that

$$u = u_{\text{hom}} + u_{\text{part}} = c_1 \exp\left(i(\omega + \sqrt{g/l})t\right) + c_2 \exp\left(i(\omega - \sqrt{g/l})t\right) + \frac{\omega^2 R}{(g/l) - \omega^2}. \quad (3.31)$$

With the initial condition $x(0) = x_0$ and $y(0) = \dot{x}(0) = \dot{y}(0) = 0$, it follows for c_1 and c_2

$$c_1 = \frac{\sqrt{g/l} - \omega}{2\sqrt{g/l}} \left(x_0 + \frac{\omega^2 R}{\omega^2 - g/l}\right),$$

$$c_2 = \frac{\sqrt{g/l} + \omega}{2\sqrt{g/l}} \left(x_0 + \frac{\omega^2 R}{\omega^2 - g/l}\right). \quad (3.32)$$

By decomposing the solution (3.31) into real and imaginary parts, the solutions for $x(t)$ and $y(t)$ can be found.

$$x = c_1 \cos\left(\omega + \sqrt{\frac{g}{l}}\right)t + c_2 \cos\left(\omega - \sqrt{\frac{g}{l}}\right)t + \frac{\omega^2 R}{g/l - \omega^2}$$

$$= \sqrt{\frac{l}{g}} \left(x_0 + \frac{\omega^2 R}{\omega^2 - g/l}\right) \left\{\sqrt{\frac{g}{l}} \cos\sqrt{\frac{g}{l}}t \cos\omega t + \omega \sin\sqrt{\frac{g}{l}}t \sin\omega t\right\} + \frac{\omega^2 R}{g/l - \omega^2}$$

$$= \left(x_0 + \frac{\omega^2 R}{\omega^2 - g/l}\right) \left\{\cos\sqrt{\frac{g}{l}}t \cos\omega t + \omega\sqrt{\frac{l}{g}} \sin\sqrt{\frac{g}{l}}t \sin\omega t\right\} + \frac{\omega^2 R}{g/l - \omega^2}, \quad (3.33)$$

$$y = c_1 \sin\left(\omega + \sqrt{\frac{g}{l}}\right)t + c_2 \sin\left(\omega - \sqrt{\frac{g}{l}}\right)t$$

$$= \left(x_0 + \frac{\omega^2 R}{\omega^2 - g/l}\right) \left\{\sin\omega t \cos\sqrt{\frac{g}{l}}t - \omega\sqrt{\frac{l}{g}} \sin\sqrt{\frac{g}{l}}t \cos\omega t\right\}. \quad (3.34)$$

Because $\omega \ll \sqrt{g/l}$, $\omega\sqrt{l/g} \ll 1$. From this it follows that

$$x = x_0 \cos\sqrt{\frac{g}{l}}t \cos\omega t,$$

$$y = x_0 \cos\sqrt{\frac{g}{l}}t \sin\omega t.$$

This describes a rotation of the pendulum plane with the frequency ω (as for Foucault's pendulum).

The pendulum period T can now be obtained from the following consideration: For $t = 0$, the brace in (3.33) equals 1. For $t = (\pi/2)\sqrt{l/g} + t'$, where $t' \ll (\pi/2)\sqrt{l/g}$, the brace vanishes for the first time, which corresponds to a quarter of T. By expanding the brace, for t' we find

$$t' = \frac{\pi}{2}\left(\frac{l}{g}\right)^{3/2} \omega^2,$$

$$T = 4\left(\frac{\pi}{2}\sqrt{\frac{l}{g}} + \frac{\pi}{2}\left(\frac{l}{g}\right)^{3/2}\omega^2\right)$$
$$= 2\pi\sqrt{\frac{l}{g}}\left(1 + \frac{\omega^2}{g/l}\right).$$

On the other hand, in part (a) we found

$$T = \frac{2\pi}{\sqrt{g/l - \omega^2}} = 2\pi\sqrt{\frac{l}{g}}\left(1 + \frac{1}{2}\frac{\omega^2}{g/l}\right),$$

which suggests the conclusion that the Coriolis force should not be neglected from the outset in this consideration.

Example 3.3: Exercise: Formation of cyclones

Explain to which directions the winds from north, east, south, and west will be deflected in the northern hemisphere. Explain the formation of cyclones.

Solution We derive the equation of motion for a parcel of air P that moves near the earth's surface. The X, Y, Z system is considered as an inertial system; i.e., we shall not take the rotation of the earth about the sun into account. Moreover, we assume the air mass is moving at constant height; i.e., there is no velocity component along the z-direction ($\dot{z} = 0$). The centrifugal acceleration shall also be neglected.

With the assumptions mentioned above, the equation of motion of the particle is defined by the differential equation

$$\ddot{\mathbf{r}} = \mathbf{g} - 2(\boldsymbol{\omega} \times \dot{\mathbf{r}}) = \mathbf{g} - 2(\boldsymbol{\omega}_\perp \times \dot{\mathbf{r}}) - 2(\boldsymbol{\omega}_\| \times \dot{\mathbf{r}}),$$

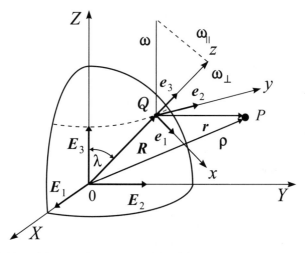

Figure 3.7. Definition of the coordinates: $0 =$ origin of the inertial systems X, Y, Z; $Q =$ origin of the moving system x, y, z; $P =$ a point with mass m; $\rho =$ position vector in the system X, Y, Z; and $\mathbf{r} =$ position vector in the x, y, z-system.

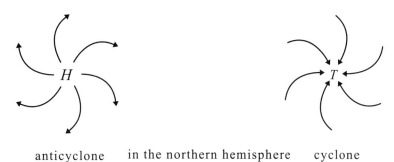

anticyclone   in the northern hemisphere   cyclone

Figure 3.8. If air flows in the northern hemisphere from a high-pressure region to a low-pressure region, a left-rotating cyclone arises in the low-pressure region, and a right-rotating anticyclone is formed in the high-pressure region.

where $\boldsymbol{\omega} \times \dot{\mathbf{r}} = (\boldsymbol{\omega}_\perp + \boldsymbol{\omega}_\parallel) \times \dot{\mathbf{r}}$. Let $\boldsymbol{\omega}_\parallel$ be the component of $\boldsymbol{\omega}$ within the tangential plane at the point Q of the earth's surface (see Figure 3.7); it points to the negative $\mathbf{e}_1$-direction. $\boldsymbol{\omega}_\perp$ points to the $\mathbf{e}_3$-direction.

We consider the dominant term $-2\boldsymbol{\omega}_\perp \times \dot{\mathbf{r}}$: An air parcel moving in x-direction (south) is accelerated toward the negative y-axis, and a motion along the y-direction causes an acceleration in x-direction. The deflection proceeds from the direction of motion toward the right side. The wind from the west is deflected toward the south, the north wind toward the west, the east wind toward the north, and the south wind toward the east.

The force $-2m\boldsymbol{\omega}_\parallel \times \dot{\mathbf{r}}$ for the north and south winds exactly equals zero. For the west or east winds the force points along $\mathbf{e}_3$ or the opposite direction. Accordingly the air masses are pushed away from or toward the ground. This force component is however very small compared to the gravitational force mg, which also points toward the negative $\mathbf{e}_3$-direction.

If we consider an air parcel moving in the southern hemisphere, then $\lambda > \pi/2$ and $\cos \lambda$ is negative. Thus, a west wind is here deflected to the north, a north wind toward the east, and a south wind toward the west.

Example 3.4:   Exercise: Movable mass in a rotating tube

A tube rotates with constant angular velocity ω (relative motion) and is inclined from the rotational axis by the angle α. A mass m inside of the tube is pulled inward with constant velocity c by a string.

(a) What forces act on the mass?

(b) What work is performed by these forces while the mass moves from x_1 to x_2? (Calculate the energy balance!) Numerical values: $m = 5\,\text{kg}$, $\alpha = 45°$, $x_1 = 1\,\text{m}$, $x_2 = 5\,\text{m}$, $\omega = 2\,\text{s}^{-1}$, $c = 5\,\text{m/s}$, $g = 9.81\,\text{m/s}^2$.

Solution   (a) The mass m within the tube performs a relative motion with constant velocity $\mathbf{c} = c(-1, 0, 0)$, and thus the resulting acceleration is composed of the guiding acceleration $\mathbf{a}_f$ in the tube, the relative acceleration $\mathbf{a}_r$, and the Coriolis acceleration $\mathbf{a}_c$:

$$\mathbf{a} = \mathbf{a}_f + \mathbf{a}_r + \mathbf{a}_c.$$

The guiding acceleration consists in general of a translational acceleration $\mathbf{b}_0$ of the vehicle, and of the accelerations $\mathbf{a}_t$ (tangential acceleration) and $\mathbf{a}_n$ (normal acceleration) due to a rotational motion.

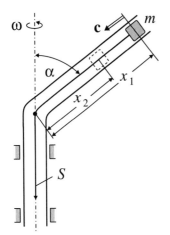 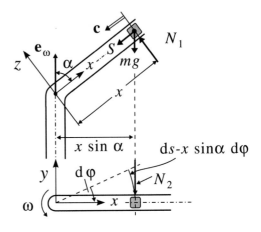

Figure 3.9.

In the present problem, we are dealing with a rotation about a fixed axis $\mathbf{e}_\omega = (\cos\alpha, 0, \sin\alpha)$ with a constant angular velocity ω, so that $\mathbf{a}_0 = \mathbf{a}_t = 0$, and the guiding acceleration is therefore

$$\mathbf{a}_f = \mathbf{a}_0 + \mathbf{a}_t + \mathbf{a}_n = \mathbf{a}_n = x\sin\alpha\,\omega^2(-\sin\alpha, 0, \cos\alpha),$$

i.e., the guiding acceleration obviously consists of the centripetal acceleration $\mathbf{b}_n$ only.

The relative acceleration $\mathbf{a}_r = 0$, since the relative velocity is constant. The Coriolis acceleration $\mathbf{a}_c$, defined by

$$\mathbf{a}_c = 2\omega \mathbf{e}_\omega \times \mathbf{c},$$

is therefore

$$\mathbf{a}_c = -2\omega c \sin\alpha\,(0, 1, 0).$$

Therefore, for the total acceleration $\mathbf{a}$ we have

$$\mathbf{a} = \mathbf{a}_f + \mathbf{a}_c = (-x\omega^2\sin^2\alpha,\ -2\omega c\sin\alpha,\ x\omega^2\sin\alpha\cos\alpha). \tag{3.35}$$

$\mathbf{a}$ is the result of the following forces acting on the mass m (see Figure 3.9):

$$\mathbf{S} = S(-1, 0, 0), \qquad \mathbf{G} = mg(-\cos\alpha, 0 - \sin\alpha),$$
$$\mathbf{N}_1 = N_1(0, 0, 1), \qquad \mathbf{N}_2 = N_2(0, -1, 0).$$

The resulting total force is therefore

$$\mathbf{F} = (-S - mg\cos\alpha,\ -N_2,\ N_1 - mg\sin\alpha). \tag{3.36}$$

With Newton's equation

$$\mathbf{F} = m \cdot \mathbf{a}, \tag{3.37}$$

DISCUSSION OF THE SOLUTION

one can determine the unknown quantities S, N_1 and N_2: From the equations (3.35), (3.36) and (3.37) it follows that

$$(-S - mg\cos\alpha, \ -N_2, \ N_1 - mg\sin\alpha) = m\left(-x\omega^2\sin^2\alpha, \ -2\omega c\sin\alpha, \ x\omega^2\sin\alpha\cos\alpha\right)$$
$$\Rightarrow S = m\left(x\omega^2\sin^2\alpha - g\cos\alpha\right)$$
$$N_1 = m\left(x\omega^2\sin\alpha\cos\alpha + g\sin\alpha\right)$$
$$N_2 = 2\omega m c\sin\alpha.$$

S becomes negative if $x < (g\cos\alpha)/(\omega^2\sin\alpha)$; i.e., the mass m would have to be decelerated additionally within the tube if a constant velocity is to be maintained.

(b) During the motion, work is performed by the string force $\mathbf{S}$, by the gravitational force $\mathbf{G}$, and by the Coriolis force $\mathbf{N}_2$; $\mathbf{N}_1$ is the normal force. The work performed by the string force is

$$W_s = \int_{x_1}^{x_2} dW_s = -\int_{x_1}^{x_2} S(x)\,dx = \frac{m}{2}\omega^2\sin^2\alpha(x_1^2 - x_2^2) - mg\cos\alpha(x_1 - x_2). \tag{3.38}$$

The work performed by the weight force is

$$W_G = \int_{x_1}^{x_2} dW_G = mg\cos\alpha(x_1 - x_2). \tag{3.39}$$

The work performed by the Coriolis force, taking into account $dx/dt = -c$, is

$$W_{N_2} = \int dW_{N_2} = -\int N_2\,ds = -\int N_2 x \sin\alpha\,d\varphi$$
$$= -\int_{x_1}^{x_2} N_2 x \sin\alpha \frac{d\varphi}{dt}\frac{dt}{dx}\,dx \tag{3.40}$$
$$= -m\omega^2\sin^2\alpha(x_1^2 - x_2^2).$$

Insertion of the numerical values given in the formulation of the problem yields

$$W_S = (3.75 - 17.34)\,\text{Nm} = -13.59\,\text{Nm},$$
$$W_G = 17.34\,\text{Nm}, \qquad W_{N_2} = 7.5\,\text{Nm}.$$

To check the results, one uses the fact that the sum of the work performed by the external forces must be equal to the difference of the kinetic energies (energy balance)

$$\Delta E = W_S + W_{N_2} + W_G,$$

where

$$\Delta E = \frac{m}{2}(c^2 + x_2^2\sin^2\alpha\omega^2) - \frac{m}{2}(c^2 + x_1^2\sin^2\alpha\omega^2)$$
$$= -\frac{m}{2}\omega^2\sin^2\alpha(x_1^2 - x_2^2),$$

and according to equations (3.38), (3.39), and (3.40)

$$\begin{aligned} W_S + W_G + W_{N_2} &= \frac{m}{2}\omega^2 \sin^2\alpha(x_1^2 - x_2^2) - mg\cos\alpha(x_1 - x_2) \\ &\quad - m\omega^2 \sin^2\alpha(x_1^2 - x_2^2) + mg\cos\alpha(x_1 - x_2) \\ &= \frac{m}{2}\omega^2 \sin^2\alpha(x_1^2 - x_2^2). \end{aligned}$$

PART II

MECHANICS OF PARTICLE SYSTEMS

So far we have considered only the mechanics of a mass point. We now proceed to describe systems of mass points. A particle system is called a *continuum* if it consists of so great a number of mass points that a description of the individual mass points is not feasible. On the other hand, a particle system is called *discrete* if it consists of a manageable number of mass points.

An idealization of a body (continuum) is the *rigid body*. The notion of a rigid body implies that the distances between the individual points of the body are fixed, so that these points cannot move relative to each other. If one considers the relative motion of the points of a body, one speaks of a *deformable medium*.

4 Degrees of Freedom

The number of degrees of freedom f of a system represents the number of coordinates that are necessary to describe the motion of the particles of the system. A mass point that can freely move in space has 3 translational degrees of freedom: (x, y, z). If there are n mass points freely movable in space, this system has $3n$ degrees of freedom:

$$(x_i, y_i, z_i), \qquad i = 1, \ldots, n.$$

Degrees of freedom of a rigid body

We look for the number of degrees of freedom of a rigid body that can freely move. To describe a rigid body in space, one must know 3 noncollinear points of it. Hence, one has 9 coordinates:

$$\mathbf{r}_1 = (x_1, y_1, z_1), \qquad \mathbf{r}_2 = (x_2, y_2, z_2), \qquad \mathbf{r}_3 = (x_3, y_3, z_3),$$

which, however, are mutually dependent. Since by definition we are dealing with a rigid body, the distances between any two points are constant. One obtains

$$(x_1 - x_2)^2 + (y_1 - y_2)^2 + (z_1 - z_2)^2 = C_1^2 = \text{constant},$$
$$(x_1 - x_3)^2 + (y_1 - y_3)^2 + (z_1 - z_3)^2 = C_2^2 = \text{constant},$$
$$(x_2 - x_3)^2 + (y_2 - y_3)^2 + (z_2 - z_3)^2 = C_3^1 = \text{constant}.$$

Three coordinates can be eliminated by means of these 3 equations. The remaining 6 coordinates represent the 6 degrees of freedom. These are the 3 *degrees of freedom of translation* and the 3 *degrees of freedom of rotation*. The motion of a rigid body can always be understood as a *translation of any of its points* relative to an inertial system and a *rotation of the body about this point* (Chasles'[1] theorem). This is illustrated by Figure 4.1:

---

[1] *Michael Chasles*, French mathematician, b. in Épernon Nov. 15, 1793–d. Dec. 18, 1880, Paris. Banker in Chartres; 1841 to 1851 professor at the École Polytechnique; after 1846 professor at the Sorbonne in Paris. Chasles is independently of J. Steiner one of the founders of the synthetical geometry. His *Aperçu historique* by far surpassed the older representations of the development of geometry and stimulated new geometrical research in his age.

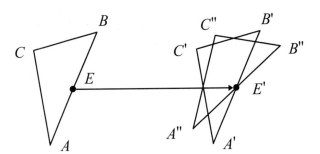

Figure 4.1. Chasles' theorem: The translation vector depends on the rotation, and vice versa.

$\triangle ABC \rightarrow \triangle A''B''C''$, namely, by translation in $\triangle A'B'C'$ and by rotation about the point E' in $\triangle A''B''C''$.

We now consider the rigid body with one point fixed in space. The motion is completely described if we know the coordinates of two points

$$\mathbf{r}_1 = (x_1, y_1, z_1) \quad \text{and} \quad \mathbf{r}_2 = (x_2, y_2, z_2)$$

and adopt the fixed point as the origin of the coordinate system. Since the body is rigid, we have

$$x_1^2 + y_1^2 + z_1^2 = \text{constant}, \qquad x_2^2 + y_2^2 + z_2^2 = \text{constant},$$
$$(x_1 - x_2)^2 + (y_1 - y_2)^2 + (z_1 - z_2)^2 = \text{constant}.$$

From these 3 equations one can eliminate 3 coordinates, so that the remaining 3 coordinates describe the *3 degrees of freedom of rotation*.

If a particle moves along a given curve in space, the number of degrees of freedom is $f = 1$. The curve can be written in the parametric form

$$x = x(s), \qquad y = y(s), \qquad z = z(s),$$

i.e., for a given curve the position of the particle is fully determined by specifying one parameter value s.

A deformable medium or a fluid has an infinite number of degrees of freedom (e.g., a vibrating string, a flexible bar, a drop of fluid).

Figure 4.2. Example of the parametric form: A caterpillar creeps on a blade of grass.

5 Center of Gravity

Definition: Let a system consist of n particles with the position vectors $\mathbf{r}_\nu$ and the masses m_ν for $\nu = (1, \ldots, n)$. The center of gravity of this system is defined as point S with the position vector $\mathbf{r}_s$:

$$\mathbf{r}_s = \frac{m_1 \mathbf{r}_1 + m_2 \mathbf{r}_2 + \cdots + m_n \mathbf{r}_n}{m_1 + m_2 + \cdots + m_n} = \frac{\sum_{\nu=1}^n m_\nu \mathbf{r}_\nu}{\sum_{\nu=1}^n m_\nu},$$

$$\mathbf{r}_s = \frac{1}{M} \sum_{\nu=1}^n m_\nu \mathbf{r}_\nu,$$

where $M = \sum_{\nu=1}^n m_\nu$ is the total mass of the system, and

$$M \mathbf{r}_s = \sum_{\nu=1}^n m_\nu \mathbf{r}_\nu$$

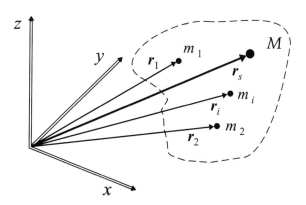

Figure 5.1. Definition of the center of gravity.

is the mass moment. For systems with uniform mass distribution over a volume V with the *volume density* ϱ, the sum $\sum_i m_i \mathbf{r}_i$ becomes an integral, and one obtains

$$\mathbf{r}_s = \frac{\int_V \mathbf{r}\varrho(\mathbf{r})\, dV}{\int_V \varrho\, dV}.$$

The individual components are

$$x_s = \frac{\sum_\nu m_\nu x_\nu}{M}, \qquad y_s = \frac{\sum_\nu m_\nu y_\nu}{M}, \qquad z_s = \frac{\sum_\nu m_\nu z_\nu}{M},$$

and for a continuous mass distribution

$$x_s = \frac{\int_V \varrho x\, dV}{M}, \qquad y_s = \frac{\int_V \varrho y\, dV}{M}, \qquad z_s = \frac{\int_V \varrho z\, dV}{M},$$

where the total mass is given by

$$M = \sum_\nu m_\nu \qquad \text{or} \qquad M = \int_V \varrho\, dV.$$

We consider three systems of masses with the centers of gravity $\mathbf{r}_1, \mathbf{r}_2, \mathbf{r}_3$ and the total masses M_1, M_2, M_3. The system 1 consists of the mass $M_1 = (m_{11} + m_{12} + m_{13} + \cdots)$ with the position vectors $\mathbf{r}_{11}, \mathbf{r}_{12}, \mathbf{r}_{13}, \ldots$; the systems 2 and 3 are analogous. Then by definition the centers of gravity are

$$\text{system 1:} \qquad \mathbf{r}_{s1} = \frac{\sum_i m_{1i} \mathbf{r}_{1i}}{\sum_i m_{1i}},$$

$$\text{system 2:} \qquad \mathbf{r}_{s2} = \frac{\sum_i m_{2i} \mathbf{r}_{2i}}{\sum_i m_{2i}},$$

$$\text{system 3:} \qquad \mathbf{r}_{s3} = \frac{\sum_i m_{3i} \mathbf{r}_{3i}}{\sum_i m_{3i}}.$$

For the center of gravity of the total system we have the same relation:

$$\mathbf{r}_s = \frac{\sum_i m_{1i} \mathbf{r}_{1i} + \sum_i m_{2i} \mathbf{r}_{2i} + \sum_i m_{3i} \mathbf{r}_{3i}}{\sum_i m_{1i} + \sum_i m_{2i} + \sum_i m_{3i}}$$

$$= \frac{M_1 \mathbf{r}_{s1} + M_2 \mathbf{r}_{s2} + M_3 \mathbf{r}_{s3}}{M_1 + M_2 + M_3}.$$

Hence, for composite systems we can determine the centers of gravity and masses of the partial systems, and from them calculate the center of gravity of the total system. The calculation can thereby be much simplified. This fact is often referred to as the *cluster property* of the center of gravity.

CENTER OF GRAVITY

The linear momentum of a particle system is the sum of the momenta of the individual particles:

$$\mathbf{P} = \sum_{\nu=1}^{n} \mathbf{p}_\nu = \sum_{\nu=1}^{n} m_\nu \dot{\mathbf{r}}_\nu.$$

If we introduce the center of gravity by $M\mathbf{r}_s = \sum_i m_i \mathbf{r}_i$, we see that $\mathbf{P} = M\dot{\mathbf{r}}_s$, i.e., the total momentum of a particle system equals the product of the total mass M united in the center of gravity and its velocity $\dot{\mathbf{r}}_s$. This means that the translation of a body can be described by the motion of the center of gravity.

Example 5.1: Exercise: Center of gravity for a system of three mass points

Find the coordinates of the center of gravity for a system of 3 mass points.

$m_1 = 1\,\text{g}, \qquad m_2 = 3\,\text{g}, \qquad m_3 = 10\,\text{g},$
$\mathbf{r}_1 = (1, 5, 7)\,\text{cm}, \qquad \mathbf{r}_2 = (-1, 2, 3)\,\text{cm}, \qquad \mathbf{r}_3 = (0, 4, 5)\,\text{cm}.$

Solution   For the center of gravity, one finds

$$\mathbf{r}_s = \frac{1}{14}(1 - 3,\; 5 + 3\cdot 2 + 10\cdot 4,\; 7 + 3\cdot 3 + 10\cdot 5)\,\text{cm}$$

or, recalculated,

$$\mathbf{r}_s = \frac{1}{14}(-2,\; 51,\; 66)\,\text{cm}.$$

Example 5.2: Exercise: Center of gravity of a pyramid

Find the center of gravity of a pyramid with edge length a and a homogeneous mass distribution.

Solution   Because of the homogeneous mass distribution, the mass density $\rho(\mathbf{r}) = \rho_0 = $ constant. The base of the pyramid is represented by the equation

$$x + y + z = a.$$

The coordinate axes are the edges, and the origin is the top. Then

$$\mathbf{r}_s = \frac{\int_V \rho_0 \mathbf{r}\, dV}{\int_V \rho_0\, dV} = \frac{\int_V \mathbf{r}\, dV}{\int_V dV}, \qquad dV = dx\, dy\, dz.$$

The integration limits are evident from Figure 5.2. The integration runs over z along the column from $z = 0$ to $z = a - x - y$; over y along the prism from $y = 0$ to $y = a - x$; and over x along the pyramid from $x = 0$ to $x = a$:

$$\mathbf{r}_s = \frac{\int_V \mathbf{r}\, dV}{\int_V dV} = \frac{\int_{x=0}^{a} \int_{y=0}^{a-x} \int_{z=0}^{a-x-y} \mathbf{r}\, dz\, dy\, dx}{\int_{x=0}^{a} \int_{y=0}^{a-x} \int_{z=0}^{a-x-y} dz\, dy\, dx},$$

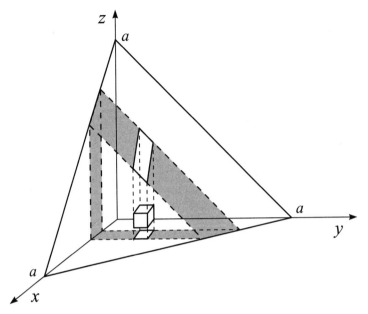

Figure 5.2.

with

$$\mathbf{r} = (x, y, z) \Rightarrow \int_V \mathbf{r}\, dV = \int_{x=0}^{a} \int_{y=0}^{a-x} \left(xy, yz, \frac{1}{2}z^2\right)\Bigg|_{z=0}^{z=a-x-y} dy\, dx,$$

$$\int_V \mathbf{r}\, dV = \int_0^a \int_0^{a-x} \left(x(a-x-y), y(a-x-y), \frac{1}{2}(a-x-y)^2\right) dy\, dx.$$

The corresponding integration over y and x yields

$$\int_V \mathbf{r}\, dV = \frac{a^4}{24}(1,1,1), \qquad \int_V dV = V = \frac{a^3}{6}.$$

Thus, the center of gravity is at

$$\mathbf{r}_s = \frac{\int_V \mathbf{r}\, dV}{\int_V dV} = \frac{a}{4}(1,1,1).$$

Example 5.3: Exercise: Center of gravity of a semicircle

Find the center of gravity of a semicircular disk of radius a. The surface density is constant.

CENTER OF GRAVITY

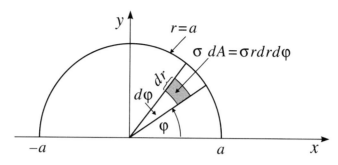

Figure 5.3.

Solution   The *surface density* σ = constant. (The surface density is defined by $\sigma(\mathbf{r}) = \lim_{\Delta A \to 0} \Delta m(x, y, z)/\Delta A$.) x_s and y_s represent the coordinates of the center of gravity. We use polar coordinates for calculating the center of gravity. The equation of the semicircle then reads

$$r = a, \qquad 0 \leq \varphi \leq \pi.$$

Because of symmetry, $x_s = 0$, and for y_s, we have

$$y_s = \frac{\int_A \sigma y \, dA}{\int_A \sigma \, dA} = \frac{\int_{\varphi=0}^{\pi} \int_{r=0}^{a} (r \sin\varphi) r \, dr \, d\varphi}{A}.$$

The evaluation of the integral yields

$$y_s = \frac{2a^3/3}{\pi a^2/2} = \frac{4a}{3\pi},$$

i.e., the center of gravity lies at $\mathbf{r}_s = (0, 4a/(3\pi))$.

Example 5.4: Exercise: Center of gravity of a circular cone

Determine the center of gravity of

(a) a homogeneous circular cone with base radius a and height h; and

(b) a circular cone as in (a), with a hemisphere of radius a set onto its base.

Solution   (a) Because of symmetry, the center of gravity is on the z-axis, i.e., $x_s = y_s = 0$. For the z-component, we have

$$z_s = \frac{\int_k z \, dV}{\int_k dV} = \frac{\int_k z \, dV}{(1/3)\pi a^2 h}.$$

We adopt cylindrical coordinates for evaluating the integral:

$$\int_k z \, dV = \int_{\varphi=0}^{2\pi} \int_{\varrho=0}^{a} \int_{z=0}^{h(1-\varrho/a)} z \varrho \, d\varrho \, d\varphi \, dz$$

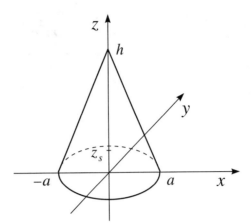

Figure 5.4.

$$= 2\pi \int_{\varrho=0}^{a} \frac{1}{2} h^2 \left(1 - \frac{\varrho}{a}\right)^2 \varrho \, d\varrho$$

$$= \pi h^2 \left[\frac{1}{2}\varrho^2 - \frac{2\varrho^3}{3a} + \frac{\varrho^4}{4a^2}\right]_0^a = \pi \frac{a^2 h^2}{12},$$

$$z_s = \frac{\pi h^2 a^2 \cdot 3}{12 \pi a^2 h} = \frac{1}{4} h.$$

Thus, the center of gravity of a circular cone is independent of the radius of the base.

(b) See Figure 5.5. One then has

$$z_s = \frac{\int_{\text{cone}} z \, dV + \int_{\text{hemisphere}} z \, dV}{V_{\text{cone}} + V_{\text{hemisphere}}} = \frac{\frac{1}{12}\pi h^2 a^2 + \int_{\text{hemisphere}} z \, dV}{(\pi/3)(h + 2a)a^2},$$

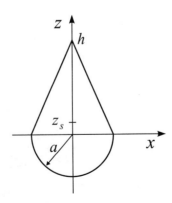

Figure 5.5. Because of symmetry the center of gravity is again on the z-axis.

CENTER OF GRAVITY

$$\int_{\text{hemisphere}} z\, dV = \int_{\varphi=0}^{2\pi}\int_{\varrho=0}^{a}\int_{z=-\sqrt{a^2-\varrho^2}}^{0} \varrho z\, d\varphi\, d\varrho\, dz$$

$$= \pi \int_{\varrho=0}^{a} (\varrho^2 - a^2)\varrho\, d\varrho$$

$$= \pi \left[\frac{\varrho^4}{4} - \frac{a^2\varrho^2}{2}\right]_0^a$$

$$= -\frac{\pi a^4}{4}.$$

Hence, the center of gravity is given by

$$z_s = \frac{\frac{1}{12}\pi a^2 h^2 - \frac{1}{4}\pi a^4}{\frac{a^2\pi}{3}(h+2a)} = \frac{1}{4}\frac{h^2 - 3a^2}{h+2a},$$

$$y_s = 0, \quad x_s = 0.$$

Example 5.5: **Exercise: Momentary center and pole path**

(a) Show that any positional variation of a rigid disk in the plane can be represented by a pure rotation about a point at a finite distance or at infinity. (*Hint*: The position of the disk is already fixed by specifying two points A and B).

(b) Show by "differential" variation of position: The planar motion of a rigid disk can be described at any moment by a pure rotation about a point varying with the motion, the so-called *momentary center*. The geometric locus of these momentary centers is called the *pole path* or the *fixed pole curve*.

(c) Calculate the fixed pole curve $r(\varphi)$ for a ladder sliding on two perpendicular walls.

(d) Calculate the fixed pole curve $r(\varphi)$ for a bar of length l that can move in the guide shown in Figure 5.6.

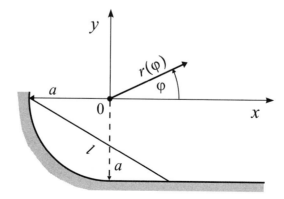

Figure 5.6.

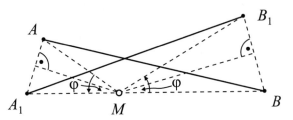

Figure 5.7.

Solution (a) For describing the motion of the disk we take the (arbitrary) straight line AB; it turns into the straight line A_1B_1. The intersection M of the mid-perpendiculars onto AA_1 and BB_1 is the desired center of rotation.

Argument: The triangles ABM and A_1B_1M are congruent. Hence the motion can be considered as a rotation of the triangle ABM (involving the straight line AB) about M by the angle φ.

(b) For an infinitely small rotation by $d\varphi$ the same considerations hold. But now the individual turning points vary. These are the so-called *momentary centers*. In a differential rotation about a momentary center M, for any point the path element $d\mathbf{r}$ and the velocity vector $\mathbf{v}$ point along the same direction and are perpendicular to the connecting lines to M (see Figure 5.8). The geometric locus of the momentary centers is called the pole curve.

(c) According to (b), one gets Figure 5.9. The straight line $l = AB$ forms a diagonal of the square $OBMA$. Since the diagonals of a square are equal, M must move along a circle of radius l.

(d) According to (b), one can construct Figure 5.10. Evidently,

$$\sin\alpha = \frac{a}{l}(1 - \sin\alpha) \quad \text{and} \quad \cos\alpha = \sqrt{1 - \frac{a^2}{l^2}(1 - \sin\varphi)^2},$$

$$\overline{AC} = l\cos\alpha - a\cos\varphi,$$

$$\overline{OM} = \frac{\overline{AC}}{\cos\varphi} = l\frac{\cos\alpha}{\cos\varphi} - a = l\sqrt{\frac{1 - (a^2/l^2)(1 - \sin\varphi)^2}{\cos^2\varphi}} - a.$$

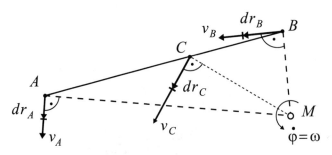

Figure 5.8.

CENTER OF GRAVITY

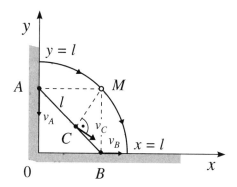

Figure 5.9.

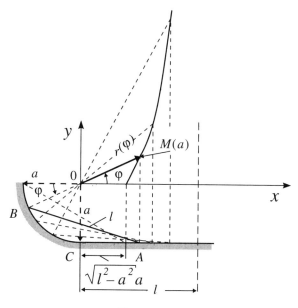

Figure 5.10.

Thus, in polar coordinates the equation of the fixed pole curve $r(\varphi)$ reads

$$r = r(\varphi) = \overline{OM} = -a + l\sqrt{\frac{1 - (a^2/l^2)(1 - \sin\varphi)^2}{\cos^2\varphi}}.$$

Example 5.6: Scattering in a central field

(1) The problem

The two-body problem appeared for the first time in recent physics in investigations of planetary motion. However, the classical formulation of the two-body problem provides information both on the bound state as well as on the unbound state (scattering state) of a system.

The study of the unbound states of a system became of great importance in modern physics. One learns about the mutual interaction of two objects by scattering them off each other and observing the path of the scattered particles as a function of the incident energy and of other path parameters. The objects studied in this way are usually molecules, atoms, atomic nuclei, and elementary particles. Scattering processes in these microscopic regions must be described by quantum mechanics. However, one can obtain information on scattering processes by means of classical mechanics which is confirmed by a quantum mechanical calculation. Moreover, one may learn the methods for describing scattering phenomena by studying the classical case.

The schematic arrangement of a scattering experiment is shown in Figure 5.11. We consider a homogeneous beam of incoming particles (projectiles) of the same mass and energy. The force acting on a particle is assumed to drop to zero at large distances from the scattering center. This guarantees that the interaction is somehow localized. Let the initial velocity v_0 of each projectile relative to the force center be so large that the system is in the unbound state, i.e., for $t \to \infty$ the distance between the two scattering particles shall become arbitrarily large. For a repulsive potential this happens for any value of v_0; this does not hold for an attractive potential.

The interaction of a projectile with the target particle manifests itself by the fact that the flight direction after the collision differs from that before the collision (the usage of the words "before" and "after" in this context presupposes a more or less finite range of the interaction potential).

(2) Definition of the cross section

Measured quantities are count rates (number of particles/s in the detector, which is assumed to be small). These count rates depend first on the physical data as kind of projectile and target, incidence energy, and scattering direction, and second on the specific experimental conditions such as detector size, distance between target and detector, number of scattering centers, or incident intensity. In order to have a quantity that is independent of the latter features, one defines the *differential cross section*

$$\frac{d\sigma}{d\Omega}(\vartheta, \varphi) := \frac{(\text{number of particles scattered to } d\Omega)\ /\text{s}}{d\Omega \cdot n \cdot I}. \tag{5.1}$$

Here, n is the number of scattering centers and I is the beam intensity, which is given by (number of projectiles)/(s·m^2). The scattering direction is represented here by ϑ and φ. ϑ is the angle between the asymptotic scattering direction and the incidence direction; it is called the scattering angle. φ is the azimuth angle. $d\Omega$ denotes the solid-angle element covered by the detector. Since we assumed the detector to be small, we have

$$d\Omega = \sin\vartheta\, d\vartheta\, d\varphi, \tag{5.2}$$

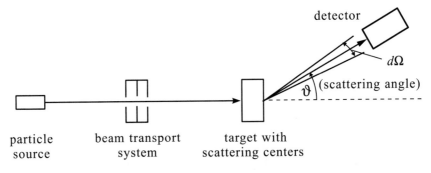

Figure 5.11. Schematic setup of a scattering experiment.

where $d\vartheta$ and $d\varphi$ specify the detector size. Note that $(d\sigma/d\Omega)(\vartheta,\varphi)$ is defined by (5.1) and is not a derivative of a quantity σ with respect to Ω. Obviously $d\sigma/d\Omega)(\vartheta,\varphi)$ has the dimension of an area. The standard unit is

$$1\,\text{b} = 1\,\text{Barn} = 100(\text{fm})^2 = 10^{-28}\,\text{m}^2. \tag{5.3}$$

Often the differential cross section will be independent of the azimuth angle φ (we shall restrict ourselves to this case), and one can define

$$\frac{d\sigma}{d\vartheta} := 2\pi \sin\vartheta \frac{d\sigma}{d\Omega}(\vartheta,\varphi); \tag{5.4}$$

see Figure 5.12.

$$\frac{d\sigma}{d\vartheta}(\vartheta) = \frac{(\text{number of particles scattered to } d\Omega)\,/\text{s}}{d\vartheta \cdot n \cdot I}. \tag{5.5}$$

Finally, we introduce the total cross section, defined by

$$\sigma_{\text{tot}} = \int d\Omega \frac{d\sigma}{d\Omega}(\vartheta,\varphi) = \int_0^{2\pi} d\varphi \int_0^{\pi} d\vartheta \sin\vartheta \frac{d\sigma}{d\Omega} = \int_0^{\pi} d\vartheta \frac{d\sigma}{d\vartheta}(\vartheta). \tag{5.6}$$

It depends only on the kinds of particles, and possibly on the incidence energy:

$$\sigma_{\text{tot}} = \frac{(\text{number of scattered particles})/\text{s}}{n \cdot I}. \tag{5.7}$$

Like $d\sigma/d\Omega$ it has the dimension of an area. It equals the size of the (fictive) area of a scattering center which must be traversed perpendicularly by the projectiles in order to be deflected at all.

(3) Introduction of the collision parameter, its relation to the scattering angle, and the formula for the differential cross section

It is clear that the scattering angle ϑ at fixed energy can depend only on the collision parameter b, since the initial position and the initial velocity of the particle are then specified. The *collision parameter* is defined as the vertical distance of the asymptotic incidence direction of the projectile from the initial position of the scatterer. Hence, for $E = $ constant the scattering angle is

$$\vartheta = \vartheta(b). \tag{5.8}$$

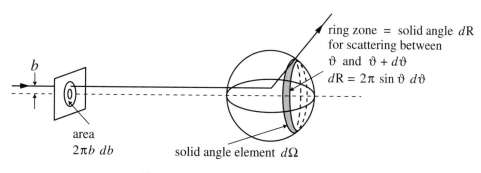

Figure 5.12. Definition of the cross section.

Since the movements in classical mechanics are determined, this connection is unambiguous. (This statement is no longer valid in quantum mechanics.) Thus,

$$b = b(\vartheta), \tag{5.9}$$

which means that, by observing an arbitrary particle at a definite scattering angle ϑ, one can determine in a straightforward manner the value of the scattering parameter b of the incident particle. This fact allows the following consideration. The number dN of projectiles per second that move with values b' of the collision parameter

$$b \leq b' \leq b + db$$

toward a scattering center is

$$dN = I \cdot 2\pi b \, db \quad \text{or} \quad dN = I \cdot 2\pi b \left| \frac{db}{d\vartheta} \right| d\vartheta.$$

The sign of absolute value stands since the number dN by definition cannot become negative. Just this number of particles are scattered into the solid angle element

$$dR = 2\pi \sin \vartheta \, d\vartheta.$$

By inserting this into (5.5), we get

$$\frac{d\sigma}{d\vartheta}(\vartheta) = 2\pi b \left| \frac{db}{d\vartheta} \right|, \tag{5.10}$$

and for the differential cross section

$$\frac{d\sigma}{d\Omega}(\vartheta) = \frac{b(\vartheta)}{\sin \vartheta} \left| \frac{db}{d\vartheta} \right|. \tag{5.11}$$

This is just the desired relation. The function $b(\vartheta)$ is determined by the force law that governs the particular case. One realizes that the knowledge of the differential cross section allows one to determine the interaction potential between the projectile and the target particle.

In general, the scattering angle will depend not only on the collision parameter but also on the incident energy. As a consequence, the differential cross section also becomes energy dependent. Hence, one can measure the differential cross section as a function of the projectile energy by observing the scattered particles at a fixed scattering angle.

(4) Transition to the center-of-mass system, and transformation of the differential cross section from the center-of-mass system to the laboratory system

The considerations of the last section are to some extent independent of the reference system. If we move from the laboratory system S to another system S' that moves with constant velocity $\mathbf{V}$ parallel to the beam axis, the scattering angle and the differential cross section (5.5) will change, but the derivation in the last section remains unchanged, so that the relation (5.11) remains valid.

This has a practical meaning inasmuch as cross sections are always measured in the laboratory system S where the target is at rest, but the calculation of $b(\vartheta')$ often simplifies in the center-of-mass system S'. We therefore derive a relation between these two cross sections. In the following the primed and nonprimed quantities shall always refer to these two systems.

First, we investigate the relation between the scattering angles ϑ and ϑ'. Let $\mathbf{v}_1^f$ and $\mathbf{v_1'}^f$ be the asymptotic final velocity (f = final) of the projectile of mass m_1 in the system S and S', respectively. $\mathbf{V}$ is the relative velocity of the two systems.

CENTER OF GRAVITY

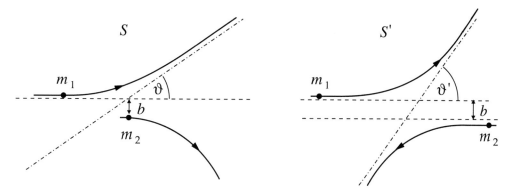

Figure 5.13. Scattering in the laboratory system (S) and in the center-of-mass system (S').

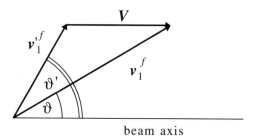

Figure 5.14.

From Figure 5.14, one immediately sees that

$$\tan \vartheta = \frac{v_1'^f \sin \vartheta'}{v_1'^f \cos \vartheta' + V} = \frac{\sin \vartheta'}{\cos \vartheta' + V/v_1'^f},$$

where V stands for the magnitude of $\mathbf{V}$, and analogously for $v_1'^f$. Furthermore,

$$m_1 v_1^i = (m_1 + m_2)V,$$

where v_1^i is the initial velocity of the projectile in the laboratory system (i = initial), and

$$v_1^i = V + v_1'^i.$$

Because $m_1 v_1'^i = m_2 v_2'^i$ and $m_1 v_1'^f = m_2 v_2'^f$ for elastic scattering ($E'^i_{\text{kin}} = E'^f_{\text{kin}}$), $v_1'^i = v_1'^f$, and therefore,

$$\frac{V}{v_1'^f} = \frac{m_1}{m_2}.$$

Hence,

$$\tan \vartheta = \frac{\sin \vartheta'}{\cos \vartheta' + m_1/m_2}. \tag{5.12}$$

This relation defines the function $\vartheta'(\vartheta)$; we will not give it explicitly. If a projectile in S is scattered into the ring dR with the "radius" ϑ and the width $d\vartheta$ (see Figure 5.12), it will in S' be scattered into a ring dR' with the "radius" $\vartheta'(\vartheta)$ and the width $d\vartheta' = (d\vartheta'/d\vartheta)d\vartheta$. The number of particles scattered to dR in S and to dR' in S' is therefore identical, and with equation (5.5), we get

$$\frac{d\sigma}{d\vartheta}(\vartheta) \cdot d\vartheta = \frac{d\sigma'}{d\vartheta'}(\vartheta') \cdot d\vartheta' = \frac{d\sigma'}{d\vartheta'}(\vartheta')\frac{d\vartheta'}{d\vartheta}d\vartheta,$$

thus,

$$\frac{d\sigma}{d\vartheta}(\vartheta) = \frac{d\sigma'}{d\vartheta'}(\vartheta')\frac{d\vartheta'}{d\vartheta}, \tag{5.13}$$

or

$$\frac{d\sigma}{d\Omega}(\vartheta) = \frac{d\sigma'}{d\Omega'}(\vartheta')\frac{\sin\vartheta'}{\sin\vartheta}\frac{d\vartheta'}{d\vartheta}. \tag{5.14}$$

This is the desired connection.

The difference between the scattering angles and the cross sections, respectively, is obviously determined by the mass ratio of projectile and target particle (see eq. (5.12)).

Example 5.7: Exercise: Rutherford scattering cross section

A particle of mass m moves from infinity with the collision parameter b toward a force center. The central force is inversely proportional to the square of the distance:

$$F = kr^{-2}.$$

(a) Calculate the scattering angle as a function of b and of the initial velocity of the particle.

(b) What are the differential and the total cross sections?

Solution (a) From the discussion of the Kepler problem, we know that the underlying force law has the form

$$F = -\frac{k}{r^2}. \tag{5.15}$$

The minus sign means that the force is attractive. The path equation reads (see *Classical Mechanics: Point Particles and Relativity*):

$$\frac{1}{r} = \frac{mk}{l^2}\left(1 + \sqrt{1 + \frac{2El^2}{mk^2}}\cos(\theta - \theta')\right) \tag{5.16}$$

(E = initial energy, l = angular momentum, m = mass of the particle, θ' = integration constant). With the standard abbreviation

$$\varepsilon = \sqrt{1 + \frac{2El^2}{mk^2}}, \tag{5.17}$$

one can write for equation (5.16)

$$\frac{1}{r} = \frac{mk}{l^2}(1 + \varepsilon\cos(\theta - \theta')). \tag{5.18}$$

The path is characterized by ε:

$$\begin{aligned} \varepsilon > 1, \quad & E > 0: \quad & \text{hyperbola,} \\ \varepsilon = 1, \quad & E = 0: \quad & \text{parabola,} \\ \varepsilon < 1, \quad & E < 0: \quad & \text{ellipse,} \\ \varepsilon = 0, \quad & E = -\frac{mk^2}{2l^2}: \quad & \text{circle.} \end{aligned} \qquad (5.19)$$

In the given problem the force law is

$$F = \frac{k}{r^2}. \qquad (5.20)$$

The force is repulsive. For illustration, we consider the scattering of charged particles by a Coulomb field (e.g., atomic nuclei by atomic nuclei, protons by nuclei, or electrons by electrons, etc.). The scattering force center is created by a fixed charge $-Ze$ and acts on the particle with the charge $-Z'e$. The force is then

$$F = \frac{ZZ'e^2}{r^2}. \qquad (5.21)$$

If we set $k = -ZZ'e^2$, we can directly take over the equations for an attractive potential. The path equation (5.18) now reads

$$\frac{1}{r} = -\frac{mZZ'e^2}{l^2}(1 + \varepsilon \cos \theta). \qquad (5.22)$$

The coordinates were rotated so that $\theta' = 0$. For ε (equation (5.17)) it follows that

$$\varepsilon = \sqrt{1 + \frac{2El^2}{m(ZZ'e^2)^2}} = \sqrt{1 + \left(\frac{2Eb}{(ZZ'e^2)^2}\right)^2}. \qquad (5.23)$$

Here, we used the relation

$$l = mbv_\infty = b\sqrt{2mE}, \qquad E = \frac{1}{2}mv_\infty^2 \qquad (5.24)$$

between angular momentum (l) and collision parameter (b).

Since $\varepsilon > 1$, equation (5.22) represents a hyperbola (see equation (5.19)). Because of the minus sign, the values of θ for the path are restricted to values for which

$$\cos \theta < -\frac{1}{\varepsilon} \qquad (5.25)$$

(see Figure 5.15). Note that the force center for repulsive forces is in the *outer* focal point (see Figure 5.16).

The change of θ that occurs if the particle comes from infinity, and is then scattered and moves to infinity again, equals the angle ϕ between the asymptotes, which is the supplement to the scattering angle θ (see Figure 5.16).

From Figure 5.15 and equation (5.25), it follows that

$$\cos\left(\frac{\pi}{2} - \frac{\theta}{2}\right) = \sin\left(\frac{\theta}{2}\right) = \cos\left(\frac{\phi}{2}\right) = \frac{1}{\varepsilon}. \qquad (5.26)$$

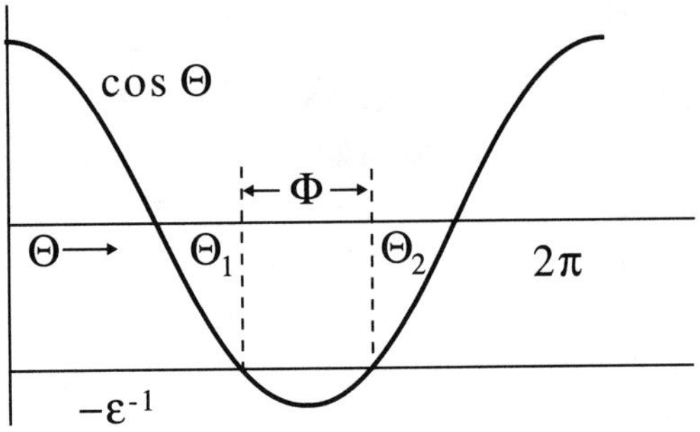

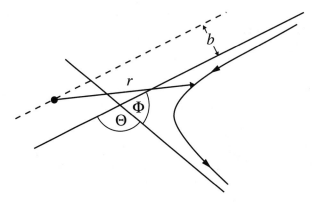

Figure 5.15. Region of θ for repulsive Coulomb scattering.

Figure 5.16. Illustration of the hyperbolic path of a particle that is pushed off by a force center. The force center lies in the outer focal point.

The relation $\cos(\phi/2) = 1/\varepsilon$ can be proved as follows: The two limiting angles θ_1 and θ_2 satisfy the condition

$$\cos\theta_1 = -\frac{1}{\varepsilon},$$
$$\cos\theta_2 = -\frac{1}{\varepsilon}.$$
(5.27)

From this it follows that (see Figure 5.17):

$$\sin\theta_1 = -\sin\theta_2,$$
$$\cos\left(\frac{\theta_1}{2}\right) = -\cos\left(\frac{\theta_2}{2}\right).$$
(5.28)

The first of these equations can be rewritten as

$$2\cos\left(\frac{\theta_1}{2}\right)\sin\left(\frac{\theta_1}{2}\right) = \sin\theta_1 = -\sin\theta_2 = -2\cos\left(\frac{\theta_2}{2}\right)\sin\left(\frac{\theta_2}{2}\right),$$
(5.29)

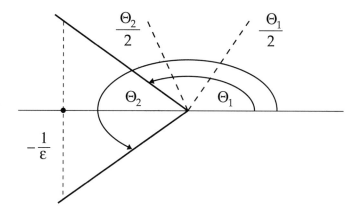

Figure 5.17. The limiting angles θ_1 and θ_2 have the same cosine.

and therefore,

$$\sin\left(\frac{\theta_1}{2}\right) = \sin\left(\frac{\theta_2}{2}\right). \tag{5.30}$$

We look for $\cos(\phi/2) = \cos((\theta_2 - \theta_1)/2)$, $\phi = \theta_2 - \theta_1$:

$$\cos\frac{\phi}{2} = \cos\left(\frac{\theta_2}{2} - \frac{\theta_1}{2}\right) = \cos\left(\frac{\theta_2}{2}\right)\cos\left(\frac{\theta_1}{2}\right) + \sin\left(\frac{\theta_2}{2}\right)\sin\left(\frac{\theta_1}{2}\right)$$
$$= -\cos^2\left(\frac{\theta_1}{2}\right) + \sin^2\left(\frac{\theta_1}{2}\right). \tag{5.31}$$

From $\cos\theta_1 = -1/\varepsilon$, it follows that

$$-\frac{1}{\varepsilon} = \cos\theta_1 = \cos^2\left(\frac{\theta_1}{2}\right) - \sin^2\left(\frac{\theta_1}{2}\right) \tag{5.32}$$

$$\Rightarrow \quad \sin^2\left(\frac{\theta_1}{2}\right) = \cos^2\left(\frac{\theta_1}{2}\right) + \frac{1}{\varepsilon}. \tag{5.33}$$

Insertion into equation (5.31) yields

$$\cos\frac{\phi}{2} = -\cos^2\left(\frac{\theta_1}{2}\right) + \cos^2\left(\frac{\theta_1}{2}\right) + \frac{1}{\varepsilon} = \frac{1}{\varepsilon}. \tag{5.34}$$

From there we find

$$\frac{1}{\sin^2(\theta/2)} = 1 + \left(\frac{2Eb}{ZZ'e^2}\right)^2$$

$$\Leftrightarrow \quad \frac{1}{\sin^2(\theta/2)} - 1 = \frac{1 - \sin^2(\theta/2)}{\sin^2(\theta/2)} = \cot^2\left(\frac{\theta}{2}\right) = \left(\frac{2Eb}{ZZ'e^2}\right)^2$$

$$\Rightarrow \quad \frac{\theta}{2} = \text{arccot}\left(\frac{2Eb}{ZZ'e^2}\right)^2. \tag{5.35}$$

(b) From equations (5.24) and (5.26) it follows that

$$b = \frac{ZZ'e^2}{2E} \cot\left(\frac{\theta}{2}\right)$$

$$\Rightarrow \quad \frac{db}{d\theta} = -\frac{ZZ'e^2}{4E} \frac{1}{\sin^2(\theta/2)}. \tag{5.36}$$

The differential cross section as a function of θ is given by

$$\frac{d\sigma}{d\Omega} = -\frac{b}{\sin\theta} \frac{db}{d\theta}. \tag{5.37}$$

Thus, one obtains

$$\frac{d\sigma}{d\Omega} = \frac{(ZZ'e^2)^2}{2E \sin\theta \, 2E} \cdot \frac{1}{\sin^2(\theta/2)} \cot\left(\frac{\theta}{2}\right)$$

$$= \frac{1}{2}\left(\frac{ZZ'e^2}{2E}\right)^2 \frac{\cot(\theta/2)}{\sin\theta \sin^2(\theta/2)}, \tag{5.38}$$

and with the identity

$$\sin\theta = 2\sin\left(\frac{\theta}{2}\right)\cos\left(\frac{\theta}{2}\right) \tag{5.39}$$

it follows that

$$\frac{d\sigma}{d\Omega} = \frac{1}{4}\left(\frac{ZZ'e^2}{2E}\right)^2 \frac{1}{\sin^4(\theta/2)}. \tag{5.40}$$

This is the well-known *Rutherford scattering formula*. The total cross section is calculated according to

$$\sigma_{\text{total}} = \int \frac{d\sigma}{d\Omega}(\Omega) \, d\Omega = 2\pi \int \frac{d\sigma}{d\Omega}(\theta) \sin\theta \, d\theta. \tag{5.41}$$

By inserting $d\sigma/d\Omega(\theta)$ from equation (5.40), one quickly realizes that the expression diverges because of the strong singularity at $\theta = 0$. This is due to the long-range nature of the Coulomb force. If one uses potentials which decrease faster than $1/r$, this singularity disappears.

Example 5.8: Exercise: Scattering of a particle by a spherical square well potential

A particle is scattered by a spherical square well potential with radius a and depth U_0:

$$U = 0 \quad (r > a),$$
$$U = -U_0 \quad (r \leq a).$$

Calculate the differential and the total cross section. *Hint:* Use the refraction law for particles at sharp surfaces which results from the following consideration: Let the velocity of the particle before scattering by a sharp potential well be $v_1 = v_\infty$ and after scattering v_2. Due to momentum conservation perpendicular to the incident normal ("transverse momentum conservation") one has

$$v_\infty \sin\alpha = v_2 \sin\beta \tag{5.42}$$

$$\Rightarrow \quad \frac{\sin\alpha}{\sin\beta} = \frac{v_2}{v_\infty}. \tag{5.43}$$

From the energy conservation law it follows that

$$E = T + U = \frac{1}{2}mv_\infty^2 + U_1 = \frac{1}{2}mv_2^2 + U_2. \tag{5.44}$$

Solving for v_2 yields

$$v_2 = \sqrt{v_\infty^2 + \frac{2}{m}(U_1 - U_2)} = \sqrt{v_\infty^2 + \frac{2}{m}U_0}. \tag{5.45}$$

Insertion into equation (5.43) finally yields

$$n = \frac{\sin\alpha}{\sin\beta} = \frac{\sqrt{v_\infty^2 + (2/m)U_0}}{v_\infty} = \sqrt{1 + \frac{2U_0}{mv_\infty^2}}. \tag{5.46}$$

Solution The straight path of the particle is broken when entering and leaving the field. We have the relation

$$\frac{\sin\alpha}{\sin\beta} = n, \tag{5.47}$$

where according to (5.46)

$$n = \sqrt{1 + \frac{2U_0}{mv_\infty^2}}.$$

The deflection angle is (see Figure 5.18)

$$\chi = 2(\alpha - \beta)$$

$$\Rightarrow \frac{\sin\alpha}{\sin\beta} = \frac{\sin(\alpha - \chi/2)}{\sin\alpha}$$

$$= \frac{\sin\alpha\cos(\chi/2) - \cos\alpha\sin(\chi/2)}{\sin\alpha}$$

$$= \cos\left(\frac{\chi}{2}\right) - \cot\alpha\sin\left(\frac{\chi}{2}\right) = \frac{1}{n}. \tag{5.48}$$

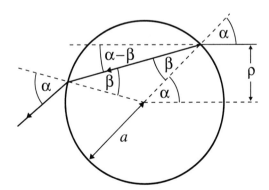

Figure 5.18. In the inner and outer region of the spherical potential well the particle moves along straight lines. When passing the surface it will be refracted.

From Figure 5.18, we have

$$a \sin \alpha = \varrho. \tag{5.49}$$

Because $\sin^2 \alpha + \cos^2 \alpha = 1$, it follows that

$$\cos \alpha = \sqrt{1 - \left(\frac{\varrho}{a}\right)^2}. \tag{5.50}$$

Now we can eliminate α from equation (5.49):

$$\frac{\cos(\chi/2) - 1/n}{\sin(\chi/2)} = \cot \alpha = \frac{\cos \alpha}{\sin \alpha} = \frac{a \cos \alpha}{\varrho} \tag{5.51}$$

and with (5.50), we get

$$\varrho = a \frac{\sqrt{1 - (\varrho/a)^2} \sin(\chi/2)}{(\cos(\chi/2) - 1/n)},$$

$$\varrho^2 = \frac{a^2 \sin^2(\chi/2) - \varrho^2 \sin^2(\chi/2)}{(\cos(\chi/2) - 1/n)^2}$$

$$= \frac{a^2 \sin^2(\chi/2)}{(\cos(\chi/2) - 1/n)^2 + \sin^2(\chi/2)} = \frac{a^2 \sin^2(\chi/2)}{1 - (2/n) \cos(\chi/2) + 1/n^2}$$

$$\Rightarrow \quad \varrho^2 = a^2 \frac{n^2 \sin^2(\chi/2)}{n^2 - 2n \cos(\chi/2) + 1}. \tag{5.52}$$

To get the cross section, we differentiate

$$\varrho = a \frac{n \sin(\chi/2)}{(n^2 - 2n \cos(\chi/2) + 1)^{1/2}} \tag{5.53}$$

with respect to χ.

$$\Rightarrow \quad \frac{d\varrho}{d\chi} = \frac{\frac{an}{2} \cos(\chi/2)}{(n^2 - 2n \cos(\chi/2) + 1)^{1/2}} - \frac{1}{2} \frac{an \sin(\chi/2) \cdot n \sin(\chi/2)}{(n^2 - 2n \cos(\chi/2) + 1)^{3/2}}$$

$$= \frac{\frac{an}{2} \cos \frac{\chi}{2} \left(n^2 + 1 - 2n \cos \frac{\chi}{2}\right) - \frac{1}{2} an^2 \sin^2 \frac{\chi}{2}}{\left(n^2 + 1 - 2n \cos \frac{\chi}{2}\right)^{3/2}}$$

$$= \frac{\frac{a}{2} n^3 \cos \frac{\chi}{2} + \frac{an}{2} \cos \frac{\chi}{2} - an^2 \cos^2 \frac{\chi}{2} - \frac{1}{2} an^2 \sin^2 \frac{\chi}{2}}{\left(n^2 + 1 - 2n \cos \frac{\chi}{2}\right)^{3/2}}$$

$$= \frac{an}{2} \frac{n^2 \cos \frac{\chi}{2} + \cos \frac{\chi}{2} - n - n \cos^2 \frac{\chi}{2}}{\left(n^2 + 1 - 2n \cos \frac{\chi}{2}\right)^{3/2}}$$

$$= \frac{an}{2} \frac{\left(n \cos \frac{\chi}{2} - 1\right)\left(n - \cos \frac{\chi}{2}\right)}{\left(n^2 + 1 - 2n \cos \frac{\chi}{2}\right)^{3/2}} \tag{5.54}$$

CENTER OF GRAVITY

$$\Rightarrow \quad \sigma(\chi) = \frac{\frac{d\sigma}{d\Omega}}{\sin\chi} \left| \frac{d\varrho}{d\chi} \right|$$

$$= \frac{a^2 n^2}{2} \frac{\sin(\chi/2)}{\sin\chi} \frac{|(n\cos(\chi/2) - 1)(n - \cos(\chi/2))|}{(n^2 + 1 - 2n\cos(\chi/2))^2}$$

$$= \frac{a^2 n^2}{4} \frac{1}{\cos(\chi/2)} \frac{|(n\cos(\chi/2) - 1)(n - \cos(\chi/2))|}{(n^2 + 1 - 2n\cos(\chi/2))^2}. \tag{5.55}$$

Here, we utilized

$$\sin\chi = 2\cos\frac{\chi}{2}\sin\frac{\chi}{2}. \tag{5.56}$$

The angle χ takes the values from zero (for $\varrho = 0$) up to the value $\chi_{\max}$ (for $\varrho = a$) which is determined by the equation

$$\cos\frac{\chi_{\max}}{2} = \frac{1}{n}. \tag{5.57}$$

The total cross section obtained by integration of $(d\sigma/d\Omega)(\chi)$ over all angles within the cone $\chi < \chi_{\max}$ of course equals the geometrical cross section πa^2.

We still want to show that the total cross section for scattering by the spherical square well potential equals the geometrical cross section πa^2. This is obvious since for $r > a$ we have $U = 0$, i.e., there is no scattering.

We start from equation (5.55)

$$\frac{d\sigma}{d\Omega}(\chi) = \frac{\varrho}{\sin\chi} \left| \frac{d\varrho}{d\chi} \right| = \frac{a^2 n^2}{4} \frac{1}{\cos(\chi/2)} \frac{[n\cos(\chi/2) - 1][n - \cos(\chi/2)]}{[n^2 + 1 - 2n\cos(\chi/2)]^2} \tag{5.58}$$

and integrate over all angles χ from 0 to $\chi_{\max}$: ($d\Omega = 2\pi \sin\chi\, d\chi$)

$$\sigma_{\text{tot}} = \int_0^{\chi_{\max}} \frac{d\sigma}{d\Omega}(\chi)\, d\Omega = \pi a^2 \int_0^{\chi_{\max}} n^2 \sin\frac{\chi}{2} \frac{[n\cos(\chi/2) - 1][n - \cos(\chi/2)]}{[n^2 + 1 - 2n\cos(\chi/2)]^2}\, d\chi$$

$$= \pi a^2 \int_0^{\chi_{\max}} \frac{n^2}{(1 + n^2 - 2n\cos(\chi/2))^2} \times$$

$$\left\{ \underbrace{(n^2 + 1)\cos\frac{\chi}{2}\sin\frac{\chi}{2}}_{\text{I}} - \underbrace{n\cos^2\frac{\chi}{2}\sin\frac{\chi}{2}}_{\text{II}} - \underbrace{n\sin\frac{\chi}{2}}_{\text{III}} \right\} d\chi. \tag{5.59}$$

Part III can be integrated at once; I and II are transformed by integrating by parts:

$$\sigma_{\text{tot}} = \left[\pi a^2 \left(n^2 + 1 - 2n\cos\frac{\chi}{2} \right)^{-1} n^2 \right]_0^{\chi_{\max}}$$

$$- \pi a^2 n(n^2 + 1) \left| \left[\frac{\cos(\chi/2)}{(1 + n^2 - 2n\cos(\chi/2))} \right] \right|_0^{\chi_{\max}}$$

$$- \pi a^2 n(n^2 + 1) \int_0^{\chi_{\max}} \frac{(1/2)\sin(\chi/2)}{(1 + n^2 - 2n\cos(\chi/2))}\, d\chi$$

$$+n^2\pi a^2 \left[\frac{\cos^2(\chi/2)}{(1+n^2-2n\cos(\chi/2))}\right]\Big|_0^{\chi_{max}}$$

$$-\pi a^2 n^2 \int_0^{\chi_{max}} \frac{-\cos(\chi/2)\sin(\chi/2)}{(1+n^2-2n\cos(\chi/2))}d\chi. \qquad (5.60)$$

In the last integral, we substitute

$$y := \cos\frac{\chi}{2},$$

$$dy = -\frac{1}{2}\sin\frac{\chi}{2}d\chi \qquad (5.61)$$

and obtain

$$\sigma_{tot} = \pi a^2 \Bigg[\left\{\frac{n^2(1+\cos^2(\chi/2)) - n(n^2+1)\cos(\chi/2)}{(1+n^2-2n\cos(\chi/2))}\right\}\Big|_0^{\chi_{max}}$$

$$-\frac{n(n^2+1)}{2}\int_0^{\chi_{max}} \frac{\sin(\chi/2)}{(1+n^2-2n\cos(\chi/2))}d\chi$$

$$-2n^2 \int_1^{\cos(\chi_{max}/2)} \frac{y}{(1+n^2-2ny)}dy \Bigg]$$

$$= \pi a^2 \Bigg[\left\{\frac{n^2(1+\cos^2(\chi/2)) - n(n^2+1)\cos(\chi/2)}{(1+n^2-2n\cos(\chi/2))}\right\}\Big|_0^{\chi_{max}}$$

$$-\left\{\left(\frac{n^2+1}{2}\right)\ln\left(1+n^2-2n\cos\frac{\chi}{2}\right)\right\}\Big|_0^{\chi_{max}}$$

$$+\left\{ny + \left(\frac{n^2+1}{2}\right)\ln(1+n^2-2ny)\right\}\Big|_1^{\cos(\chi_{max}/2)} \Bigg];$$

and finally, with $\chi_{max} = 2\arccos(1/n)$,

$$\sigma_{tot} = \pi a^2 \Bigg\{ \underbrace{\frac{n^2(1+1/n^2) - n(n^2+1)(1/n)}{(1+n^2-2)}}_{=0} - \frac{n^2(1+1) - n(n^2+1)\cdot 1}{(1+n^2-2n)} + 1 - n \Bigg\}$$

$$= \pi a^2 \left\{\frac{n - 2n^2 + n^3 - n^3 + 2n^2 - n + n^2 - 2n + 1}{(n-1)^2}\right\} = \pi a^2. \qquad (5.62)$$

Example 5.9: Exercise: Scattering of two atoms

A hydrogen atom moves along the x-axis with a velocity $v_H = 1.78 \cdot 10^2$ m $\cdot$ s^{-1}. It reacts with a chlorine atom that moves perpendicular to the x-axis with $v_{Cl} = 3.2 \cdot 10^1$ m $\cdot$ s^{-1}. Calculate the angle and the velocity of the HCl–molecule. The atomic weights are H = 1.00797 and Cl = 35.453.

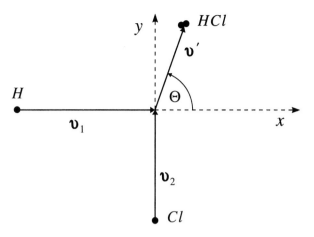

Figure 5.19.

Solution   We utilize momentum conservation. The initial momenta are

$$\mathbf{P}_1 = m_1 v_1 \mathbf{e}_x, \quad m_1 = A_1 \cdot 1 \text{ amu},$$
$$\mathbf{P}_2 = m_2 v_2 \mathbf{e}_y, \quad m_2 = A_2 \cdot 1 \text{ amu}. \tag{5.63}$$

Here, A_1, A_2 mean the atomic weights, and 1 amu ("atomic mass unit") $= 1/12\, m\, (^{12}\text{C})$. We require

$$\mathbf{P}' = \mathbf{P} = (m_1 v_1, m_2 v_2) \quad \text{with} \quad \mathbf{P}' = (m_1 + m_2)\mathbf{v}', \tag{5.64}$$

from which we get

$$\mathbf{v}' = \frac{1}{m_1 + m_2}(m_1 v_1, m_2 v_2) = \frac{m_1 m_2}{m_1 + m_2}\left(\frac{v_1}{m_2}, \frac{v_2}{m_1}\right) = \mu\left(\frac{v_1}{m_2}, \frac{v_2}{m_1}\right). \tag{5.65}$$

Here, μ is the reduced mass. It is calculated as

$$\mu = \frac{m_1 m_2}{m_1 + m_2} = 0.9801 \text{ amu}. \tag{5.66}$$

Thus, one obtains

$$\mathbf{v}' = (4.9208, 31.1154) \text{ m} \cdot \text{s}^{-1},$$
$$\Rightarrow \quad v' = 31.502 \text{ m} \cdot \text{s}^{-1}. \tag{5.67}$$

The angle θ is found from $\tan\theta = v'_y/v'_x$ to be $\theta = 81.013°$.

6 Mechanical Fundamental Quantities of Systems of Mass Points

Linear momentum of the many-body system

If we consider a system of mass points, for the total force acting on the νth particle we have

$$\mathbf{F}_\nu + \sum_\lambda \mathbf{f}_{\nu\lambda} = \dot{\mathbf{p}}_\nu. \tag{6.1}$$

The force $\mathbf{f}_{\nu\lambda}$ is the force of the particle λ on the particle ν; $\mathbf{F}_\nu$ is the force acting on the particle ν from the outside of the system; $\sum_\lambda \mathbf{f}_{\nu\lambda}$ is the resulting internal force of all other particles on the particle ν.

The resulting force acting on the system is obtained by summing over the individual forces:

$$\sum_\nu \dot{\mathbf{p}}_\nu = \sum_\nu \mathbf{F}_\nu + \sum_\nu \sum_\lambda \mathbf{f}_{\nu\lambda} = \dot{\mathbf{P}}.$$

Since force equals (−) counter force (here Newton's third law becomes operative), it follows that $\mathbf{f}_{\nu\lambda} + \mathbf{f}_{\lambda\nu} = 0$, so that the terms of the above double sum cancel pairwise. One thus obtains for the total force acting on the system

$$\dot{\mathbf{P}} = \mathbf{F} = \sum_\nu \mathbf{F}_\nu.$$

If no external force acts on the system, one has

$$\mathbf{F} = \dot{\mathbf{P}} = 0, \quad \text{i.e.,} \quad \mathbf{P} = \text{constant}.$$

The total momentum $\mathbf{P} = \sum_\nu \mathbf{p}_\nu$ of the particle system is thus conserved if the sum of the external forces acting on the system vanishes.

Angular momentum of the many-body system

The situation is similar for the angular momentum if the internal forces are assumed to be central forces.

The angular momentum of the νth particle with respect to the coordinate origin is

$$\mathbf{l}_\nu = \mathbf{r}_\nu \times \mathbf{p}_\nu.$$

The angular momentum of a single particle is defined with respect to the origin. The same holds for the total angular momentum. The angular momentum of the system then equals the sum over all individual angular momenta,

$$\mathbf{L} = \sum_\nu \mathbf{l}_\nu.$$

Analogously, the torque acting on the νth particle is

$$\mathbf{d}_\nu = \mathbf{r}_\nu \times \mathbf{F}_\nu,$$

and the total torque is given by

$$\mathbf{D} = \sum_\nu \mathbf{d}_\nu.$$

The internal forces $\mathbf{f}_{\nu\lambda}$ do not perform a torque, since we assumed them to be central forces. This can be seen as follows: For the force acting on the νth particle, according to (6.1) we have

$$\mathbf{F}_\nu + \sum_\lambda \mathbf{f}_{\nu\lambda} = \frac{d}{dt}\mathbf{p}_\nu.$$

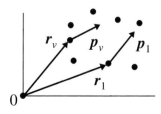

Figure 6.1.

By vectorial multiplication of the equation from the left by $\mathbf{r}_\nu$, we obtain

$$\mathbf{r}_\nu \times \mathbf{F}_\nu + \sum_\lambda \mathbf{r}_\nu \times \mathbf{f}_{\nu\lambda} = \mathbf{r}_\nu \times \frac{d}{dt}\mathbf{p}_\nu = \frac{d}{dt}(\mathbf{r}_\nu \times \mathbf{p}_\nu) = \dot{\mathbf{l}}_\nu.$$

The differentiation can be moved to the left, because $\dot{\mathbf{r}}_\nu \times \mathbf{p}_\nu = 0$. Summation over ν yields

$$\underbrace{\sum_\nu \mathbf{r}_\nu \times \mathbf{F}_\nu}_{\mathbf{D}} + \underbrace{\sum_\lambda \sum_\nu \mathbf{r}_\nu \times \mathbf{f}_{\nu\lambda}}_{0} = \dot{\mathbf{L}},$$

$$\mathbf{D} = \dot{\mathbf{L}} = \sum \dot{\mathbf{l}}_\nu.$$

Here, $\sum_\nu \sum_\lambda \mathbf{r}_\nu \times \mathbf{f}_{\nu\lambda} = 0$, since the terms of the double sum cancel pairwise, e.g.,

$$\mathbf{r}_\nu \times \mathbf{f}_{\nu\lambda} + \mathbf{r}_\lambda \times \mathbf{f}_{\lambda\nu} = (\mathbf{r}_\nu - \mathbf{r}_\lambda) \times \mathbf{f}_{\nu\lambda}.$$

Since for central forces $(\mathbf{r}_\nu - \mathbf{r}_\lambda)$ is parallel to $\mathbf{f}_{\nu\lambda}$, the vector product vanishes.

The total torque on a system is given by the sum of the external torques

$$\mathbf{D} = \dot{\mathbf{L}}.$$

For $\mathbf{D} = 0$, it follows that $\mathbf{L} = $ constant. If no external torques act on a system, the total angular momentum is conserved.

Example 6.1: Conservation of the total angular momentum of a many-body system: Flattening of a galaxy

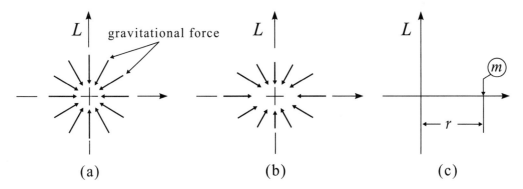

Figure 6.2. Formation of a galaxy from a cloud of gas with angular momentum **L**: (a) The gas contracts due to the mutual gravitational attraction between its constituents. (b) The gas contracts faster along the direction of the angular momentum **L** than in the plane perpendicular to **L**, since the angular momentum must be conserved. In this way a flattening appears. (c) The galaxy in equilibrium: In the plane perpendicular to **L**, the gravitational force balances the centrifugal force due to the rotational motion.

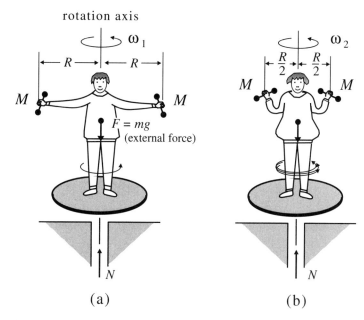

Figure 6.3. Demonstration of angular momentum conservation in the absence of external torques. A person stands on a platform that rotates about a vertical axis.

Example 6.2: Conservation of angular momentum of a many-body problem: The pirouette

(a) The person holds two weights and is set into uniform circular motion with angular velocity ω. The arms are stretched out, so that the angular momentum is large.

(b) If the person pulls the arms towards the body, the moment of inertia (see Chapter 11) decreases. Since angular momentum is conserved, the angular velocity ω significantly increases. Skaters exploit this effect when performing a pirouette.

Energy law of the many-body system

Let $\mathbf{f}_{\nu\lambda}$ be the force of the λth particle on the νth particle. According to equation (6.1), we have

$$\mathbf{F}_\nu + \sum_\lambda \mathbf{f}_{\nu\lambda} = \frac{d}{dt}(m_\nu \dot{\mathbf{r}}_\nu).$$

Scalar multiplication of the equation by $\dot{\mathbf{r}}_\nu$, with

$$\dot{\mathbf{r}}_\nu \cdot \frac{d}{dt}(m_\nu \dot{\mathbf{r}}_\nu) = \frac{d}{dt}\left(\frac{1}{2} m_\nu \dot{\mathbf{r}}_\nu^2\right),$$

leads to

$$\mathbf{F}_\nu \cdot \dot{\mathbf{r}}_\nu + \sum_\lambda \mathbf{f}_{\nu\lambda} \cdot \dot{\mathbf{r}}_\nu = \frac{d}{dt}\left(\frac{1}{2}m_\nu \dot{\mathbf{r}}_\nu^{\,2}\right).$$

$(1/2)m_\nu \dot{\mathbf{r}}_\nu^{\,2}$ is however the kinetic energy T_ν of the νth particle. By summation over ν, we obtain

$$\sum_\nu \mathbf{F}_\nu \cdot \dot{\mathbf{r}}_\nu + \sum_\lambda \sum_\nu \mathbf{f}_{\nu\lambda} \cdot \dot{\mathbf{r}}_\nu = \sum_\nu \frac{d}{dt}\left(\frac{1}{2}m_\nu \dot{\mathbf{r}}_\nu^2\right) = \sum_\nu \dot{T}_\nu = \frac{d}{dt}\sum_\nu T_\nu.$$

$\sum_\nu \dot{T}_\nu$ is the time derivative of the total kinetic energy of the system. By integration from t_1 to t_2, with

$$\dot{\mathbf{r}}_\nu \, dt = d\mathbf{r}_\nu,$$

we get

$$T(t_2) - T(t_1) = \underbrace{\sum_\nu \int_{t_1}^{t_2} \mathbf{F}_\nu \cdot d\mathbf{r}_\nu}_{A_a} + \underbrace{\sum_{\nu\lambda} \int_{t_1}^{t_2} \mathbf{f}_{\nu\lambda} \cdot d\mathbf{r}_\nu}_{A_i}. \tag{6.2}$$

T is the total kinetic energy, A_a is the work performed by external forces, and A_i is the work performed by internal forces over the time interval $t_2 - t_1$.

If we assume that the forces can be derived from a potential, we can express the performed internal and external work by potential differences.

For the external work, we have

$$A_a = \sum_\nu \int \mathbf{F}_\nu \cdot d\mathbf{r}_\nu = -\sum_\nu \int \nabla_\nu V^a \cdot d\mathbf{r}_\nu = -\sum_\nu \int_{t_1}^{t_2} dV_\nu^a$$
$$= -\sum_\nu \left[V_\nu^a(t_2) - V_\nu^a(t_1)\right],$$
$$A_a = V^a(t_1) - V^a(t_2).$$

V_ν^a is the potential of the particle ν in an external field. By summing over all particles, one obtains the total external potential $V^a = \sum_\nu V_\nu^a$.

The force acting between two particles ν and λ is assumed to be a central force. For the "internal" potential, we set

$$V_{\lambda\nu}^i(\mathbf{r}_{\lambda\nu}) = V_{\lambda\nu}^i(r_{\lambda\nu}) = V_{\nu\lambda}^i(r_{\nu\lambda}).$$

The mutual potential depends only on the absolute value of the distance:

$$r_{\nu\lambda} = |\mathbf{r}_\nu - \mathbf{r}_\lambda| = \sqrt{(x_\nu - x_\lambda)^2 + (y_\nu - y_\lambda)^2 + (z_\nu - z_\lambda)^2}.$$

Thus, the principle of action and reaction is satisfied, since from this it follows automatically that the force $\mathbf{f}_{\nu\lambda}$ is equal and opposite to the counterforce $\mathbf{f}_{\lambda\nu}$:

$$\mathbf{f}_{\nu\lambda} = -\nabla_\nu V^i_{\nu\lambda} = +\nabla_\lambda V^i_{\nu\lambda} = -\mathbf{f}_{\lambda\nu}.$$

The index ν on the gradient indicates that the gradient is to be calculated with respect to the components of the position vector $\mathbf{r}_\nu$ of the particle ν. Hence,

$$\nabla_\nu = \left\{ \frac{\partial}{\partial x_\nu}, \frac{\partial}{\partial y_\nu}, \frac{\partial}{\partial z_\nu} \right\}, \quad \nabla_\lambda = \left\{ \frac{\partial}{\partial x_\lambda}, \frac{\partial}{\partial y_\lambda}, \frac{\partial}{\partial z_\lambda} \right\}.$$

Hence, for the internal work we can write

$$A_i = \sum_{\nu,\lambda} \int \mathbf{f}_{\nu\lambda} \cdot d\mathbf{r}_\nu = \frac{1}{2} \left(\sum_{\nu,\lambda} \int \mathbf{f}_{\nu\lambda} \cdot d\mathbf{r}_\nu + \sum_{\lambda,\nu} \int \mathbf{f}_{\lambda\nu} \cdot d\mathbf{r}_\lambda \right)$$

$$= \frac{1}{2} \sum_{\nu,\lambda} \int \mathbf{f}_{\nu\lambda} (d\mathbf{r}_\nu - d\mathbf{r}_\lambda).$$

We now replace the difference of the position vectors by the vector $\mathbf{r}_{\nu\lambda} = \mathbf{r}_\nu - \mathbf{r}_\lambda$ and introduce the operator $\nabla_{\nu\lambda}$ which forms the gradient with respect to this difference. We get

$$A_i = -\frac{1}{2} \sum_{\nu,\lambda} \int \nabla_{\nu\lambda} V^i_{\nu\lambda} \cdot d\mathbf{r}_{\nu\lambda} = -\frac{1}{2} \sum_{\nu,\lambda} \int dV^i_{\nu\lambda} = -\frac{1}{2} \sum_{\nu,\lambda} \left(V^i_{\nu\lambda}(t_2) - V^i_{\nu\lambda}(t_1) \right),$$

where

$$\nabla_{\nu\lambda} = \left\{ \frac{\partial}{\partial(x_\nu - x_\lambda)}, \frac{\partial}{\partial(y_\nu - y_\lambda)}, \frac{\partial}{\partial(z_\nu - z_\lambda)} \right\}.$$

Hence, the internal work is the difference of the internal potential energy. This quantity is significant for deformable media (deformation energy).

For rigid bodies where the differences (distances) $|\mathbf{r}_\nu - \mathbf{r}_\lambda|$ are invariant, the internal work vanishes. Changes $d\mathbf{r}_{\nu\lambda}$ can occur only perpendicular to $\mathbf{r}_\nu - \mathbf{r}_\lambda$ and hence perpendicular to the direction of force, i.e., the scalar products $\mathbf{f}_{\nu\lambda} \cdot d\mathbf{r}_{\nu\lambda}$ vanish.

If we set for the total potential energy

$$V = \sum_\nu V^a_\nu + \frac{1}{2} \sum_{\nu,\lambda} V^i_{\nu\lambda},$$

for equation (6.2) we find

$$T(t_2) - T(t_1) = V(t_1) - V(t_2)$$

or

$$V(t_1) + T(t_1) = V(t_2) + T(t_2); \tag{6.3}$$

the sum of potential and kinetic energy for the total system remains conserved. Since energy can be transferred by the interaction of the particles (e.g., collisions between gas molecules), energy conservation must not hold for the individual particle but must hold for all particles together, i.e., for the entire system.

Transformation to center-of-mass coordinates

When investigating the motion of particle systems, one often disregards the common translation of the system in space, since only the motions of particles relative to the center of gravity of the system are of interest. One therefore transforms the quantities characterizing the particles to a system whose origin is the center of gravity.

According to the Figure 6.4, the origin of the primed coordinate system is the center of gravity; the position, velocity, and mass $\mathbf{R}$, $\mathbf{V}$, and M of the center of gravity are denoted by capital letters. One has

$$\mathbf{r}_\nu = \mathbf{R} + \mathbf{r}'_\nu, \quad \dot{\mathbf{r}}_\nu = \mathbf{V} + \mathbf{v}'_\nu = \dot{\mathbf{R}} + \dot{\mathbf{r}}'_\nu.$$

According to the definition of the center of gravity, we have

$$M \cdot \mathbf{R} = \sum_\nu m_\nu \mathbf{r}_\nu = \sum_\nu m_\nu (\mathbf{R} + \mathbf{r}'_\nu),$$

$$M \cdot \mathbf{R} = M \cdot \mathbf{R} + \sum_\nu \mathbf{r}'_\nu,$$

where $M = \sum_\nu m_\nu$ is the total mass of the system.

From the last equation, it follows that

$$\sum_\nu m_\nu \mathbf{r}'_\nu = 0. \tag{6.4}$$

Thus, the sum of the mass moments relative to the center of gravity vanishes. If there acts a constant external force, as for example the gravity $\mathbf{F}_\nu = m_\nu \mathbf{g}$, then it also follows that

$$\mathbf{D} = \sum_\nu \mathbf{r}'_\nu \times \mathbf{F}_\nu = \left(\sum_\nu m_\nu \mathbf{r}'_\nu \right) \times \mathbf{g} = 0.$$

A body in the earth's field is therefore in equilibrium if it is supported in the center of gravity.

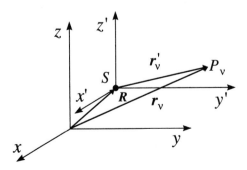

Figure 6.4.

Differentiation of the equation (6.4) with respect to time yields

$$\sum_\nu m_\nu \mathbf{v}_\nu' = 0, \qquad (6.5)$$

i.e., in the center-of-mass system the sum of the momenta vanishes. In relativistic physics this statement is often used as definition of the "center-of-momentum" system; there it is not possible to introduce the notion of the center of mass, – as defined above – in a consistent way. Only the "center-of-momentum" system can be formulated in a relativistically consistent way.

The equivalent transformation of the angular momentum leads to

$$\mathbf{L} = \sum_\nu m_\nu (\mathbf{r}_\nu \times \mathbf{v}_\nu) = \sum_\nu m_\nu (\mathbf{R} + \mathbf{r}_\nu') \times (\mathbf{V} + \mathbf{v}_\nu'),$$

$$\mathbf{L} = \sum_\nu m_\nu (\mathbf{R} + \mathbf{V}) + \sum_\nu m_\nu (\mathbf{R} \times \mathbf{v}_\nu') + \sum_\nu m_\nu (\mathbf{r}_\nu' \times \mathbf{V}) + \sum_\nu m_\nu (\mathbf{r}_\nu' \times \mathbf{v}_\nu').$$

By appropriate grouping, one obtains

$$\mathbf{L} = M(\mathbf{R} \times \mathbf{V}) + \mathbf{R} \times \left(\sum_\nu m_\nu \mathbf{v}_\nu'\right) + \left(\sum_\nu m_\nu \mathbf{r}_\nu'\right) \times \mathbf{V} + \sum_\nu m_\nu (\mathbf{r}_\nu' \times \mathbf{v}_\nu')$$

and sees that the two middle terms disappear, because of the definition (6.4) of the center-of-mass coordinates. Hence,

$$\mathbf{L} = M(\mathbf{R} \times \mathbf{V}) + \sum_\nu m_\nu (\mathbf{r}_\nu' \times \mathbf{v}_\nu') = \mathbf{L}_s + \sum_\nu \mathbf{l}_\nu'. \qquad (6.6)$$

Thus, the angular momentum $\mathbf{L}$ can be decomposed into the angular momentum of the center of gravity $\mathbf{L}_s = M\mathbf{R} \times \mathbf{V}$ with the total mass M, and the sum of angular momenta of the individual particles about the center of gravity.

For the torque as the derivative of the angular momentum, the same decomposition holds:

$$\mathbf{D} = \mathbf{D}_s + \sum_\nu \mathbf{d}_\nu'. \qquad (6.7)$$

Transformation of the kinetic energy

We have

$$T = \frac{1}{2} \sum_\nu m_\nu \mathbf{v}_\nu^2 = \frac{1}{2} \sum_\nu m_\nu \mathbf{V}^2 + \mathbf{V} \cdot \sum_\nu m_\nu \mathbf{v}_\nu' + \frac{1}{2} \sum_\nu m_\nu \mathbf{v}_\nu'^2.$$

Because $\sum m_\nu \mathbf{v}_\nu' = 0$, the middle term again vanishes, and we find

$$T = \frac{1}{2} M \mathbf{V}^2 + \frac{1}{2} \sum_\nu m_\nu \mathbf{v}_\nu'^2 = T_s + T'. \qquad (6.8)$$

The total kinetic energy T is thus composed of the kinetic energy of a virtual particle of mass M with the position vector $\mathbf{R}(t)$ (the center of gravity), and the kinetic energy of the individual particles relative to the center of gravity. Mixed terms, e.g. of the form $\mathbf{V} \cdot \mathbf{v}_\nu'^2$, do not appear! This is the remarkable property of the center-of-mass coordinates, the foundation of their meaning.

Example 6.3: Exercise: Reduced mass

Show that the kinetic energy of two particles with the masses m_1, m_2 splits into the energy of the center of gravity and the kinetic energy of relative motion.

Solution The total kinetic energy is

$$T = \frac{1}{2} m_1 \mathbf{v}_1^2 + \frac{1}{2} m_2 \mathbf{v}_2^2. \tag{6.9}$$

The center of gravity is defined by

$$\mathbf{R} = \frac{m_1 \mathbf{r}_1 + m_2 \mathbf{r}_2}{m_1 + m_2},$$

and its velocity is

$$\dot{\mathbf{R}} = \frac{1}{m_1 + m_2}(m_1 \mathbf{v}_1 + m_2 \mathbf{v}_2). \tag{6.10}$$

The velocity of relative motion is denoted by $\mathbf{v}$. We have

$$\mathbf{v} = \mathbf{v}_1 - \mathbf{v}_2. \tag{6.11}$$

We now express the particle velocity by the center of gravity and relative velocity, respectively.

By inserting $\mathbf{v}_2$ from (6.11) into equation (6.10), we have

$$(m_1 + m_2)\dot{\mathbf{R}} = m_1 \mathbf{v}_1 + m_2 \mathbf{v}_1 - m_2 \mathbf{v}.$$

From this, it follows that

$$\mathbf{v}_1 = \dot{\mathbf{R}} + \frac{m_2}{m_1 + m_2} \mathbf{v}.$$

Analogously, we get

$$\mathbf{v}_2 = \dot{\mathbf{R}} - \frac{m_1}{m_1 + m_2} \mathbf{v}.$$

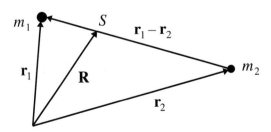

Figure 6.5. Center of gravity and relative coordinates of two masses.

Inserting the two particle velocities into equation (6.9), we obtain

$$T = \frac{1}{2}m_1\left(\dot{\mathbf{R}} + \frac{m_2}{m_1+m_2}\mathbf{v}\right)^2 + \frac{1}{2}m_2\left(\dot{\mathbf{R}} - \frac{m_1}{m_1+m_2}\mathbf{v}\right)^2$$

or

$$T = \frac{1}{2}M\dot{\mathbf{R}}^2 + \frac{1}{2}\frac{m_1 m_2^2 \mathbf{v}^2}{(m_1+m_2)^2} + \frac{1}{2}\frac{m_2 m_1^2 \mathbf{v}^2}{(m_1+m_2)^2},$$

$$T = \frac{1}{2}M\dot{\mathbf{R}}^2 + \frac{1}{2}\mu v^2.$$

The mixed terms cancel. The mass related to the center-of-mass motion is the total mass $M = m_1 + m_2$; the mass related to the relative motion is the reduced mass

$$\mu = \frac{m_1 m_2}{m_1 + m_2}.$$

The reduced mass is often written in the form

$$\frac{1}{\mu} = \frac{1}{m_1} + \frac{1}{m_2}.$$

It is remarkable that the kinetic energy for two bodies decomposes into the kinetic energies of the motion of the center of gravity and of the relative motion. There are no mixed terms, e.g., of the form $\dot{\mathbf{R}} \cdot \mathbf{v}$, which considerably simplifies the solution of the two-body problem (see the next problem).

Example 6.4: Exercise: Movement of two bodies under the action of mutual gravitation

Two bodies of masses m_1 and m_2 move under the action of their mutual gravitation. Let $\mathbf{r}_1$ and $\mathbf{r}_2$ be the position vectors in a space-fixed coordinate system, and $\mathbf{r} = \mathbf{r}_1 - \mathbf{r}_2$. Find the equations of motion for $\mathbf{r}_1, \mathbf{r}_2$, and $\mathbf{r}$ in the center-of-gravity system. How do the trajectories in the space-fixed system and in the center-of-mass system look like?

Solution Newton's gravitational law immediately yields

$$\ddot{\mathbf{r}}_1 = -\frac{Gm_2 \mathbf{r}}{r^3}, \quad \ddot{\mathbf{r}}_2 = \frac{Gm_1 \mathbf{r}}{r^3}.$$

With the relative coordinate $\mathbf{r} = \mathbf{r}_1 - \mathbf{r}_2$, it follows that

$$\ddot{\mathbf{r}}_1 = -\frac{Gm_2(\mathbf{r}_1 - \mathbf{r}_2)}{r^3} \quad \text{and} \quad \ddot{\mathbf{r}}_2 = \frac{Gm_1(\mathbf{r}_1 - \mathbf{r}_2)}{r^3}.$$

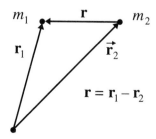

Figure 6.6. Laboratory system.

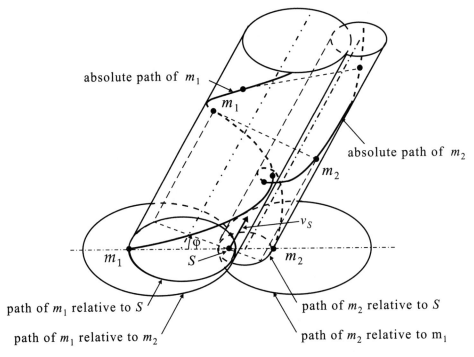

Figure 6.7.

In the center-of-mass system, we have $m_1 \mathbf{r}_1 = -m_2 \mathbf{r}_2$

$$\Rightarrow \quad \ddot{\mathbf{r}}_1 = \frac{-G(m_1 + m_2)\mathbf{r}_1}{r^3} \quad \text{and} \quad \ddot{\mathbf{r}}_2 = \frac{-G(m_1 + m_2)\mathbf{r}_2}{r^3}.$$

Subtraction yields

$$\ddot{\mathbf{r}} = \ddot{\mathbf{r}}_1 - \ddot{\mathbf{r}}_2 = -\frac{G(m_1 + m_2)\mathbf{r}}{r^3}.$$

Since

$$r_1 = \frac{m_2}{m_1 + m_2} r \quad \text{and} \quad r_2 = \frac{m_1}{m_1 + m_2} r,$$

it follows that

$$\ddot{\mathbf{r}}_1 = \frac{-G m_2^3 \mathbf{r}_1}{(m_1 + m_2)^2 r_1^3} \quad \text{and} \quad \ddot{\mathbf{r}}_2 = \frac{-G m_1^3 \mathbf{r}_2}{(m_1 + m_2)^2 r_2^3}.$$

Hence, Newton's gravitational law holds with respect to the center of gravity, but with modified mass factors. This means that the trajectories are conic sections as before (relative path with respect to S). Because of the superimposed translation of the center of gravity, the trajectories become spirals in space.

Example 6.5: Exercise: Atwoods fall machine

Two masses ($m_1 = 2$ kg and $m_2 = 4$ kg) are connected by a massless rope (without sliding) via a frictionless disk of mass $M = 2$ kg and radius $R = 0.4$ m (Atwoods machine). Find the acceleration of the mass $m_2 = 4$ kg if the system moves under the influence of gravitation.

Solution

For the given masses $m_1 = 2$ kg, $m_2 = 4$ kg and the tension forces at the rope ends $\mathbf{N}_1$ and $\mathbf{N}_2$, it follows that

$$m_1 a_1 = N_1 - m_1 g, \qquad m_2 a_2 = m_2 g - N_2, \tag{6.12}$$

and for the torques acting on the disk, we get

$$D_1 + D_2 = -N_1 R + N_2 R = R(N_2 - N_1) = \dot{\omega}\theta_s, \tag{6.13}$$

since the disk is accelerated. θ_s is the moment of inertia of the disk. From this, it follows that $N_2 \neq N_1$; otherwise, there is no motion at all. For the accelerations, we have

$$a = a_1 = a_2 = \dot{\omega} R, \tag{6.14}$$

since the rope is tight and does not slide, i.e., it adheres to the disk.

Inserting the moment of inertia of the disk $\theta_s = MR^2/2$ (see example 11.7) into equation (6.13) and using equation (6.14) yields for the acceleration

$$a = \frac{N_1}{m_1} - g = g - \frac{N_2}{m_2} = \dot{\omega} R = \frac{R^2}{MR^2/2}(N_2 - N_1). \tag{6.15}$$

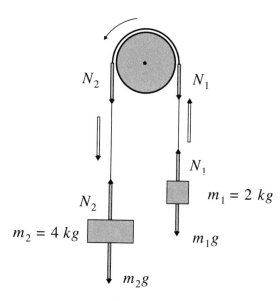

Figure 6.8.

Inserting equation (6.12) and performing the algebraic steps yields

$$a = \frac{2}{M}(N_2 - N_1) = \frac{g(m_2 - m_1) - m_2 a_2 - m_1 a_1}{M/2}$$

and, because $a = a_1 = a_2$,

$$0 = \frac{aM/2 - g(m_2 - m_1) + a(m_2 + m_1)}{M/2} = \frac{a(m_1 + m_2 + M/2) - g(m_2 - m_1)}{M/2}$$

$$\Rightarrow \quad a = \frac{g(m_2 - m_1)}{m_1 + m_2 + M/2}.$$

The Atwoods machine serves as a transparent and easily controllable demonstration of the laws of free fall. By varying the difference of the masses $(m_2 - m_1)$, the acceleration a can be varied.

Example 6.6: **Exercise: Our solar system in the Milky Way**

Our solar system is about $r_0 \approx 5 \cdot 10^{20}$ m away from the center of the Milky Way, and its orbital velocity relative to the galactic center v_0 is $\approx 3 \cdot 10^5$ m/s. This is schematically shown in Figure 6.9.

(a) Determine the mass M of our galaxy.

(b) Discuss the hypothesis that the motion of our solar system is a consequence of the contraction of our Milky Way (see Figure 6.9), and then verify, $r_0 = GM/v_0^2$. Here $G = 6.7 \cdot 10^{-11} \mathrm{m^3 s^2\, kg^{-1}}$ is the gravitational constant.

Solution   (a) If a mass point moves on a circular path, then according to Newton the force per unit mass equals the acceleration. Since our sun (mass m) is at the periphery of our Milky Way, the attractive force toward the center can approximately be represented by

$$F = G\frac{mM}{r_0^2}, \qquad (6.16)$$

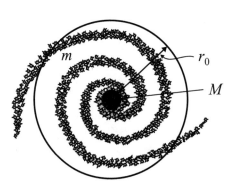

Figure 6.9.

where m is the solar mass and M is the mass of the Milky Way. The acceleration points toward the center,

$$a = \frac{v_0^2}{r_0} = \frac{F}{m}, \tag{6.17}$$

from which it follows that

$$\frac{v_0^2}{r_0} = \frac{GM}{r_0^2} \quad \text{or} \quad r_0 = \frac{GM}{v_0^2}. \tag{6.18}$$

Using the numbers given in the formulation of the problem, one gets from equation (6.18) the mass of our Milky Way:

$$M = \frac{r_0 v_0^2}{G} \approx \frac{5 \cdot 10^{20} \cdot 9 \cdot 10^{10}}{6.7 \cdot 10^{-11}} \text{kg} = 6.7 \cdot 10^{41} \text{ kg}.$$

This means that the mass of the Milky Way is

$$M \approx 3 \cdot 10^{11} m,$$

where m is the solar mass.

(b) If r, v are the initial values for the distance and velocity of our sun, for the available energies we have

$$V_{\text{pot}} = -\frac{GMm}{r} \quad \text{and} \quad T_{\text{kin}} = \frac{1}{2}mv^2, \tag{6.19}$$

where M is the mass of the Milky Way, and G is the gravitational constant. If the sun moves with decreasing radius about the center of the Milky Way, the angular momentum about the center remains constant; however, the orbital velocity increases. Hence, the kinetic energy T_{kin} can be given as a function of the radius

$$T = \frac{1}{2}m\frac{l^2}{m^2 r^2} = \frac{1}{2}\frac{l^2}{m}\frac{1}{r^2}, \tag{6.20}$$

where we used $l = (mr^2)\omega = mvr = $ constant.

The assumption is now that at the present distance r the increase in the kinetic energy ΔT_{kin} is balanced by the decrease in the potential energy if r is reduced by Δr. Differentiation of equations (6.19) and (6.20) with respect to r yields:

$$\Delta T_{\text{kin}} = \left(\frac{dT_{\text{kin}}}{dr}\right)\Delta r = -\frac{l^2}{m}\frac{1}{r^3}\Delta r, \quad \Delta T_{\text{kin}} > 0, \quad \text{if } \Delta r < 0,$$

$$\Delta V_{\text{pot}} = \left(\frac{dV_{\text{pot}}}{dr}\right)\Delta r = \frac{GMm}{r^2}\Delta r, \quad \Delta V_{\text{pot}} < 0, \quad \text{if } \Delta r < 0.$$

In the equilibrium, however, $\Delta T_{\text{kin}} + \Delta V_{\text{pot}} = 0$. Replacing l by $l = mv_0 r_0$ yields

$$\frac{m^2 v_0^2 r_0^2}{m r_0^3} = G\frac{Mm}{r_0^2} \quad \text{or} \quad r_0 v_0^2 = MG. \tag{6.21}$$

Equation (6.21) again corresponds exactly to the result of problem (a).

PART III

VIBRATING SYSTEMS

7 Vibrations of Coupled Mass Points

As the first and most simple system of vibrating mass points, we consider the free vibration of two mass points, fixed to two walls by springs of equal spring constant, as is shown in the Figure 7.1.

The two mass points shall have equal masses. The displacements from the rest positions are denoted by x_1 and x_2, respectively. We consider only vibrations along the line connecting the mass points.

When displacing the mass 1 from the rest position, there acts the force $-kx_1$ by the spring fixed to the wall, and the force $+k(x_2 - x_1)$ by the spring connecting the two mass points. Thus, the mass point 1 obeys the equation of motion

$$m\ddot{x}_1 = -kx_1 + k(x_2 - x_1). \tag{7.1a}$$

Analogously, for the mass point 2 we have

$$m\ddot{x}_2 = -kx_2 - k(x_2 - x_1). \tag{7.1b}$$

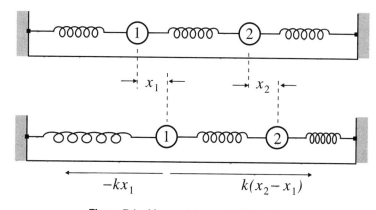

Figure 7.1. Mass points coupled by springs.

We first determine the possible frequencies of common vibration of the two particles. *The frequencies that are equal for all particles are called eigenfrequencies.* The related vibrational states are called eigen- or normal vibrations. These definitions are correspondingly generalized for a N-particle system. We use the *ansatz*

$$x_1 = A_1 \cos \omega t, \qquad x_2 = A_2 \cos \omega t, \tag{7.2}$$

i.e., both particles shall vibrate with the same frequency ω. The specific type of the *ansatz*, be it a sine or cosine function or a superposition of both, is not essential. We would always get the same condition for the frequency, as can be seen from the following calculation.

Insertion of the *ansatz* into the equations of motion yields two linear homogeneous equations for the amplitudes:

$$\begin{aligned} A_1(-m\omega^2 + 2k) - A_2 k &= 0, \\ -A_1 k + A_2(-m\omega^2 + 2k) &= 0. \end{aligned} \tag{7.3}$$

The system of equations has nontrivial solutions for the amplitudes only if the determinant of coefficients D vanishes:

$$D = \begin{vmatrix} -m\omega^2 + 2k & -k \\ -k & -m\omega^2 + 2k \end{vmatrix} = (-m\omega^2 + 2k)^2 - k^2 = 0.$$

We thus obtain an equation for determining the frequencies:

$$\omega^4 - 4\frac{k}{m}\omega^2 + 3\frac{k^2}{m^2} = 0.$$

The positive solutions of the equation are the frequencies

$$\omega_1 = \sqrt{\frac{3k}{m}} \quad \text{and} \quad \omega_2 = \sqrt{\frac{k}{m}}.$$

These frequencies are called *eigenfrequencies* of the system; the corresponding vibrations are called *eigenvibrations* or *normal vibrations*. To get an idea about the type of the normal vibrations, we insert the eigenfrequency into the system (7.3). For the amplitudes, we find

$$A_1 = -A_2 \quad \text{for} \quad \omega_1 = \sqrt{\frac{3k}{m}}$$

and

$$A_1 = A_2 \quad \text{for} \quad \omega_2 = \sqrt{\frac{k}{m}}.$$

The two mass points vibrate in-phase with the lower frequency ω_2, and with the higher frequency ω_1 against each other. The two vibration modes are illustrated by Figure 7.2.

The number of normal vibrations equals the number of coordinates (degrees of freedom) which are necessary for a complete description of the system. This is a consequence of the fact that for N degrees of freedom there appear N equations of the kind (7.2) and N equations of motion of the kind (7.1a),(7.1b). This leads to a determinant of rank N for ω^2, and therefore in general to N normal frequencies. Since we have restricted ourselves in the

VIBRATIONS OF COUPLED MASS POINTS

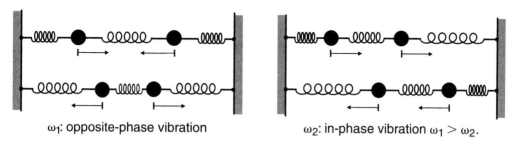

ω_1: opposite-phase vibration ω_2: in-phase vibration $\omega_1 > \omega_2$.

Figure 7.2.

example to the vibrations along the x-axis, the two coordinates x_1 and x_2 are sufficient to describe the system, and we obtain the two eigenvibrations with the frequencies ω_1, ω_2.

In our example, the normal vibrations mean in-phase or opposite-phase (= in-phase with different sign of the amplitudes) oscillations of the mass points. The amplitudes of equal size are related to the equality of masses ($m_1 = m_2$). The general motion of the mass points corresponds to a superposition of the normal modes with different phase and amplitude.

The differential equations (7.1a),(7.1b) are linear. The general form of the vibration is therefore the superposition of the normal modes. It reads

$$\begin{aligned} x_1(t) &= C_1 \cos(\omega_1 t + \varphi_1) + C_2 \cos(\omega_2 t + \varphi_2), \\ x_2(t) &= -C_1 \cos(\omega_1 t + \varphi_1) + C_2 \cos(\omega_2 + \varphi_2). \end{aligned} \quad (7.4)$$

Here, we already utilized the result that x_1 and x_2 have opposite-equal amplitudes for a pure ω_1-vibration, and equal amplitudes for pure ω_2-vibrations. This ensures that the special cases of the pure normal vibrations with $C_2 = 0$, $C_1 \neq 0$ and $C_1 = 0$, $C_2 \neq 0$ are included in the *ansatz* (7.4). Equation (7.4) is the most general *ansatz* since it involves 4 free constants. Thus one can incorporate any initial values for $x_1(0)$, $x_2(0)$, $\dot{x}_1(0)$, $\dot{x}_2(0)$.

For example, the initial conditions are

$$x_1(0) = 0, \quad x_2(0) = a, \quad \dot{x}_1(0) = \dot{x}_2(0) = 0.$$

To determine the 4 free constants C_1, C_2, φ_1, φ_2, we insert the initial conditions into the equations (7.4) and their derivatives:

$$\begin{aligned} x_1(0) &= C_1 \cos \varphi_1 + C_2 \cos \varphi_2 = 0, & (7.5) \\ x_2(0) &= -C_1 \cos \varphi_1 + C_2 \cos \varphi_2 = a, & (7.6) \\ \dot{x}_1(0) &= -C_1 \omega_1 \sin \varphi_1 - C_2 \omega_2 \sin \varphi_2 = 0, & (7.7) \\ \dot{x}_2(0) &= C_1 \omega_1 \sin \varphi_1 - C_2 \omega_2 \sin \varphi_2 = 0. & (7.8) \end{aligned}$$

Addition of (7.7) and (7.8) yields

$$C_2 \sin \varphi_2 = 0.$$

Subtraction of (7.7) and (7.8) yields

$$C_1 \sin \varphi_1 = 0.$$

From addition and subtraction of (7.5) and (7.6), it follows that

$$2C_2 \cos \varphi_2 = a \quad \text{and} \quad 2C_1 \cos \varphi_1 = -a.$$

Thus, one obtains

$$\varphi_1 = \varphi_2 = 0, \qquad C_1 = -\frac{a}{2}, \qquad C_2 = \frac{a}{2}.$$

The overall solution therefore reads

$$x_1(t) = \frac{a}{2}(-\cos \omega_1 t + \cos \omega_2 t) = a \sin\left(\frac{\omega_1 - \omega_2}{2}\right) t \sin\left(\frac{\omega_1 + \omega_2}{2}\right) t,$$

$$x_2(t) = \frac{a}{2}(\cos \omega_1 t + \cos \omega_2 t) = a \cos\left(\frac{\omega_1 - \omega_2}{2}\right) t \cos\left(\frac{\omega_1 + \omega_2}{2}\right) t.$$

For $t = 0$: $x_1(0) = 0$, $x_2(0) = a$, as required. The second mass plucks at the first one and causes it to vibrate. These are *beat vibrations* (see Example 7.2).

Example 7.1: Exercise: Two equal masses coupled by two equal springs

Two equal masses move without friction on a plate. They are connected to each other and to the wall by two springs, as is indicated by Figure 7.3. The two spring constants are equal, and the motion shall be restricted to a straight line (one-dimensional motion).

Find

(a) the equations of motion,

(b) the normal frequencies, and

(c) the amplitude ratios of the normal vibrations and the general solution.

Solution (a) Let x_1 and x_2 be the displacements from the rest positions. The equations of motion then read

$$m\ddot{x}_1 = -kx_1 + k(x_2 - x_1), \tag{7.9}$$
$$m\ddot{x}_2 = -k(x_2 - x_1). \tag{7.10}$$

(b) For determining the normal frequencies, we use the *ansatz*

$$x_1 = A_1 \cos \omega t, \qquad x_2 = A_2 \cos \omega t$$

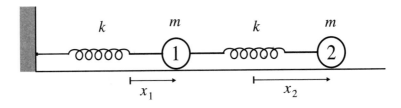

Figure 7.3. Two equal masses coupled by two equal springs.

VIBRATIONS OF COUPLED MASS POINTS

and thereby get from (7.9) and (7.10) the equations

$$(2k - m\omega^2)A_1 - kA_2 = 0,$$
$$-kA_1 + (k - m\omega^2)A_2 = 0. \tag{7.11}$$

From the requirement for nontrivial solutions of the system of equations, it follows that the determinant of coefficients vanishes:

$$D = \begin{vmatrix} 2k - m\omega^2 & -k \\ -k & k - m\omega^2 \end{vmatrix} = 0.$$

From this follows the determining equation for the eigenfrequencies,

$$\omega^4 - 3\frac{k}{m}\omega^2 + \frac{k^2}{m^2} = 0,$$

with the positive solutions

$$\omega_1 = \frac{\sqrt{5}+1}{2}\sqrt{\frac{k}{m}} \quad \text{and} \quad \omega_2 = \frac{\sqrt{5}-1}{2}\sqrt{\frac{k}{m}}, \quad \omega_1 > \omega_2.$$

(c) By inserting the eigenfrequencies in (7.11) one sees that the higher frequency ω_1 corresponds to the opposite-phase mode, and the lower frequency ω_2 to the equal-phase normal vibration:

with $\omega_1^2 = \frac{1}{2}(3 + \sqrt{5})\frac{k}{m}$, it follows from (7.11) that $A_2 = -\frac{\sqrt{5}-1}{2}A_1$,

with $\omega_2^2 = \frac{1}{2}(3 - \sqrt{5})\frac{k}{m}$, it follows from (7.11) that $A_2 = \frac{\sqrt{5}+1}{2}A_1$.

Since the two mass points are fixed in different ways, we find amplitudes of different magnitudes.

The general solution is obtained as a superposition of the normal vibrations, using the calculated amplitude ratios:

$$x_1(t) = C_1 \cos(\omega_1 t + \varphi_1) + C_2 \cos(\omega_2 t + \varphi_2),$$
$$x_2(t) = -\frac{\sqrt{5}-1}{2}C_1 \cos(\omega_1 t + \varphi_1) + \frac{\sqrt{5}+1}{2}C_2 \cos(\omega_2 t + \varphi_2).$$

The 4 free constants are determined from the initial conditions of the specific case.

Example 7.2: Exercise: Coupled pendulums

Two pendulums of equal mass and length are connected by a spiral spring. They vibrate in a plane. The coupling is weak (i.e., the two eigenmodes are not very different). Find the motion with small amplitudes.

Solution The initial conditions are

$$x_1(0) = 0, \quad x_2(0) = A, \quad \dot{x}_1(0) = \dot{x}_2(0) = 0.$$

We start from the vibrational equation of the simple pendulum:

$$ml\ddot{\alpha} = -mg \sin\alpha.$$

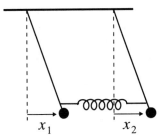
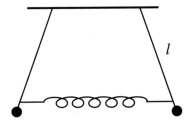

Figure 7.4.

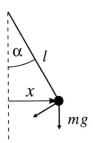

Figure 7.5.

For small amplitudes, we set $\sin\alpha = \alpha = x/l$ and obtain

$$m\ddot{x} = -m\frac{g}{l}x.$$

For the coupled pendulums, the force $\mp k(x_1 - x_2)$ caused by the spring still enters, which leads to the equations

$$\ddot{x}_1 = -\frac{g}{l}x_1 - \frac{k}{m}(x_1 - x_2),$$
$$\ddot{x}_2 = -\frac{g}{l}x_2 + \frac{k}{m}(x_1 - x_2). \quad (7.12)$$

This coupled set of differential equations can be decoupled by introducing the coordinates

$$u_1 = x_1 - x_2 \quad \text{and} \quad u_2 = x_1 + x_2.$$

Subtraction and addition of the equations (7.12) yield

$$\ddot{u}_1 = -\frac{g}{l}u_1 - 2\frac{k}{m}u_1 = -\left(\frac{g}{l} + 2\frac{k}{m}\right)u_1,$$
$$\ddot{u}_2 = -\frac{g}{l}u_2.$$

These two equations can be solved immediately:

$$u_1 = A_1 \cos\omega_1 t + B_1 \sin\omega_1 t,$$
$$u_2 = A_2 \cos\omega_2 t + B_2 \sin\omega_2 t, \quad (7.13)$$

where $\omega_1 = \sqrt{g/l + 2(k/m)}$, $\omega_2 = \sqrt{g/l}$ are the eigenfrequencies of the two vibrations. The coordinates u_1, u_2 are called *normal coordinates*. Normal coordinates are often introduced to decouple a coupled system of differential equations. The coordinate $u_1 = x_1 - x_2$ describes the opposite-phase and $u_2 = x_1 + x_2$ the equal-phase normal vibration. The equal-phase normal mode proceeds as if the coupling were absent.

For sake of simplicity, we incorporate the initial conditions in the system (7.13). For the normal coordinates we then have

$$u_1(0) = -A, \qquad u_2(0) = A, \qquad \dot{u}_1(0) = \dot{u}_2(0) = 0.$$

Insertion into (7.13) yields

$$A_1 = -A, \qquad A_2 = A, \qquad B_1 = B_2 = 0,$$

and thus,

$$u_1 = -A \cos \omega_1 t, \qquad u_2 = A \cos \omega_2 t.$$

Returning to the coordinates x_1 and x_2:

$$x_1 = \frac{1}{2}(u_1 + u_2) = \frac{A}{2}(-\cos \omega_1 t + \cos \omega_2 t),$$
$$x_2 = \frac{1}{2}(u_2 - u_1) = \frac{A}{2}(\cos \omega_1 t + \cos \omega_2 t).$$

After transforming the angular functions, one has

$$x_1 = A \sin\left(\frac{\omega_1 - \omega_2}{2} t\right) \sin\left(\frac{\omega_1 + \omega_2}{2} t\right),$$
$$x_2 = A \cos\left(\frac{\omega_1 - \omega_2}{2} t\right) \cos\left(\frac{\omega_1 + \omega_2}{2} t\right).$$

We have presupposed the coupling of the two pendulums to be weak, i.e.,

$$\omega_2 = \sqrt{\frac{g}{l}} \approx \omega_1 = \sqrt{\frac{g}{l} + 2\frac{k}{m}},$$

hence, the frequency $\omega_1 - \omega_2$ is small. The vibrations $x_1(t)$ and $x_2(t)$ can then be interpreted as follows: The amplitude factor of the pendulum vibrating with the frequency $\omega_1 + \omega_2$ is slowly modulated by the frequency $\omega_1 - \omega_2$. This process is called *beat vibration*. Figure 7.6 illustrates the process. The two pendulums exchange their energy with the amplitude modulation frequency $\omega_1 - \omega_2$. If one pendulum reaches its maximum amplitude (energy), the other pendulum comes to rest. This complete energy transfer occurs only for identical pendulums. If the pendulums differ in mass or length, the energy transfer becomes incomplete; the pendulums vary in amplitudes but without coming to rest.

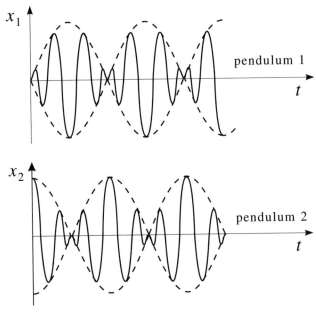

Figure 7.6.

The vibrating chain[1]

We consider another vibrating mass system: the vibrating chain. The "chain" is a massless thread set with N mass points. All mass points have the mass m and are fixed to the thread at equal distances a. The points 0 and $N+1$ at the ends of the thread are tightly fixed and do not participate in the vibration. The displacement from the rest position in y-direction is assumed to be relatively small, so that the minor displacement in x-direction is negligible. The total string tension T is only due to the clamping of the end points and is constant over the entire thread.

If one picks out the νth particle, the forces acting on this particle are due to the displacements of the particles $(\nu - 1)$ and $(\nu + 1)$. According to Figure 7.7 the backdriving forces are given by

$$\mathbf{F}_{\nu-1} = -(T \cdot \sin\alpha)\mathbf{e}_2,$$
$$\mathbf{F}_{\nu+1} = -(T \cdot \sin\beta)\mathbf{e}_2.$$

Since the displacement in y-direction is small by definition, α and β are small angles, and

---

[1] It is recommended that the reader go through Chapter 8 ("The Vibrating String") before studying this section. The concepts presented here will be more easily understood, and the mathematical approaches will be more transparent in their physical motivation.

THE VIBRATING CHAIN

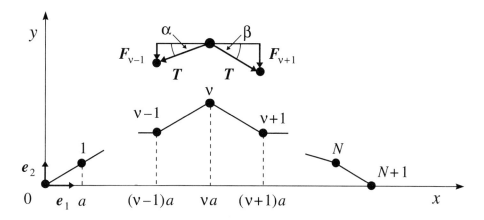

Figure 7.7.

hence, one has to a good approximation

$$\sin\alpha = \tan\alpha \quad \text{and} \quad \sin\beta = \tan\beta.$$

From Figure 7.7, one sees that

$$\tan\alpha = \frac{y_\nu - y_{\nu-1}}{a} \quad \text{and} \quad \tan\beta = \frac{y_\nu - y_{\nu+1}}{a}.$$

Hence, the forces are given by

$$\mathbf{F}_{\nu-1} = -T\left(\frac{y_\nu - y_{\nu-1}}{a}\right)\mathbf{e}_2,$$

$$\mathbf{F}_{\nu+1} = -T\left(\frac{y_\nu - y_{\nu+1}}{a}\right)\mathbf{e}_2.$$

The total backdriving force is the sum $\mathbf{F}_{\nu-1} + \mathbf{F}_{\nu+1}$, i.e., the equation of motion for the particle reads

$$m\frac{d^2 y_\nu}{dt^2}\mathbf{e}_2 = -T\left(\frac{y_\nu - y_{\nu-1}}{a}\right)\mathbf{e}_2 - T\left(\frac{y_\nu - y_{\nu+1}}{a}\right)\mathbf{e}_2$$

or

$$\frac{d^2 y_\nu}{dt^2} = \frac{T}{ma}(y_{\nu-1} - 2y_\nu + y_{\nu+1}). \tag{7.14}$$

Since the index ν runs from $\nu = 1$ to $\nu = N$, one obtains a system of N coupled differential equations. Considering that the endpoints are fixed, by setting for the indices $\nu = 0$ and $\nu = N + 1$

$$y_0 = 0 \quad \text{and} \quad y_{N+1} = 0 \quad \text{(boundary condition)},$$

one obtains from the differential equation (7.14) with the indices $\nu = 1$ and $\nu = N$ the differential equation for the first and last particle that can participate in the vibration:

$$m\frac{d^2 y_1}{dt^2} = \frac{T}{a}(-2y_1 + y_2),$$
$$m\frac{d^2 y_N}{dt^2} = \frac{T}{a}(y_{N-1} - 2y_N). \tag{7.15}$$

We now look for the eigenfrequencies of the particle system, i.e., the frequencies of vibration common to all particles. To get a determining equation for the eigenfrequency ω_n, we introduce in equation (7.14) the *ansatz*

$$y_\nu = A_\nu \cos \omega t. \tag{7.16}$$

We obtain

$$-m\omega^2 \cdot A_\nu \cdot \cos \omega t = \frac{T}{a}(A_{\nu-1} - 2A_\nu + A_{\nu+1})\cos \omega t,$$

and after rewriting,

$$-A_{\nu-1} + \left(2 - \frac{ma\omega^2}{T}\right) A_\nu - A_{\nu+1} = 0, \qquad \nu = 2, \ldots, N-1. \tag{7.17a}$$

By insertion of (7.16) into (7.15), we get the equations for the first and the last vibrating particle:

$$\left(2 - \frac{ma\omega^2}{T}\right) A_1 - A_2 = 0,$$
$$-A_{N-1} + \left(2 - \frac{ma\omega^2}{T}\right) A_N = 0. \tag{7.17b}$$

With the abbreviation

$$\frac{2T - ma\omega^2}{T} = c, \tag{7.18}$$

the equations (7.17a) and (7.17b) can be rewritten as follows:

$$\begin{aligned}
cA_1 - A_2 &= 0, \\
-A_1 + cA_2 - A_3 &= 0, \\
-A_2 + cA_3 - A_4 &= 0, \\
&\vdots \\
-A_{N-1} + cA_N &= 0
\end{aligned}$$

This is a system of homogeneous linear equations for the coefficients A_ν. For any nontrivial solution of the equation system (not all $A_\nu = 0$) the determinant of coefficients must vanish. This determinant has the form

$$D_N = \begin{vmatrix} c & -1 & 0 & 0 & 0 & \cdots & 0 & 0 & 0 \\ -1 & c & -1 & 0 & 0 & \cdots & 0 & 0 & 0 \\ 0 & -1 & c & -1 & 0 & \cdots & 0 & 0 & 0 \\ \vdots & \vdots & \vdots & \vdots & \vdots & \ddots & \vdots & \vdots & \vdots \\ 0 & 0 & 0 & 0 & 0 & \cdots & -1 & c & -1 \\ 0 & 0 & 0 & 0 & 0 & \cdots & 0 & -1 & c \end{vmatrix}.$$

It has N rows and N columns. The eigenfrequencies are obtained as solution of the equation

$$D_N = 0.$$

Expanding D_N with respect to the first row, we get

$$D_N = c \cdot \begin{vmatrix} c & -1 & 0 & \cdots & 0 & 0 & 0 \\ -1 & c & -1 & \cdots & 0 & 0 & 0 \\ 0 & -1 & c & \cdots & 0 & 0 & 0 \\ 0 & 0 & -1 & \cdots & 0 & 0 & 0 \\ \vdots & \vdots & \vdots & \ddots & \vdots & \vdots & \vdots \\ 0 & 0 & 0 & \cdots & -1 & 0 & 0 \\ 0 & 0 & 0 & \cdots & c & -1 & 0 \\ 0 & 0 & 0 & \cdots & -1 & c & -1 \\ 0 & 0 & 0 & \cdots & 0 & -1 & c \end{vmatrix}$$

$$+ \begin{vmatrix} -1 & -1 & 0 & 0 & \cdots & 0 & 0 \\ 0 & c & -1 & 0 & \cdots & 0 & 0 \\ 0 & -1 & -c & -1 & \cdots & 0 & 0 \\ 0 & 0 & -1 & c & \cdots & 0 & 0 \\ \vdots & \vdots & \vdots & \vdots & \ddots & \vdots & \vdots \\ 0 & 0 & 0 & 0 & \cdots & -1 & 0 \\ 0 & 0 & 0 & 0 & \cdots & c & -1 \\ 0 & 0 & 0 & 0 & \cdots & -1 & c \end{vmatrix}$$

The left-hand determinant has exactly the same form as D_N, but is lower by one order ($N-1$ rows, $N-1$ columns). It would be the determinant of coefficients for a similar system with one mass point less, i.e., D_{N-1}. The right-hand determinant is now expanded with respect to the first column, which leads to

$$D_N = cD_{N-1} + (-1) \cdot \begin{vmatrix} c & -1 & 0 & \cdots & 0 & 0 \\ -1 & c & -1 & \cdots & 0 & 0 \\ 0 & -1 & c & \cdots & 0 & 0 \\ \vdots & \vdots & \vdots & \ddots & \vdots & \vdots \\ 0 & 0 & 0 & \cdots & -1 & 0 \\ 0 & 0 & 0 & \cdots & c & -1 \\ 0 & 0 & 0 & \cdots & -1 & c \end{vmatrix}.$$

The last determinant is just D_{N-2}. Hence we get the determinant recursion equation

$$D_N = cD_{N-1} - D_{N-2}, \quad \text{if} \quad N \geq 2. \tag{7.19}$$

Moreover,

$$D_1 = |c| = c \quad \text{and} \quad D_2 = \begin{vmatrix} c & -1 \\ -1 & c \end{vmatrix} = c^2 - 1. \tag{7.20}$$

By setting $N = 2$ in (7.19), we recognize that (7.19) combined with (7.20) is satisfied only if we formally set

$$D_0 = 1. \tag{7.21}$$

Our problem is now to solve the determinant equation (7.19). We use the *ansatz*

$$D_N = p^N,$$

where the constant p must be determined. Insertion into (7.19) yields

$$p^N = cp^{N-1} - p^{N-2},$$

and after division by p^{N-2},

$$p^2 - cp + 1 = 0 \quad \text{or} \quad p = \frac{c \pm \sqrt{c^2 - 4}}{2}.$$

The mathematical possibility $p^{N-2} = 0$ that leads to $p \equiv 0$ does not obey the boundary condition $D_0 = 1$ and is therefore inapplicable. Substituting $c = 2\cos\Theta$, we obtain for p

$$p = \cos\Theta \pm \sqrt{\cos^2\Theta - 1} = \cos\Theta \pm i\sin\Theta = e^{\pm i\Theta}.$$

The solutions of equation (7.19) are then

$$D_N = p^N = (e^{i\Theta})^N = e^{iN\Theta} = \cos N\Theta + i\sin N\Theta$$

and

$$D_N = (e^{-i\Theta})^N = e^{-iN\Theta} = \cos N\Theta - i \sin N\Theta.$$

Since the equation system (7.19) is homogeneous and linear, the general solution is a linear combination of $\cos N\Theta$ and $\sin N\Theta$:

$$D_N = G \cos N\Theta + H \sin N\Theta. \tag{7.22}$$

Since $D_0 = 1$ and $D_1 = c = 2 \cos \Theta$ (see above), G and H are determined as

$$G = 1, \qquad H = \cot \Theta,$$

so that

$$D_N = \cos N\Theta + \frac{\sin N\Theta \cos \Theta}{\sin \Theta} = \frac{\sin(N+1)\Theta}{\sin \Theta},$$

because $\sin \Theta \cos N\Theta + \sin N\Theta \cos \Theta = \sin(N+1)\Theta$.

For any nontrivial solution of the equation system we must have $D_N = 0$, i.e., D_N must vanish for all N; it follows that

$$\sin((N+1)\Theta) = 0,$$

or

$$\Theta = \Theta_n = \frac{n\pi}{N+1}, \qquad n = 1, \ldots, N. \tag{7.23}$$

$n = 0$ drops out since it leads to the solution $\Theta_0 = 0$, and hence to $D_N = N + 1 \neq 0$, and thus does not lead to a solution of the equation $D_N = 0$. For c we then get according to (7.18):

$$c = 2 - \frac{\omega^2 ma}{T} = 2 \cos \frac{n\pi}{N+1},$$

and ω is calculated from

$$\omega^2 = \omega_{(n)}^2 = \frac{2T}{ma}\left(1 - \cos \frac{n\pi}{N+1}\right) \tag{7.24a}$$

as

$$\omega_{(n)} = \sqrt{\frac{2T}{ma}}\sqrt{1 - \cos \frac{n\pi}{N+1}}. \tag{7.24b}$$

These are the *eigenfrequencies of the system*; the fundamental frequency is obtained for $n = 1$ as the lowest eigenfrequency. There are exactly N eigenfrequencies, as is seen from (7.23): For $n \geq N + 1$, we set $n = (N + 1) + \tau$ and find

$$\Theta_n = \pi + \frac{\tau \pi}{N+1}.$$

If one inserts the above expression into equation (7.17a) and (7.17b) for ω and c, respectively, one obtains for the amplitudes of the normal vibration

$$-A_{\nu-1}^{(n)} + 2A_{\nu}^{(n)} \cos \frac{n\pi}{N+1} - A_{\nu+1}^{(n)} = 0,$$

$$2A_1^{(n)} \cos \frac{n\pi}{N+1} = A_2^{(n)}, \tag{7.25}$$

$$2A_N^{(n)} \cos \frac{n\pi}{N+1} = A_{N-1}^{(n)}$$

where the A_ν depend on n ($A_\nu = A_\nu^{(n)}$). The system of equations (7.25) for the A_ν is the same as that for the determinants D_N (equation (7.19)), with the same coefficient $c = 2\cos n\pi/(N+1) = 2\cos\Theta_n$. Only the boundary conditions (7.25) do not correspond to those for the D_N (see equations (7.20) and (7.21)). The general solution for the coefficients A_ν is therefore obtained from equation (7.22) with at first arbitrary coefficients $E^{(n)}$:

$$A_\nu^{(n)} = E_1^{(n)} \cos \nu\Theta_n + E_2^{(n)} \sin \nu\Theta_n,$$

or, in detail,

$$A_\nu^{(n)} = E_1^{(n)} \cos \frac{n\pi\nu}{N+1} + E_2^{(n)} \sin \frac{n\pi\nu}{N+1}. \tag{7.26}$$

Since the points $\nu = 0$ and $\nu = N+1$ are tightly clamped, for all eigenmodes n we have $y_0 = y_{N+1} = 0$, or

$$A_0^{(n)} = A_{N+1}^{(n)} = 0 \quad \text{(boundary condition)}.$$

Then one obtains for $\nu = 0$ in (7.26):

$$E_1^{(n)} = 0, \quad \text{i.e.,} \quad A_\nu^{(n)} = E_2^{(n)} \sin \frac{n\pi\nu}{N+1}.$$

After insertion into equation (7.16), one gets

$$y_\nu^{(n)} = E_2^{(n)} \sin \frac{n\pi\nu}{N+1} \cos \omega_{(n)} t. \tag{7.27}$$

If one inserts $y_\nu = B_\nu \sin \omega t$ instead of equation (7.16) into equation (7.14), one determines B_ν by the same method as A_ν and obtains

$$B_\nu^{(n)} = E_4^{(n)} \sin \frac{n\pi\nu}{N+1}, \quad (E_3^{(n)} = 0);$$

hence, the solutions for the y_ν read

$$y_\nu^{(n)} = E_2^{(n)} \sin \frac{n\pi\nu}{N+1} \cos \omega_{(n)} t \tag{7.28a}$$

and

$$y_\nu^{(n)} = E_4^{(n)} \sin \frac{n\pi\nu}{N+1} \sin \omega_{(n)} t. \tag{7.28b}$$

THE VIBRATING CHAIN

The sum of these individual solutions yields the general solution, which therefore reads

$$y_\nu = \sum_{n=1}^{N} \sin \frac{n\pi \nu}{N+1} \left(E_4^{(n)} \sin \omega_{(n)} t + E_2^{(n)} \cos \omega_{(n)} t \right)$$

$$= \sum_{n=1}^{N} \sin \frac{n\pi \nu}{N+1} (a_n \sin \omega_{(n)} t + b_n \cos \omega_{(n)} t), \tag{7.29}$$

where the constants $E_2^{(n)}$ and $E_4^{(n)}$ were renamed b_n and a_n, respectively. They are determined from the initial conditions.

The equation of the vibrating chord must follow from the limit for $N \to \infty$ and $a \to 0$ (continuous mass distribution):

$$\sin \frac{n\pi \nu}{N+1} = \sin \frac{n\pi a\nu}{(N+1)a} \quad (x_\nu = a\nu \text{ takes only discrete values})$$

$$= \sin \frac{\pi n(a\nu)}{l+a} \quad (l = Na \text{ is the length of the chord})$$

$$\lim_{\substack{N \to \infty \\ a \to 0}} \left(\sin \frac{\pi n x}{l+a} \right) = \sin \frac{\pi n x}{l} \quad (x \text{ continuous}).$$

$\omega_{(n)}^2$ becomes (expansion of the cosine in (7.24a) in a Taylor series):

$$\omega_{(n)}^2 = \frac{2T}{ma}\left(1 - 1 + \frac{1}{2}\left(\frac{n\pi}{N+1}\right)^2 - \cdots \right) \approx \frac{T(n\pi)^2}{(m/a)(N+1)^2 a^2},$$

and with $\sigma = m/a =$ mass density of the chord,

$$\lim_{\substack{N \to \infty \\ a \to 0}} \left(\frac{T(n\pi)^2}{\sigma(N+1)^2 a^2} \right) = \frac{T(n\pi)^2}{\sigma l^2},$$

i.e.,

$$\omega_{(n)} = \sqrt{\frac{T}{\sigma}} \frac{n\pi}{l}.$$

Hence, one has as a limit

$$y_n(x) = \sin\left(\frac{n\pi x}{l}\right) \left[a_n \sin\left(\sqrt{\frac{T}{\sigma}} \cdot \frac{n\pi}{l} t\right) + b_n \cos\left(\sqrt{\frac{T}{\sigma}} \frac{n\pi}{l} t\right) \right]. \tag{7.30}$$

This is the equation for the nth eigenmode of the vibrating chord (l is the chord length). It will be derived once again in the next chapter in a different way and will then be discussed in more detail.

Example 7.3: Exercise: Eigenfrequencies of the vibrating chain

When solving the determinant equation (7.19), we have made a mathematical restriction for c by setting $c = 2\cos\Theta$.

Show that for the cases

(a) $|c| = 2$,

(b) $c < -2$

the eigenvalue equation $D_N = 0$ cannot be satisfied. Clarify that thereby the special choice of the constant c is justified.

Solution (a)

$$D_n = cD_{n-1} - D_{n-2}, \qquad D_1 = c = \pm 2, \qquad D_0 = 1. \tag{7.31}$$

We assert and prove by induction

$$|D_n| \geq |D_{n-1}|. \tag{7.32}$$

Induction start: $n = 2$, $|D_0| = 1$, $|D_1| = 2$, $|D_2| = 3$.
Induction conclusion from $n-1$, $n-2$ to n:

$$|D_n|^2 = 4|D_{n-1}|^2 \pm 4|D_{n-1}||D_{n-2}| + |D_{n-2}|^2$$
$$\geq 4|D_{n-1}|^2 + |D_{n-2}|^2 - 4|D_{n-1}||D_{n-2}|$$

$$\Rightarrow \qquad |D_n|^2 - |D_{n-1}|^2 \geq 3|D_{n-1}|^2 + |D_{n-2}|^2 - 4|D_{n-1}||D_{n-2}|.$$

According to the induction condition,

$$|D_{n-1}| = |D_{n-2}| + \epsilon \quad \text{with} \quad \epsilon \geq 0.$$

From this, it follows that

$$|D_n|^2 - |D_{n-1}|^2 \geq 4|D_{n-2}|^2 + 6\epsilon|D_{n-2}| + 3\epsilon^2 - 4\epsilon|D_{n-2}| - 4|D_{n-2}|^2$$
$$\geq 2\epsilon|D_{n-2}|$$
$$\geq 0$$

$$\Rightarrow \qquad |D_n| \geq |D_{n-1}|. \tag{7.33}$$

Since $|D_n|$ monotonically increases in n, and $|D_1| = 2 > 0$, we have $|D_N| > 0$. Therefore $D_N = 0$ cannot be satisfied. $\omega = 0$ and $\omega = \sqrt{2T/ma}$ are not eigenfrequencies of the vibrating chain.

(b) By inserting the *ansatz* $D_n = Ap^n$, $p \neq 0$, we also find the solution of the recursion formula $D_n = cD_{n-1} - D_{n-2}$, $D_1 = c$, $D_0 = 1$:

$$\left. \begin{array}{l} p_1 = \frac{1}{2}\left(c + (c^2 - 4)^{1/2}\right) < 0 \\ p_2 = \frac{1}{2}\left(c - (c^2 - 4)^{1/2}\right) < 0 \end{array} \right\} \quad 0 > p_1 > p_2. \tag{7.34}$$

The general solution for incorporating the boundary conditions $D_0 = 1$, $D_1 = c$ reads

$$D_n = A_1 p_1^n + A_2 p_2^n. \tag{7.35}$$

THE VIBRATING CHAIN

With $D_0 = 1$, $D_1 = c$, it follows that

$$A_1 + A_2 = 1,$$

$$\frac{A_1}{2}\left(c + (c^2 - 4)^{1/2}\right) + \frac{A_2}{2}\left(c - (c^2 - 4)^{1/2}\right) = c,$$

$$A_1 = \frac{c + (c^2 - 4)^{1/2}}{2(c^2 - 4)^{1/2}}, \quad \Leftrightarrow \quad A_2 = \frac{-c + (c^2 - 4)^{1/2}}{2(c^2 - 4)^{1/2}}. \tag{7.36}$$

One then has

$$D_n = \frac{1}{2}\frac{c + (c^2 - 4)^{1/2}}{(c^2 - 4)^{1/2}} p_1^n + \frac{1}{2}\frac{(c^2 - 4)^{1/2} - c}{(c^2 - 4)^{1/2}} p_2^n$$

$$= \frac{1}{(c^2 - 4)^{1/2}}\left(p_1^{n+1} - p_2^{n+1}\right). \tag{7.37}$$

To determine the physically possible vibration modes, we had required that $D_N = 0$:

$$D_N = 0 \quad \Rightarrow \quad \left(\frac{p_2}{p_1}\right)^{N+1} = 1. \tag{7.38}$$

But now $0 > p_1 > p_2$, hence $(p_2/p_1)^{N+1} > 1$. Thus, for the case $c < -2$ eigenfrequencies do not exist too.

These supplementary investigations can be summarized as follows: The possible eigenfrequencies of the vibrating chain lie between 0 and $\sqrt{2T/ma}$:

$$0 < |\omega| < \sqrt{\frac{2T}{ma}}. \tag{7.39}$$

Example 7.4:   Exercise: Vibration of two coupled mass points, two dimensional

Two mass points (equal mass m) lie on a frictionless horizontal plane and are fixed to each other and to two fixed points A and B by means of springs (spring tension T, length l).

(a) Establish the equation of motion.

(b) Find the normal vibrations and frequencies and describe the motions.

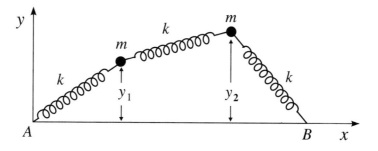

Figure 7.8.

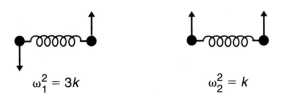

Figure 7.9.

Solution (a) For the vibrating chain with n mass points, which are equally spaced by the distance l, the equations of motion

$$\frac{d^2 y_N}{dt^2} = \frac{T}{ml}(y_{N-1} - 2y_N + y_{N+1}) \quad (N = 1, \ldots, n)$$

were established. For the first and second mass point, we have

$$\ddot{y}_1 = k(y_0 - 2y_1 + y_2) = k(y_2 - 2y_1),$$
$$\ddot{y}_2 = k(y_1 - 2y_2 + y_3) = k(y_1 - 2y_2)$$ (7.40)

with $k = T/ml$; the chain is clamped at the points A and B, i.e., $y_0 = y_3 = 0$.

(b) Solution *ansatz*: $y_1 = A_1 \cos \omega t$, $y_2 = A_2 \cos \omega t$ (ω = eigenfrequency). Insertion into (7.40) yields

$$(2k - \omega^2) A_1 - k A_2 = 0,$$
$$(2k - \omega^2) A_2 - k A_1 = 0.$$ (7.41)

To get the nontrivial solution, the determinant of coefficients must vanish, i.e.,

$$D = \begin{vmatrix} 2k - \omega^2 & -k \\ -k & 2k - \omega^2 \end{vmatrix} = 0;$$

i.e., $\omega^4 + 3k^2 - 4k\omega^2 = 0$, from which it follows that $\omega_1^2 = 3k$, $\omega_2^2 = k$.

Insertion in (7.41) yields $A_1 = A_2$ for ω_2 and $A_1 = -A_2$ for ω_1. This is an opposite-phase and an equal-phase vibration, respectively. We note that the vibration with the higher frequency has opposite phases and a "node," while the vibration with lower frequency has equal phases and a "vibration antinode."

Example 7.5: Exercise: Three masses on a string

Three mass points are fixed equidistantly on a string that is fixed at its endpoints.

(a) Determine the eigenfrequencies of this system if the string tension T can be considered constant (this holds for small amplitudes).

(b) Discuss the eigenvibrations of the system. *Hint*: Note Problems 8.1 and 8.2 in Chapter 8.

Solution (a) For the equations of motion of the system, one finds straightaway

$$m\ddot{x}_1 + \left(\frac{2T}{L}\right) x_1 - \left(\frac{T}{L}\right) x_2 = 0,$$

THE VIBRATING CHAIN

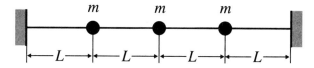

Figure 7.10.

$$m\ddot{x}_2 + \left(\frac{2T}{L}\right)x_2 - \left(\frac{T}{L}\right)x_3 - \left(\frac{T}{L}\right)x_1 = 0, \tag{7.42}$$

$$m\ddot{x}_3 + \left(\frac{2T}{L}\right)x_3 - \left(\frac{T}{L}\right)x_2 = 0.$$

Assuming periodic oscillations, i.e., solutions of the form

$$x_1 = A\sin(\omega t + \psi) \qquad \ddot{x}_1 = -\omega^2 A\sin(\omega t + \psi),$$
$$x_2 = B\sin(\omega t + \psi) \qquad \ddot{x}_2 = -\omega^2 B\sin(\omega t + \psi),$$
$$x_3 = C\sin(\omega t + \psi) \qquad \ddot{x}_3 = -\omega^2 C\sin(\omega t + \psi),$$

we get after insertion into equation (7.42)

$$\left(\frac{2T}{L} - \omega^2 m\right)A - \left(\frac{T}{L}\right)B = 0,$$
$$-\left(\frac{T}{L}\right)A + \left(\frac{2T}{L} - \omega^2 m\right)B - \left(\frac{T}{L}\right)C = 0, \tag{7.43}$$
$$-\left(\frac{T}{L}\right)B + \left(\frac{2T}{L} - \omega^2 m\right)C = 0.$$

As in Problem 8.2, one gets the equation for the frequencies of the system from the expansion of the determinant of coefficients:

$$\left(\frac{Lm}{T}\right)^3 \omega^6 - 6\left(\frac{Lm}{T}\right)^2 \omega^4 + \frac{10Lm}{T}\omega^2 - 4 = 0$$

or

$$\left(\frac{Lm}{T}\right)^3 \Omega^3 - 6\left(\frac{Lm}{T}\right)^2 \Omega^2 + \frac{10Lm}{T}\Omega - 4 = 0 \tag{7.44}$$

with $\Omega \triangleq \omega^2$. This cubic equation with the coefficients

$$a = \left(\frac{Lm}{T}\right)^3, \qquad b = -6\left(\frac{Lm}{T}\right)^2, \qquad c = \frac{10Lm}{T}, \qquad d = -4$$

can be solved by Cardano's method.

With the substitutions

$$y = \Omega + \frac{b}{3a}, \qquad 3p = -\frac{1}{3}\frac{b^2}{a^2} + \frac{c}{a} = -2\frac{T^2}{L^2 m^2}, \qquad 2q = \frac{2}{27}\frac{b^3}{a^3} - \frac{1}{3}\frac{bc}{a^2} + \frac{d}{a} = 0$$

we get $q^2 + p^3 < 0$, i.e., there are three real solutions which by using the auxiliary quantities

$$\cos\varphi = -\frac{q}{\sqrt{-p^3}} = 0, \quad y_1 = -2\sqrt{-p}\cos\left(\frac{\varphi}{3} - \frac{\pi}{3}\right) = -\sqrt{2}\frac{T}{Lm},$$

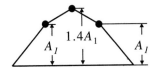

Figure 7.11.

Figure 7.12.

$$y_2 = -2\sqrt{-p} \cos\left(\frac{\varphi}{3} + \frac{\pi}{3}\right) = 0,$$

$$y_3 = 2\sqrt{-p} \cos\frac{\varphi}{3} = \sqrt{2}\frac{T}{Lm}$$

can be calculated as

$$\omega_1 = \sqrt{0.6\frac{T}{Lm}}, \quad \omega_2 = \sqrt{\frac{2T}{Lm}}, \quad \omega_3 = \sqrt{3.4\frac{T}{Lm}}.$$

(b) From the first and third equation of (7.43), one finds for the amplitude ratios

$$\frac{B}{A} = \frac{B}{C} = 2 - \frac{mL\omega^2}{T}. \tag{7.45}$$

Discussion of the modes:

(1) $\omega = \omega_1 = (0.6T/Lm)^{1/2}$ inserted into (7.45) $\Rightarrow$ $B_1/A_1 = B_1/C_1 = 1.4$ or $B_1 = 1.4A_1 = 1.4C_1$.

All three masses are deflected in the same direction, where the first and third mass have equal amplitudes, and the second mass has a larger amplitude.

(2) $\omega = \omega_2 = (2T/Lm)^{1/2}$ inserted into (7.45) $\Rightarrow$ $B_2/A_2 = B_2/C_2 = 0$ and $A_2 = -C_2$ from the second equation of (7.43). The central mass is at rest, while the first and third mass are vibrating in opposite directions but with equal amplitude.

(3) $\omega = \omega_3 = (3.4T/Lm)^{1/2}$ inserted into (7.45) $\Rightarrow$ $B_3/A_3 = B_3/C_3 = -1.4$, i.e., $A_3 = C_3 = -1.4B_3$. The first and the last mass are deflected in the same direction, while the central mass vibrates with different amplitude in the opposite direction.

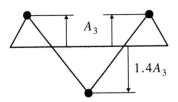

Figure 7.13.

THE VIBRATING CHAIN

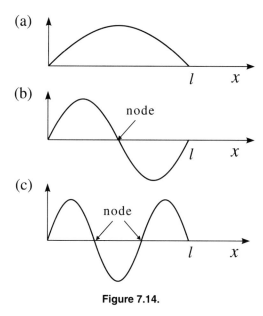

Figure 7.14.

The system discussed here has three vibration modes with 0, 1, and 2 nodes, respectively. For a system with n mass points, both the number of modes as well as the number of possible nodes $(n-1)$ increases. A system with $n \to \infty$ is called a "vibrating string."

A comparison of the figures clearly shows the approximation of the vibrating string by the system of three mass points.

Example 7.6: Exercise: Eigenvibrations of a three-atom molecule

Discuss the eigenvibrations of a three-atom molecule. In the equilibrium state of the molecule, the two atoms of mass m are in the same distance from the atom of mass M. For simplicity one should consider only vibrations along the molecule axis connecting the three atoms, where the complicated interatomic potential is approximated by two strings (with spring constant k).

(a) Establish the equation of motion.

(b) Calculate the eigenfrequencies and discuss the eigenvibrations of the system.

Solution (a) Let x_1, x_2, x_3 be the displacements of the atoms from the equilibrium positions at time t. From Newton's equations and Hooke's law then it follows that

$$m\ddot{x}_1 = -k(x_1 - x_2),$$

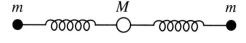

Figure 7.15.

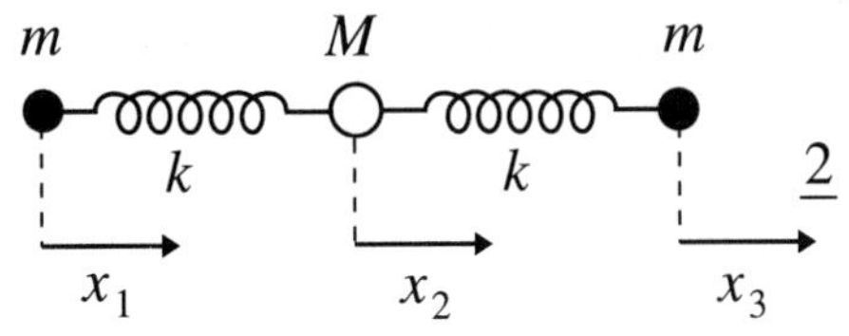

Figure 7.16.

$$M\ddot{x}_2 = -k(x_2 - x_3) - k(x_2 - x_1) = k(x_3 + x_1 - 2x_2), \quad (7.46)$$
$$m\ddot{x}_3 = -k(x_3 - x_2).$$

(b) By inserting the *ansatz* $x_1 = a_1 \cos\omega t$, $x_2 = a_2 \cos\omega t$, and $x_3 = a_3 \cos\omega t$ into equation (7.46), one obtains

$$\begin{aligned}
(m\omega^2 - k)a_1 + ka_2 &= 0, \\
ka_1 + (M\omega^2 - 2k)a_2 + ka_3 &= 0, \\
ka_2 + (m\omega^2 - k)a_3 &= 0.
\end{aligned} \quad (7.47)$$

The eigenfrequencies of this system are obtained by setting the determinant of coefficients equal to zero:

$$\begin{vmatrix} m\omega^2 - k & k & 0 \\ k & M\omega^2 - 2k & k \\ 0 & k & m\omega^2 - k \end{vmatrix} = 0. \quad (7.48)$$

From this, it follows that

$$(m\omega^2 - k)[\omega^4 mM - \omega^2(kM + 2km)] = 0 \quad (7.49)$$

or

$$\omega^2(m\omega^2 - k)[\omega^2 mM - k(M + 2m)] = 0.$$

By factorization of equation (7.49) with respect to ω, one obtains for the eigenvibrations of the system:

$$\omega_1 = 0, \quad \omega_2 = \sqrt{\frac{k}{m}}, \quad \omega_3 = \sqrt{\frac{k}{m}\left(1 + \frac{2m}{M}\right)}.$$

Discussion of the vibration modes:

(1) Insertion of $\omega = \omega_1 = 0$ into (7.47) yields $a_1 = a_2 = a_3$. The eigenfrequency $\omega_1 = 0$ does not correspond to a vibrational motion, but represents only a uniform translation of the entire molecule: •→ ○→ •→.

(2) Inserting $\omega = \omega_2 = (k/m)^{1/2}$ into (7.47) yields $a_1 = -a_3$, $a_2 = 0$; i.e., the central atom is at rest, while the outer atoms vibrate against each other: ←• ○ •→.

(3) Inserting $\omega = \omega_3 = \{k/m(1 + 2m/M)\}^{1/2}$ into (7.47) yields $a_1 = a_3$, $a_2 = -(2m/M)a_1$, i.e., the two outer atoms vibrate in phase, while the central atom vibrates with opposite phase and with another amplitude: •→ ←○ •→.

8 The Vibrating String

A string of length l is fixed at both ends. Thereby appear forces T that are constant in time and independent of the position. The string tension acts as a backdriving force when the string is displaced out of the rest position. A string element Δs at the position x experiences the force

$$F_y(x) = -T \sin \Theta(x)$$

in y-direction. At the position $x + \Delta x$ there acts in y-direction the force

$$F_y(x + \Delta x) = T \sin \Theta(x + \Delta x).$$

In y-direction, the string element Δs experiences the total force

$$F_y = T \sin \Theta(x + \Delta x) - T \sin \Theta(x). \tag{8.1}$$

Accordingly, along the x-direction the string element Δs is pulled by the force

$$F_x = T \cos \Theta(x + \Delta x) - T \cos \Theta(x).$$

In a first approximation we assume that the displacement in x-direction shall be zero. A displacement of the string in y-direction causes only a very small motion in the x-direction. This displacement is negligible compared to the displacement in y-direction, i.e.,

$$F_x = 0.$$

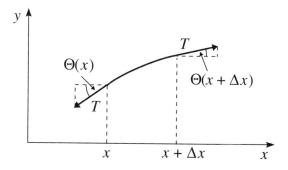

Figure 8.1. The string tension T.

Since we neglect the displacement in x-direction, the only acceleration component of the string element is given by $\partial^2 y/\partial t^2$. The mass of the element is $m = \sigma \Delta s$, where σ represents the line density. From that and by means of equation (8.1) we obtain the equation of motion:

$$F_y = \sigma \Delta s \frac{\partial^2 y}{\partial t^2} = T \sin \Theta(x + \Delta x) - T \sin \Theta(x). \tag{8.2}$$

Both sides are divided by Δx:

$$\frac{\sigma \Delta s d^2 y}{\Delta x dt^2} = \frac{T \sin \Theta(x + \Delta x) - T \sin \Theta(x)}{\Delta x}. \tag{8.3}$$

Inserting for Δs in the left-hand side of equation (8.3)

$$\Delta s = \sqrt{\Delta x^2 + \Delta y^2},$$

one has

$$\frac{\sigma \sqrt{\Delta x^2 + \Delta y^2}}{\Delta x} \frac{\partial^2 y}{\partial t^2} = \sigma \sqrt{1 + \left(\frac{\Delta y}{\Delta x}\right)^2} \frac{\partial^2 y}{\partial t^2}$$

$$= \frac{T \sin \Theta(x + \Delta x) - T \sin \Theta(x)}{\Delta x}. \tag{8.4}$$

By forming the limit for $\Delta x, \Delta y \to 0$ on both sides of equation (8.4), we obtain

$$\sigma \sqrt{1 + \left(\frac{\partial y}{\partial x}\right)^2} \frac{\partial^2 y}{\partial t^2} = T \frac{\partial}{\partial x}(\sin \Theta). \tag{8.5}$$

For $\sin \Theta$ we have $\sin \Theta = \tan \Theta / \sqrt{1 + \tan^2 \Theta}$. Since $\tan \Theta = \partial y/\partial x$ (inclination of the curve), we write

$$\sin \Theta = \frac{\partial y/\partial x}{\sqrt{1 + (\partial y/\partial x)^2}}. \tag{8.6}$$

By means of relation (8.6) the equation (8.5) can be transformed as follows:

$$\sigma \sqrt{1 + \left(\frac{\partial y}{\partial x}\right)^2} \frac{\partial^2 y}{\partial t^2} = T \frac{\partial}{\partial x} \left(\frac{\partial y/\partial x}{\sqrt{1 + (\partial y/\partial x)^2}} \right). \tag{8.7}$$

In order to simplify the equation, we again consider only small displacements of the string in y-direction. Then $\partial y/\partial x \ll 1$, and $(\partial y/\partial x)^2$ can be neglected too.

Thus, we obtain

$$\sigma \frac{\partial^2 y}{\partial t^2} = T \frac{\partial}{\partial x}\left(\frac{\partial y}{\partial x}\right) \tag{8.8}$$

or

$$\sigma \frac{\partial^2 y}{\partial t^2} = T \frac{\partial^2 y}{\partial x^2}. \tag{8.9}$$

We set $c^2 = T/\sigma$ (c has the dimension of a velocity). The desired differential equation (also called the *wave equation*) then reads

$$\frac{\partial^2 y}{\partial t^2} = c^2 \frac{\partial^2 y}{\partial x^2} \quad \text{or} \quad \left(\frac{\partial^2}{\partial x^2} - \frac{1}{c^2}\frac{\partial^2}{\partial t^2}\right) y(x,t) = 0. \tag{8.10}$$

Solution of the wave equation

The wave equation (8.10) is solved with given definite boundary conditions and initial conditions. The *boundary conditions* state that the string is tightly clamped at both ends $x = 0$ and $x = l$, i.e.,

$$y(0,t) = 0, \quad y(l,t) = 0 \quad \text{(boundary conditions)}.$$

The *initial conditions* specify the state of the string at the time $t = 0$ (initial excitation).
The excitation is performed by a displacement of the form $f(x)$,

$$y(x,0) = f(x) \quad \text{(first initial condition)},$$

and the velocity of the string is zero,

$$\left. \frac{\partial}{\partial t} y(x,t) \right|_{t=0} = 0 \quad \text{(second initial condition)}.$$

For solving the partial differential equation (PDE), we use the product ansatz $y(x,t) = X(x) \cdot T(t)$. Such an approach is obvious, since we are looking for eigenvibrations. These are defined so that all mass points (i.e., any string element at any position x) vibrate with the same frequency. By the ansatz $y(x,t) = X(x) \cdot T(t)$, the time behavior is decoupled from the spatial one. Thus we try to split the partial differential equation into a function of the position $X(x)$ and a function of the time $T(t)$. Inserting $y(x,t) = X(x) \cdot T(t)$ into the differential equation (8.10) yields

$$X(x)\ddot{T}(t) = c^2 X''(x) T(t),$$

where $\partial^2 T/\partial t^2 = \ddot{T}$ and $\partial^2 X/\partial x^2 = X''$. The above equation can be rewritten as

$$\frac{\ddot{T}(t)}{T(t)} = c^2 \frac{X''(x)}{X(x)}.$$

Since one side depends only on x and the other side depends on t, while x and t are independent of each other, there is only one possible solution: Both sides are constant. The constant will be denoted by $-\omega^2$.

$$\frac{\ddot{T}}{T} = -\omega^2 \quad \text{or} \quad \ddot{T} + \omega^2 T = 0, \tag{8.11}$$

or

$$\frac{X''}{X} = -\frac{\omega^2}{c^2} \quad \text{or} \quad X'' + \frac{\omega^2}{c^2} X = 0. \tag{8.12}$$

The solutions of the differential equations (continuous harmonic vibrations) have the form

$$T(t) = A \sin \omega t + B \cos \omega t,$$
$$X(x) = C \sin \frac{\omega}{c} x + D \cos \frac{\omega}{c} x.$$

The general solution then reads

$$y(x,t) = (A \sin \omega t + B \cos \omega t) \cdot \left(C \sin \frac{\omega}{c} x + D \cos \frac{\omega}{c} x \right). \tag{8.13}$$

The constants A, B, C, and D are determined from the boundary and initial conditions.
From the boundary conditions, it follows for (8.11) that

$$y(0,t) = 0 = D(A \sin \omega t + B \cos \omega t).$$

Since the expression in brackets differs from zero, we must have $D = 0$. Then (8.13) simplifies to

$$y(x,t) = C \sin \frac{\omega}{c} x (A \sin \omega t + B \cos \omega t).$$

With the second boundary condition, we get

$$y(l,t) = 0 = C \sin \frac{\omega}{c} l (A \sin \omega t + B \cos \omega t)$$
$$\Rightarrow \quad 0 = C \sin \frac{\omega}{c} l.$$

This equation will be satisfied if either of the following holds:

(a) $\quad C = 0,\quad$ which means that the entire string is not displaced,

or

(b) $\quad \sin(\omega l/c) = 0.\quad$ The sine equals zero if $(\omega/c)l = n\pi$, i.e., if $\omega = \omega_n = n\pi c/l$, where $n = 1, 2, 3, \ldots$ ($n = 0$ would lead to case (a)).

From the boundary conditions, we thus obtain the *eigenfrequencies* $\omega_n = n\pi c/l$ of the string. Since the string is a continuous system, there are *infinitely many eigenfrequencies*. The solution for an eigenfrequency, the normal vibration, was marked by the index n. The equation (8.11) becomes

$$y_n(x,t) = C \cdot \sin \frac{n\pi}{l} x \left(A_n \sin \frac{n\pi c}{l} t + B_n \cos \frac{n\pi c}{l} t \right),$$
$$y_n(x,t) = \sin \frac{n\pi}{l} x \left(a_n \sin \frac{n\pi c}{l} t + b_n \cos \frac{n\pi c}{l} t \right),$$

where we set $C \cdot A_n = a_n$ and $C \cdot B_n = b_n$.
From the initial conditions, we have

$$\left. \frac{\partial}{\partial t} y_n(x,t) \right|_{t=0} = 0 = \frac{n\pi c}{l} \sin \frac{n\pi}{l} x \left(a_n \cos \frac{n\pi c}{l} t - b_n \sin \frac{n\pi c}{l} t \right) \bigg|_{t=0}.$$

Then

$$a_n \cdot \frac{n\pi c}{l} \cdot \sin\frac{n\pi}{l}x = 0$$

is satisfied for all x only if $a_n = 0$. Thus, the solution of the differential equation is

$$y_n(x,t) = b_n \cdot \sin\frac{n\pi}{l}x \cos\frac{n\pi c}{l}t. \tag{8.14}$$

The parameter n describes the excitation states of a system, in this case those of the vibrating string. *In quantum physics such a discrete parameter n is called a quantum number.*

Interjection: If we had selected a negative separation constant in equation (8.11), i.e., $+\omega^2$ instead of $-\omega^2$, we would have arrived at the solution

$$y(x,t) = (Ae^{\omega t} + Be^{-\omega t})\left(Ce^{\frac{\omega}{c}x} + De^{-\frac{\omega}{c}x}\right).$$

The boundary conditions $y(0,t) = y(l,t) = 0$ would have led to the conditions

$$C + D = 0; \qquad Ce^{\frac{\omega}{c}l} + De^{-\frac{\omega}{c}l} = 0$$

with the solutions $C = D = 0$. The string would have remained at rest. But this is not the desired solution.

Since the one-dimensional wave equation is a linear differential equation, one can obtain the most general solution, according to the superposition principle, by the superposition (addition) of the particular solutions:

$$y(x,t) = \sum_{n=1}^{\infty} b_n \sin\frac{n\pi x}{l} \cos\frac{n\pi c}{l}t = \sum_{n=1}^{\infty} b_n \sin k_n x \cos \omega_n t.$$

The coefficients b_n can be calculated from the given initial curve by using the considerations on the Fourier series (see the next chapter):

$$y(x,0) \equiv f(x) = \sum_{n=1}^{\infty} b_n \sin\frac{n\pi x}{l}.$$

The calculation of the Fourier coefficients b_n will be shown in the next chapter. One then gets the following general solution of the differential equation:

$$y(x,t) = \sum_{n=1}^{\infty} \left(\frac{2}{l}\int_0^l f(x')\sin\frac{n\pi x'}{l}dx'\right) \sin\frac{n\pi x}{l}\cos\frac{n\pi ct}{l}. \tag{8.15}$$

Normal vibrations

Normal vibrations are described by the following equation:

$$y_n(x,t) = C_n \sin(k_n x)\cos(\omega_n t). \tag{8.16}$$

For a fixed time t, the spatial variation (positional dependence) of the normal vibration depends on the expression $\sin(n\pi x/l)$ (for $n > 1$, $\sin(n\pi x/l)$ has exactly $n-1$ nodes). All mass points (position x) vibrate with the same frequency ω_n.

At a definite position x, the time dependence of the normal vibration is represented by the expression $\cos(n\pi c/l)t$. The *wave number* k_n is defined as

$$k_n \equiv \frac{\omega_n}{c} = \frac{n\pi}{l} = \frac{2\pi}{\lambda_n}, \tag{8.17}$$

where $\lambda_n = 2l/n$ is the *wavelength*.

The *angular frequency* is defined as follows:

$$\omega_n \equiv \frac{n\pi c}{l} = 2\pi \nu_n. \tag{8.18}$$

Solving the equation (8.18) for ν_n, we obtain for the *frequency*

$$\nu_n = \frac{nc}{2l}, \tag{8.19}$$

i.e., the frequencies increase with increasing index n. By definition,

$$c = \sqrt{\frac{T}{\sigma}}; \tag{8.20}$$

c can be interpreted as *"sound" velocity* in the string, as we shall see below. T is the tension in the string, σ is the mass density. From the equations (8.19) and (8.20) we find

$$\nu_n = \frac{n}{2l}\sqrt{\frac{T}{\sigma}}, \tag{8.21}$$

i.e., the longer and thicker a string is, the smaller the frequency. The frequency increases with the string tension T. This agrees with our experience that long, thick strings sound deeper than short, thin ones. With increasing string tension the frequency increases. This property is utilized when tuning up a violin.

Multiplication of the wavelength by the frequency yields a constant c which has the dimension of a velocity:

$$\lambda_n \nu_n = \frac{2l}{n}\frac{nc}{2l} = c \quad \text{(dispersion law)}. \tag{8.22}$$

c is the velocity (phase velocity) by which the wave propagates in a medium. This can be seen as follows: If an initial perturbation $y(x, 0) = f(x)$ is given as in Figure 8.2, $f(x-ct)$ is also a solution of the wave equation (8.10), because with $z = x - ct$ we have

$$\frac{\partial f}{\partial t} = \frac{\partial f}{\partial z}\frac{\partial z}{\partial t} = -c\frac{\partial f}{\partial z}, \quad \frac{\partial^2 f}{\partial t^2} = -c\frac{\partial^2 f}{\partial z^2}\frac{\partial z}{\partial t} = c^2\frac{\partial^2 f}{\partial z^2},$$

and

$$\frac{\partial f}{\partial x} = \frac{\partial f}{\partial z}, \quad \frac{\partial^2 f}{\partial x^2} = \frac{\partial^2 f}{\partial z^2}.$$

NORMAL VIBRATIONS

Figure 8.2. Propagation of a perturbation f(x) along a long string: After the time t, the perturbation has moved away by ct; it is then described by f(x − ct).

Hence,

$$\frac{1}{c^2}\frac{\partial^2}{\partial t^2}f(x-ct) = \frac{c^2}{c^2}\frac{\partial^2 f}{\partial z^2} = \frac{\partial^2 f}{\partial z^2} = \frac{\partial^2 f(x-ct)}{\partial x^2}.$$

$f(x - ct)$ thus satisfies the wave equation (8.10).

Let the maximum of the perturbation $f(x)$ be at x_0. After the time t, it lies at

$$x - ct = x_0.$$

It thus propagates with the velocity

$$\frac{dx}{dt} = c$$

along the string, namely to the right (positive x-direction). One can say that the perturbation $f(x)$ moves along the string with the velocity

$$\frac{dx}{dt} = c. \tag{8.23}$$

The propagation velocity of small perturbations is called the *sound velocity*. One easily realizes as above that $f(x + ct)$ is also a solution of the wave equation and represents a perturbation that moves to the left (negative x-direction). We are dealing here with *running* waves, while for the tightly clamped string we have *standing* waves.

If a string is excited with an arbitrary normal frequency, there are points on the string that remain at rest at any time (*nodes*).

The wavelength, the number of nodes, and the shape of normal vibrations can be represented as a function of the index n (see Figure 8.3).

n	Wavelength	Number of nodes	Figure
1	$2l$	0	(a)
2	l	1	(b)
3	$\frac{2}{3}l$	2	(c)
⋮	⋮	⋮	
n	$\frac{2}{n}l$	$n-1$	

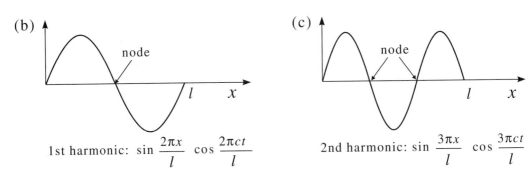

Figure 8.3. The lowest normal vibrations of a string.

Example 8.1: Exercise: Kinetic and potential energy of a vibrating string

Consider a string of density σ that is stretched between two points and is excited with small amplitudes.

(a) Calculate in general the kinetic and potential energy of the string.

(b) Calculate the kinetic and potential energy for waves of the form

$$y = C \cos\left(\frac{\omega(x - ct)}{c}\right)$$

with $T_0 = 500$ N, $C = 0.01$ m, and $\lambda = 0.1$ m.

Solution (a) The part $\overline{PQ}$ of the string has the mass $\sigma \Delta x$ and the velocity $\partial y/\partial t$. Its kinetic energy is then

$$\Delta T = \frac{1}{2}\sigma \Delta x \left(\frac{\partial y}{\partial t}\right)^2. \tag{8.24}$$

The total kinetic energy of the string between $x = a$ and b is

$$T = \frac{1}{2}\sigma \int_a^b \left(\frac{\partial y}{\partial t}\right)^2 dx. \tag{8.25}$$

The work which is needed to elongate the string from Δx to Δl is

$$dP = T_0(\Delta l - \Delta x), \quad \frac{\Delta l}{\Delta x} \sim 1. \tag{8.26}$$

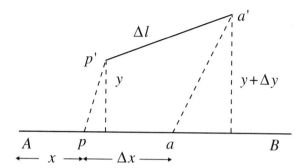

Figure 8.4. Displacement and deformation (elongation compression) of the string element Δx.

For small displacements, we have

$$\Delta l = (\Delta x^2 + \Delta y^2)^{1/2} = \Delta x \left[1 + \left(\frac{\partial y}{\partial x}\right)^2\right]^{1/2} \simeq \Delta x \left[1 + \frac{1}{2}\left(\frac{\partial y}{\partial x}\right)^2\right]. \quad (8.27)$$

The potential energy for the region $x = a$ to $x = b$ is then

$$P = \frac{1}{2} T_0 \int_a^b \left(\frac{\partial y}{\partial x}\right)^2 dx. \quad (8.28)$$

For a wave $y = F(x - ct)$ propagating in a direction, we have

$$T = P = \frac{1}{2} T_0 \int_a^b \left[F'(x - ct)\right]^2 dx, \quad c^2 = \frac{T_0}{\sigma}. \quad (8.29)$$

Hence, the kinetic and potential energy are equal. If a, b are fixed points, then T and P vary with time. But if we admit that a and b can propagate with the sound velocity c, so that

$$a = A + ct \quad \text{and} \quad b = B + ct, \quad (8.30)$$

then P and T are constant:

$$T = P = \frac{1}{2} T_0 \int_A^B (F'(x))^2 dx. \quad (8.31)$$

(b)

$$\frac{\partial y}{\partial t} = C \sin\left(\frac{\omega}{c}x - \omega t\right) \omega$$

$$\Rightarrow \quad \left(\frac{\partial y}{\partial t}\right)^2 = C^2 \sin^2\left(\frac{\omega}{c}x - \omega t\right) \omega^2 \quad (8.32)$$

Insertion into equation (8.25) yields ($a = 0, b = \lambda$)

$$T = \frac{1}{2}\frac{T_0}{c^2}C^2\omega^2 \int_0^\lambda \sin^2\left(\frac{\omega}{c}x - \omega t\right) dx = \frac{1}{2}\frac{T_0}{c^2}C^2\omega^2 \cdot I. \tag{8.33}$$

With the substitution $z = (\omega/c)x - \omega t$ for the integral I, we find

$$I = \frac{c}{\omega}\int_{-\omega t}^{(\omega/c)\lambda - \omega t} \sin^2 z\, dz = \frac{c}{\omega}\int_0^{(\omega/c)\lambda} \sin^2 z\, dz = \frac{c}{\omega}\int_0^{2\pi} \sin^2 z\, dz \tag{8.34}$$

$$= \frac{c}{\omega}\left[\frac{1}{2}z - \frac{1}{4}\sin(2z)\right]_0^{2\pi} = \frac{c}{\omega}\pi$$

$$\Rightarrow \quad T = \frac{1}{2}\frac{T_0}{c^2}C^2\omega^2 \frac{c}{\omega}\pi = \frac{\pi^2 C^2 T_0}{\lambda}, \quad \lambda = 2\pi\frac{c}{\omega}. \tag{8.35}$$

One gets the same expression for the potential energy. Insertion of the numerical values yields

$$T = P = (0.01)^2 \cdot \pi^2 \frac{500\,\text{N}}{0.1\,\text{m}}\text{m}^2 \sim 5\,\text{Nm}.$$

Example 8.2: Exercise: Three different masses equidistantly fixed on a string

Calculate the eigenfrequencies of the system of three different masses that are fixed equidistantly on a stretched string, as is shown in Figure 8.5. (*Hint*: For small amplitudes, the string tension T does not change!)

Solution From Figure 8.6, we extract for the equations of motion

$$2m\ddot{x}_1 = T\left[\frac{(x_2 - x_1)}{L}\right] - T\left[\frac{x_1}{L}\right],$$

$$m\ddot{x}_2 = -T\left[\frac{(x_2 - x_1)}{L}\right] - T\left[\frac{(x_2 - x_3)}{L}\right],$$

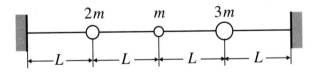

Figure 8.5.

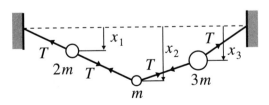

Figure 8.6.

NORMAL VIBRATIONS

$$3m\ddot{x}_3 = T\left[\frac{(x_2 - x_3)}{L}\right] - T\left[\frac{x_3}{L}\right]. \tag{8.36}$$

We look for the eigenvibrations. All mass points must then vibrate with the same frequency. We therefore start with

$$x_1 = A\sin(\omega t + \psi), \qquad \ddot{x}_1 = -\omega^2 A\sin(\omega t + \psi),$$
$$x_2 = B\sin(\omega t + \psi), \qquad \ddot{x}_2 = -\omega^2 B\sin(\omega t + \psi),$$
$$x_3 = C\sin(\omega t + \psi), \qquad \ddot{x}_3 = -\omega^2 C\sin(\omega t + \psi).$$

Hence, after insertion into equation (8.36) one gets

$$\left(\frac{2T}{L} - 2m\omega^2\right)A - \left(\frac{T}{L}\right)B = 0,$$
$$-\left(\frac{T}{L}\right)A + \left(\frac{2T}{L} - m\omega^2\right)B - \left(\frac{T}{L}\right)C = 0, \tag{8.37}$$
$$-\left(\frac{T}{L}\right)B + \left(\frac{2T}{L} - 3m\omega^2\right)C = 0.$$

For evaluating the eigenfrequencies of the system, i.e., for solving equation (8.37), the determinant of coefficients must vanish:

$$\begin{vmatrix} \left(2\frac{T}{L} - 2m\omega^2\right) & -\frac{T}{L} & 0 \\ -\frac{T}{L} & \left(2\frac{T}{L} - m\omega^2\right) & -\frac{T}{L} \\ 0 & -\frac{T}{L} & \left(2\frac{T}{L} - 3m\omega^2\right) \end{vmatrix} = 0.$$

Expansion of the determinant leads to

$$0 = 6m^3\omega^6 - \left(\frac{22Tm^2}{L}\right)\omega^4 + \left(\frac{19T^2m}{L^2}\right)\omega^2 - \left(\frac{4T^3}{L^3}\right)$$

or

$$0 = 6m^3\Omega^3 + \left(\frac{-22Tm^2}{L}\right)\Omega^2 + \left(\frac{19T^2m}{L^2}\right)\Omega + \left(\frac{-4T^3}{L^3}\right), \tag{8.38}$$

where we substituted $\Omega = \omega^2$. This leads to the cubic equation

$$a\Omega^3 + b\Omega^2 + c\Omega + d = 0,$$

where

$$a = 6m^3, \qquad b = \frac{-22Tm^2}{L}, \qquad c = \frac{19T^2m}{L^2}, \qquad d = \frac{-4T^3}{L^3}.$$

It can be transformed to the representation (reduction of the cubic equation)

$$y^3 + 3py + 2q = 0, \tag{8.39}$$

where

$$y = \Omega + \frac{b}{3a} = \Omega - \frac{11}{9}\frac{T}{Lm}$$

and

$$3p = -\frac{1}{3}\frac{b^2}{a^2} + \frac{c}{a} \quad \text{and} \quad 2q = \frac{2}{27}\frac{b^3}{a^3} - \frac{1}{3}\frac{bc}{a^2} + \frac{d}{a}.$$

Insertion leads to

$$3p = -\frac{71}{54}\frac{T^2}{L^2 m^2}, \quad 2q = -\frac{653}{1458}\frac{T^3}{L^3 m^3}.$$

From this, it follows that

$$q^2 + p^3 < 0,$$

i.e., there exist 3 real solutions of the cubic equation (8.39).

For the case $q^2 + p^3 \leq 0$, the solutions y_1, y_2, y_3 can be calculated using tabulated auxiliary quantities (see mathematical supplement 8.4). Direct application of Cardano's formula would lead to complex expressions for the real roots, hence the above method is convenient.

After insertion one obtains for the auxiliary quantities

$$\cos\varphi = \frac{-q}{\sqrt{-p^3}}, \quad y_1 = 2\sqrt{-p}\cos\frac{\varphi}{3},$$

$$y_2 = -2\sqrt{-p}\cos\left(\frac{\varphi}{3} + \frac{\pi}{3}\right),$$

$$y_3 = -2\sqrt{-p}\cos\left(\frac{\varphi}{3} - \frac{\pi}{3}\right),$$

and finally, for the eigenfrequencies of the system

$$\omega_1 = 0.563\sqrt{\frac{T}{Lm}}, \quad \omega_2 = 0.916\sqrt{\frac{T}{Lm}}, \quad \omega_3 = 1.585\sqrt{\frac{T}{Lm}}.$$

Example 8.3: Exercise: Complicated coupled vibrational system

Determine the eigenfrequencies of the system of three equal masses suspended between springs with the spring constant k, as is shown in Figure 8.7. *Hint*: Consider the solution method of the preceding problem 8.2 and the mathematical supplement 8.4.

Solution From Figure 8.7, we extract for the equations of motion

$$m\ddot{x}_1 = -kx_1 - k(x_1 - x_2) - k(x_1 - x_3),$$
$$m\ddot{x}_2 = -kx_2 - k(x_2 - x_1) - k(x_2 - x_3), \quad (8.40)$$
$$m\ddot{x}_3 = -kx_3 - k(x_3 - x_1) - k(x_3 - x_2),$$

or

$$m\ddot{x}_1 + 3kx_1 - kx_2 - kx_3 = 0,$$
$$m\ddot{x}_2 + 3kx_2 - kx_3 - kx_1 = 0, \quad (8.41)$$
$$m\ddot{x}_3 + 3kx_3 - kx_1 - kx_2 = 0.$$

We look for the eigenvibrations. All mass points must vibrate with the same frequency. Thus, we adopt the *ansatz*

$$x_1 = A\cos(\omega t + \psi), \quad \ddot{x}_1 = -\omega^2 A\cos(\omega t + \psi),$$

NORMAL VIBRATIONS

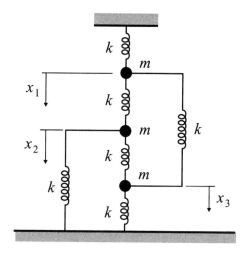

Figure 8.7. Vibrating coupled masses.

$$x_2 = B\cos(\omega t + \psi), \quad \ddot{x}_2 = -\omega^2 B \cos(\omega t + \psi),$$
$$x_3 = C\cos(\omega t + \psi), \quad \ddot{x}_3 = -\omega^2 C \cos(\omega t + \psi),$$

and after insertion into equation (8.41), we get

$$(3k - m\omega^2)A - kB - kC = 0,$$
$$-kA + (3k - m\omega^2)B - kC = 0, \quad \quad \text{(8.42)}$$
$$-kA - kB + (3k - m\omega^2)C = 0.$$

To get a nontrivial solution of equation (8.42), the determinant of coefficients must vanish:

$$\begin{vmatrix} (3k - m\omega^2) & -k & -k \\ -k & (3k - m\omega^2) & -k \\ -k & -k & (3k - m\omega^2) \end{vmatrix} = 0.$$

Expansion of the determinant leads to

$$0 = \omega^6 - \frac{9k}{m}\omega^4 + \frac{24k^2}{m^2}\omega^2 - \frac{16k^3}{m^3}$$

or

$$0 = \Omega^3 - \frac{9k}{m}\Omega^2 + \frac{24k^2}{m^2}\Omega - \frac{16k^3}{m^3},$$

where we substituted $\Omega = \omega^2$ (see Example 8.2). The general cubic equation $a\Omega^3 + b\Omega^2 + c\Omega + d = 0$ (in our case $a = 1, b = -9k/m, c = 24k^2/m^2, d = -16k^3/m^3$) can according to mathematical supplement 8.4 be reduced to

$$y^3 + 3py + 2q = 0,$$

where

$$y = \Omega + \frac{b}{3a}, \quad 3p = -\frac{1}{3}\frac{b^2}{a^2} + \frac{c}{a}, \quad 2q = \frac{2}{27}\frac{b^3}{a^3} - \frac{1}{3}\frac{bc}{a^2} + \frac{d}{a}.$$

Insertion leads to

$$3p = -3\frac{k^2}{m^2}, \quad 2q = 2\frac{k^3}{m^3} \quad \Rightarrow \quad q^2 + p^3 = 0,$$

i.e., there exist 3 solutions (the real roots); 2 of them coincide. Hence, the vibrating system being treated here is *degenerate*. As in problem 8.2, the solutions can be calculated using tabulated auxiliary quantities. For these, we obtain

$$\cos\varphi = \frac{-q}{\sqrt{-p^3}}, \quad y_1 = 2\sqrt{-p}\cos\frac{\varphi}{3},$$

$$y_2 = -2\sqrt{-p}\cos\left(\frac{\varphi}{3} + \frac{\pi}{3}\right),$$

$$y_3 = -2\sqrt{-p}\cos\left(\frac{\varphi}{3} - \frac{\pi}{3}\right),$$

and, after insertion, for the eigenfrequencies of the system

$$\omega_3 = \sqrt{\frac{k}{m}}, \quad \omega_1 = \omega_2 = 2\sqrt{\frac{k}{m}}.$$

Example 8.4: Mathematical supplement: The Cardano formula[1]

In theoretical physics, one often meets the problem of solving a cubic equation, just as in the Examples 8.2 and 8.3. We now will clarify this problem.

Reduction of the general cubic equation: If the general cubic equation

$$x^3 + ax^2 + bx + c = 0 \tag{8.43}$$

with nonvanishing coefficients a, b, and c is to be solved, one must first eliminate the quadratic term of the equation, i.e., *reduce* the equation. If the unknown x is replaced by $y + \lambda$, where y and λ are new, unknown quantities, equation (8.43) turns into

$$(y^3 + 3y^2\lambda + 3y\lambda^2 + \lambda^3) + (ay^2 + 2ay\lambda + a\lambda^2) + (by + b\lambda) + c = 0,$$
$$y^3 + (3\lambda + a)y^2 + (3\lambda^2 + 2a\lambda + b)y + (\lambda^3 + a\lambda^2 + b\lambda + c) = 0. \tag{8.44}$$

Since we have replaced one unknown quantity x by two unknown ones, y and λ, we can freely dispose of one of the two unknown quantities. This freedom is exploited so as to let the quadratic term of the equation disappear. This is achieved by setting the coefficient of y^2, that is, $3\lambda + a$, equal to zero, i.e., $\lambda = -a/3$. By inserting this value the equation (8.44) changes to

$$y^3 + \left(-\frac{a^2}{3} + b\right)y + \left(\frac{2a^3}{27} - \frac{ab}{3} + c\right) = 0. \tag{8.45}$$

---

[1]We follow the exposition of E. v. Hanxleben and R. Hentze, *Lehrbuch der Mathematik*, Friedrich Vieweg & Sohn 1952, Braunschweig-Berlin-Stuttgart.

NORMAL VIBRATIONS

If we set the expressions determined by the known coefficients a, b, and c of the cubic equation,

$$-\frac{a^2}{3} + b = p \quad \text{and} \quad \frac{2a^3}{27} - \frac{ab}{3} + c = q, \tag{8.46}$$

the cubic equation takes the form

$$y^3 + py + q = 0 \quad \text{(reduced cubic equation)}. \tag{8.47}$$

Result: To reduce the cubic equation given in the normal form, one sets $x = y - a/3$. Then equation (8.47) follows from equation (8.43).

Example: $x^3 - 9x^2 + 33x - 65 = 0$.

(1) *Solution*: Set $x = y - (-3) = y + 3$.

$$(y+3)^3 - 9(y+3)^2 + 33(y+3) - 65 = 0,$$
$$(y^3 + 9y^2 + 27y + 27) - 9(y^2 + 6y + 9) + 33(y+3) - 65 = 0,$$
$$y^3 + 6y - 20 = 0.$$

(2) *Solution*: Insert the values calculated from equation (8.46) into equation (8.47).

Special case: If in the general cubic equation, the linear term is missing ($b = 0$), i.e., the cubic equation is given in the form

$$x^3 + ax^2 + c = 0, \tag{8.48}$$

the reduction can also be performed by inserting

$$x = \frac{c}{y}. \tag{8.49}$$

From equations (8.48) and (8.49), we obtain the reduced equation

$$\frac{c^3}{y^3} + a\frac{c^2}{y^2} + c = 0 \quad \text{or} \quad y^3 + acy + c^2 = 0. \tag{8.50}$$

Solution of the reduced cubic equation: If one sets in the reduced cubic equation

$$y^3 + py + q = 0,$$
$$y = u + v, \tag{8.51}$$

one obtains

$$u^3 + 3u^2v + 3uv^2 + v^3 + p(u+v) + q = 0,$$
$$(u^3 + v^3 + q) + 3uv(u+v) + p(u+v) = 0,$$
$$(u^3 + v^3 + q) + (3uv + p)(u+v) = 0. \tag{8.52}$$

Since one can freely dispose of one of the unknown quantities u or v (justification?), these are suitably chosen so that the coefficient of $(u+v)$ vanishes. We therefore set

$$3uv + p = 0, \quad \text{i.e.,} \quad uv = -\frac{p}{3}. \tag{8.53}$$

Then equation (8.52) simplifies to

$$u^3 + v^3 + q = 0 \quad \text{or} \quad u^3 + v^3 = -q. \tag{8.54}$$

u and v are determined by equations (8.53) and (8.54). The quantities u and v can no longer be arbitrarily chosen. By raising equation (8.54) to the second power and equation (8.53) to the third power, one obtains

$$u^6 + 2u^3v^3 + v^6 = q^2,$$
$$4u^3v^3 = -4\left(\frac{p}{3}\right)^3.$$

Subtraction of the two equations yields

$$(u^3 - v^3)^2 = q^2 + 4\left(\frac{p}{3}\right)^3,$$
$$u^3 - v^3 = \pm\sqrt{q^2 + 4\left(\frac{p}{3}\right)^3}. \tag{8.55}$$

By addition and subtraction of equations (8.54) and (8.55), one obtains

$$u^3 = \frac{1}{2}\left[-q \pm \sqrt{q^2 + 4\left(\frac{p}{3}\right)^3}\right] \quad \text{and} \quad v^3 = \frac{1}{2}\left[-q \mp \sqrt{q^2 + 4\left(\frac{p}{3}\right)^3}\right],$$

$$u = \sqrt[3]{-\frac{q}{2} + \sqrt{\left(\frac{q}{2}\right)^2 + \left(\frac{p}{3}\right)^3}} \quad \text{and} \quad v = \sqrt[3]{-\frac{q}{2} \mp \sqrt{\left(\frac{q}{2}\right)^2 + \left(\frac{p}{3}\right)^3}}. \tag{8.56}$$

If one sets

$$\sqrt[3]{-\frac{q}{2} + \sqrt{\left(\frac{q}{2}\right)^2 + \left(\frac{p}{3}\right)^3}} = m \quad \text{and} \quad \sqrt[3]{-\frac{q}{2} - \sqrt{\left(\frac{q}{2}\right)^2 + \left(\frac{p}{3}\right)^3}} = n,$$

one gets

$$u_1 = m, \quad u_2 = m\epsilon_2, \quad u_3 = m\epsilon_3,$$
$$v_1 = n, \quad v_2 = n\epsilon_2, \quad v_3 = n\epsilon_3.$$

Here, the ϵ_i are the unit roots of the cubic equation $x^3 = 1$ which, as is evident, read

$$\epsilon_1 = 1, \quad \epsilon_2 = -\frac{1}{2} + i\frac{\sqrt{3}}{2}, \quad \epsilon = -\frac{1}{2} - i\frac{\sqrt{3}}{2}.$$

Since now $y = u + v$, one can actually form 9 values for y (why?). But since the quantities u and v must satisfy the determining equation (8.53), the number of possible connections between u and v is restricted to 3, namely,

$$y_1 = u_1 + v_1, \quad y_2 = u_2 + v_3, \quad y_3 = u_3 + v_2;$$

hence,

$$y_1 = m + n = \sqrt[3]{-\frac{q}{2} + \sqrt{\left(\frac{q}{2}\right)^2 + \left(\frac{p}{3}\right)^3}} + \sqrt[3]{-\frac{q}{2} - \sqrt{\left(\frac{q}{2}\right)^2 + \left(\frac{p}{3}\right)^3}},$$
$$y_2 = m\epsilon_2 + n\epsilon_3 = -\frac{m+n}{2} + \frac{m-n}{2}i\sqrt{3}, \tag{8.57}$$
$$y_3 = m\epsilon_3 + n\epsilon_2 = -\frac{m+n}{2} - \frac{m-n}{2}i\sqrt{3}.$$

The real root of the cubic equation, i.e., the root

$$y_1 = \sqrt[3]{-\frac{q}{2} + \sqrt{\left(\frac{q}{2}\right)^2 + \left(\frac{p}{3}\right)^3}} + \sqrt[3]{-\frac{q}{2} - \sqrt{\left(\frac{q}{2}\right)^2 + \left(\frac{p}{3}\right)^3}}$$

is known as the *"Cardano formula."* It was named in honor of the Italian Hieronimo Cardano[2] to whom the discovery of the formula was falsely ascribed. Actually, the formula is due to the Bolognesian professor of mathematics Scipione del Ferro,[3] who found this ingenious algorithm.

Example: $y^3 - 15y - 126 = 0$. Here,

$$p = -15, \quad q = -126,$$
$$\frac{p}{3} = -5, \quad \frac{q}{2} = -63.$$

By inserting into the Cardano formula, one obtains

$$\begin{aligned}
y_1 &= \sqrt[3]{63 + \sqrt{63^2 - 5^3}} &&+ \sqrt[3]{63 - \sqrt{63^2 - 5^3}} \\
&= \sqrt[3]{63 + \sqrt{3844}} &&+ \sqrt[3]{63 - \sqrt{3844}}c \\
&= \sqrt[3]{63 + 62} &&+ \sqrt[3]{63 - 62} \\
&= \sqrt[3]{125} &&+ \sqrt[3]{1} \quad (= m+n) \\
&= 6, \\
y_2 &= -\frac{5+1}{2} + \frac{5-1}{2}i\sqrt{3} = -3 + 2i\sqrt{3}, \\
y_3 &= -\frac{5+1}{2} - \frac{5-1}{2}i\sqrt{3} = -3 - 2i\sqrt{3}.
\end{aligned}$$

Check the validity of the roots by insertion!

Discussion of Cardano's formula: The square root appearing in the Cardano formula only yields a real value if the radicand $(q/2)^2 + (p/3)^3 \geq 0$. If the radicand is negative, the three values for y yield complex numbers. We consider the possible cases:

---

[2]*Hieronimo Cardano*, Italian physicist, mathematician, and astrologer, b. Sept. 24, 1501, Pavia–d. Sept. 20, 1576, Rome. Cardano was the illegitimate son of Fazio (Bonifacius) Cardano, a friend of Leonardo da Vinci. He studied at the universities of Pavia and Padua, and in 1526 he graduated in medicine. In 1532, he went to Milan, where he lived in deep poverty, until he got a position teaching in mathematics. In 1539, he worked at a high school of physics, where he soon became the director. In 1543, he accepted a professorship for medicine in Pavia.

As a mathematician, Cardano was the most prominent personality of his age. In 1539, he published two books on arithmetic methods. At this time, the discovery of a solution method for the cubic equation became known. Nicolo Tartaglia, a Venetian mathematician, was the owner. Cardano tried in vain to get permission to publish it. Tartaglia left the method to him under the condition that he keeps it secret. In 1545, Cardano's book *Artis magnae sive de regulis algebraicis*, one of the cornerstones of the history of algebra, was published. The book contained, besides many other new facts, the method of solving cubic equations. The publication caused a serious controversy with Tartaglia.

[3]*Scipione del Ferro*, b. 1465(?)–d. 1526 (?). About his life we know only that he lectured from 1496 to 1526 at the university of Bologna. By 1500, he discovered the method of solving the cubic equation but did not publish it. Tartaglia rediscovered the method in 1535.

		$\sqrt{\left(\frac{q}{2}\right)^2 + \left(\frac{p}{3}\right)^3}$	Form of the roots
(1)	$p > 0$	Real	A real value, two complex conjugate values
(2)	$p < 0$, namely,		
(a)	$\left|\left(\frac{p}{3}\right)^3\right| < \left(\frac{q}{2}\right)^2$	Real	As in 1.
(b)	$\left|\left(\frac{p}{3}\right)^3\right| = \left(\frac{q}{2}\right)^2$	$= 0$	Three real values, among them a double root
(c)	$\left|\left(\frac{p}{3}\right)^3\right| > \left(\frac{q}{2}\right)^2$	Imaginary	All three roots by the form imaginary

The case (2c) was of particular interest to the mathematicians of the Middle Ages. Since any cubic equation has at least one real root, but they could not find it by means of Cardano's formula, the case was called the *casus irreducibilis*.[4] The first to solve this case was the French politician and mathematician Vieta.[5] He proved by using trigonometry that this case was solvable too, and that in this case the equation has three real roots.

Trigonometric solution of the irreducible case: Since p is negative in this case, one starts from the reduced cubic equation

$$y^3 - py + q = 0, \tag{8.58}$$

where p must now be kept fixed as absolute numerical value. According to the trigonometric formulae we have

$$\begin{aligned}
\cos 3\alpha &= \cos(2\alpha + \alpha) = \cos 2\alpha \cos \alpha - \sin 2\alpha \sin \alpha \\
&= (\cos^2 \alpha - \sin^2 \alpha) \cos \alpha - 2 \sin^2 \alpha \cos \alpha \\
&= \cos^3 \alpha - \sin^2 \alpha \cos \alpha - 2 \sin^2 \alpha \cos \alpha \\
&= \cos^3 \alpha - (1 - \cos^2 \alpha) \cos \alpha - 2(1 - \cos^2 \alpha) \cos \alpha \\
&= \cos^3 \alpha - \cos \alpha + \cos^3 \alpha - 2 \cos \alpha + 2 \cos^3 \alpha \\
&= 4 \cos^3 \alpha - 3 \cos \alpha,
\end{aligned}$$

---

[4] *Casus irreducibilis* (Lat.) = "the nonreducible case."

[5] *Francois Vieta*, French mathematician, b. 1540, Fontenay-le-Comte–d. Dec. 13, 1603, Paris. Advocate and advisor of Parliament in the Bretagne. His greatest achievements were in the theory of equations and algebra, where he introduced and systematically used letter notations. He established the rules for the rectangular spherical triangle which are often ascribed to Neper. In his *Canon mathematicus*, a table of angular functions (1571), he emphasized the advantages of decimal notation. [BR]

NORMAL VIBRATIONS

thus,

$$\cos^3 \alpha - \frac{3}{4} \cos \alpha - \frac{1}{4} \cos 3\alpha = 0. \tag{8.59}$$

If one considers $\cos \alpha$ to be unknown, equation (8.59) coincides with the form of equation (8.58). But since the value of the cosine varies only between the limits -1 and $+1$, while y, according to the values of p and q, can take any values, one cannot simply set $\cos \alpha = y$. By multiplying equation (8.59) by a still uncertain positive factor ϱ^3, one obtains

$$\varrho^3 \cos^3 \alpha - \frac{3}{4} \varrho^2 \cdot \varrho \cos \alpha - \frac{1}{4} \varrho^3 \cos 3\alpha = 0. \tag{8.60}$$

By setting $\varrho \cdot \cos \alpha = y$, $p = (3/4)\varrho^2$, and $q = -(1/4)\varrho^3 \cos 3\alpha$, equation (8.60) turns into (8.58). From this, we find

$$\varrho = 2 \cdot \sqrt{\frac{p}{3}} \tag{8.61}$$

and

$$\cos 3\alpha = -\frac{4q}{\varrho^3} = \frac{-4q}{8 \cdot (p/3)\sqrt{p/3}} = -\frac{q/2}{\sqrt{(p/3)^3}}. \tag{8.62}$$

Equation (8.62) is ambiguous, since the cosine is a periodic function. One has

$$3\alpha = \varphi + k \cdot 360°, \quad \text{where} \quad k = 0, 1, 2, 3, \ldots. \tag{8.63}$$

From this, we find for α

$$\alpha_1 = \frac{\varphi}{3}, \qquad \alpha_2 = \frac{\varphi}{3} + 120°, \qquad \alpha_3 = \frac{\varphi}{3} + 240°.$$

Compare this consideration with the problem of cyclotomy! Which values are obtained for α if $k = 3, 4, \ldots$?

For y, one obtains

$$y_1 = 2\sqrt{\frac{p}{3}} \cos \frac{\varphi}{3}, \qquad y_2 = 2\sqrt{\frac{p}{3}} \cos\left(\frac{\varphi}{3} + 120°\right), \qquad y_3 = 2\sqrt{\frac{p}{3}} \cos\left(\frac{\varphi}{3} + 240°\right).$$

Now

$$\cos\left(\frac{\varphi}{3} + 120°\right) = -\cos\left(60° - \frac{\varphi}{3}\right)$$

and

$$\cos\left(\frac{\varphi}{3} + 240°\right) = -\cos\left(60° + \frac{\varphi}{3}\right),$$

so that the roots of the cubic equations are

$$y_1 = 2\sqrt{\frac{p}{3}} \cos \frac{\varphi}{3},$$
$$y_2 = -2\sqrt{\frac{p}{3}} \cos\left(60° - \frac{\varphi}{3}\right), \tag{8.64}$$
$$y_3 = -2\sqrt{\frac{p}{3}} \cos\left(60° + \frac{\varphi}{3}\right).$$

Comment: The formulas of the *casus irreducibilis* can also be derived by means of the Moivre's theorem.

Example: Calculate the roots of the equation

$$y^3 - 981y - 11340 = 0.$$

Solution: Since $p < 0$ and

$$\left|\left(\frac{p}{3}\right)^3\right| = 327^3, \qquad \log\left|\left(\frac{p}{3}\right)^3\right| = 3 \cdot \log 327 = 7.5436,$$

$$\left(\frac{q}{2}\right)^2 = 5670^2, \qquad \log\left(\frac{q}{2}\right)^2 = 2 \cdot \log 5670 = 7.5072,$$

by comparing the logarithms it follows that $|(p/3)^3| > (q/2)^2$. Thus, the condition of the *casus irreducibilis* is fulfilled. According to equation (8.62)

$$\cos 3\alpha = +\frac{5670}{\sqrt{327^3}},$$

$$\log \cos 3\alpha = 3.7536 - 3.7718 = 9.9818 - 10,$$

$$\varphi = 3\alpha \approx 16°30', \qquad \text{hence,} \qquad \frac{\varphi}{3} = \alpha = 5°30'.$$

From equation (8.64), we obtain $y_1 = 36$, $y_2 = -21$, $y_3 = -15$. Check the root values by insertion!

9 Fourier Series

When setting the initial conditions for the problem of the vibrating string, a trigonometric series was set equal to a given function $f(x)$. The expansion coefficients of the series had to be determined. To solve the problem, the function $f(x)$ should also be represented by a trigonometric series. These trigonometric series are called Fourier series.[1] The conditions that allow an expansion of a function into a Fourier series are summarized as follows:

(1) $f(x)$ is defined in the interval $a \leq x < a + 2l$,

(2) $f(x)$ and $f'(x)$ are piecewise continuous on $a \leq x < a + 2l$,

(3) $f(x)$ has a finite number of discontinuities which are finite jump discontinuities, and

(4) $f(x)$ has the period $2l$, i.e., $f(x + 2l) = f(x)$.

---

[1] *Jean Baptiste Joseph Fourier*, b. March 21, 1768, Auxerre, son of a tailor–d. May 16, 1830, Paris. Fourier attended the home École Militaire. Because of his origin he was excluded from an officer's career. Fourier decided to join the clergy, but did not take a vow because of the outbreak of the revolution of 1789. Fourier first took a teaching position in Auxerre. Soon he turned to politics and was arrested several times. In 1795, he was sent to Paris to study at the École Normale. He soon became member of the teaching staff of the newly founded École Polytechnique. In 1798, he became director of the Institut d'Egypte in Cairo. Only in 1801 did he return to Paris, where he was appointed by Napoleon as a prefect of the departement Isère. During his term of office from 1802 to 1815, he arranged the drainage of the malaria-infested marshes of Bourgoin. After the downfall of Napoleon, Fourier was dismissed from all posts by the Bourbons. However, in 1817 the king had to agree to Fourier's election to the Academy of Sciences, where he became permanent secretary in 1822. Fourier's most important mathematical achievement was his treatment of the notion of the function. The problem of the vibrating string that had been treated already by D'Alembert, Euler, and Lagrange, and had been solved in 1755 by D. Bernoulli by a trigonometric series. The subsequent question of whether an "arbitrary" function can be represented by such a series was answered 1807/12 by Fourier in the affirmative. The question about the conditions for such a representation could be answered only by his friend Dirichlet. Fourier became known mainly by his *Théorie analytique de la chaleur* (1822) which deals mainly with the discussion of the equation of heat propagation in terms of Fourier-series. This work represents the starting point for treating partial differential equations with boundary conditions by means of trigonometric series. Fourier also made import contributions to the theory of solving equations and to the probability calculus.

These conditions (Dirichlet conditions) are sufficient to represent $f(x)$ by a Fourier series:

$$f(x) = \frac{a_0}{2} + \sum_{n=1}^{\infty}\left(a_n \cos\frac{n\pi x}{l} + b_n \sin\frac{n\pi x}{l}\right). \tag{9.1}$$

The Fourier coefficients a_n, b_n, and a_0 are determined as follows:

$$a_n = \frac{1}{l}\int_a^{a+2l} f(x) \cos\frac{n\pi x}{l}\,dx,$$

$$b_n = \frac{1}{l}\int_a^{a+2l} f(x) \sin\frac{n\pi x}{l}\,dx, \tag{9.2}$$

$$a_0 = \frac{1}{l}\int_a^{a+2l} f(x)\,dx.$$

To prove these formulas, one needs the so-called orthogonality relations of the trigonometric functions:

$$\int_0^{2l} \cos\frac{n\pi x}{l} \cos\frac{m\pi x}{l}\,dx = l\,\delta_{nm},$$

$$\int_0^{2l} \sin\frac{n\pi x}{l} \sin\frac{m\pi x}{l}\,dx = l\,\delta_{nm}, \tag{9.3}$$

$$\int_0^{2l} \sin\frac{n\pi x}{l} \cos\frac{m\pi x}{l}\,dx = 0.$$

The first relation can be proven by means of the theorem

$$\cos A \cos B = \frac{1}{2}\Big(\cos(A+B) + \cos(A-B)\Big),$$

$$\int_0^{2l} \cos\frac{n\pi x}{l} \cos\frac{m\pi x}{l}\,dx = \frac{1}{2}\int_0^{2l}\left(\cos\frac{(n+m)\pi x}{l} + \cos\frac{(n-m)\pi x}{l}\right)dx = 0,$$

if $n \neq m$. The integral of the cosine function over a full period vanishes. For $n = m$ we have

$$\int_0^{2l} \cos\frac{n\pi x}{l} \cos\frac{m\pi x}{l}\,dx = \frac{1}{2}\int_0^{2l}\left(1 + \cos\frac{2n\pi x}{l}\right)dx = l.$$

The other relations can be proved in an analogous way.

FOURIER SERIES

The formula (9.2) for calculating the Fourier coefficients can be proved by means of the orthogonality relations.

To determine the a_n, one multiplies the equation

$$f(x) = \frac{a_0}{2} + \sum_{n=1}^{\infty} a_n \cos \frac{n\pi x}{l} + \sum_{n=1}^{\infty} b_n \sin \frac{n\pi x}{l}$$

by $\cos(m\pi x/l)$ and then integrates over the interval 0 to $2l$:

$$\int_0^{2l} f(x) \cos \frac{m\pi x}{l} dx = \frac{a_0}{2} \int_0^{2l} \cos \frac{m\pi x}{l} dx + \sum_{n=1}^{\infty} a_n \int_0^{2l} \cos \frac{n\pi x}{l} \cos \frac{m\pi x}{l} dx$$

$$+ \sum_{n=1}^{\infty} b_n \int_0^{2l} \sin \frac{n\pi x}{l} \cos \frac{m\pi x}{l} dx$$

$$= \sum_{n=1}^{\infty} a_n l \delta_{nm} = l a_m,$$

and therefore,

$$a_m = \frac{1}{l} \int_0^{2l} f(x) \cos \frac{m\pi x}{l} dx, \qquad (9.4)$$

as is given by the equations (9.2).

The analogous relation for the b_m can be confirmed by multiplication of equation (9.1) by $\sin(m\pi x/l)$ and integration from 0 to $2l$; the same holds for the calculation of a_0.

Functions that satisfy

$$f(x) = f(-x)$$

are called *even functions*; functions with the property

$$f(x) = -f(-x)$$

are called *odd functions*. For instance, $f(x) = \cos x$ evidently is an even function and $f(x) = \sin x$ an odd function. The part of (9.1)

$$\frac{a_0}{2} + \sum_{n=1}^{\infty} a_n \cos \frac{n\pi x}{l}$$

is obviously even, while

$$\sum_{n=1}^{\infty} b_n \sin \frac{n\pi x}{l}$$

represents the odd part of the series expansion (9.1). Therefore, for even functions all $b_n = 0$, for odd functions a_0 and all a_n are equal to zero.

Any function $f(x)$ can be decomposed into an even and an odd part. Thus, $(f(x) + f(-x))/2$ is the even part and $(f(x) - f(-x))/2$ the odd part of $f(x) = [(f(x) + f(-x))/2 + (f(x) - f(-x))/2]$.

Example 9.1: Inclusion of the initial conditions for the vibrating string by means of the Fourier expansion

A string is fixed at both ends. The center is displaced from the equilibrium position by the distance H and then released. From Figure 9.1 we see that the initial displacement is given by

$$y(x,0) = f(x) = \begin{cases} 2\dfrac{Hx}{l}, & 0 \le x \le \dfrac{l}{2}, \\ \dfrac{2H(l-x)}{l}, & \dfrac{l}{2} \le x \le l. \end{cases}$$

If we assume $f(x)$ is an odd function (dashed line), we then obtain

$$b_n = \frac{2}{l} \int_0^l f(x) \sin \frac{n\pi x}{l} dx$$

$$= \frac{2}{l} \left(\int_0^{l/2} \frac{2Hx}{l} \sin \frac{n\pi x}{l} dx + \int_{l/2}^l \frac{2H}{l}(l-x) \sin \frac{n\pi x}{l} dx \right),$$

$$\int_0^{l/2} \frac{2Hx}{l} \sin \frac{n\pi x}{l} dx = \frac{2H}{l} \left[-x \frac{l}{n\pi} \cos \frac{n\pi x}{l} + \frac{l^2}{n^2\pi^2} \sin \frac{n\pi x}{l} \right]_0^{l/2}$$

$$= \frac{2lH}{n^2\pi^2} \sin \frac{n\pi}{2} - \frac{Hl}{n\pi} \cos \frac{n\pi}{2},$$

$$\int_{l/2}^l \frac{2H}{l}(l-x) \sin \frac{n\pi x}{l} dx = \frac{2H}{l} \left(\int_{l/2}^l l \sin \frac{n\pi x}{l} dx - \int_{l/2}^l x \sin \frac{n\pi x}{l} dx \right)$$

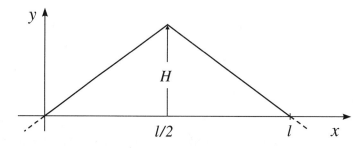

Figure 9.1.

$$= \frac{2H}{l}\left[-\frac{l^2}{n\pi}\cos\frac{n\pi x}{l}+\frac{xl}{n\pi}\cos\frac{n\pi x}{l}-\frac{l^2}{n^2\pi^2}\sin\frac{n\pi x}{l}\right]_{l/2}^{l}$$

$$= \frac{2lH}{n^2\pi^2}\sin\frac{n\pi}{2}+\frac{lH}{n\pi}\cos\frac{n\pi}{2},$$

$$b_n = \frac{2}{l}\left(\frac{2lH}{n^2\pi^2}\sin\frac{n\pi}{2}+\frac{2lH}{n^2\pi^2}\sin\frac{n\pi}{2}\right)$$

$$= \frac{8H}{n^2\pi^2}\sin\frac{n\pi}{2}.$$

By inserting the solution for the Fourier coefficient b_n into the general solution of the differential equation (8.15), we get the equation that describes the vibrations of a string:

$$y(x,t) = \sum_{n=1}^{\infty}\left(\frac{8H}{n^2\pi^2}\sin\frac{n\pi}{2}\right)\sin\frac{n\pi x}{l}\cos\frac{n\pi ct}{l}$$

$$= \frac{8H}{\pi^2}\left(\frac{1}{1^2}\sin\frac{\pi x}{l}\cos\frac{\pi ct}{l}-\frac{1}{3^2}\sin\frac{3\pi x}{l}\cos\frac{3\pi ct}{l}\right.$$

$$\left.+\frac{1}{5^2}\sin\frac{5\pi x}{l}\cos\frac{5\pi ct}{l}-\cdots\right).$$

Thus, by plucking the string in the center one essentially excites the fundamental mode (lowest eigenvibration) $\sin(\pi x/l)\cos(\pi ct/l)$. Several overtones are admixed with small amplitude. The initial displacement obviously corresponds to the fundamental vibration. If one wants to excite pure overtones, the initial displacement must be selected according to the desired higher harmonic vibration (compare the figures after equation (8.23)).

Example 9.2: Exercise: Fourier series of the sawtooth function

Find the Fourier series of the function

$$f(x) = 4x, \qquad 0 \le x \le 10, \qquad \text{with period} \quad 2l = 10, \qquad l = 5.$$

Solution The Fourier coefficients are

$$a_0 = \frac{1}{5}\int_0^{10} 4x\,dx = \frac{2}{5}x^2\Big|_0^{10} = 40,$$

$$a_n = \frac{1}{5}\int_0^{10} 4x\cos\frac{n\pi x}{5}\,dx = \frac{4x}{n\pi}\cos\frac{n\pi x}{5}\Big|_0^{10}-\frac{4}{n\pi}\int_0^{10}\sin\frac{n\pi x}{5}\,dx$$

$$= 0+\frac{20}{n^2\pi^2}\cos\frac{n\pi x}{5}\Big|_0^{10} = 0,$$

$$b_n = \frac{4}{5}\int_0^{10} x\sin\frac{n\pi x}{5}\,dx = -\frac{4x}{n\pi}\cos\frac{n\pi x}{5}\Big|_0^{10}+\frac{4}{n\pi}\int_0^{10}\cos\frac{n\pi x}{5}\,dx$$

$$= -\frac{40}{n\pi}+\frac{20}{n^2\pi^2}\sin\frac{n\pi x}{5}\Big|_0^{10} = -\frac{40}{n\pi}.$$

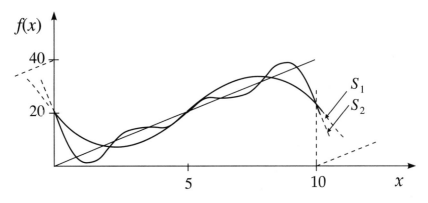

Figure 9.2.

Hence, the Fourier series reads

$$f(x) = 20 - \frac{40}{\pi} \sum_{n=1}^{\infty} \frac{1}{n} \sin \frac{n \pi x}{5}.$$

The first partial sums S_n of this series are drawn in Figure 9.2. A comparison of this series with the starting curve $f(x)$ illustrates the convergence of this Fourier series.

Example 9.3:   Exercise: Vibrating string with a given velocity distribution

Find the transverse displacement of a vibrating string of length l with fixed endpoints if the string is initially in its rest position and has a velocity distribution $g(x)$.

Solution   We look for the solution of the boundary value problem

$$\frac{\partial^2 y}{\partial t^2} = c^2 \frac{\partial^2 y}{\partial x^2}, \tag{9.5}$$

where $y = y(x, t)$, with

$$y(0, t) = 0, \qquad y(l, t) = 0,$$
$$y(x, 0) = 0, \qquad \left.\frac{\partial}{\partial t} y(x, t)\right|_{t=0} = g(x). \tag{9.6}$$

We use the separation *ansatz* $y = X(x) \cdot T(t)$. By inserting it into equation (9.5), one obtains

$$X \cdot \ddot{T} = c^2 X'' T \quad \text{or} \quad \frac{X''}{X}(x) = \frac{\ddot{T}}{c^2 T}(t). \tag{9.7}$$

Since the left-hand side of equation (9.7) depends only on x, the right side only on t, and x and t are independent of each other, the equation is satisfied only then if both sides are constant. The constant is denoted by $-\lambda^2$.

$$\frac{X''}{X} = -\lambda^2 \quad \text{and} \quad \frac{\ddot{T}}{c^2 T} = -\lambda^2,$$

or, transformed,

$$X'' + \lambda^2 X = 0 \quad \text{and} \quad \ddot{T} + \lambda^2 c^2 T = 0. \tag{9.8}$$

The two equations have the solutions

$$X = A_1 \cos \lambda x + B_1 \sin \lambda x, \qquad T = A_2 \cos \lambda ct + B_2 \sin \lambda ct.$$

Since $y = X \cdot T$, we have

$$y(x, t) = (A_1 \cos \lambda x + B_1 \sin \lambda x)(A_2 \cos \lambda ct + B_2 \sin \lambda ct). \tag{9.9}$$

From the condition $y(0, t) = 0$, it follows that $A_1(A_2 \cos \lambda ct + B_2 \sin \lambda ct) = 0$. This condition is satisfied by $A_1 = 0$. Then

$$y(x, t) = B_1 \sin \lambda x (A_2 \cos \lambda ct + B_2 \sin \lambda ct).$$

We now set

$$B_1 A_2 = a, \qquad B_1 B_2 = b,$$

and it follows that

$$y(x, t) = \sin \lambda x (a \cos \lambda ct + b \sin \lambda ct). \tag{9.10}$$

From the condition $y(l, t) = 0$, it follows that $\sin \lambda l = 0$. This happens if

$$\lambda l = n\pi \quad \text{or} \quad \lambda = \frac{n\pi}{l}. \tag{9.11}$$

Here, $n = 1, 2, 3, \ldots$. The value $n = 0$ which seems possible at first sight leads to $y(x, t) \equiv 0$ and must be excluded. The relation (9.11) is inserted into (9.10). The normal vibration will be labeled by the index n:

$$y_n(x, t) = \sin \frac{n\pi x}{l} \left(a_n \cos \frac{n\pi ct}{l} + b_n \sin \frac{n\pi ct}{l} \right). \tag{9.12}$$

Because $y(x, 0) = 0$, all $a_n = 0$, we have

$$y_n(x, t) = b_n \sin \frac{n\pi x}{l} \sin \frac{n\pi ct}{l}. \tag{9.13}$$

By differentiation of (9.13), we get

$$\frac{\partial y_n}{\partial t} = b_n \frac{n\pi c}{l} \sin \frac{n\pi x}{l} \cos \frac{n\pi ct}{l}. \tag{9.14}$$

For linear differential equations, the superposition principle holds, so that the entire solution looks as follows:

$$\frac{\partial y}{\partial t} = \sum_{n=1}^{\infty} \frac{n\pi c b_n}{l} \sin \frac{n\pi x}{l} \cos \frac{n\pi ct}{l}. \tag{9.15}$$

Because

$$\left. \frac{\partial}{\partial t} y(x, t) \right|_{t=0} = g(x),$$

it follows that

$$g(x) = \sum_{n=1}^{\infty} \frac{n\pi c b_n}{l} \sin \frac{n\pi x}{l}. \tag{9.16}$$

The Fourier coefficients then follow by

$$\frac{n\pi c b_n}{l} = \frac{2}{l} \int_0^l g(x) \sin \frac{n\pi x}{l} dx \tag{9.17}$$

or

$$b_n = \frac{2}{n\pi c} \int_0^l g(x) \sin \frac{n\pi x}{l} dx. \tag{9.18}$$

By inserting (9.18) into (9.13), we obtain the final solution for $y(x,t)$:

$$y(x,t) = \sum_{n=1}^{\infty} \left(\frac{2}{n\pi c} \int_0^l g(x') \sin \frac{n\pi x'}{l} dx' \right) \sin \frac{n\pi x}{l} \sin \frac{n\pi c t}{l}. \tag{9.19}$$

Example 9.4: Exercise: Fourier series for a step function

Given the function

$$f(x) = \begin{cases} 0, & \text{for } -5 \le x \le 0, \\ 3, & \text{for } 0 \le x \le 5 \end{cases} \quad \text{period } 2l = 10.$$

(a) Sketch the function.

(b) Determine its Fourier series.

Solution (a)

$$f(x) = \begin{cases} 0, & \text{for } -5 \le x \le 0, \\ 3, & \text{for } 0 \le x \le 5 \end{cases} \quad \text{period } 2l = 10.$$

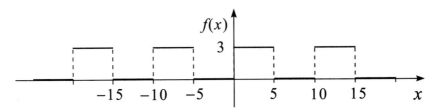

Figure 9.3.

(b) For period $2l = 10$ and $l = 5$, we choose the interval a to $a + 2l$ to be -5 to 5, i.e., $a = -5$:

$$a_n = \frac{1}{l}\int_a^{a+2l} f(x)\cos\frac{n\pi x}{l}dx = \frac{1}{5}\int_{-5}^{5} f(x)\cos\frac{n\pi x}{l}dx$$

$$= \frac{1}{5}\left\{\int_{-5}^{0}(0)\cos\frac{n\pi x}{5}dx + \int_0^5 3\cos\frac{n\pi x}{5}dx\right\} = \frac{3}{5}\int_0^5\cos\frac{n\pi x}{5}dx$$

$$= \frac{3}{5}\left\{\frac{5}{n\pi}\sin\frac{n\pi x}{5}\right\}\bigg|_0^5 = 0 \quad \text{for} \quad n \neq 0.$$

For $n = 0$, one has $a_n = a_0 = (3/5)\int_0^5 \cos(0\pi x/5)\,dx = (3/5)\int_0^5 dx = 3$. Furthermore,

$$b_n = \frac{1}{l}\int_a^{a+2l} f(x)\sin\frac{n\pi x}{l}dx = \frac{1}{5}\int_{-5}^{5} f(x)\sin\frac{n\pi x}{l}dx$$

$$= \frac{1}{5}\left\{\int_{-5}^{0}(0)\sin\frac{n\pi x}{5}dx + \int_0^5 3\sin\frac{n\pi x}{5}dx\right\} = \frac{3}{5}\int_0^5\sin\frac{n\pi x}{5}dx$$

$$= \frac{3}{5}\left(-\frac{5}{n\pi}\cos\frac{n\pi x}{5}\right)\bigg|_0^5 = \frac{3}{n\pi}(1 - \cos n\pi).$$

Thus,

$$f(x) = \frac{3}{2} + \sum_{n=1}^{\infty}\frac{3}{n\pi}(1 - \cos n\pi)\sin\left(\frac{n\pi x}{5}\right),$$

i.e.,

$$f(x) = \frac{3}{2} + \frac{6}{\pi}\left(\sin\frac{\pi x}{5} + \frac{1}{3}\sin\frac{3\pi x}{5} + \frac{1}{5}\sin\frac{5\pi x}{5} + \cdots\right).$$

Example 9.5: Exercise: On the unambiguousness of the tautochrone problem

Which trajectory of the mass of a mathematical pendulum yields a pendulum period that is independent of the amplitude?

Solution   We consider the Figure 9.4. From energy conservation, we have

$$\frac{m}{2}\dot{s}^2(y) + gmy = mgh \tag{9.20}$$

or

$$\dot{s}(y) = \sqrt{2g(h-y)}. \tag{9.21}$$

From this, one can calculate the period by separation of the variables:

$$\frac{1}{4}T = \int_0^{T/4}dt = \int_0^{s(h)}\frac{ds}{\sqrt{2g(h-y)}} = \int_0^h\frac{(ds/dy)dy}{\sqrt{2g(h-y)}}. \tag{9.22}$$

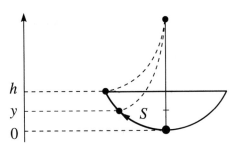

Figure 9.4.

Using the variable $u = y/h$, (9.22) changes to

$$\frac{T}{4} = \int_0^1 \frac{(ds/dy)\sqrt{h}\,du}{\sqrt{2g(1-u)}}. \qquad (9.23)$$

We now require that T be independent of the maximum height h:

$$\frac{dT}{dh} = 0 \quad \text{for all } h. \qquad (9.24)$$

Thus, we get from (9.23) ($s' \equiv ds/dy$)

$$\frac{d}{dh} \int_0^1 \frac{s'\sqrt{h}\,du}{\sqrt{2g(1-u)}} = \int_0^1 \frac{du}{\sqrt{2g(1-u)}} \left(\frac{1}{2}h^{-1/2}s' + \sqrt{h}\frac{ds'}{dh}\right) = 0 \quad \text{for all } h. \qquad (9.25)$$

With the condition that we keep the dimensionless variable $u = y/h$ constant, we can rewrite the derivative with respect to h as a derivative with respect to y,

$$\frac{ds'}{dh} = \frac{uds'}{d(uh)} = u\frac{ds'}{dy} = us'', \qquad (9.26)$$

and thus, we can transform (9.25) into

$$\int_0^1 \frac{du}{\sqrt{8g(1-u)}} (s' + 2ys'') \frac{1}{\sqrt{h}} = 0 \quad \text{for all } h. \qquad (9.27)$$

Any periodic function $f(u)$ satisfying $\int_0^1 f(u)\,du = 0$ can generally be expanded into a Fourier series:

$$f(u) = \sum_{m=1}^{\infty} [a_m \sin(2\pi mu) + b_m \cos(2\pi mu)]. \qquad (9.28)$$

Therefore, from (9.27) it follows that

$$s'' + \frac{1}{2y}s' = \frac{\sqrt{8gh(1-u)}}{2y} \sum_{m=1}^{\infty} \left(a_m \sin(2\pi mu) + b_m \cos(2\pi mu) \right)$$

FOURIER SERIES

$$= \frac{\sqrt{8gh(h-y)}}{2y} \sum_{m=1}^{\infty} \left(a_m \sin\left(2\pi m \frac{y}{h}\right) + b_m \cos\left(2\pi m \frac{y}{h}\right) \right). \tag{9.29}$$

This holds for all values of h. The left-hand side of (9.29) does not contain h; therefore, the right-hand side must be independent of h too. This holds only for $a_m = b_m = 0$ (for all m), as we shall prove now.

To have the right-hand side of (9.29) independent of h, we must have

$$\sum_{m=1}^{\infty} \left[a_m \sin\left(2\pi m \frac{y}{h}\right) + b_m \cos\left(2\pi m \frac{y}{h}\right) \right] = \frac{\text{constant} \cdot (y/h) h^{1/2}}{\sqrt{8g(1-y/h)}} \tag{9.30}$$

or

$$\sum_{m=1}^{\infty} [a_m \sin(2\pi m u) + b_m \cos(2\pi m u)] = \frac{u}{\sqrt{1-u}} \frac{h^{1/2}}{\sqrt{8g}} C. \tag{9.31}$$

By integrating (9.31) from 0 to 1, we obtain

$$0 = \frac{h^{1/2}}{\sqrt{8g}} C \int_0^1 \frac{u}{\sqrt{1-u}} \, du = \frac{4}{3} \frac{h^{1/2}}{\sqrt{8g}} C, \tag{9.32}$$

thus, $C = 0$. (This reflects the fact that $u/\sqrt{1-u}$ cannot be expanded into a Fourier series à la (9.31).)

Inserting this result $C = 0$ again into (9.30), we have $a_m = b_m = 0 \, \forall m$, and thus, from (9.29)

$$s'' + \frac{s'}{2y} = 0. \tag{9.33}$$

From this, one finds by integrating once

$$\frac{s''}{s'} = -\frac{1}{2y} \quad \Rightarrow \quad s' \equiv \frac{ds}{dy} = \tilde{C} e^{-(1/2)\ln y} = \frac{\tilde{C}}{\sqrt{y}}. \tag{9.34}$$

The constant is usually denoted by

$$\tilde{C} = \sqrt{\frac{l}{2}}, \tag{9.35}$$

so that we have to solve

$$\frac{ds}{dy} = \sqrt{\frac{l}{2}} \frac{1}{\sqrt{y}}. \tag{9.36}$$

This is the differential equation of a cycloid (see *Classical Mechanics: Point Particles and Relativity*, Exercise 24.4).

10 The Vibrating Membrane

We consider a two-dimensional system: the vibrating membrane. We shall see that the methods applied for the treatment of a vibrating string can be simply transferred in many respects.

The membrane is a skin without an elasticity of its own. The stretching of the membrane along the edge leads to a tension force that acts as a backdriving force on a deformed membrane.

Let the tangential tension in the membrane be spatially constant and time independent. We consider only vibrations with amplitudes so small that displacements within the membrane plane can be neglected.

Derivation of the differential equation

We introduce the following notations: σ is the surface density of the membrane, and the membrane tension is T (force per unit length). Let the coordinate system be oriented so that the membrane lies in the x,y-plane. The displacements perpendicular to this plane are denoted by $u = u(x, y, t)$.

To set up the equation of motion, we imagine a cut of length Δx through the membrane parallel to the x-axis, and a cut Δy parallel to the y-axis. The force acting on the membrane element $\Delta x \Delta y$ in the x-direction is the product of the tension and the length of the cut: $F_x = T \Delta y$. Analogously for the y-component we have $F_y = T \Delta x$.

The surface element $\Delta x \Delta y$ is pulled by the sum of the two forces. If the membrane is displaced, the u-component of this sum acts on it.

From Figure 10.1, we see

$$F_u = T\Delta x(\sin\varphi(y+\Delta y) - \sin\varphi(y)) + T\Delta y(\sin\vartheta(x+\Delta x) - \sin\vartheta(x)). \tag{10.1}$$

DERIVATION OF THE DIFFERENTIAL EQUATION

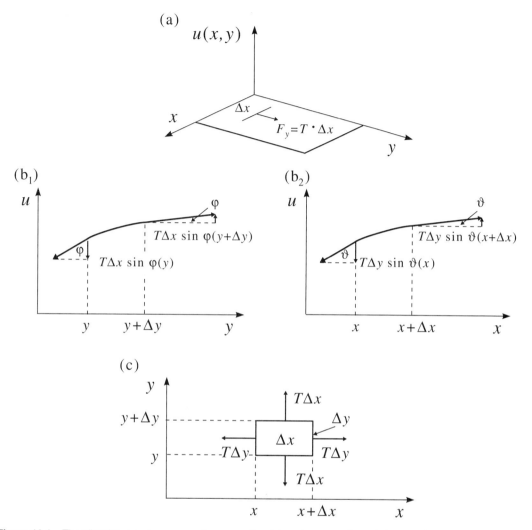

Figure 10.1. The vibrating membrane seen in perspective (a), various cuts through the membrane (b), and from the top (c).

Since we restrict ourselves to small amplitudes and angles, the sine can be replaced by the tangent. For the tangent we then insert the differential quotient, e.g.,

$$\tan \varphi(x, y + \Delta y) = \frac{\partial u}{\partial y}(x, y + \Delta y),$$

i.e., the partial derivative with respect to y at the point $y + \Delta y$.

Equation (10.1) then takes the form

$$F_u = T\Delta x \left(\frac{\partial u}{\partial y}(x, y + \Delta y) - \frac{\partial u}{\partial y}(x, y)\right) + T\Delta y \left(\frac{\partial u}{\partial x}(x + \Delta x) - \frac{\partial u}{\partial x}(x, y)\right).$$

Moving the product $T\Delta x \Delta y$ to the left side, one has

$$F_u = T\Delta x \Delta y \left(\frac{\frac{\partial u}{\partial y}(x, y + \Delta y) - \frac{\partial u}{\partial y}(x, y)}{\Delta y} + \frac{\frac{\partial u}{\partial x}(x + \Delta x, y) - \frac{\partial u}{\partial x}(x, y)}{\Delta x}\right).$$

We replace the area $\Delta x \Delta y$ of the membrane element by $\Delta m/\sigma$, where Δm is its mass, and $\sigma = \Delta m/\Delta x \Delta y$ is the mass density per unit surface. Turning now to the differentials, $\Delta x, \Delta y \to 0$, we find

$$\lim_{\Delta x \to 0} \frac{\frac{\partial u}{\partial x}(x + \Delta x, y) - \frac{\partial u}{\partial x}(x, y)}{\Delta x} = \frac{\partial^2 u}{\partial x^2}(x, y)$$

or

$$F_u = T\frac{\Delta m}{\sigma}\left(\frac{\partial^2 u}{\partial x^2} + \frac{\partial^2 u}{\partial y^2}\right).$$

With this force, we arrive at the equation of motion

$$\Delta m \frac{\partial^2 u}{\partial t^2} = T\frac{\Delta m}{\sigma}\left(\frac{\partial^2 u}{\partial x^2} + \frac{\partial^2 u}{\partial y^2}\right).$$

With the abbreviation $T/\sigma = c^2$ and the Laplace operator, one obtains

$$\Delta u - \frac{1}{c^2}\frac{\partial^2 u}{\partial t^2} = 0. \tag{10.2}$$

This form of the wave equation is independent of the dimension of the vibrating medium. If we insert the three-dimensional Laplace operator and set $u = u(x, y, z, t)$, equation (10.2) also holds for sound vibrations (u then represents the density variation of the air). c is the propagation velocity of small perturbations (velocity of sound) – similar to the case of the vibrating string.

Solution of the differential equation: Rectangular membrane

We will now solve the two-dimensional wave equation (10.2) for the example of the *rectangular membrane*.

We have the *boundary conditions* which mean that the membrane cannot vibrate at the boundary: $u(0, y, t) = u(a, y, t) = u(x, 0, t) = u(x, b, t) = 0$. To solve the equation, we again use the product *ansatz*

$$u(x, y, t) = V(x, y) \cdot Z(t),$$

SOLUTION OF THE DIFFERENTIAL EQUATION

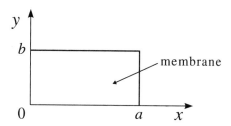

Figure 10.2. A rectangular membrane.

by means of which we first of all separate the space variables from the time variables. Normal vibrations are of this type. All points x, y (mass points) then have the same time behavior. This is typical for eigenvibrations. By insertion into the wave equation, we obtain

$$\frac{\ddot{Z}(t)}{Z(t)} = c^2 \frac{\Delta V(x, y)}{V(x, y)}.$$

Here, one has a function of *only* the position equal to a function that depends *only* on the time. Thus, this identity is only valid if both functions are constants, i.e., unchanging with respect to space and time. The constant that equals these functions is denoted by $-\omega^2$, the quotient ω^2/c^2 by k^2.

One then has

$$\frac{\ddot{Z}}{Z} = -\omega^2, \tag{10.3}$$

$$\frac{\Delta V(x, y)}{V(x, y)} = -k^2, \qquad k^2 = \frac{\omega^2}{c^2}. \tag{10.4}$$

We can at once write down the general solution of (10.3):

$$Z(t) = A \sin(\omega t + \delta).$$

If we had selected a positive separation constant, i.e., $+\omega^2$ in equation (10.3), the solution would have been $Z(t) = e^{\pm \omega t}$. This means that the solution would either explode with the time ($e^{+\omega t}$) or fade away ($e^{-\omega t}$). The negative separation constant in equation (10.3) obviously guarantees harmonic solutions.

In order to separate the two space variables, we use a further separation *ansatz*:

$$V(x, y) = X(x) \cdot Y(y).$$

Insertion into (10.4) yields

$$Y \frac{\partial^2 X}{\partial x^2} + X \frac{\partial^2 Y}{\partial y^2} + k^2 XY = 0.$$

From this, it follows after division by $X(x)Y(y)$ that

$$\frac{1}{X(x)} \frac{\partial^2 X(x)}{\partial x^2} + \frac{1}{Y(y)} \frac{\partial^2 Y(y)}{\partial y^2} + k^2 = 0, \qquad k^2 = \frac{\omega^2}{c^2}.$$

Here again, a function of x equals a function of y only if both are constants. We split the constant k^2 into

$$k^2 = k_x^2 + k_y^2$$

and thus obtain

$$\frac{1}{X}\frac{\partial^2 X}{\partial x^2} = -k_x^2, \qquad \frac{1}{Y}\frac{\partial^2 Y}{\partial y^2} = -k_y^2.$$

Therefore, one has

$$\frac{\partial^2 X}{\partial x^2} + k_x^2 X = 0, \quad \text{solution: } X(x) = A_1 \sin(k_x x + \delta_1),$$

$$\frac{\partial^2 Y}{\partial y^2} + k_y^2 Y = 0, \quad \text{solution: } Y(y) = A_2 \sin(k_y y + \delta_2).[1]$$

By multiplying the partial solutions and combining the constants, one obtains the complete solution of the two-dimensional wave equation:

$$u(x, y, t) = B \sin(k_x x + \delta_1) \sin(k_y y + \delta_2) \sin(\omega t + \delta).$$

Inclusion of the boundary conditions

With the given boundary conditions for u, we obtain

$$u(0, y, t) = B \sin \delta_1 \sin(k_y y + \delta_2) \sin(\omega t + \delta) = 0,$$
$$u(x, 0, t) = B \sin(k_x x + \delta_1) \sin \delta_2 \sin(\omega t + \delta) = 0.$$

Both equations are only satisfied for all values of the variables x, y, t if

$$\sin \delta_1 = \sin \delta_2 = 0,$$

which is for example correct for $\delta_1 = \delta_2 = 0$.

From this, we obtain the other boundary conditions:

$$u(a, y, t) = B \sin(k_x a) \sin(k_y y) \sin(\omega t + \delta) = 0,$$
$$u(x, b, t) = B \sin(k_x x) \sin(k_y b) \sin(\omega t + \delta) = 0.$$

From considerations similar to those above, we find

$$\sin(k_x a) = \sin(k_y b) = 0,$$

---

[1] One of the two separation constants k_x^2 or k_y^2 could in principle be chosen to be negative, so that e.g., $k_x^2 - k_y^2 = k^2$. In this case we would get $Y = Ae^{k_y \cdot y} + Be^{-k_y \cdot y}$, and the boundary conditions $u(x, 0, t) = u(x, b, t) = 0$ could be satisfied only by $A = B = 0$.

from which we get

$$k_x a = n_x \pi, \quad k_y b = n_y \pi, \quad \text{with} \quad n_x, n_y = 1, 2, \ldots.$$

The values $n_x = n_y = 0$ must be excluded, since they lead to $u(x, y, t) = 0$–as for the vibrating string.

Now we have

$$k^2 = k_x^2 + k_y^2 = n_x^2 \left(\frac{\pi}{a}\right)^2 + n_y^2 \left(\frac{\pi}{b}\right)^2,$$

and because $\omega = k \cdot c$, we find for the eigenfrequency

$$\omega_{n_x n_y} = c\pi \sqrt{\frac{n_x^2}{a^2} + \frac{n_y^2}{b^2}}.$$

Eigenfrequencies

Thus, the eigenfrequencies of the rectangular membrane are

$$\omega_{n_x n_y} = c\pi \sqrt{\frac{n_x^2}{a^2} + \frac{n_y^2}{b^2}},$$

where the lowest frequency is the *fundamental harmonic*:

$$\omega_{11} = c\pi \sqrt{\frac{1}{a^2} + \frac{1}{b^2}}.$$

For the string, we have $\omega_n = n\omega_1$, i.e., the higher harmonics are integer multiples of the fundamental frequency. This is no longer valid in the two-dimensional case. Contrary to the *harmonic frequency spectrum* ($\omega_n = n\omega_1$) of the string, the membrane has an *anharmonic spectrum* ($\omega_{n_x n_y} \neq n\omega_{11}$).

Degeneracy

If in the special case of a square membrane, the edges have equal length, $a = b$, then it follows that

$$\omega_{n_x n_y} = \frac{\sqrt{n_x^2 + n_y^2}}{\sqrt{2}} \omega_{11}, \qquad \omega_{11} = \frac{c\pi \sqrt{2}}{a}.$$

Table 10.1.

The ratio $\omega_{n_x n_y}/\omega_{11}$ as a function of n_x and n_y				
$n_y \backslash n_x$	1	2	3	4
1	1.00	1.58	2.24	2.92
2	1.58	2.00	2.55	3.16
3	2.24	2.55	3.00	3.54
4	2.92	3.16	3.54	4.00

The table of the ratios $\omega_{n_x n_y}/\omega_{11}$ for several values of the "quantum numbers" n_x, n_y of a square membrane shows that for different pairs of "quantum numbers" there exist the same eigenvalues, i.e., there are different possible eigenvibrations with the same frequency. *Such states are called degenerate.* For a square membrane which is symmetric with respect to the meaning of the x- and y-coordinate, all states $n_x n_y$ arranged symmetrically with respect to the main diagonal of the table are degenerate.

The degeneracy is removed at once if $a \neq b$. Generally *degeneracies appear only in systems with definite symmetries.*

We further recognize that the square membrane contains a fraction of harmonic overtones (diagonal elements of the table).

Nodal lines

At the points where the position-dependent part of the wave motion vanishes, the string has a node, and the membrane correspondingly has a *nodal line*.

The position-dependent part reads

$$\sin \frac{n_x \pi x}{a} \sin \frac{n_y \pi y}{b}.$$

Then for $n_x = 2$ and $n_y = 1$ we have

$$\sin \frac{2\pi x}{a} \sin \frac{\pi y}{b} = 0$$

as the condition for a nodal line.

Away from the edges this condition is still satisfied for the straight line $x = a/2$, which represents a nodal line for $(n_x, n_y) = (2, 1)$. In general all straight lines of the form

$$x = \frac{ma}{n_x}; \qquad y = \frac{nb}{n_y} \quad (m = 1, 2, \ldots,), \quad m < n_x; \quad (n = 1, 2, \ldots,), \quad n < n_y$$

are nodal lines.

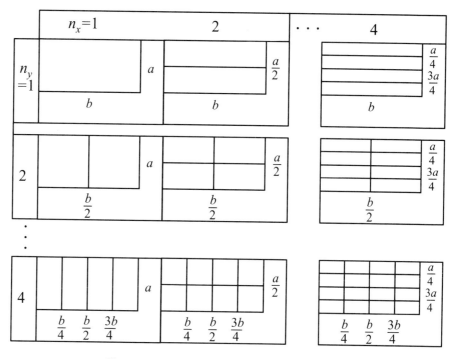

Figure 10.3. Nodal lines of several eigenvibrations.

General solution (inclusion of the initial conditions)

The general solution of the wave equation for the rectangular membrane, since it is a linear differential equation, is obtained as a sum of the particular solutions (superposition principle):

$$u(x, y, t) = \sum_{n_x=1}^{\infty} \sum_{n_y=1}^{\infty} c_{n_x n_y} \sin \frac{n_x \pi x}{a} \sin \frac{n_y \pi y}{b} \sin(\omega_{n_x n_y} t + \varphi_{n_x n_y}).$$

We now can evaluate the $c_{n_x n_y}$ and the $\varphi_{n_x n_y}$ from the initial conditions

$$u(x, y, t = 0) = u_0(x, y),$$
$$\dot{u}(x, y, t = 0) = v_0(x, y).$$

For $t = 0$, the general solution and its time derivative read as follows:

$$u_0(x, y) = \sum_{n_x, n_y=1}^{\infty} c_{n_x n_y} \sin \varphi_{n_x n_y} \cdot \sin \frac{n_x \pi x}{a} \cdot \sin \frac{n_y \pi y}{b},$$

$$v_0(x, y) = \sum_{n_x, n_y=1}^{\infty} \omega_{n_x n_y} c_{n_x n_y} \cos \varphi_{n_x n_y} \cdot \sin \frac{n_x \pi x}{a} \cdot \sin \frac{n_y \pi y}{b}.$$

We redefine the constants:

$$A_{n_x n_y} = C_{n_x n_y} \sin \varphi_{n_x n_y}, \tag{10.5}$$

$$B_{n_x n_y} = \omega_{n_x n_y} C_{n_x n_y} \cos \varphi_{n_x n_y}. \tag{10.6}$$

The above equations then change to

$$u_0(x, y) = \sum_{n_x, n_y=1}^{\infty} A_{n_x n_y} \sin \frac{n_x \pi x}{a} \sin \frac{n_y \pi y}{b}, \tag{10.7}$$

$$v_0(x, y) = \sum_{n_x, n_y=1}^{\infty} B_{n_x n_y} \sin \frac{n_x \pi x}{a} \sin \frac{n_y \pi y}{b}. \tag{10.8}$$

The coefficients $A_{n_x n_y}$ and $B_{n_x n_y}$ can be determined by means of the orthogonality relations. These read

$$\int_{-a}^{a} \sin \frac{\bar{n}_x \pi x}{a} \sin \frac{n_x \pi x}{a} dx = a \delta_{\bar{n}_x n_x},$$

$$\int_{-b}^{b} \sin \frac{\bar{n}_y \pi y}{b} \sin \frac{n_y \pi y}{b} dy = b \delta_{\bar{n}_y n_y}. \tag{10.9}$$

We assume (10.7) to be continued across the borders as an odd function, multiply (10.7) by $\sin(\bar{n}_x \pi x/a)$, and integrate over x from $-a$ to a. Next we multiply by $\sin(\bar{n}_y \pi y/b)$ and integrate over y from $-b$ to b:

$$\int_{-a}^{a} \int_{-b}^{b} u_0(x, y) \sin \frac{\bar{n}_x \pi x}{a} \sin \frac{\bar{n}_y \pi y}{b} dx dy = 4 \int_{0}^{a} \int_{0}^{b} u_0(x, y) \sin \frac{\bar{n}_x \pi x}{a} \sin \frac{\bar{n}_y \pi y}{b} dx dy$$

$$= \sum_{n_x, n_y}^{\infty} A_{n_x n_y} \int_{-a}^{a} \sin \frac{n_x \pi x}{a} \sin \frac{\bar{n}_x \pi x}{a} dx \int_{-b}^{b} \sin \frac{\bar{n}_y \pi y}{b} \sin \frac{n_y \pi y}{b} dy$$

$$= \sum_{n_x, n_y}^{\infty} A_{n_x n_y} \delta_{\bar{n}_x n_x} a \delta_{\bar{n}_y n_y} b = ab A_{\bar{n}_x \bar{n}_y}.$$

Likewise, we treat equation (10.8) to evaluate the coefficients $B_{n_x n_y}$. One then obtains

$$A_{n_x n_y} = \frac{4}{ab} \int_{0}^{a} \int_{0}^{b} u_0(x, y) \sin \frac{n_x \pi x}{a} \sin \frac{n_y \pi y}{b} dx dy,$$

$$B_{n_x n_y} = \frac{4}{ab} \int_{0}^{a} \int_{0}^{b} v_0(x, y) \sin \frac{n_x \pi x}{a} \sin \frac{n_y \pi y}{b} dx dy. \tag{10.10}$$

Superposition of node line figures

With the knowledge of the $A_{n_x n_y}$ and $B_{n_x n_y}$, one now can calculate the $c_{n_x n_y}$ and $\varphi_{n_x n_y}$ from (10.5) and (10.6).

Superposition of node line figures

In the case of degenerate vibrations of the membrane, there can also appear node lines that arise by superposition of the node line figures of the *degenerate* normal vibrations.

As an example we consider the position dependence of the degenerate normal vibrations of the quadratic membrane

$$u_{12} = \sin\frac{\pi x}{a} \sin\frac{2\pi y}{a} \sin\omega_{12} t \quad \text{and} \quad u_{21} = \sin\frac{2\pi x}{a} \sin\frac{\pi y}{a} \sin\omega_{21} t. \tag{10.11}$$

For the superposition of the two normal vibrations, we write

$$u = u_{12} + C u_{21}.$$

The constant C specifies the particular kind of superposition. The equation of the nodal line is obtained from $u = 0$. The common numerical factor $\sin\omega_{12}t = \sin\omega_{21}t$ obviously factors out. For the special case $C = \pm 1$, we find

$$\sin\frac{\pi x}{a} \sin\frac{2\pi y}{a} \pm \sin\frac{2\pi x}{a} \sin\frac{\pi y}{a} = 0$$

or, rewritten,

$$\sin\frac{\pi x}{a} \sin\frac{\pi y}{a} \left(\cos\frac{\pi y}{a} \pm \cos\frac{\pi x}{a} \right) = 0. \tag{10.12}$$

By setting the bracket equal to zero, we get the equations for the two nodal lines:

$$y = x \quad \text{for} \quad C = -1 \quad \text{and} \quad y = a - x \quad \text{for} \quad C = +1.$$

Figure 10.4 illustrates the nodal lines.

We recognize that new vibrations with new kinds of nodal lines can be constructed by superposing appropriate normal vibrations. One can excite such specific superpositions of normal vibrations by stretching wires along the nodal lines (right figure) so that the membrane remains at rest along these lines.

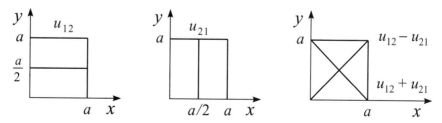

Figure 10.4. Nodal lines for degenerate eigenvibrations.

The circular membrane

In the case of the circular membrane, it is convenient to change from the Cartesian coordinates to polar coordinates, i.e., from $u = f(x, y, t)$ to $u = \psi(r, \varphi, t)$.

For this recalculation, we have

$$x = r \cos \varphi, \qquad y = r \sin \varphi,$$
$$\tan \varphi = \frac{y}{x}, \qquad r = \sqrt{x^2 + y^2}. \tag{10.13}$$

For the transformation of the Laplace operator, we need the derivatives

$$\frac{\partial r}{\partial x} = \frac{x}{r} = \cos \varphi, \qquad \frac{\partial r}{\partial y} = \frac{y}{r} = \sin \varphi. \tag{10.14}$$

By differentiating the tangent, we get

$$\frac{\partial \tan \varphi}{\partial x} = \frac{\partial \tan \varphi}{\partial \varphi} \frac{\partial \varphi}{\partial x} = \frac{1}{\cos^2 \varphi} \frac{\partial \varphi}{\partial x} = -\frac{y}{x^2}. \tag{10.15}$$

By inserting the polar representations for x and y, one gets $\partial \varphi / \partial x = -(\sin \varphi)/r$. The corresponding differentiation of $\tan \varphi$ with respect to y yields $\partial \varphi / \partial y = (\cos \varphi)/r$. To get the two-dimensional vibration equation in polar coordinates, we first transform the Laplace operator $\Delta(x, y)$ to polar coordinates $\Delta(r, \varphi)$. The differential quotients are interpreted as operators.

We demonstrate the calculation for the x-component; the recalculation of the y-component then runs likewise. According to the chain rule, we have

$$\frac{\partial}{\partial x} = \frac{\partial}{\partial r} \frac{\partial r}{\partial x} + \frac{\partial}{\partial \varphi} \frac{\partial \varphi}{\partial x}. \tag{10.16}$$

After insertion of the above results, we obtain

$$\frac{\partial}{\partial x} = \cos \varphi \frac{\partial}{\partial r} - \frac{\sin \varphi}{r} \frac{\partial}{\partial \varphi}. \tag{10.17}$$

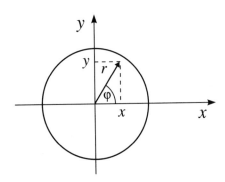

Figure 10.5. Circular membrane (drum).

THE CIRCULAR MEMBRANE

We square this result, taking into account that the terms act on each other as operators. (The square of an operator means double application.)

$$\frac{\partial^2}{\partial x^2} = \left(\cos\varphi \frac{\partial}{\partial r} - \sin\varphi \frac{1}{r}\frac{\partial}{\partial \varphi}\right)\left(\cos\varphi \frac{\partial}{\partial r} - \sin\varphi \frac{1}{r}\frac{\partial}{\partial \varphi}\right). \quad (10.18)$$

By multiplying out, one first gets the four terms

$$\frac{\partial^2}{\partial x^2} = \left(\cos\varphi \frac{\partial}{\partial r} \cdot \cos\varphi \frac{\partial}{\partial r}\right) + \left(\frac{\sin\varphi}{r}\frac{\partial}{\partial \varphi} \cdot \frac{\sin\varphi}{r}\frac{\partial}{\partial \varphi}\right)$$
$$- \left(\cos\varphi \frac{\partial}{\partial r} \cdot \frac{\sin\varphi}{r}\frac{\partial}{\partial \varphi}\right) - \left(\frac{\sin\varphi}{r}\frac{\partial}{\partial \varphi} \cdot \cos\varphi \frac{\partial}{\partial \varphi}\right). \quad (10.19)$$

We now treat the individual terms according to the product rule:

$$\cos\varphi\left(\frac{\partial}{\partial r} \cdot \cos\varphi \frac{\partial}{\partial r}\right) = \cos^2\varphi \frac{\partial^2}{\partial r^2},$$

$$\frac{\sin\varphi}{r}\left(\frac{\partial}{\partial \varphi} \cdot \frac{\sin\varphi}{r}\frac{\partial}{\partial \varphi}\right) = \frac{\sin\varphi\cos\varphi}{r^2}\frac{\partial}{\partial \varphi} + \frac{\sin^2\varphi}{r^2}\frac{\partial^2}{\partial \varphi^2},$$

$$\cos\varphi\left(\frac{\partial}{\partial r} \cdot \frac{\sin\varphi}{r}\frac{\partial}{\partial \varphi}\right) = -\frac{\cos\varphi\sin\varphi}{r^2}\frac{\partial}{\partial \varphi} + \frac{\cos\varphi\sin\varphi}{r}\frac{\partial}{\partial r}\frac{\partial}{\partial \varphi},$$

$$\frac{\sin\varphi}{r}\left(\frac{\partial}{\partial \varphi} \cdot \cos\varphi \frac{\partial}{\partial r}\right) = -\frac{\sin^2\varphi}{r}\frac{\partial}{\partial r} + \frac{\sin\varphi\cos\varphi}{r}\frac{\partial}{\partial \varphi}\frac{\partial}{\partial r}.$$

From this, one obtains

$$\frac{\partial^2}{\partial x^2} = \cos^2\varphi \frac{\partial^2}{\partial r^2} + \frac{\sin^2\varphi}{r^2}\left(r\frac{\partial}{\partial r} + \frac{\partial^2}{\partial \varphi^2}\right) + \frac{2\sin\varphi\cos\varphi}{r^2}\left(\frac{\partial}{\partial \varphi} - r\frac{\partial}{\partial \varphi}\frac{\partial}{\partial r}\right).$$

Analogously, one gets for the y-component

$$\frac{\partial^2}{\partial y^2} = \sin^2\varphi \frac{\partial^2}{\partial r^2} + \frac{\cos^2\varphi}{r^2}\left(r\frac{\partial}{\partial r} + \frac{\partial^2}{\partial \varphi^2}\right) - \frac{2\sin\varphi\cos\varphi}{r^2}\left(\frac{\partial}{\partial \varphi} - r\frac{\partial}{\partial \varphi}\frac{\partial}{\partial r}\right).$$

By adding both expressions, we obtain the Laplace operator in polar coordinates:

$$\frac{\partial^2}{\partial x^2} + \frac{\partial^2}{\partial y^2} = \Delta = \frac{\partial^2}{\partial r^2} + \frac{1}{r}\frac{\partial}{\partial r} + \frac{1}{r^2}\frac{\partial^2}{\partial \varphi^2}. \quad (10.20)$$

The vibration equation then takes the following form:

$$\frac{\partial^2 u(r,\varphi,t)}{\partial r^2} + \frac{1}{r}\frac{\partial u(r,\varphi,t)}{\partial r} + \frac{1}{r^2}\frac{\partial^2 u(r,\varphi,t)}{\partial \varphi^2} = \frac{1}{c^2}\frac{\partial^2 u(r,\varphi,t)}{\partial t^2}. \quad (10.21)$$

The equation of motion is solved again by separation of the variables. We use a product *ansatz* for separating the position and time functions:

$$u(r,\varphi,t) = V(r,\varphi) \cdot Z(t). \quad (10.22)$$

By insertion into the wave equation, we obtain

$$Z(t)\left(\frac{\partial^2 V}{\partial r^2} + \frac{1}{r}\frac{\partial V}{\partial r} + \frac{1}{r^2}\frac{\partial^2 V}{\partial \varphi^2}\right) = \frac{1}{c^2}V\frac{\partial^2 Z}{\partial t^2}. \tag{10.23}$$

We divide both sides by $V(r, \varphi) \cdot Z(t)$:

$$\frac{\frac{\partial^2 V}{\partial r^2} + \frac{1}{r}\frac{\partial V}{\partial r} + \frac{1}{r^2}\frac{\partial^2 V}{\partial \varphi^2}}{V(r, \varphi)} = \frac{1}{c^2}\frac{\ddot{Z}(t)}{Z(t)}. \tag{10.24}$$

As a separation constant, we choose

$$\frac{1}{c^2}\frac{\ddot{T}}{T} = -k^2 \tag{10.25}$$

and introduce the angular frequency ω by

$$\omega = ck. \tag{10.26}$$

From this, we get

$$\ddot{Z} + \omega^2 Z = 0 \tag{10.27}$$

with the solution

$$Z(t) = C\sin(\omega t + \delta). \tag{10.28}$$

By insertion of the constant $-k^2$, the equation of motion takes the form

$$\frac{\partial^2 V}{\partial r^2} + \frac{1}{r}\frac{\partial V}{\partial r} + \frac{1}{r^2}\frac{\partial^2 V}{\partial \varphi^2} + k^2 V = 0. \tag{10.29}$$

We separate the radial and angular functions by a second product *ansatz*:

$$V(r, \varphi) = R(r) \cdot \phi(\varphi). \tag{10.30}$$

Hence, we obtain

$$\frac{\frac{d^2 R}{dr^2} + \frac{1}{r}\frac{dR}{dr}}{R(r)} + \frac{\frac{1}{r^2}\frac{d^2\phi}{d\varphi^2}}{\phi(\varphi)} + k^2 = 0. \tag{10.31}$$

We separate the variables by multiplying by r^2:

$$\frac{r^2(d^2 R/dr^2) + r(dR/dr)}{R(r)} + k^2 r^2 + \frac{d^2\phi/d\varphi^2}{\phi(\varphi)} = 0. \tag{10.32}$$

Here again, the equation is valid only then if both functions are constants. Hence, we choose

$$\frac{1}{\phi}\frac{d^2\phi}{d\varphi^2} = -\sigma, \tag{10.33}$$

SOLUTION OF BESSEL'S DIFFERENTIAL EQUATION

from which one obtains as a solution for $\phi(\varphi)$

$$\phi(\varphi) = A e^{i\sqrt{\sigma}\varphi} + B e^{-i\sqrt{\sigma}\varphi}$$
$$= C \sin(m\varphi + \delta) \quad \text{with} \quad m = \pm\sqrt{\sigma}, \quad m = 0, 1, 2, 3, \ldots. \quad (10.34)$$

m must take only integer values to get the periodicity of the solution. At the angle $2\pi + \varphi$, the solution must be identical with that for the angle φ. This fact is often described by the phrase *periodic boundary conditions*.

Now we can admit—without restricting the problem—only positive m, since with negative m only the sense of rotation angle is inverted.

Thus, the equation of motion for the radial function R looks as follows:

$$r^2 \frac{d^2 R}{dr^2} + r \frac{dR}{dr} + k^2 r^2 R - \sigma R = 0$$

or

$$\frac{d^2 R}{dr^2} + \frac{1}{r}\frac{dR}{dr} + \left(k^2 - \frac{m^2}{r^2}\right) R = 0. \quad (10.35)$$

We substitute $z = kr$, $dr = dz/k$. Then we get

$$k^2 \frac{d^2 R}{dz^2} + \frac{k^2}{z}\frac{dR}{dz} + \left(k^2 - \frac{m^2 k^2}{z^2}\right) R = 0,$$

$$\frac{d^2 R}{dz^2} + \frac{1}{z}\frac{dR}{dz} + \left(1 - \frac{m^2}{z^2}\right) R = 0. \quad (10.36)$$

In this form, the equation is called *Bessel's differential equation*. This differential equation and its solutions appear in many problems of mathematical physics.

Solution of Bessel's differential equation[2]

The solution of our differential equation

$$\frac{d^2 g(z)}{dz^2} + \frac{1}{z}\frac{dg(z)}{dz} + \left(1 - \frac{m^2}{z^2}\right) g(z) = 0 \quad (10.37)$$

cannot be found by integration. Approaches using elementary functions also fail. We therefore try with the most general power series expansion:

$$g(z) = z^\mu \left(\sum_{n=0}^{\infty} a_n z^n\right). \quad (10.38)$$

---

[2]*Friedrich Wilhelm Bessel*, b. July 22, 1784, Minden–d. March 17, 1846, Königsberg (Kaliningrad). Bessel was first a trade apprentice in Bremen, then until 1809 an assistant at the observatory in Lilienthal, and then professor of astronomy in Königsberg and director of the observatory there. As a mathematician Bessel was best known for his investigations on differential equations and on Bessel functions.

The separation of a power factor is not necessary, but will prove to be very convenient.

Since in the center of our membrane the vibration remains always finite, $g(z)$ must not have a singularity at $z = 0$. But since for $z \to 0$ we have

$$g(z) \approx a_0 z^{\mu}, \tag{10.39}$$

for these physical reasons we must have $\mu \geq 0$. To get a more general statement, we consider the asymptotic behavior of Bessel's differential equation for $z \to 0$ for at first arbitrary μ.

We then can set as above

$$g(z) \approx a_0 z^{\mu} \tag{10.40}$$

and obtain by inserting:

$$\mu(\mu - 1)z^{\mu-2} + \mu z^{\mu-2} + z^{\mu} - m^2 z^{\mu-2} = (\mu(\mu - 1) + \mu + z^2 - m^2)z^{\mu-2}$$
$$\approx (\mu^2 - m^2)z^{\mu-2} = 0, \tag{10.41}$$

since for $z \to 0$ we also have $z^2 \to 0$. We thus have the condition

$$\mu^2 - m^2 = 0. \tag{10.42}$$

For the above-mentioned reasons, which are of a purely physical nature, it follows that

$$\mu = m, \quad m \in \mathbf{N}_0. \tag{10.43}$$

The constant m is itself an integer. To see this, we remind ourselves of the angular dependence of the total solution, namely,

$$f(\varphi) = \sin(m\varphi + \delta). \tag{10.44}$$

Since after a full revolution we return again to the same point of the membrane, the solution function must have the period 2π. But this holds only then if m is an integer!

We now try to to determine the coefficients of our *ansatz*

$$g_m(z) = z^m(a_0 + a_1 z + a_2 z^2 \ldots), \quad m = 0, 1, 2, \ldots . \tag{10.45}$$

For this purpose, we insert the *ansatz* in the Bessel equation. The individual terms of this equation then have the following form:

$$\frac{d^2 g}{dz^2} = z^{m-2}(a_0 m(m-1) + a_1(m+1)mz + a_2(m+2)(m+1)z^2$$
$$+ a_3(m+3)(m+2)z^3 + \cdots),$$

$$\frac{1}{z}\frac{dg}{dz} = z^{m-2}(a_0 m + a_1(m+1)z + a_2(m+2)z^2 + a_3(m+3)z^3 + \cdots),$$

$$g(z) = z^{m-2}(a_0 z^2 + a_1 z^3 + \cdots),$$

$$-\frac{m^2}{z^2}g(z) = z^{m-2}(-a_0 m^2 - a_1 m^2 z - a_2 m^2 z^2 - a_3 m^2 z^3 - \cdots).$$

SOLUTION OF BESSEL'S DIFFERENTIAL EQUATION

The sum of the coefficients for each power of z must vanish, i.e., $a_0(m(m-1) + m - m^2) = 0$. Since the bracket vanishes, a_0 can be arbitrary.

For a_1, we get

$$a_1(m(m+1) + (m+1) - m^2) = 0,$$
$$a_1(2m+1) = 0, \quad \text{i.e.,} \quad a_1 = 0. \tag{10.46}$$

From the coefficient of z^m, it follows that

$$a_2((m+2)(m+1) + (m+2) - m^2) + a_0 = 0$$

or

$$a_2(4m+4) = -a_0. \tag{10.47}$$

Furthermore, we get

$$a_3((m+3)(m+2) + (m+3) - m^2) + a_1 = 0,$$
$$a_3(6m+9) = -a_1, \quad \text{i.e.,} \quad a_3 = 0. \tag{10.48}$$

Generally, we find the condition equation

$$a_{p+2}((m+p+2)(m+p+1) + (m+p+2) - m^2) + a_p = 0,$$

$$a_{p+2}((m+p+2)^2 - m^2) = -a_p,$$

$$a_{p+2} = \frac{-a_p}{(m+p+2)^2 - m^2} = \frac{-a_p}{(p+2)(2m+p+2)}. \tag{10.49}$$

This recursion formula allows one to determine the coefficient a_{p+2} from the preceding a_p. Because $a_1 = 0$, it follows that all a_{2n-1} vanish, i.e., in the series expansion of the solution function there appear only even exponents. For these one obtains with $a_0 \neq 0$:

$$a_{2n} = \frac{-a_{2n-2}}{2n(2m+2n)} = \frac{-a_{2n-2}}{2n \cdot 2(m+n)}. \tag{10.50}$$

In the next step, we replace a_{2n-2} by a_{2n-4} and obtain

$$a_{2n} = \frac{+a_{2n-4}}{2n(2n-2)(2m+2n)(2m+2n-2)}$$
$$= \frac{a_{2n-4}}{2^2 n(n-1) 2^2 (m+n)(m+n-1)}. \tag{10.51}$$

By continuing this way, we can relate a_{2n} back to a_0. We obtain

$$a_{2n} = \frac{(-1)^n a_0}{2^n n(n-1) \cdots 1 \cdot 2^n (m+n)(m+n-1) \cdots (m+1)}$$
$$= \frac{(-1)^n a_0}{2^{2n} n! (m+n)!/m!}. \tag{10.52}$$

Thereby, we obtain the following solution functions:

$$g_m(z) = a_0 z^m m! \sum_{n=0}^{\infty} \frac{(-1)^n}{n!(m+n)!} \frac{z^{2n}}{2^{2n}}. \tag{10.53}$$

With the special choice $a_0 \cdot m! = 2^{-m}$, we obtain the *Bessel functions:*

$$\begin{aligned} J_m(z) &= \left(\frac{z}{2}\right)^m \sum_{n=0}^{\infty} \frac{(-1)^n}{n!(m+n)!} \left(\frac{z}{2}\right)^{2n} \\ &= \sum_{n=0}^{\infty} \frac{(-1)^n}{n!(m+n)!} \left(\frac{z}{2}\right)^{2n+m}. \end{aligned} \tag{10.54}$$

The graph of the first Bessel functions is given in Figure 10.6. We see that for large arguments the Bessel functions vary like the trigonometric functions sine or cosine.

Now we can immediately write down the solutions of our differential equation:

$$V_m(r, \varphi) = c_m J_m(kr) \sin(m\varphi + \delta_m). \tag{10.55}$$

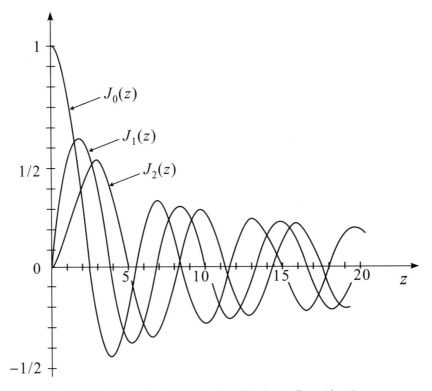

Figure 10.6. Graphical representation of the lowest Bessel functions.

SOLUTION OF BESSEL'S DIFFERENTIAL EQUATION

The membrane cannot vibrate at the border $r = a$, i.e., the boundary condition reads

$$V(a, \varphi) = 0 \quad \text{for all} \quad \varphi.$$

From this, we obtain the condition

$$J_m(k \cdot a) = 0,$$

from which the eigenfrequencies can be determined. For this purpose we must find the zeros of the Bessel function:

$$J_0(z) = 1 - \frac{z^2}{4} + \frac{z^4}{64} - + \cdots = 0,$$

$$J_1(z) = \frac{z}{2} - \frac{z^3}{16} + \frac{z^5}{384} - + \cdots = 0, \quad \text{etc.} \tag{10.56}$$

These zeros – except for the trivial ones for $z = 0$ – can in general not be determined exactly; they must be calculated by numerical methods. If we denote the nth node of the function $J_m(z)$ by $z_n^{(m)}$, we obtain the following table for the values of the first $z_n^{(m)}$:

Table 10.2. Zeros of the Bessel functions.

n	m = 0	1	2	3	4	5
1	2.41	3.83	5.14	6.38	7.59	8.77
2	5.52	7.02	8.42	9.76	11.06	12.34
3	8.65	10.17	11.62	13.02	14.37	15.70
4	11.79	13.32	14.80	16.22	17.62	18.98
5	14.93	16.47	17.96	19.41	20.83	22.22
6	18.07	19.62	21.12	22.51	24.02	25.43
7	21.21	22.76	24.27	25.75	27.20	28.63
8	24.35	25.90	27.42	28.91	30.37	31.81
9	27.49	29.05	30.57	32.07	33.51	34.99

Useful approximate solutions may also be obtained by considering the asymptotes of the Bessel functions for $z \to \infty$. Then

$$J_m(z) \to \sqrt{\frac{2}{mz}} \cos\left(z - \frac{m\pi}{2} - \frac{\pi}{4}\right). \tag{10.57}$$

We give this without proof. A look to the graphical variation of the Bessel functions shows the close analogy with the cosine function for large arguments.

From this one can determine zeros:

$$\cos\left(z_n^{(m)} - \frac{m\pi}{2} - \frac{\pi}{4}\right) = 0$$

$$\Rightarrow \quad z_n^{(m)} - \frac{m\pi}{2} - \frac{\pi}{4} = n\pi - \frac{\pi}{2},$$

$$z_n^{(m)} = n\pi + \frac{m\pi}{2} - \frac{\pi}{4} = (4n + 2m - 1)\frac{\pi}{4}. \tag{10.58}$$

A comparison of these values with the exact ones from Table 10.2 shows that particularly for n large compared to m one obtains good approximate values:

Table 10.3. Comparison of the exact zeros of the Bessel functions with those obtained from the asymptotic approximation.

	$m = 0$		$m = 5$	
	$z_n^{(0)}$	$\bar{z}_n^{(0)}$	$z_n^{(5)}$	$\bar{z}_n^{(5)}$
$n = 1$	2.41	2.36	8.77	10.21
$n = 2$	5.52	5.49	12.34	13.35
...	...	...	...	...
$n = 9$	27.49	27.49	34.99	35.34

With the exact solutions $z_n^{(m)}$, the boundary condition is

$$k_n^{(m)} \cdot a = z_n^{(m)}, \qquad k_n^{(m)} = \frac{1}{a} \cdot z_n^{(m)}.$$

For the eigenfrequencies, we get

$$\omega_n^{(m)} = k_n^{(m)} \cdot c = \frac{c}{a} \cdot z_n^{(m)} = \omega_0 \cdot z_n^{(m)}. \tag{10.59}$$

Thus, Table 10.2 also shows the values for the ratio $\omega_n^{(m)}/\omega_0$. By drawing all these eigenfrequencies along an axis, one arrives at Figure 10.7. The distances between the individual eigenfrequencies are fully chaotic. Thus, we are dealing with extremely anharmonic overtones. This is the reason why drums are badly suited as melodic instruments!

The general solution of the vibration equation is the superposition of the normal vibrations. It now reads

$$u(r, \varphi, t) = \sum_{m,n} c_n^{(m)} J_m(k_n^{(m)} r) \cdot \sin(m\varphi + \delta_m) \cdot \sin(\omega_n^{(m)} t + \delta_n^{(m)}). \tag{10.60}$$

In analogy to the Fourier analysis, the $c_n^{(m)}$ can be found so that $u(r, \varphi, t)$ can be adjusted to any given initial condition $u(r, \varphi, 0)$ or $\dot{u}(r, \varphi, 0)$.

Finally, we want to get a survey of the nodal lines of the vibrating membrane. On these lines we must have

$$u_{m,n}(r, \varphi) = c_n^{(m)} J_m(k_n^{(m)} r) \cdot \sin(m\varphi + \delta_m) = 0. \tag{10.61}$$

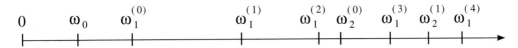

Figure 10.7. Linear representation of the eigenfrequencies of the circular membrane.

SOLUTION OF BESSEL'S DIFFERENTIAL EQUATION

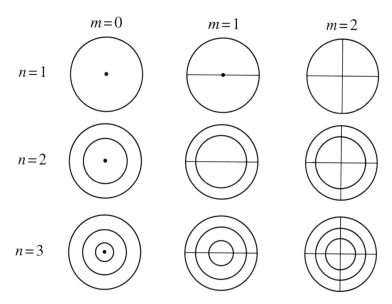

Figure 10.8. Nodal lines of the circular membrane.

Then we get nodal lines if either

$$J_m(k_n^{(m)} r) = 0; \tag{10.62}$$

this is realized for

$$r = \frac{z_i^{(m)}}{k_n^{(m)}}, \qquad i = 1, 2, \ldots, n-1; \tag{10.63}$$

or if

$$\sin(m\varphi + \delta_m) = 0, \tag{10.64}$$

i.e., for angles

$$\varphi = \frac{\nu \pi - \delta_m}{m}, \qquad \nu = 1, 2, \ldots, m. \tag{10.65}$$

For the first nodal lines, we get Figure 10.8 (with $\delta_m = 0$).

Example 10.1: The longitudinal chain: Poincaré recurrence time

The equations of motion for a system with n vibrating mass points which are connected by $n+1$ springs of equal spring constant k read

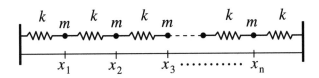

Figure 10.9.

$$\begin{aligned}
m\ddot{x}_1 &= -kx_1 + k(x_2 - x_1) \\
m\ddot{x}_2 &= -k(x_2 - x_1) + k(x_3 - x_2) \\
m\ddot{x}_3 &= -k(x_3 - x_2) + k(x_4 - x_3) \\
&\vdots \\
m\ddot{x}_{n-1} &= -k(x_{n-1} - x_{n-2}) + k(x_n - x_{n-1}) \\
m\ddot{x}_n &= -k(x_n - x_{n-1}) - kx_n.
\end{aligned} \quad (10.66)$$

With $\mathbf{r} = \begin{pmatrix} x_1 \\ x_2 \\ \vdots \\ x_n \end{pmatrix}$, this can be written succinctly as

$$m\ddot{\mathbf{r}} = \hat{C}k\mathbf{r}, \quad (10.67)$$

where

$$\hat{C} = \begin{pmatrix} -2 & 1 & & & & \\ 1 & -2 & 1 & & & \\ & 1 & -2 & 1 & & \\ & & & \ddots & & \\ & & & 1 & -2 & 1 \\ & & & & 1 & -2 \end{pmatrix} \quad \text{and} \quad \mathbf{r} = \begin{pmatrix} x_1 \\ x_2 \\ \vdots \\ x_n \end{pmatrix}. \quad (10.68)$$

With the *ansatz* $\mathbf{r} = \mathbf{a}\cos\omega t$, $\mathbf{a} = \begin{pmatrix} a_1 \\ a_2 \\ \vdots \\ a_n \end{pmatrix}$, we look for the normal modes:

$$\underbrace{(k\hat{C} + m\omega^2 \hat{E}_n)}_{=\hat{D}_n(\omega)} \mathbf{a}\cos\omega t = 0,$$

SOLUTION OF BESSEL'S DIFFERENTIAL EQUATION

$$\hat{D}_n(\omega) = \begin{pmatrix} m\omega^2 - 2k & k & & \\ k & m\omega^2 - 2k & k & \\ & & \ddots & \end{pmatrix}. \tag{10.69}$$

Here, $\hat{E}_n = \begin{pmatrix} 1 & 0 & \cdots & 0 \\ 0 & 1 & \cdots & 0 \\ \vdots & \vdots & & \vdots \\ 0 & 0 & \cdots & 1 \end{pmatrix}$ represents the unit matrix.

For nontrivial solutions for $\mathbf{a}$, the determinant $D_n(\omega)$ of the matrix $\hat{D}_n(\omega)$ vanishes. Furthermore, we use for the coefficients $\mathbf{a}$ an *ansatz* with phase δ and γ to be determined.

$$\mathbf{a} := (a_j = \sin(j\gamma - \delta), \ j = 1, 2, \ldots, n). \tag{10.70}$$

The evaluation of the line j yields

$$ka_{j-1} + (m\omega^2 - 2k)a_j + ka_{j+1} = 0,$$
$$k\sin((j-1)\gamma - \delta) + (m\omega^2 - 2k)\sin(j\gamma - \delta) + k\sin((j+1)\gamma - \delta) = 0,$$
$$\Rightarrow k\cos\gamma + (m\omega^2 - 2k) + k\cos\gamma = 0,$$
$$\Leftrightarrow \omega^2 = 2\frac{k}{m}(1 - \cos\gamma), \quad \omega = 2\sqrt{\frac{k}{m}}\sin\frac{\gamma}{2}. \tag{10.71}$$

We know that the characteristic polynomial has n zeros:

$$\omega_i = 2\sqrt{\frac{k}{m}}\sin\frac{\gamma_i}{2}, \quad i = 1, 2, \ldots, n. \tag{10.72}$$

The boundary conditions $a_0 = a_{n+1} = 0$ must be satisfied. The first one requires that

$$\sin\delta = 0, \quad \text{hence} \quad \delta = l\pi, \ l \in \mathbb{Z}, \quad \text{and therefore w.l.o.g.} \ l = 0. \tag{10.73}$$

From the second boundary condition, it further follows that

$$\sin((n+1)\gamma_i) = 0 \quad \Rightarrow \quad \gamma_i = \frac{i\pi}{n+1}, \quad \text{for all} \ i \in \{1, \ldots, n\}. \tag{10.74}$$

We summarize the result for the ith eigenmodes:

$$\mathbf{r}_i(t) = \left(\sin\left(j\frac{i\pi}{n+1}\right) \cdot \cos\omega_i t, \ j = 1, 2, \ldots, n\right) \tag{10.75}$$

with

$$\omega_i = 2\sqrt{\frac{k}{m}}\sin\frac{i\pi}{2n+2}. \tag{10.76}$$

The general solution of (10.66) is a superposition of the various eigenmodes, i.e., a vector $\mathbf{r}(t)$ with the components $x_j(t)$:

$$x_j(t) = \sum_{i=1}^{n}(c_i \cdot \cos\omega_i t + b_i \cdot \sin\omega_i t) \cdot \sin\left(j \cdot \frac{i\pi}{n+1}\right), \quad \text{for} \ j = 1, 2, \ldots, n. \tag{10.77}$$

The coefficients $\sin(ji\pi/(n+1))$ are, according to (10.70), (10.73) and (10.76), the components of the eigenvector to the ith mode, and since $\hat{\mathbf{D}}_n(\omega)$ is symmetric, the latter ones represent an orthogonal basis in $\mathbb{R}^n$.

$$\sum_{j=1}^{n} \sin\left(l\frac{i\pi}{n+1}\right) \sin\left(j\frac{l\pi}{n+1}\right) = \frac{n+1}{2}\delta_{il}. \tag{10.78}$$

We explicitly check this relation in the following problem 10.2. We define the *orthonormal eigenmodes* $\mathbf{a}_i$:

$$\mathbf{a}_i = \sqrt{\frac{2}{n+1}} \left\{ \sin\left(j\frac{i\pi}{n+1}\right), \; j = 1, 2, \ldots, n, \right\}, \tag{10.79}$$

or in detail

$$\mathbf{a}_i = \sqrt{\frac{2}{n+1}} \left\{ \sin\frac{\pi i}{n+1}, \sin\frac{2\pi i}{n+1}, \ldots, \sin\frac{n\pi i}{n+1} \right\}.$$

The general solution can then be written as follows:

$$\mathbf{r}(t) = \sum_{i=1}^{n} (c_i \cos \omega_i t + b_i \sin \omega_i t)\mathbf{a}_i. \tag{10.80}$$

The following interesting question arises: Let the system of n mass points (degrees of freedom) at the time t_0 be at $\mathbf{r}(t_0) = \mathbf{r}_0$ with the velocity $\dot{\mathbf{r}}(t_0) = \dot{\mathbf{r}}_0$. The system moves away from this configuration, but after a certain time τ it can closely approach the initial configuration and possibly return exactly into the initial configuration. We call this time τ the *Poincaré recurrence time*. One looks for the difference between the actual time-dependent state vector in the phase space $(\mathbf{r}(t), \dot{\mathbf{r}}(t))$ and the start vector $(\mathbf{r}_0, \dot{\mathbf{r}}|_{t=0})$:

$$\varepsilon(t) =: \sqrt{\|\mathbf{r}(t) - \mathbf{r}_0\|^2 + \|\dot{\mathbf{r}}(t) - \dot{\mathbf{r}}|_{t=0}\|_\Omega^2}. \tag{10.81}$$

The index Ω at the second scalar product for the velocities indicates a diagonal weight matrix Ω which is suitably included into this normalization:

$$\Omega = \begin{pmatrix} \frac{1}{\omega_1^2} & & & \\ & \frac{1}{\omega_2^2} & & \\ & & \ddots & \\ & & & \frac{1}{\omega_n^2} \end{pmatrix}; \tag{10.82}$$

ω_i are the eigenfrequencies. In this way the factors ω_i obtained by differentiation of $\mathbf{r}$ in (10.80) cancel out. So it is guaranteed that both terms under the root in (10.81) have the same dimension. We formulate with (10.80) the following initial value problem:

$$\mathbf{r}(0) = \mathbf{r}_0 = \sum_{i=1}^{n} c_i \mathbf{a}_i,$$

$$\dot{\mathbf{r}}(0) = \dot{\mathbf{r}}_0 = 0 = \sum_{i=1}^{n} b_i \mathbf{a}_i \omega_i \quad \Rightarrow \quad b_i = 0 \quad \text{for all} \quad i.$$

For this choice, the distance $\varepsilon(t)$ in phase space, given by equation (10.81), is

$$\varepsilon(t) = \sqrt{\mathbf{r}_0 \cdot \mathbf{r}_0 - 2\mathbf{r}_0 \cdot \sum_i c_i \cos \omega_i t \mathbf{a}_i + \sum_i c_i^2 \cos^2 \omega_i t + \sum_i c_i^2 \sin^2 \omega_i t} \,. \tag{10.83}$$

Because $\mathbf{r}_0 = \sum_i c_i \mathbf{a}_i$, this turns into

$$\varepsilon(t) = \sqrt{\sum_{i=1}^n c_i^2 (1 - 2\cos \omega_i t + \underbrace{\cos^2 \omega_i t + \sin^2 \omega_i t}_{=1})}$$

$$= 2\sqrt{\sum_{i=1}^n c_i^2 \sin^2 \frac{\omega_i t}{2}} \,. \tag{10.84}$$

It is easily seen that this expression after $t = 0$ vanishes again only if the eigenfrequencies ω_i are related by rational fractions. In the general case $\varepsilon(t)$ is only *conditionally periodic*. This notion will be explained more precisely as follows: We first consider the purely periodic case. The period is then determined by the lowest of the frequencies that are related by rational fractions:

$$\omega_i = 2\sqrt{\frac{k}{m}} \sin \frac{i\pi}{2n+2}. \tag{10.85}$$

We denote it by $\tilde{\omega} = q \cdot \omega_1$ ($q \in \mathbf{Z}_+$),

$$\tilde{\omega} = 2q\sqrt{\frac{k}{m}} \sin \frac{\pi}{2n+2} \approx 2q\sqrt{\frac{k}{m}} \frac{\pi}{2n+2}. \tag{10.86}$$

The last approximation holds for $n \gg 1$ (many mass points). To this frequency corresponds the time

$$\tau = \frac{2\pi}{q\omega_1} = 2\sqrt{\frac{k}{m}} \cdot \frac{n+1}{q}. \tag{10.87}$$

This is the *Poincaré recurrence time*, since after this time $\varepsilon(t)$ vanishes again, i.e., the initial configuration in the phase space is reached again after the time τ. For very many mass points ($n \to \infty$), this time tends to infinity

$$\tau \to \infty, \quad \text{for} \quad n \to \infty. \tag{10.88}$$

This is an important and physically plausible result: After preparation of an initial configuration, the system develops with time away from this configuration. At some time, just after the Poincaré recurrence time, the state of motion of the system returns to the initial state (or in the general case very close to it). However, in the case of a great many degrees of freedom, the system "escapes" and the recurrence time becomes ∞. For example, if one of the n masses coupled by strings—say, the first one—is being pushed (this corresponds to setting of $\mathbf{r}_0$ and $\dot{\mathbf{r}}_0$), the energy of this motion will spread more and more over the other masses. After the Poincaré time τ the first mass will have regained the entire energy. Only in the case of infinitely many coupled masses will this no longer happen, since $\tau \to \infty$. This is of great importance for the statistical behavior of systems of particles.

Addendum: Periodic systems with several degrees of freedom: Conditionally periodic systems.
Let us define a periodic system with several degrees of freedom as a system for which according to equation (10.83) the orthogonal coordinates used in the description are periodic functions:

$$\mathbf{r}_i = \mathbf{a}_i \cos \omega_i t. \tag{10.89}$$

The quantities $\tau_i = 2\pi/\omega_i$ are the periods belonging to the coordinates x_i. In analogy to (10.77) and (10.80), we expand the general configuration vector $\mathbf{r}(t)$ in a Fourier series

$$\mathbf{r}(t) = \sum_i (c_i \cos \omega_i t + b_i \sin \omega_i t) \mathbf{a}_i. \tag{10.90}$$

The Fourier series (10.90) is in general no longer a periodic function with respect to the time t, although every individual term is periodic. The periodicity is assured only for such degrees of freedom whose frequencies $\omega_1, \omega_2, \ldots$ are related by rational fractions. Therefore systems with several degrees of freedom are called *conditionally periodic systems*.

The number of frequencies which are related by rational fractions determines the *degree of degeneracy of the system*. If there are no relations of this kind, the system is *non-degenerate*. If all frequencies are rationally related to each other, the system is called *fully degenerate*. In this case we observe a *periodic time function*.

The Kepler problem treated earlier is an example for a system with two degrees of freedom (r, φ) which is degenerate and thus has only one frequency. By inventing a perturbation, say a quadrupole-like potential with the typical variation $1/r^3$, the degeneracy can be removed, which causes a rosette-like motion.

As an example of a conditionally periodic motion, we note the anisotropic linear harmonic oscillator, which is a mass point with different spring constants in the various Cartesian directions. The trajectory of the mass point is a *Lissajous figure* which never turns into itself and in the course of time tightly covers the area given by the amplitudes. Only in the case of degeneracy there are periodicities in the motion.

In the discussion of the Poincaré recurrence time, we assumed a periodic motion. In the case of a conditionally periodic motion, the situation is—as was expected—completely analogous. In this case, after the Poincaré recurrence time τ the configuration vectors $\mathbf{r}(t)$, $\dot{\mathbf{r}}(t)$ come very close to the initial configuration $\mathbf{r}_0$, $\dot{\mathbf{r}}_0$. The initial configuration will not be reached again, but will be reached "nearly" again after the time τ. For further discussion, we refer to the literature.

Example 10.2: Exercise: Orthogonality of the eigenmodes

Prove the orthogonality relation for the eigenmodes

$$\sum_{i=1}^{n} \sin\left(j\frac{i\pi}{n+1}\right) \sin\left(j\frac{l\pi}{n+1}\right) = \left(\frac{n+1}{2}\right) \delta_{il}, \tag{10.91}$$

which is explicitly used in equation (10.78) of the last example.

Solution Proof of orthogonality of the eigenvectors:

$$d_{il} = \sum_{j=1}^{n} \sin\left(i\frac{\pi}{n+1}j\right) \sin\left(l\frac{\pi}{n+1}j\right)$$

$$= \frac{1}{2} \sum_{j=1}^{n} \left\{ \cos\frac{(k-l)\pi}{n+1} j - \cos\frac{(k+l)\pi}{n+1} \right\}.$$

Before continuing the exposition, we evaluate the sum of the following series:

$$\sum_{k=1}^{n} \cos kx = \frac{\sin(xn/2) \cos x(n+1)/2}{\sin x/2} = \frac{\cos(xn/2) \sin x(n+1)/2}{\sin x/2} - 1. \tag{10.92}$$

This result is easily obtained by writing the cosine in terms of exponential functions and then evaluating the sum as a geometrical series.

The case $k = l$ yields

$$d_{kk} = \frac{1}{2}\left(n - \sum_{j=1}^{n} \cos\frac{2k\pi}{n+1}j\right) = \frac{1}{2}\left(n+1 - \frac{\cos\frac{kn\pi}{n+1}\sin k\pi}{\sin\frac{k\pi}{n+1}}\right) = \frac{1}{2}(n+1), \qquad (10.93)$$

since $\sin k\pi = 0$ for all k.

In the case $k \neq l$, we calculate the sums in both alternatives given in (10.92):

$$d_{kl} = \frac{1}{2}\left\{\frac{\cos\frac{(k-l)\pi n}{(n+1)2}\sin\frac{(k-l)\pi}{2}}{\sin\frac{(k-l)\pi}{2(n+1)}} - \frac{\cos\frac{(k+l)\pi n}{(n+1)2}\sin\frac{(k+l)\pi}{2}}{\sin\frac{(k+l)\pi}{2(n+1)}}\right\}, \qquad (10.94)$$

$$d_{kl} = \frac{1}{2}\left\{\frac{\sin\frac{(k-l)\pi n}{(n+1)2}\cos\frac{(k-l)\pi}{2}}{\sin\frac{(k-l)\pi}{2(n+1)}} - \frac{\sin\frac{(k+l)\pi n}{2(n+1)}\cos\frac{(k+l)\pi}{2}}{\sin\frac{(k+l)\pi}{2(n+1)}}\right\}. \qquad (10.95)$$

The vanishing of $d_{kl}(k \neq l)$ is immediately seen from (10.94) for even $k - l$ and $k + l$, and from (10.95) for odd $k - l$ and $k + l$.

Supplement: Everything happened already—A physical theorem?

Near the end of the nineteenth century, many physicists discussed the hypothesis that the course of the world repeats in eternal cycles. This interest was stimulated mainly by the works of Henri Poincaré (1854–1912). Also a philosopher like Friedrich Nietzsche (1844–1900) was tempted by this theorem to a short guest performance in physics. The first speculations along these lines originated from almost nonscientific attempts to explain the phenomenon of heat. The Lord of Verulam (1561–1626), Francis Bacon, had already identified heat as a form of motion but had failed to construct a quantitative theory. For lack of systematic investigations he had included the development of heat in dung hills in his considerations. The topic attracted more and more actors from physics, metaphysics, philosophy, politics and theology. We reproduce here two quotations from Poincaré's works from this era. Henri Poincaré in 1893 wrote in "Review of metaphysics and morality":

Everybody knows the mechanistic world view that tempted so many good people, and the various forms in which it comes up. Some people imagine the material world as being composed of atoms which move along straight lines because of their inertia, and change their velocity only if two atoms collide. Other people assume that the atoms perform an attraction or repulsion on each other, which depends on their distance. The following considerations will meet both points of view.

It would possibly be appropriate to dispute here the metaphysical difficulties that are related to these opinions, but I don't have the necessary expert knowledge. Therefore I will deal here only with the difficulties the mechanists met when they tried to reconcile their system with the experimental facts, and with the efforts they made to overcome or to elude these difficulties.

According to the mechanistic hypothesis all phenomena must be reversible; the stars for example could move along their orbits also in the opposite sense, without conflicting with Newton's laws. Reversibility is a consequence of all mechanistic hypotheses.

A theorem that can easily be proved tells us that a restricted world, which is governed only by the laws of mechanics, will pass again and again a state that is very close to its initial state. On the other hand, according to the assumed experimental laws the universe tends towards a certain final state which it never will leave. In this final state which represents some kind of death, all bodies will be at rest at the same temperature.

The doubts provoked this way, accompanying the developing theory of heat based upon an irreversible motion of atomic particles, have not yet been clearly removed.

A classical illustrative example of Poincaré's "recurrence objection" is the partitioned box, one half filled with gas which uniformly distributes over the entire box after removal of the membrane. Experience tells us what happens, and an inversion of this "irreversible process" is never observed in practice. But Poincaré did not think at all of an inversion, but rather of the chance that brought the particles into the initially empty half. This chance—after some "appropriate time"—should also bring them back again to the initial half.

In 1955, Enrico Fermi, John Pasta, and Stanislaw Ulam considered a problem which corresponds to our example 10.1, except for the additional inclusion of a nonlinear coupling term. Their interest focussed on finding, by means of the first computers, recurring processes such as we looked for in our purely linear problem. Surprisingly, they found an almost perfect recurrence of the initial conditions after large numbers of oscillations. The investigations and reflections of such properties of nonlinear wave equations continue to this day and have been introduced into the theory of elementary particles (solitons).

PART IV

MECHANICS OF RIGID BODIES

11 Rotation About a Fixed Axis

As we have seen in Chapter 4, a rigid body has 6 degrees of freedom, 3 of translation and 3 of rotation. The most general motion of a rigid body can be separated into the translation of a body point and the rotation about an axis through this point (Chasles' theorem). In the general case the rotation axis will change its orientation too. The meaning of the 6 degrees of freedom becomes clear once again: The 3 translational degrees of freedom give the coordinates of the particular body point, 2 of the rotational degrees of freedom determine the orientation of the rotation axis, and the third one fixes the rotation angle about this axis.

If a point of the rigid body is kept fixed, then any displacement corresponds to a rotation of the body about an axis through this fixed point (Euler's theorem). Hence, there exists an axis (through the fixed point) such that the result of several consecutive rotations can be replaced by a *single* rotation about this axis.

For an extended body the vanishing of the sum of all acting forces is no longer sufficient as an equilibrium condition.

Two oppositely oriented equal forces $-\mathbf{F}$ and $\mathbf{F}$ that act at two points of a body separated by the distance vector $\mathbf{l}$ are called a *couple*. A couple causes, independent of the reference point, the torque

$$\mathbf{D} = \mathbf{l} \times \mathbf{F}.$$

While the torque on a mass point is always related to a fixed point, the torque of a couple is completely free and can be shifted in space.

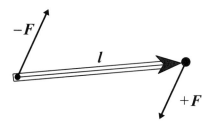

Figure 11.1. A couple causes a torque.

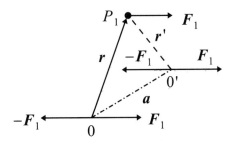

Figure 11.2. The forces acting on a rigid body are equivalent to a total force and a couple.

The forces acting on a rigid body can always be replaced by a total force acting on an arbitrary point, and a couple. This can easily be shown by the following example: At the point P_1, the force $\mathbf{F}_1$ acts. Nothing is changed if we let the forces $-\mathbf{F}_1$ and $\mathbf{F}_1$ act at O'. The force $\mathbf{F}_1$ acting on P_1 and the force $-\mathbf{F}_1$ acting on O' represent a couple, and there remains the force $\mathbf{F}_1$ acting on O'.

If there are several forces acting, we combine them into the resultant force $\mathbf{F} = \sum_i \mathbf{F}_i$. The torque is then given by $\mathbf{D} = \sum_i \mathbf{r}_i' \times \mathbf{F}_i$.

An extended body is in equilibrium if both the total force and the total torque vanish:

$$\sum_i \mathbf{F}_i = 0$$

and

$$\sum_i \mathbf{r}_i' \times \mathbf{F}_i = 0$$

(equilibrium condition at the point O').

For the calculation of the equilibrium condition, the origin of the vectors $\mathbf{r}_i$ (reference point of the moments) is arbitrary. Actually, for the point O it follows that (see Figure 11.2)

$$\sum_i \mathbf{F}_i = 0$$

and

$$\sum_i \mathbf{r}_i \times \mathbf{F}_i = \sum_i (\mathbf{a} + \mathbf{r}_i') \times \mathbf{F}_i = \mathbf{a} \times \sum_i \mathbf{F}_i + \sum_i \mathbf{r}_i' \times \mathbf{F}_i = 0,$$

i.e., the condition that the sum of all forces and the sum of all torques must vanish.

Moment of inertia (elementary consideration)

A rigid body rotates about a rotation axis z fixed in space. By substituting the angular velocity $v_i = \omega \cdot r_i$ for the velocity in the kinetic energy, one obtains

MOMENT OF INERTIA

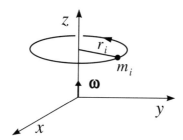

Figure 11.3. Rotation about the fixed z-axis with the angular velocity ω.

$$T = \sum_i \frac{1}{2} m_i v_i^2 = \frac{1}{2}\omega^2 \sum_i m_i r_i^2 = \frac{1}{2}\Theta\omega^2.$$

Analogously, for the angular momentum in z-direction we have

$$L_z = \sum_i m_i r_i v_i = \omega \sum_i m_i r_i^2 = \Theta\omega.$$

Here, r_i is the distance of the ith mass element from the z-axis.

The sum appearing in both relations is called the *moment of inertia* with respect to the rotation axis. One has

$$\Theta = \sum_i m_i r_i^2.$$

To calculate the moments of inertia of extended continuous systems, we change from the sum to the integral, i.e.,

$$\Theta = \int_{\text{body}} r^2 \, dm = \int_{\text{body}} r^2 \varrho \, dV,$$

with ϱ representing the density.

For a spatially extended, not axially symmetric rigid body which rotates about the z-axis, there can also appear components of the angular momentum perpendicular to the z-axis:

$$\begin{aligned}
\mathbf{L} &= \sum_\nu m_\nu \mathbf{r}_\nu \times \mathbf{v}_\nu = \sum_\nu m_\nu \mathbf{r}_\nu \times (\boldsymbol{\omega} \times \mathbf{r}_\nu) \\
&= \sum_\nu m_\nu \omega (x_\nu, y_\nu, z_\nu) \times (-y_\nu, x_\nu, 0) \\
&= \omega \sum_\nu (-x_\nu z_\nu, -y_\nu z_\nu, x_\nu^2 + y_\nu^2) m_\nu.
\end{aligned}$$

Since the body is supported in such a way that the rotation axis is constantly fixed, in the bearings appear torques (bearing moments) $\mathbf{D} = \dot{\mathbf{L}}$. They can be compensated by "balancing," i.e., by attaching additional masses so that the *deviation moments*

$$-\sum_\nu x_\nu z_\nu m_\nu \quad \text{and} \quad -\sum_\nu y_\nu z_\nu m_\nu$$

vanish.

Example 11.1:   Moment of inertia of a homogeneous circular cylinder

We determine the moment of inertia of a homogeneous circular cylinder with density ϱ about its symmetry axis. Adapted to the problem, we use cylindrical coordinates. The volume element then reads $dV = r\,dr\,d\varphi\,dz$, and $dm = \varrho\,dV$. The moment of inertia about the z-axis is then given by

$$\Theta = \int_{\text{cylinder}} r^2\,dm = \varrho \int_0^{2\pi} d\varphi \int_0^h dz \int_0^R r^3\,dr;$$

integration over the angle and the z-coordinate yields

$$\Theta = 2\pi h \varrho \int_0^R r^3\,dr.$$

Integration over the radius yields

$$\Theta = \frac{\pi}{2} h \varrho R^4 = \varrho \pi R^2 h \frac{R^2}{2} = \frac{1}{2} M R^2.$$

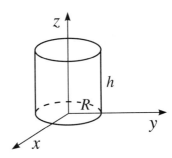

Figure 11.4. A homogeneous cylinder rotates about its axis.

Steiner's theorem[1]

If the moment of inertia Θ_s with respect to an axis through the center of gravity S of a rigid body is known, the moment of inertia Θ for an arbitrary *parallel* axis with the distance b from the center of gravity is given by the relation

$$\Theta = \Theta_s + Mb^2.$$

If AB is the axis through the center of gravity and $A'B'$ the parallel one with the unit vector $\mathbf{e}$ along the axis, this can be shown as follows:

$$\Theta_{AB} = \sum_\nu m_\nu (\mathbf{r}_\nu \times \mathbf{e})^2, \qquad \Theta_{A'B'} = \sum_\nu m_\nu (\mathbf{r}_\nu' \times \mathbf{e})^2.$$

The relation between $\mathbf{r}_\nu$ and $\mathbf{r}_\nu'$ is given by Figure 11.5. Obviously $\mathbf{r}_\nu' = -\mathbf{b} + \mathbf{r}_\nu$, and therefore,

$$\begin{aligned}
\Theta_{A'B'} &= \sum_\nu m_\nu ((-\mathbf{b} + \mathbf{r}_\nu) \times \mathbf{e})^2 \\
&= \sum_\nu m_\nu [(-\mathbf{b} \times \mathbf{e}) + (\mathbf{r}_\nu \times \mathbf{e})]^2 \\
&= \sum_\nu m_\nu (-\mathbf{b} \times \mathbf{e})^2 + 2 \sum_\nu m_\nu (-\mathbf{b} \times \mathbf{e}) \cdot (\mathbf{r}_\nu \times \mathbf{e}) + \sum_\nu m_\nu (\mathbf{r}_\nu \times \mathbf{e})^2 \\
&= Mb^2 + \Theta_{AB}.
\end{aligned}$$

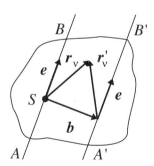

Figure 11.5. On Steiner's theorem.

---

[1] *Jacob Steiner*, b. March 18, 1796, Utzenstorf–d. April 1, 1863, Bern. Steiner was son of a peasant and grew up without education. He received his first education from Pestalozzi in Yverdon. Subsequently Steiner studied in Heidelberg, and then he served as a teacher of mathematics in Berlin; in 1834, he became an associate professor at the university there. Steiner is considered the founder of synthetic geometry, which was systematically developed by him. He worked on geometric constructions and isoperimetric problems. A peculiar feature of his work is that he almost completely avoided analytic and algebraic methods in geometric investigations.

The middle term vanishes because

$$2(-\mathbf{b} \times \mathbf{e}) \cdot \left(\sum_{\nu} m_{\nu} \mathbf{r}_{\nu}\right) \times \mathbf{e} = 0,$$

since S is the center of gravity and hence $\sum_{\nu} m_{\nu} \mathbf{r}_{\nu} = 0$.

If for a *planar* mass distribution the moments of inertia Θ_{xx}, Θ_{yy} in the x,y-plane are known, for the moment of inertia Θ_{zz} with respect to the z-axis we have

$$\Theta_{zz} = \Theta_{xx} + \Theta_{yy}.$$

If $r_{\nu} = \sqrt{x_{\nu}^2 + y_{\nu}^2}$ is the distance of the mass element from the z-axis, we have

$$\Theta_{zz} = \sum_{\nu} m_{\nu} r_{\nu}^2 = \sum_{\nu} m_{\nu} x_{\nu}^2 + \sum_{\nu} m_{\nu} y_{\nu}^2,$$

i.e.,

$$\Theta_{zz} = \Theta_{yy} + \Theta_{xx}.$$

Example 11.2: Moment of inertia of a thin rectangular disk

We consider the moment of inertia of a thin rectangular disk of density ϱ. For the calculation of the moment of inertia about the x-axis, we take as the mass element $dm = \varrho a \, dy$. We then obtain

$$\Theta_{xx} = \int_0^b y^2 a \varrho \, dy = a \varrho \frac{b^3}{3} = \frac{1}{3} M b^2.$$

The moment about the y-axis follows likewise:

$$\Theta_{yy} = \frac{1}{3} M a^2.$$

From $\Theta_{zz} = \Theta_{xx} + \Theta_{yy}$, we then get

$$\Theta_{zz} = \frac{1}{3} M (a^2 + b^2).$$

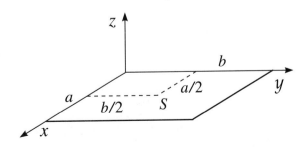

Figure 11.6. A rectangular planar mass distribution.

The moment of inertia about a perpendicular axis through the center of gravity is found, according to Steiner's theorem, from the moment of inertia about the z-axis:

$$\Theta_{zz} = \Theta_s + M\left(\sqrt{\left(\frac{a}{2}\right)^2 + \left(\frac{b}{2}\right)^2}\right)^2 = \Theta_s + M\frac{a^2+b^2}{4},$$

$$\Theta_s = \Theta_{zz} - \frac{M}{4}(a^2+b^2) = M(a^2+b^2)\left(\frac{1}{3}-\frac{1}{4}\right),$$

$$\Theta_s = \frac{M}{12}(a^2+b^2).$$

The physical pendulum

An arbitrary rigid body with the center of gravity S is suspended revolving on an axis through the point P. The distance vector $\vec{PS}$ is $\mathbf{r}$. Let Θ_0 be the moment of inertia of the body about a horizontal axis through P, and let M be the total mass. If the body in the gravitation field is now displaced from its rest position, it performs pendulum motions.

If the body is displaced, there is a torque

$$\mathbf{D} = \sum_\nu \mathbf{r}_\nu \times m_\nu \mathbf{g} = \sum_\nu m_\nu \mathbf{r}_\nu \times \mathbf{g} = M\mathbf{r} \times \mathbf{g} = -aMg\sin\varphi\,\mathbf{k},$$

where $\mathbf{k}$ is a unit vector pointing out of the page in Figure 11.7, and $|\mathbf{r}| = a$. The angular velocity is then

$$\boldsymbol{\omega} = +\mathbf{k}\frac{d\varphi}{dt}.$$

From the relation $\mathbf{D} = \dot{\mathbf{L}} = \Theta_0\dot{\boldsymbol{\omega}}$, we then obtain

$$-aMg\sin\varphi = \Theta_0\frac{d^2\varphi}{dt^2} \quad \text{or} \quad \frac{d^2\varphi}{dt^2} + \frac{aMg}{\Theta_0}\sin\varphi = 0.$$

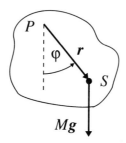

Figure 11.7. A body of mass M is suspended at the point P.

For small amplitudes, we replace $\sin \varphi$ by φ. With the abbreviation $\Omega = \sqrt{aMg/\Theta_0}$, we obtain the differential equation

$$\frac{d^2\varphi}{dt^2} + \Omega^2 \varphi = 0,$$

with the solution

$$\varphi = A \sin(\Omega t + \delta).$$

So one also obtains the period of the physical pendulum:

$$T = \frac{2\pi}{\Omega} = 2\pi \sqrt{\frac{\Theta_0}{Mag}}.$$

Since for the thread pendulum (mathematical pendulum) we have $T = 2\pi \sqrt{l/g}$, it follows that both periods coincide if the thread pendulum has the length $l = \Theta_0/Ma$.

If we replace the moment of inertia Θ_0 by the moment of inertia Θ_s about the center of gravity, then according to Steiner's theorem we have

$$T = T(a) = 2\pi \sqrt{\frac{\Theta_s + Ma^2}{Mag}} = 2\pi \sqrt{\frac{\Theta_s}{Mag} + \frac{a}{g}}.$$

From this, it follows that the period becomes a minimum if the vibration axis is a distance $a = \sqrt{\Theta_s/M}$ from the center of gravity. From this relation one can experimentally determine the moment of inertia Θ_s.

Example 11.3: Exercise: Moment of inertia of a sphere

Find the moment of inertia of a sphere about an axis through its center. The radius of the sphere is a, and the homogeneous density is ϱ.

Solution We use cylindrical coordinates (r, φ, z). The z-axis is the rotation axis. For the corresponding moment of inertia, we have

$$\Theta = \varrho \int_{\text{sphere}} r^2 \, dV.$$

The center of the sphere is at $z = 0$. The equation for the spherical surface then reads

$$x^2 + y^2 + z^2 = a^2 \quad \text{or} \quad r^2 + z^2 = a^2.$$

We write out the integration limits:

$$\Theta = \varrho \int_0^{2\pi} d\varphi \int_{-a}^{a} dz \int_0^{\sqrt{a^2-z^2}} r^3 \, dr$$

THE PHYSICAL PENDULUM

or

$$\Theta = 2\pi\varrho \int_{-a}^{a} \left[\frac{1}{4}r^4\right]_0^{\sqrt{a^2-z^2}} dz = \frac{\pi}{2}\varrho \int_{-a}^{a} (a^2 - z^2)^2 dz.$$

Integration over z yields

$$\Theta = \pi a^5 \varrho \frac{8}{15} = \frac{4}{3}\pi a^3 \varrho \frac{2}{5} a^2.$$

Since the total mass of the sphere is given by $M = (4/3)\pi a^3 \varrho$, it follows that

$$\Theta = \frac{2}{5} M a^2.$$

Example 11.4: Exercise: Moment of inertia of a cube

Calculate the moment of inertia of a homogeneous massive cube about one of its edges.

Solution Let ϱ be the density and s the edge length of the cube. A mass element is then given by

$$dm = \varrho \, dV = \varrho \, dx \, dy \, dz.$$

The moment of inertia about AB (see Figure 11.8) is evaluated as

$$\Theta_{AB} = \varrho \int_0^s \int_0^s \int_0^s (x^2 + y^2) \, dx \, dy \, dz = \frac{2}{3}\varrho s^5 = \frac{2}{3} M s^2.$$

Example 11.5: Exercise: Vibrations of a suspended cube

A cube of edge length s and mass M hangs vertically down from one of its edges. Find the period for small vibrations about the equilibrium position. How long is the equivalent thread pendulum?

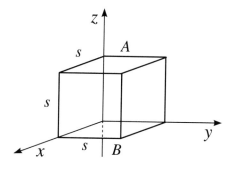

Figure 11.8. Calculation of the moment of inertia of a cube.

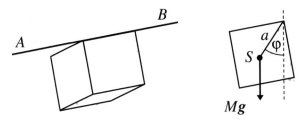

Figure 11.9. Rotation axes of the suspended cube.

Solution The moment of inertia of the cube about AB is (see Example 11.4)

$$\Theta_{AB} = \frac{2}{3} M s^2.$$

The center of gravity is in the center of the cube, i.e., for the distance a of the center of gravity S from the axis AB we have

$$a = \frac{1}{2} s \sqrt{2}.$$

The equation of motion of the physical pendulum for small angle amplitudes was

$$\ddot{\varphi} + \frac{Mga}{\Theta_{AB}} \varphi = 0$$

with the angular frequency

$$\omega = \sqrt{\frac{Mga}{\Theta_{AB}}}$$

and the period

$$T = \frac{2\pi}{\omega} = 2\pi \sqrt{\frac{\Theta_{AB}}{Mga}} = 2\pi \sqrt{\frac{2Ms^2 \cdot 2}{3Mgs\sqrt{2}}} = 2\pi \sqrt[4]{2} \sqrt{\frac{2s}{3g}}.$$

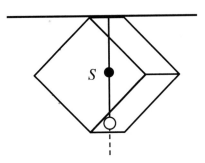

Figure 11.10. Physical pendulum and reduced pendulum length.

The length of the equivalent thread pendulum is calculated as

$$T = T' = 2\pi \sqrt{\frac{l}{g}},$$

which just defines the equivalence of the pendulums. By insertion of T one obtains

$$2\pi \sqrt[4]{2} \sqrt{\frac{2}{3}\frac{s}{g}} = 2\pi \sqrt{\frac{l}{g}},$$

or resolved,

$$l = \frac{2}{3}\sqrt{2}s.$$

This equivalent length of the thread pendulum is also called the *reduced pendulum length*.

Example 11.6: Roll off of a cylinder: Rolling pendulum

We consider a cylinder with a horizontal axis that can roll down an inclined plane. The system has one degree of freedom; hence an energy consideration is sufficient. The velocity of each point of the cylinder may be thought as being composed of the velocity $\mathbf{v}_1$ due to the translational motion and of the velocity $\mathbf{v}_{2\nu}$ due to the rotation. The energy of motion is then given by

$$\sum \frac{m_\nu}{2} \mathbf{v}_\nu^2 = \frac{1}{2} \mathbf{v}_1^2 \sum m_\nu + \sum \frac{m_\nu}{2} \mathbf{v}_{2\nu}^2 + \mathbf{v}_1 \cdot \sum m_\nu \mathbf{v}_{2\nu}. \tag{11.1}$$

For a symmetric mass distribution, the last term drops out, and we have

$$T = \frac{M}{2} \mathbf{v}_1^2 + \frac{\Theta}{2} \dot{\varphi}^2, \tag{11.2}$$

i.e., the energy of motion is additive in translational and rotation energy. For the cylinder (with symmetric mass distribution) on the inclined plane we have

$$\frac{M}{2} \dot{s}^2 + \frac{\Theta}{2} \dot{\varphi}^2 - Mgs \sin\alpha = E \tag{11.3}$$

(s measures the distance along the inclined plane). "Rolling off" without gliding means that the axis always moves just as much as corresponds to the rotation of the cylinder surface:

$$\dot{s} = R\dot{\varphi},$$

where R is the cylinder radius. We thus obtain the equation

$$\frac{1}{2}\left(M + \frac{\Theta}{R^2}\right)\dot{s}^2 - Mgs \sin\alpha = E,$$

$$\ddot{s} = \frac{1}{1 + \Theta/MR^2} g \sin\alpha. \tag{11.4}$$

The acceleration of the cylinder rolling off is smaller than that of a gliding mass point.

If the total mass of the cylinder is (approximately) concentrated on the axis, then

$$\frac{\Theta}{MR^2} = 0, \qquad \ddot{s} = g \sin\alpha,$$

and the acceleration is the same as for a gliding mass point. For a homogeneous cylinder, we have

$$\frac{\Theta}{MR^2} = \frac{1}{2}, \qquad \ddot{s} = \frac{2}{3} g \sin\alpha.$$

For a hollow cylinder with all mass on the surface, we have

$$\frac{\Theta}{MR^2} = 1, \qquad \ddot{s} = \frac{1}{2} g \sin\alpha;$$

the acceleration is only half of that for a gliding mass point. If we fix a circular disk concentric onto the cylinder, which extends beyond the base (like a wheel rim over the rail), then $\Theta/MR^2 > 1$, i.e., the acceleration can be even lower.

An investigation of the force balance lets us elucidate this problem once again from another point of view. At the point S, gravity acts and performs a torque with respect to the point A (see Figure 11.11)

$$D_A = |\mathbf{D}_A| = R \cdot Mg \sin\alpha, \tag{11.5}$$

while the constraints do not create a torque. The angular acceleration at the point A is therefore

$$\ddot{\varphi} = \dot{\omega} = \frac{D_A}{\Theta_A} = \frac{RMg \sin\alpha}{(3/2)MR^2} = \frac{2}{3} \frac{g}{R} \sin\alpha. \tag{11.6}$$

The moment of inertia Θ_A of a homogeneous cylinder is easily found by means of Steiner's theorem. Since the moment of inertia with respect to the center of gravity is $\Theta_s = MR^2/2$, it follows immediately that

$$\Theta_A = \Theta_s + MR^2 = \frac{3}{2} MR^2.$$

If the cylinder rolls without gliding, for the linear acceleration of the center of gravity, we find

$$|a_s| = |\dot{\omega} \times \mathbf{r}_A| = \dot{\omega} R = \frac{2}{3} g \sin\alpha. \tag{11.7}$$

The cylinder gets only 2/3 of the acceleration which it would get when gliding. Equation (11.8) is found from simple considerations: Since the instantaneous velocity of the contact point A equals zero, one can consider A as instantaneously at rest. But this means that the rigid body instantaneously

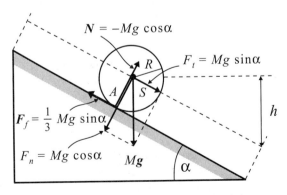

Figure 11.11. Rolling cylinder on an inclined plane.

THE PHYSICAL PENDULUM

performs a rotation about the contact point A, with an angular velocity ω. The velocity of an arbitrary point of the body is then given by (see Figure 11.11)

$$\mathbf{v} = \boldsymbol{\omega} \times \mathbf{r}_A.$$

Besides the gravitation force there acts the reaction force $\mathbf{N}$ (to balance the normal component of $M\mathbf{g}$)

$$|\mathbf{N}| = |Mg \cos \alpha|, \tag{11.8}$$

and the friction force F_f. The latter one is calculated from the balance

$$Mg \sin \alpha + F_f = Ma_s \tag{11.9}$$

and with equation (11.7),

$$-F_f = Mg \sin \alpha - \frac{2}{3} Mg \sin \alpha = \frac{1}{3} Mg \sin \alpha. \tag{11.10}$$

Thus, the friction force acts opposite to the direction of motion. The condition for a rolling motion of the cylinder is

$$|F_f| \leq \mu N, \tag{11.11}$$

where μ is the friction coefficient.
Since

$$N = Mg \cos \alpha \quad \text{and} \quad |F_f| = \frac{1}{3} Mg \sin \alpha,$$

we have

$$|F_f| \leq \mu Mg \cos \alpha \quad \text{or} \quad \frac{1}{3} Mg \sin \alpha \leq \mu Mg \cos \alpha. \tag{11.12}$$

That means that a rolling motion exists only for $\tan \alpha \leq 3\mu$.

A cylinder with asymmetrical mass distribution, which under the influence of gravitation can vibrate by rolling on a horizontal base, is called a *rolling pendulum*. It represents a system with one degree of freedom; the position of the rolling pendulum can be specified by the rotation angle φ or by the coordinate x of the cylinder axis (measured perpendicularly to the axis, see Figure 11.12). "Rolling off" means that

$$\dot{x} = R\dot{\varphi}. \tag{11.13}$$

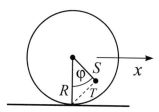

Figure 11.12. Rolling pendulum.

Since there is only one degree of freedom, the energy law is sufficient for the description. The motion is composed of a translational and a rotational motion. When applying (11.1), we have to account for the asymmetrical mass distribution. The expression $\sum m_\nu \mathbf{v}_{2\nu}$, as a momentum due to the rotational motion, can be calculated by assuming that the total mass M is concentrated at the center of gravity, which is located off the axis by the distance s: $|s\dot\varphi|$ is then the velocity $|\mathbf{v}_{2\nu}|$ of this mass on rotation, and $\pi - \varphi$ is the angle between $\mathbf{v}_1$ and $\mathbf{v}_{2\nu}$. According to (11.1), we then have

$$T = \frac{M}{2}\dot{x}^2 + \frac{\Theta}{2}\dot\varphi^2 - \dot{x}\cdot Ms\dot\varphi\cos\varphi,$$

where Θ is the moment of inertia about the cylinder axis. With the condition (11.13) for rolling follows

$$T = \frac{1}{2}(MR^2 + \Theta - 2MRs\cos\varphi)\dot\varphi^2. \tag{11.14}$$

This expression can be interpreted also in a different way: If Θ_s is the moment of inertia about an axis through the center of gravity which is parallel to the cylinder axis, then according to Steiner's theorem we have

$$\Theta = \Theta_s + Ms^2.$$

Equation (11.14) thus turns into

$$T = \frac{1}{2}[M(R^2 + s^2 - 2Rs\cos\varphi) + \Theta_s]\dot\varphi^2$$

or

$$T = \frac{1}{2}(Mr^2 + \Theta_s)\dot\varphi^2,$$

where r is the distance of the center of gravity from the contact line of the cylinder with the base. According to Steiner's theorem,

$$\Theta_u = Mr^2 + \Theta_s$$

is the moment of inertia about the contact line which changes with time, and (11.14) takes the form

$$T = \frac{\Theta_u}{2}\dot\varphi^2.$$

If we now abbreviate (11.14) by

$$T = \frac{1}{2}(B - 2MRs\cos\varphi)\dot\varphi^2$$

and add the potential energy

$$U = Mgs(1 - \cos\varphi),$$

we then get the energy law

$$\frac{1}{2}(B - 2MRs\cos\varphi)\dot\varphi^2 + Mgs(1 - \cos\varphi) = E. \tag{11.15}$$

THE PHYSICAL PENDULUM

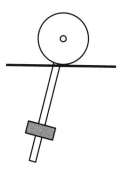

Figure 11.13. Transition from the rolling pendulum to the physical pendulum.

The equation differs from that for the physical pendulum (compare the section on the physical pendulum). For small angles φ, we obtain

$$\Theta_u \dot{\varphi}^2 + Mgs\varphi^2 = Mgs\alpha^2, \tag{11.16}$$

where we replaced the arbitrary constant E by the arbitrary constant α. Equation (11.16) is solved by

$$\varphi = \alpha \cos(\omega t + \delta), \tag{11.17}$$

with

$$\omega^2 = \frac{Mgs}{\Theta_u} = \frac{Mgs}{MR^2 + \Theta - 2MRs}. \tag{11.18}$$

In the limit of a symmetrical mass distribution ($s = 0$), one has $\omega = 0$. If the center of gravity moves to the cylinder surface ($s \to R$), then

$$\omega^2 = \frac{MgR}{\Theta_s}.$$

If the mass is limited to a more restricted region, ω becomes very large. If we imagine that a part of the mass is shifted by an appropriate device to the outside of the rolling cylinder (see Figure 11.13) and that s is large compared to R, then the vibration turns into the vibration of a physical pendulum.

Example 11.7: Moments of inertia of several rigid bodies about selected axes

Figure 11.14 shows the moments of inertia of (a) a disk, (b) a cylinder, (c) a rectangular plate, (d) a spherical shell, (e) a solid sphere, and (f) a cube about different selected axes.

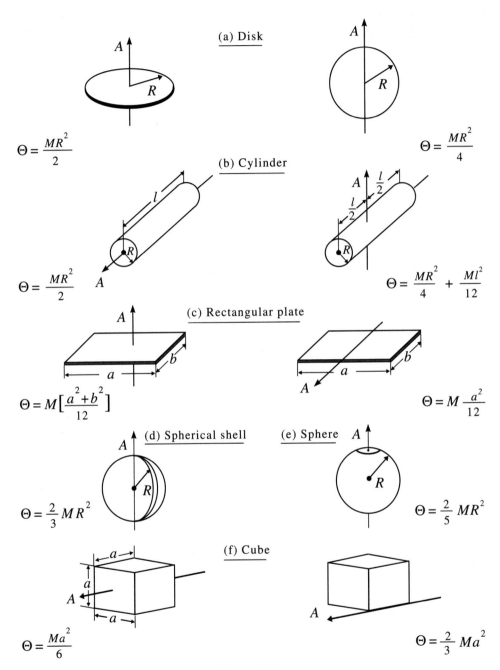

Figure 11.14.

Example 11.8: Exercise: Cube tilts over the edge of a table

A cube with the edge length $2a$ and mass M glides with constant velocity v_0 on a frictionless plate. At the end of the plate, it bumps against an obstacle and tilts over the edge (see Figure 11.15). Find the minimum velocity v_0 for which the cube still falls from the plate!

Solution We look for the velocity v_0 for which the cube can tilt over its edge, as is represented in (c). If it bumps into the obstacle at the edge of the plate, it is set into rotation about the axis A. At the time of collision all external forces act along this axis, and the angular momentum of the cube is conserved. Before hitting the obstacle, the cube has—due to the translational motion—the angular momentum

$$L = |\mathbf{r} \times \mathbf{p}| = p \cdot a = M v_0 a. \tag{11.19}$$

Immediately after the collision, the angular momentum appears as rotational motion of the cube

$$L = \Theta_A \omega_0 = M v_0 a,$$

or

$$\omega_0 = \frac{M v_0 a}{\Theta_A}. \tag{11.20}$$

If the cube begins to lift off, the gravitational force causes a torque about the axis A that counteracts the lifting process.

For the kinetic energy of the cube immediately after the collision, one has, for given ω_0,

$$T_0 = \frac{1}{2} \Theta_A \omega_0^2 = \frac{1}{2} \frac{M^2 v_0^2 a^2}{\Theta_A}. \tag{11.21}$$

The potential energy difference between position a and position c is

$$\Delta V = M(h_2 - h_1)g = M(\sqrt{2}a - a)g = Mag(\sqrt{2} - 1), \tag{11.22}$$

and from the energy conservation law, immediately it follows that

$$Mag(\sqrt{2} - 1) = \frac{1}{2} \frac{M^2 v_0^2 a^2}{\Theta_A}. \tag{11.23}$$

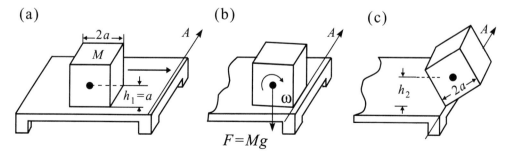

Figure 11.15. Cube tilting over an edge.

The moment of inertia of the cube Θ_A is easy to calculate:

$$\Theta_A = \varrho \int_V r^2 \, dV = \varrho \int_0^{2a}\int_0^{2a}\int_0^{2a} (x^2+y^2) \, dx \, dy \, dz = \frac{8}{3} M a^2. \tag{11.24}$$

From equation (11.23), it follows that

$$ag(\sqrt{2}-1) = \frac{1}{2} \frac{Ma^2}{(8/3)Ma^2} v_0^2,$$

and from this, we find

$$v_0 = \sqrt{ag \frac{16}{3}(\sqrt{2}-1)}. \tag{11.25}$$

This is the correct result. We emphasize this because one could easily come to another result by a *false consideration*: The kinetic energy is $(1/2)Mv_0^2$, and from the energy conservation combined with (11.22) it follows that

$$\frac{1}{2}Mv_0^2 = Mag(\sqrt{2}-1).$$

This leads to

$$v_0 = \sqrt{2ag(\sqrt{2}-1)}, \tag{11.26}$$

i.e., a value that is smaller than the correct result (11.25) by the factor $\sqrt{3/8}$. The result (11.26) is wrong since the cube loses part of its kinetic energy in the collision because of its inelasticity. The correct result (11.25) is based upon the conservation of angular momentum, which acts "more strongly" than the conservation of energy.

Example 11.9: Exercise: Hockey puck hits a bar

A thin bar of length l and mass M lies on a frictionless plate (the x,y-plane in the figure). A hockey puck of mass m and velocity v knocks the bar elastically under $90°$ at the distance d from the center of gravity. After the collision the puck is at rest.

(a) Determine the motion of the bar.

(b) Calculate the ratio m/M, accounting for the fact that the puck is at rest.

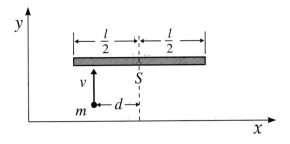

Figure 11.16. A hockey puck hits a bar on a frictionless plate.

THE PHYSICAL PENDULUM

Solution (a) Since the collision is elastic, momentum and energy conservation hold, where momentum conservation refers both to linear and angular momentum. The bar acquires both a translational and a rotational motion from the collision with the puck. Conservation of the linear momentum immediately leads to

$$P_s = M v_s = m v, \tag{11.27}$$

and the velocity of the center of gravity is

$$v_s = \frac{mv}{M}. \tag{11.28}$$

Likewise, from the conservation of angular momentum it follows that

$$L_s = \Theta_s \omega = m v d = D, \tag{11.29}$$

and for the angular velocity of the bar relative to the center of gravity,

$$\omega = \frac{mvd}{Ml^2/12}, \tag{11.30}$$

where

$$\Theta_s = \int_0^{l/2} \varrho r^2 \, dV = \frac{1}{12} M l^2.$$

Thus, the center of gravity of the bar moves uniformly with v_s along the y-axis, while the bar rotates with the angular velocity ω about the center of gravity. Figure 11.17 illustrates several stages of the motion.

(b) The kinetic energy of the bar can be determined by means of the energy conservation law. Before the collision, the kinetic energy of the puck is

$$T = \frac{1}{2} m v^2, \tag{11.31}$$

while the kinetic energy after the collision consists of two components:

$$T_t = \frac{1}{2} M v_s^2 \quad \text{``translation energy of the center of gravity''}$$

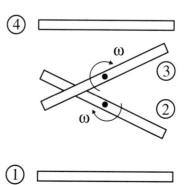

Figure 11.17. The motion of the bar.

and

$$T_r = \frac{1}{2}\Theta_s \omega^2 \quad \text{"rotation energy about the center of gravity."}$$

Since the potential energy remains unchanged, it immediately follows that

$$T = \frac{1}{2}mv^2 = T_t + T_r = \frac{1}{2}(Mv_s^2 + \Theta_s \omega^2)$$

or

$$mv^2 = Mv_s^2 + \frac{Ml^2}{12}\omega^2. \tag{11.32}$$

Insertion of equations (11.28) and (11.30) into equation (11.32) finally yields

$$mv^2 = \frac{m^2 v^2}{M} + \frac{m^2 v^2 d^2 (12)^2}{M^2 l^4} \frac{Ml^2}{12},$$

$$1 = \frac{m}{M} + 12\frac{m}{M}\frac{d^2}{l^2},$$

or

$$\frac{m}{M} = \frac{1}{1 + 12(d/l)^2}$$

for the mass ratio. If the puck kicks the bar at the center of gravity, $d = 0$, no rotation appears. In order to make the collision elastic, $m = M$ must be satisfied.

If the puck kicks the bar at the point $d = l/2$, the collision is elastic only if $M = 4m$. In this case the rotation velocity is $\omega = 6mv/Ml = 6v_s/l$.

Example 11.10: Exercise: Cue pushes a billiard ball

A billiard ball of mass M and radius R is pushed by a cue so that the center of gravity of the ball gets the velocity v_0. The momentum direction passes through the center of gravity. The friction coefficient between table and ball is μ. How far does the ball move before the initial gliding motion changes to a pure rolling motion?

Solution Since the momentum direction passes through the center of gravity, the angular momentum with respect to the center of gravity at the time $t = 0$ equals zero. The friction force **f** points opposite to the direction of motion (see Figure 11.18) and causes a torque about the center of gravity

$$D_s = f \cdot R = \mu MgR. \tag{11.33}$$

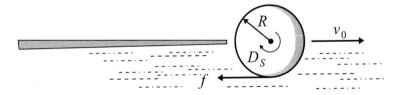

Figure 11.18. A cue pushes a billiard ball.

The result is an angular acceleration of the ball, so that

$$\dot{\omega} = \frac{\mu M g R}{\Theta_s} = \frac{\mu M g R}{(2/5) M R^2} = \frac{5}{2} \frac{\mu g}{R}. \tag{11.34}$$

Moreover, the friction force causes a deceleration of the center of gravity, i.e.,

$$M a_s = -f \quad \text{or} \quad a_s = -\frac{f}{M} = -\frac{\mu g M}{M}. \tag{11.35}$$

a_s is the acceleration of the center of gravity.

For the rotation velocity of the ball, one gets from equation (11.34)—after performing the integration—

$$\omega = \int_0^t \dot{\omega} \, dt = \frac{5}{2} \frac{\mu g}{R} t. \tag{11.36}$$

The linear velocity of the center of gravity follows from equation (11.35)—again after integration—as

$$v_s = \int a_s \, dt = v_0 - \mu g t. \tag{11.37}$$

The billiard ball begins rolling when $v_s = \omega R$, or

$$\frac{5}{2} \mu \frac{g}{R} t R = v_0 - g \mu t \tag{11.38}$$

or when

$$v_0 = \frac{7}{2} \mu g t \quad \text{and} \quad t = \frac{2}{7} \frac{v_0}{\mu g}. \tag{11.39}$$

The distance passed before rolling starts is obtained—by integrating equation (11.37)—as

$$s = \int_0^t v_s \, dt = v_0 t - \frac{\mu g t^2}{2} \tag{11.40}$$

and with t from equation (11.39) finally as

$$s = \frac{2}{7} \frac{v_0^2}{\mu g} - \frac{v_0^2}{2 \mu g} \left(\frac{2}{7}\right)^2 = \frac{12}{49} \frac{v_0^2}{\mu g}. \tag{11.41}$$

If the ball is kicked at a distance h above the center of gravity, besides the linear motion there appears a rotational motion with the angular velocity

$$\omega = \frac{M v_0 h}{\Theta} = \frac{5}{2} \frac{v_0 h}{R^2}. \tag{11.42}$$

If $h = (2/5) R$, the rolling motion of the ball starts immediately. For $h < (2/5) R$, one has $\omega < v_0/R$, and for $h > (2/5) R$ correspondingly $\omega > v_0/R$; in the second case the friction force points forward.

Figure 11.19 shows the change of v_s and ωR as a function of time for $h = 0$. If $v_s = \omega R$, the rolling motion begins, the friction vanishes, and then v_s and ω remain constant.

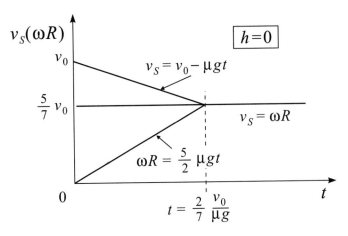

Figure 11.19.

Example 11.11: Exercise: Motion with constraints

A bar of length $2l$ and mass M is fixed at point A, so that it can rotate only in the vertical plane (see Figure 11.20). The external force $\mathbf{F}$ acts on the center of gravity. Calculate the reaction force $\mathbf{F}_r$ at the point A!

Solution In order to determine $\mathbf{F}_r$, one calculates the torque D_A with respect to the center of gravity of the bar, caused by $\mathbf{F}_r$.

The torque with respect to the fixed point A is

$$D_A = -Fl = \Theta_A \dot{\omega}, \tag{11.43}$$

since the constraints do not contribute to D_A. The angular acceleration of the bar $\dot{\omega}$ then follows from (11.43):

$$\dot{\omega} = \frac{D_A}{\Theta_A} = -\frac{Fl}{\Theta_A}, \tag{11.44}$$

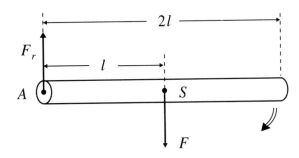

Figure 11.20. The bar rotates about the point A.

where Θ_A is the moment of inertia of the bar with respect to A. Since the moment of inertia Θ_s with respect to the center of gravity S is easily calculated as

$$\Theta_s = \int_{-l}^{l} \varrho r^2\, dV = \frac{1}{3} M l^2, \qquad (11.45)$$

one immediately gets for Θ_A by means of Steiner's theorem

$$\Theta_A = \Theta_s + M l^2 = \frac{1}{3} M l^2 + M l^2 = \frac{4}{3} M l^2. \qquad (11.46)$$

Equation (11.46) inserted into equation (11.44) leads to

$$\dot{\omega} = -\frac{Fl}{\Theta_A} = -\frac{3}{4}\frac{F}{Ml}. \qquad (11.47)$$

Since equation (11.47) must be correct, independent of the point from which the torque is being calculated, from the knowledge of the torque with respect to the center of gravity S,

$$D_s = -F_r l, \qquad (11.48)$$

and hence of the angular acceleration

$$\dot{\omega} = \frac{D_s}{\Theta_s} = -\frac{3 F_r l}{M l^2} = -\frac{3 F_r}{M l}, \qquad (11.49)$$

one can calculate the reaction force F_r, by comparing the equations (11.47) and (11.49):

$$-\frac{3}{4}\frac{F}{Ml} = -\frac{3 F_r}{Ml} \quad \Rightarrow \quad F_r = \frac{1}{4} F.$$

Example 11.12: Exercise: Bar vibrates on springs

(a) Find the moment of inertia of a thin homogeneous bar of length L with respect to an axis perpendicular to the bar.

(b) A homogeneous bar of length L and mass m is supported at the ends by identical springs (spring constant k). The bar is moved at one end by a small displacement a and then released.

Solve the equation of motion and determine the normal frequencies and normal vibrations. Sketch the normal vibrations.

Solution (a) If the bar is divided into small segments of length dx with the cross section f, we have elementary

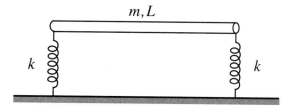

Figure 11.21. A bar is supported by two identical springs.

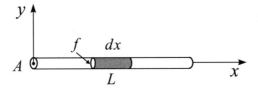

Figure 11.22.

volumes $dV = f\,dx$. Let ϱ be the constant density of the bar; then we have

$$\Theta_A = \int_0^L \varrho x^2 (f\,dx) = \varrho f \int_0^L x^2\,dx = \frac{1}{3}\varrho f L^3.$$

Since $m = \varrho f L$ is the total mass of the bar, it follows that

$$\Theta_A = \frac{1}{3}mL^2.$$

According to Steiner's theorem, the moment of inertia about an axis through the center of gravity is

$$\Theta_A = \Theta_s + m\left(\frac{L}{2}\right)^2 \quad\Rightarrow\quad \Theta_s = \frac{1}{12}mL^2.$$

(b) Let b be the length of the spring before the motion (b is not the natural length of the spring, because of the existence of the gravitation field), and x_1, x_2, x be the lengths of the first and second spring, and the height of the center of gravity of the bar at the time t. Since the bar is rigid, we have $x_1 + x_2 = 2x$. Newton's second law leads to

$$m\ddot{x} = -k(x_1 - b) - k(x_2 - b)$$

or

$$m\ddot{x} = -k(x_1 + x_2) + 2kb.$$

The constraint condition leads to

$$m\ddot{x} = -2kx + 2kb \quad\Rightarrow\quad \ddot{x} = -\frac{2k}{m}(x-b). \tag{11.50}$$

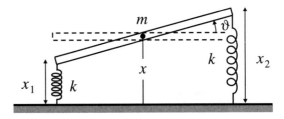

Figure 11.23.

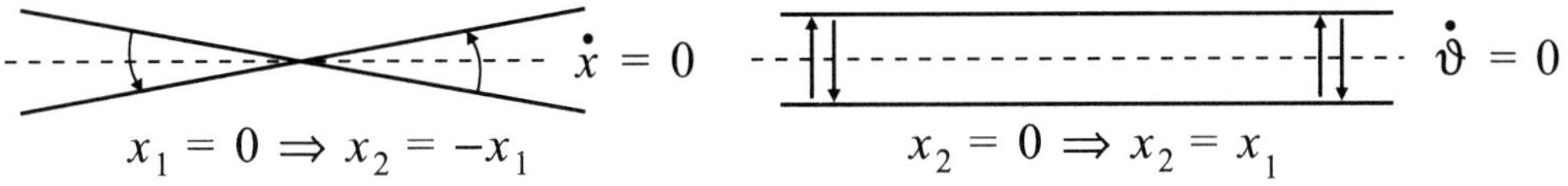

THE PHYSICAL PENDULUM

$$\dot{x} = 0$$
$$x_1 = 0 \Rightarrow x_2 = -x_1$$

$$\dot{\vartheta} = 0$$
$$x_2 = 0 \Rightarrow x_2 = x_1$$

Figure 11.24. The normal vibrations.

We assume that there are only small displacements, so that $\sin \vartheta \approx \vartheta$. Then

$$x_2 = x + \frac{L}{2}\vartheta, \qquad x_1 = x - \frac{L}{2}\vartheta.$$

For the torque, we get

$$\Theta\ddot{\vartheta} = -\frac{k}{2}L(x_2 - x_1) = -\frac{1}{2}kL^2\vartheta, \qquad \text{since} \quad x_2 - x_1 = L\vartheta.$$

From (a) $\Theta = (1/12)mL^2$, we conclude

$$\ddot{\vartheta} = -\frac{6k}{m}\vartheta. \tag{11.51}$$

The solutions of (11.50) and (11.51) are

$$x = A\cos(\omega_1 t + B) + b$$

and

$$\vartheta = C\cos(\omega_2 t + D)$$

with

$$\omega_1 = \sqrt{\frac{2k}{m}} \quad \text{and} \quad \omega_2 = \sqrt{\frac{6k}{m}}.$$

The initial conditions at the time $t = 0$ are

$$x = b - \frac{a}{2}, \qquad \vartheta = \frac{a}{L}, \qquad \dot{x} = 0, \qquad \dot{\vartheta} = 0.$$

Thus follows

$$B = D = 0,$$
$$A = -\frac{a}{2},$$
$$C = \frac{a}{L},$$

$$\begin{cases} b - \dfrac{a}{2} = A\cos(B) + b, \\ 0 = -A\omega_1 \sin(B), \\ \dfrac{a}{L} = C\cos(D), \\ 0 = -C\omega_2 \sin(D), \end{cases}$$

and, hence,

$$x = b - \frac{a}{2}\cos\sqrt{\frac{2k}{m}}t, \qquad \vartheta = \frac{a}{L}\cos\sqrt{\frac{6k}{m}}t.$$

The normal modes are

$$X_1 = x_1 + x_2 = 2b - a\cos\sqrt{\frac{2k}{m}}t,$$

$$X_2 = x_1 - x_2 = -a\cos\sqrt{\frac{6k}{m}}t.$$

12 Rotation About a Point

The general motion of a rigid body can be described as a translation and a rotation about a point of the body. This is just the content of Chasles' theorem, discussed at the begin of Chapter 4. If the origin of the body-fixed coordinate system is set at the center of gravity of the body, one can separate the center-of-mass motion and the rotation in all practical cases (compare Chapter 6, equations (6.4) to (6.8). For this reason, the rotation of a rigid body about a fixed point is of particular significance.

Tensor of inertia

We first consider the angular momentum of a rigid body that rotates with angular velocity ω about the fixed point 0 (see Figure 12.1):

$$\mathbf{L} = \sum_\nu m_\nu(\mathbf{r}_\nu \times \mathbf{v}_\nu) = \sum_\nu m_\nu(\mathbf{r}_\nu \times (\boldsymbol{\omega} \times \mathbf{r}_\nu))$$
$$= \sum_\nu m_\nu(\boldsymbol{\omega} r_\nu^2 - \mathbf{r}_\nu(\mathbf{r}_\nu \cdot \boldsymbol{\omega}));$$

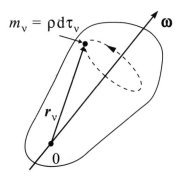

Figure 12.1. A rigid body rotates with ω about the fixed point 0.

TENSOR OF INERTIA

the latter relation holds according to the expansion rule. We decompose $\mathbf{r}_\nu$ and $\boldsymbol{\omega}$ into components and insert

$$\mathbf{L} = \sum_\nu m_\nu ((x_\nu^2 + y_\nu^2 + z_\nu^2)(\omega_x, \omega_y, \omega_z) - (x_\nu \omega_x + y_\nu \omega_y + z_\nu \omega_z)(x_\nu, y_\nu, z_\nu)).$$

Ordering by components leads to

$$\begin{aligned}\mathbf{L} = \sum_\nu m_\nu \Big(& ((x_\nu^2 + y_\nu^2 + z_\nu^2)\omega_x - x_\nu^2 \omega_x - x_\nu y_\nu \omega_y - x_\nu z_\nu \omega_z)\mathbf{e}_x \\
& + ((x_\nu^2 + y_\nu^2 + z_\nu^2)\omega_y - y_\nu^2 \omega_y - x_\nu y_\nu \omega_x - z_\nu y_\nu \omega_z)\mathbf{e}_y \\
& + ((x_\nu^2 + y_\nu^2 + z_\nu^2)\omega_z - z_\nu^2 \omega_z - x_\nu z_\nu \omega_x - y_\nu z_\nu \omega_y)\mathbf{e}_z \Big).\end{aligned}$$

For the components of the angular momentum one thus obtains

$$L_x = \left(\sum_\nu m_\nu(y_\nu^2 + z_\nu^2)\right)\omega_x + \left(-\sum_\nu m_\nu x_\nu y_\nu\right)\omega_y + \left(-\sum_\nu m_\nu x_\nu z_\nu\right)\omega_z,$$

$$L_y = \left(-\sum_\nu m_\nu x_\nu y_\nu\right)\omega_x + \left(\sum_\nu m_\nu(x_\nu^2 + z_\nu^2)\right)\omega_y + \left(-\sum_\nu m_\nu y_\nu z_\nu\right)\omega_z,$$

$$L_z = \left(-\sum_\nu m_\nu x_\nu z_\nu\right)\omega_x + \left(-\sum_\nu m_\nu y_\nu z_\nu\right)\omega_y + \left(\sum_\nu m_\nu(x_\nu^2 + y_\nu^2)\right)\omega_z.$$

The individual sums are abbreviated by

$$L_x = \Theta_{xx}\omega_x + \Theta_{xy}\omega_y + \Theta_{xz}\omega_z,$$
$$L_y = \Theta_{yx}\omega_x + \Theta_{yy}\omega_y + \Theta_{yz}\omega_z,$$
$$L_z = \Theta_{zx}\omega_x + \Theta_{zy}\omega_y + \Theta_{zz}\omega_z,$$

or

$$L_\mu = \sum_\nu \Theta_{\mu\nu}\omega_\nu,$$

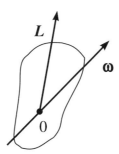

Figure 12.2. The angular momentum is in general not parallel to the angular velocity ω.

or, written in terms of vectors (matrix notation),

$$\mathbf{L} = \widehat{\Theta} \cdot \boldsymbol{\omega}.$$

The quantities $\Theta_{\mu\nu}$ are the elements of the *tensor of inertia* $\widehat{\Theta}$ that can be written as a 3×3-matrix:

$$\widehat{\Theta} = \begin{pmatrix} \Theta_{xx} & \Theta_{xy} & \Theta_{xz} \\ \Theta_{yx} & \Theta_{yy} & \Theta_{yz} \\ \Theta_{zx} & \Theta_{zy} & \Theta_{zz} \end{pmatrix}.$$

The elements in the main diagonal are called *moments of inertia*, the remaining ones are called *deviation moments*. The matrix is *symmetric*, i.e., $\Theta_{\nu\mu} = \Theta_{\mu\nu}$. Thus the tensor of inertia has 6 independent components. If the mass is continuously distributed, one changes from summation to integration for calculating the matrix elements. For example,

$$\Theta_{xy} = -\int_V \varrho(\mathbf{r}) x y \, dV,$$

$$\Theta_{xx} = \int_V \varrho(\mathbf{r})(y^2 + z^2) \, dV,$$

where $\varrho(\mathbf{r})$ is the space-dependent density.

At each point $0_0, 0_1, 0_2, \ldots$ the tensor of inertia $\Theta_{\mu\nu}$ is different. At a fixed point 0 $\Theta_{\mu\nu}$ also depends on the coordinate system.

As follows from their definition, the $\Theta_{\mu\nu}$ are constants if one selects a *body-fixed* coordinate system. The tensor of inertia is however dependent on the position of the coordinate system relative to the body and will change if the origin is shifted or the orientation of the axes is changed. *The* tensor of inertia is usually understood as the tensor in a coordinate system with the origin in the center of gravity (center-of-mass system). The corresponding principal moments of inertia (see below) are correspondingly *the* moments of inertia.

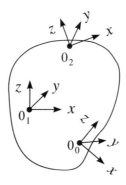

Figure 12.3. Possible rotation points and coordinate systems of the rigid body.

Kinetic energy of a rotating rigid body

Quite generally the kinetic energy of a system of mass points is

$$T = \frac{1}{2} \sum_\nu m_\nu v_\nu^2.$$

We decompose the motion of the rigid body into the translation of a point and the rotation about this point, so that $\mathbf{v}_\nu = \mathbf{V} + \boldsymbol{\omega} \times \mathbf{r}_\nu$, and we obtain

$$T = \frac{1}{2} \sum_\nu m_\nu (\mathbf{V} + \boldsymbol{\omega} \times \mathbf{r}_\nu)^2$$

$$= \frac{1}{2} M V^2 + \mathbf{V} \cdot \left(\boldsymbol{\omega} \times \sum_\nu m_\nu \mathbf{r}_\nu \right) + \frac{1}{2} \sum_\nu m_\nu (\boldsymbol{\omega} \times \mathbf{r}_\nu)^2.$$

The first and the last term correspond to pure translational and rotational energy, respectively. The mixed term can be made to vanish in two different ways.

If one point is fixed, and if we put it at the origin of the body-fixed coordinate system, then $\mathbf{V} = 0$. Otherwise the origin is put at the center of gravity, so that

$$\sum_\nu m_\nu \mathbf{r}_\nu = 0.$$

The rotation point is in this case the center of gravity. We now consider the pure rotation energy

$$T = \frac{1}{2} \sum_\nu m_\nu (\boldsymbol{\omega} \times \mathbf{r}_\nu) \cdot (\boldsymbol{\omega} \times \mathbf{r}_\nu) = \frac{1}{2} \sum_\nu m_\nu \boldsymbol{\omega} \cdot (\mathbf{r}_\nu \times (\boldsymbol{\omega} \times \mathbf{r}_\nu))$$

$$= \frac{1}{2} \boldsymbol{\omega} \cdot \sum_\nu m_\nu (\mathbf{r}_\nu \times \mathbf{v}_\nu) = \frac{1}{2} \boldsymbol{\omega} \cdot \sum_\nu \mathbf{r}_\nu \times \mathbf{p}_\nu = \frac{1}{2} \boldsymbol{\omega} \cdot \sum_\nu \mathbf{l}_\nu.$$

Hence,

$$T = \frac{1}{2} \boldsymbol{\omega} \cdot \mathbf{L}.$$

We can substitute the angular momentum $L_\mu = \sum_\nu \Theta_{\mu\nu} \omega_\nu$ ($\mu, \nu = 1, 2, 3$):

$$T = \frac{1}{2} \boldsymbol{\omega} \cdot \mathbf{L} = \frac{1}{2} \sum_\mu \omega_\mu \sum_\nu \Theta_{\mu\nu} \omega_\nu = \frac{1}{2} \sum_{\mu,\nu} \Theta_{\mu\nu} \omega_\mu \omega_\nu. \tag{12.1a}$$

Because $\Theta_{\mu\nu} = \Theta_{\nu\mu}$, the sum reads

$$T = \frac{1}{2} (\Theta_{xx} \omega_x^2 + \Theta_{yy} \omega_y^2 + \Theta_{zz} \omega_z^2 + 2\Theta_{xy} \omega_x \omega_y + 2\Theta_{xz} \omega_x \omega_z + 2\Theta_{yz} \omega_y \omega_z). \tag{12.1b}$$

Using tensor notation, the rotation energy reads

$$T = \frac{1}{2} \boldsymbol{\omega}^T \cdot \widehat{\Theta} \cdot \boldsymbol{\omega}. \tag{12.1c}$$

The vector $\boldsymbol{\omega}$ on the right-hand side of the tensor $\widehat{\Theta}$ must be given as a column vector, and on the left-hand side as a row vector:

$$T = \frac{1}{2}(\omega_x, \omega_y, \omega_z)\widehat{\Theta}\begin{pmatrix} \omega_x \\ \omega_y \\ \omega_z \end{pmatrix}. \tag{12.1d}$$

The principal axes of inertia

The elements of the tensor of inertia depend on the position of the origin and on the orientation of the (body-fixed) coordinate system. It is now possible for a fixed origin to orient the coordinate system in such a way that the deviation moments vanish. Such a special coordinate system is called a *system of principal axes*. The tensor of inertia then has diagonal form with respect to this system of axes:

$$\widehat{\Theta} = \begin{pmatrix} \Theta_1 & 0 & 0 \\ 0 & \Theta_2 & 0 \\ 0 & 0 & \Theta_3 \end{pmatrix} \quad \text{or} \quad \Theta_{\mu\nu} = \Theta_\mu \delta_{\mu\nu}. \tag{12.2}$$

For angular momenta and rotation energy in the *system of principal axes*, we have the especially simple relations (ω_ν are the components of the angular velocity $\boldsymbol{\omega}$ with respect to the principal axes)

$$L_\mu = \sum_\nu \Theta_{\mu\nu} \omega_\nu = \sum_\nu \Theta_\mu \delta_{\mu\nu} \omega_\nu = \Theta_\mu \omega_\mu, \tag{12.3}$$

$$T = \frac{1}{2}\boldsymbol{\omega} \cdot \mathbf{L} = \frac{1}{2}\sum_\mu \omega_\mu L_\mu = \frac{1}{2}\sum_\mu \Theta_\mu \omega_\mu^2, \tag{12.4a}$$

or written out,

$$T = \frac{1}{2}\left(\Theta_1 \omega_1^2 + \Theta_2 \omega_2^2 + \Theta_3 \omega_3^2\right). \tag{12.4b}$$

Because of the tensorial relation $\mathbf{L} = \widehat{\Theta}\boldsymbol{\omega}$, the angular momentum and the angular velocity have different orientations.

If the body rotates about one of the principal axes of inertia, e.g. about the μ-axis, $\boldsymbol{\omega} = \omega \mathbf{e}_\mu$, then (because in this example $\boldsymbol{\omega} = \omega \mathbf{e}_\mu$) according to (12.3) the angular momentum $\mathbf{L}$ and the angular velocity $\boldsymbol{\omega}$ have the same orientation. The vector $\boldsymbol{\omega}$ then has only one component, $\boldsymbol{\omega} = (0, \omega_2, 0)$, if the rotation is about the second principal axis. The same holds also for the angular momentum: $\mathbf{L} = (0, L_2, 0)$. This property of parallelism between the angular momentum and the angular velocity allows one to determine the principal axes. The question is namely *how to choose* $\boldsymbol{\omega} = \{\omega_1, \omega_2, \omega_3\}$ *(about which axis must the body*

THE PRINCIPAL AXES OF INERTIA

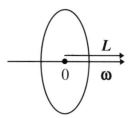

Figure 12.4. Special case: If ω is parallel to a principal axis, then **L** is parallel to ω.

rotate), in order to get the angular momentum $\mathbf{L} = \widehat{\Theta}\boldsymbol{\omega}$ and the angular velocity parallel to each other, i.e., $\mathbf{L} = \Theta\boldsymbol{\omega}$, with Θ a scalar.

From the combination of the relations $\mathbf{L} = \widehat{\Theta}\boldsymbol{\omega}$ ($\widehat{\Theta}$ is a tensor) and $\mathbf{L} = \Theta\boldsymbol{\omega}$ (Θ is a scalar), we obtain the equation

$$\mathbf{L} = \widehat{\Theta} \cdot \boldsymbol{\omega} = \Theta\boldsymbol{\omega}, \tag{12.5}$$

which is an eigenvalue equation. In this equation, the scalar Θ and the related components $\omega_x, \omega_y, \omega_z$, i.e., the rotation axis, are unknown. The equation physically states that the angular momentum **L** and the rotation velocity $\boldsymbol{\omega}$ are parallel to each other. This is fulfilled for certain directions $\boldsymbol{\omega}$ that—as stated above—must be determined. All values Θ that satisfy the equation (12.5) are called *eigenvalues* of the tensor $\widehat{\Theta}$; the corresponding vectors $\boldsymbol{\omega} \neq 0$ are *eigenvectors*.

Equation (12.5) is a shortened notation for the system of equations

$$\begin{aligned}
\Theta_{xx}\omega_x + \Theta_{xy}\omega_y + \Theta_{xz}\omega_z &= \Theta\omega_x, \\
\Theta_{yx}\omega_x + \Theta_{yy}\omega_y + \Theta_{yz}\omega_z &= \Theta\omega_y, \\
\Theta_{zx}\omega_x + \Theta_{zy}\omega_y + \Theta_{zz}\omega_z &= \Theta\omega_z,
\end{aligned} \tag{12.6}$$

or

$$\begin{aligned}
(\Theta_{xx} - \Theta)\omega_x + \Theta_{xy}\omega_y + \Theta_{xz}\omega_z &= 0, \\
\Theta_{yx}\omega_x + (\Theta_{yy} - \Theta)\omega_y + \Theta_{yz}\omega_z &= 0, \\
\Theta_{zx}\omega_x + \Theta_{zy}\omega_y + (\Theta_{zz} - \Theta)\omega_z &= 0.
\end{aligned} \tag{12.7}$$

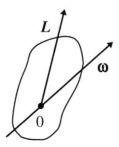

Figure 12.5. General case: The angular momentum **L** is not parallel to the rotation velocity ω.

This system of homogeneous linear equations has nontrivial solutions if its determinant of coefficients vanishes:

$$\begin{vmatrix} \Theta_{xx} - \Theta & \Theta_{xy} & \Theta_{xz} \\ \Theta_{yx} & \Theta_{yy} - \Theta & \Theta_{yz} \\ \Theta_{zx} & \Theta_{zy} & \Theta_{zz} - \Theta \end{vmatrix} = 0. \tag{12.8}$$

The expansion of the determinant leads to an equation of third order in Θ, the *characteristic equation*. Its three roots are the desired principal moments of inertia (eigenvalues) Θ_1, Θ_2, and Θ_3. By inserting Θ_i into the system of equations (12.5), one can calculate the ratio $\omega_x^{(i)} : \omega_y^{(i)} : \omega_z^{(i)}$ of the components of the vector $\boldsymbol{\omega}^{(i)}$. Thereby the orientation of the ith principal axis is determined.

Since one can find a tensor of inertia for *any possible* position of the body-fixed coordinate system, there exists also a system of principal axes at each point of the body. The orientations of these axes will however not coincide in general.

Existence and orthogonality of the principal axes

In principle, it would be possible for the cubic equation (12.8) to have two complex solutions. We therefore have to prove that a system of real orthogonal principal axes generally exists.

In order to apply a shortened summation notation, we number the coordinates ($x = 1$, $y = 2$, $z = 3$) and denote them by Latin letters. Greek letters are indices for the three different *eigenvalues*. We multiply the eigenvalue equation (12.5) for Θ_λ by the complex conjugated of $\omega_i^{(\mu)}$ and sum over i.

The equation for the component i reads

$$\sum_k \Theta_{ik} \omega_k^{(\lambda)} = L_i^{(\lambda)} = \Theta_\lambda \omega_i^{(\lambda)}. \tag{12.9}$$

This leads to

$$\sum_{i,k} \Theta_{ik} \omega_k^{(\lambda)} \omega_i^{(\mu)*} = \Theta_\lambda \sum_i \omega_i^{(\lambda)} \omega_j^{(\mu)*} = \Theta_\lambda \boldsymbol{\omega}^{(\lambda)} \cdot \boldsymbol{\omega}^{(\mu)*}. \tag{12.10}$$

In the same way, we form the complex conjugated of the equation corresponding to (12.9) for Θ_μ, multiply by $\omega_k^{(\lambda)}$, and sum over k:

$$\sum_i \Theta_{ki} \omega_i^{(\mu)} = \Theta_\mu \omega_k^{(\mu)}, \qquad \sum_i \Theta_{ki}^* \omega_i^{(\mu)*} = \Theta_\mu^* \omega_k^{(\mu)*}, \tag{12.11}$$

$$\sum_{i,k} \Theta_{ki}^* \omega_i^{(\mu)*} \omega_k^{(\lambda)} = \Theta_\mu^* \sum_k \omega_k^{(\mu)*} \omega_k^{(\lambda)} = \Theta_\mu^* \boldsymbol{\omega}^{(\mu)*} \cdot \boldsymbol{\omega}^{(\lambda)}. \tag{12.12}$$

Now we utilize the property of the tensor of inertia to be *real and symmetric*. We have $\Theta_{ik} = \Theta_{ki} = \Theta_{ki}^*$, and the left-hand sides of the equations (12.10) and (12.12) are equal to each other. We subtract equation (12.12) from equation (12.10):

$$(\Theta_\lambda - \Theta_\mu^*)\boldsymbol{\omega}^{(\lambda)} \cdot \boldsymbol{\omega}^{(\mu)*} = 0. \tag{12.13}$$

This equation allows two conclusions:

(1) Setting $\lambda = \mu$, then for the eigenvalues of

$$(\Theta_\lambda - \Theta_\lambda^*)\boldsymbol{\omega}^{(\lambda)} \cdot \boldsymbol{\omega}^{(\lambda)*} = 0 \tag{12.14}$$

follows the relation $\Theta_\lambda = \Theta_\lambda^*$, since the scalar product of two complex conjugated quantities is positively definite.

We thus proved that Θ_λ is real. Hence, any body always has three real principal moments of inertia and therefore also three real principal axes $\boldsymbol{\omega}^{(\lambda)}$. This is of course physically clear from the outset, since the principal moments of inertia are nothing else but the moments of inertia about the principal axes, and therefore they are always real.

(2) We now consider the case $\lambda \neq \mu$: Since all Θ_ν and therefore also all ω_ν are real, equation (12.13) reads

$$(\Theta_\lambda - \Theta_\mu)\boldsymbol{\omega}^{(\lambda)} \cdot \boldsymbol{\omega}^{(\mu)} = 0. \tag{12.15}$$

(a) If $\Theta_\lambda \neq \Theta_\mu$, then $\boldsymbol{\omega}^{(\lambda)} \cdot \boldsymbol{\omega}^{(\mu)} = 0$, and therefore, $\boldsymbol{\omega}^{(\lambda)}$ and $\boldsymbol{\omega}^{(\mu)}$ are *orthogonal*.

(b) If, e.g., $\Theta_1 = \Theta_2 = \Theta$, i.e., if two of the three eigenvalues are equal, then besides $\boldsymbol{\omega}^{(1)}$ and $\boldsymbol{\omega}^{(2)}$ all linear combinations of these two vectors are eigenvectors, too:

$$\widehat{\Theta} \cdot \boldsymbol{\omega}^{(1)} = \Theta \boldsymbol{\omega}^{(1)}, \qquad \widehat{\Theta} \cdot \boldsymbol{\omega}^{(2)} = \Theta \boldsymbol{\omega}^{(2)}$$
$$\Rightarrow \quad \widehat{\Theta} \cdot (\alpha \boldsymbol{\omega}^{(1)} + \beta \boldsymbol{\omega}^{(2)}) = \Theta(\alpha \boldsymbol{\omega}^{(1)} + \beta \boldsymbol{\omega}^{(2)}).$$

Thus, we can arbitrarily select two orthogonal vectors from the plane spanned this way and consider them as directions of principal axes. The third principal axis is by (12.15) fixed orthogonally to the two other axes. If two principal moments of inertia with respect to the center of gravity as rotation point are equal, the body is called a *symmetric top*.

(c) If all three moments of inertia are equal ($\Theta_1 = \Theta_2 = \Theta_3$), then any arbitrary orthogonal set of axes is a system of principal axes. If this holds with respect to the center of gravity, the body is called a *spherical top*.

If a body has rotational *symmetry* about one axis, then we are dealing with case (b), and the rotation axis is a principal axis. For other kinds of symmetries the symmetry axis also coincides with the principal axis.

Example 12.1: Tensor of inertia of a square covered with mass

We calculate the tensor of inertia and the principal axes of inertia of a square covered with mass for a corner of the square. We put the square in the x,y-plane of the coordinate system, as is shown in Figure 12.6. The components of the tensor of inertia are obtained with $z = 0$ by integration over the area:

$$\Theta_{xx} = \sigma \int_{y=0}^{a} \int_{x=0}^{a} y^2 \, dx \, dy = \frac{1}{3} M a^2,$$

$$\Theta_{yy} = \sigma \int_{y=0}^{a} \int_{x=0}^{a} x^2 \, dx \, dy = \frac{1}{3} M a^2,$$

$$\Theta_{zz} = \sigma \int_{y=0}^{a} \int_{x=0}^{a} (x^2 + y^2) \, dx \, dy = \frac{2}{3} M a^2.$$

Likewise,

$$\Theta_{xy} = \Theta_{yx} = -\sigma \int_{y=0}^{a} \int_{x=0}^{a} xy \, dx \, dy = -\frac{1}{4} M a^2.$$

The remaining deviation moments contain the factor z in the integrand and therefore vanish:

$$\Theta_{yz} = \Theta_{zy} = \Theta_{xz} = \Theta_{zx} = 0.$$

Thus, in the selected coordinate system the plate has the following tensor of inertia:

$$\widehat{\Theta} = \begin{pmatrix} \frac{1}{3} M a^2 & -\frac{1}{4} M a^2 & 0 \\ -\frac{1}{4} M a^2 & \frac{1}{3} M a^2 & 0 \\ 0 & 0 & \frac{2}{3} M a^2 \end{pmatrix}.$$

We now calculate the orientations of the principal axes.

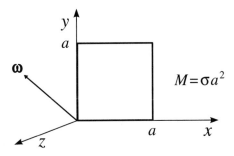

Figure 12.6. The angular velocity ω is arbitrary; however, it passes through the coordinate origin.

In accordance with the described approach, we first determine the eigenvalues of the tensor of inertia. We introduce the abbreviation $\Theta_0 = Ma^2$. Then we have the determinant

$$\begin{vmatrix} \frac{1}{3}\Theta_0 - \Theta & -\frac{1}{4}\Theta_0 & 0 \\ -\frac{1}{4}\Theta_0 & \frac{1}{3}\Theta_0 - \Theta & 0 \\ 0 & 0 & \frac{2}{3}\Theta_0 - \Theta \end{vmatrix} = 0$$

or

$$\left(\Theta^2 - \frac{2}{3}\Theta_0\Theta + \frac{7}{144}\Theta_0^2\right)\left(\frac{2}{3}\Theta_0 - \Theta\right) = 0.$$

The roots of this characteristic equation

$$\Theta_1 = \frac{1}{12}\Theta_0, \qquad \Theta_2 = \frac{7}{12}\Theta_0, \qquad \Theta_3 = \frac{2}{3}\Theta_0$$

are the principal moments of inertia with respect to the origin.

For the principal moment of inertia Θ_ν, the orientation of the axis $\boldsymbol{\omega}^{(\nu)}$ results from the eigenvalue equation $\widehat{\Theta}\boldsymbol{\omega}^{(\nu)} = \Theta_\nu \boldsymbol{\omega}^{(\nu)}$.

Written out for $\nu = 1$,

$$\begin{pmatrix} \frac{1}{3}\Theta_0 & -\frac{1}{4}\Theta_0 & 0 \\ -\frac{1}{4}\Theta_0 & \frac{1}{3}\Theta_0 & 0 \\ 0 & 0 & \frac{2}{3}\Theta_0 \end{pmatrix} \begin{pmatrix} \omega_x^{(1)} \\ \omega_y^{(1)} \\ \omega_z^{(1)} \end{pmatrix} = \frac{1}{12}\Theta_0 \begin{pmatrix} \omega_x^{(1)} \\ \omega_y^{(1)} \\ \omega_z^{(1)} \end{pmatrix}.$$

By multiplying out, we get a vector equation; after splitting into the three components, we obtain the three equations

$$\frac{1}{3}\Theta_0\omega_x^{(1)} - \frac{1}{4}\Theta_0\omega_y^{(1)} = \frac{1}{12}\Theta_0\omega_x^{(1)},$$

$$-\frac{1}{4}\Theta_0\omega_x^{(1)} + \frac{1}{3}\Theta_0\omega_y^{(1)} = \frac{1}{12}\Theta_0\omega_y^{(1)},$$

$$\frac{2}{3}\Theta_0\omega_z^{(1)} = \frac{1}{12}\Theta_0\omega_z^{(1)}.$$

From this, it follows that

$$\omega_y^{(1)} = \omega_x^{(1)}, \qquad \omega_z^{(1)} = 0,$$

and thus, the *orientation* of the first principal axis is

$$\mathbf{e}_1' = \frac{\boldsymbol{\omega}^{(1)}}{|\boldsymbol{\omega}^{(1)}|} = \frac{1}{\sqrt{2}}\begin{pmatrix} 1 \\ 1 \\ 0 \end{pmatrix}.$$

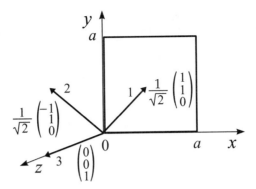

Figure 12.7. The principal axes for rotations about the point 0.

Analogously, we obtain for the two other directions

$$\mathbf{e}_2' = \frac{\boldsymbol{\omega}^{(2)}}{|\boldsymbol{\omega}^{(2)}|} = \frac{1}{\sqrt{2}}\begin{pmatrix}-1\\1\\0\end{pmatrix} \quad \text{and} \quad \mathbf{e}_3' = \frac{\boldsymbol{\omega}^{(3)}}{|\boldsymbol{\omega}^{(3)}|} = \begin{pmatrix}0\\0\\1\end{pmatrix}.$$

Evidently, the principal axes are orthogonal to each other, as is demanded by the general theory. For a rotation about the point 0 around one of the principal axes, the angular momentum **L** is parallel to **ω**, but in general the center of gravity then also moves. Such a motion can be forced only by the action of a force. Thus it is no free motion. Force-free rotations (shortly: free rotation) take place only about the center of gravity. The principal axes moments or principal moments of inertia about the center of gravity are the principal moments of inertia or principal axes of the body. In our example the orientations of the principal axes coincide with those at the point 0.

Transformation of the tensor of inertia

We investigate how the elements of the tensor $\widehat{\Theta}$ behave under a rotation of the coordinate system. The transformation of a vector under rotation of the coordinate system is described by (compare *Classical Mechanics: Point Particles and Relativity*, Chapter 30)

$$\mathbf{x}' = \widehat{A}\mathbf{x}$$

or

$$x_i' = \sum_j a_{ij} x_j, \tag{12.16}$$

or for the basis vectors

$$\mathbf{e}_i' = \sum_j a_{ij} \mathbf{e}_j, \tag{12.17}$$

TRANSFORMATION OF THE TENSOR OF INERTIA

where the components a_{ij} of the rotation matrix $\widehat{A}$ are the direction cosines between the rotated and the old axes. The inverse of this transformation reads

$$\mathbf{x} = \widehat{A}^{-1}\mathbf{x}' \quad \text{or} \quad x_i = \sum_j a_{ji} x_j'. \tag{12.18}$$

The inverse rotation matrix $(a^{-1})_{ij} = (a_{ji})$ is found by exchanging rows and columns (transposition), since the rotation is an orthogonal transformation which satisfies

$$\sum_j a_{ij} a_{kj} = \delta_{ik} \quad \text{or} \quad \sum_i a_{ij} a_{ik} = \delta_{jk}. \tag{12.19}$$

We require for the tensor of inertia that a vector equation of the form

$$L_k = \sum_l \Theta_{kl} \omega_l \tag{12.20}$$

exists also in the rotated system:

$$L_i' = \sum_j \Theta_{ij}' \omega_j'. \tag{12.21}$$

Thus, we can determine the transformation behavior of the tensor from the behavior of the vectors. The vectors $\mathbf{L}$ and $\boldsymbol{\omega}$ obey the transformation equation (12.18). If we replace L_k and ω_l in equation (12.20) by the primed quantities, we obtain

$$\sum_l \Theta_{kl} \left(\sum_j a_{jl} \omega_j' \right) = \sum_j a_{jk} L_j'.$$

Multiplication by a_{ik} and summation over k yields

$$\sum_j \left(\sum_{k,l} a_{ik} a_{jl} \Theta_{kl} \right) \omega_j' = \sum_j \left(\sum_k a_{jk} a_{ik} \right) L_j' = \sum_i \delta_{ij} L_j' = L_i'. \tag{12.22}$$

For the components of $\widehat{\Theta}'$ follows by comparison with (12.21)

$$\Theta_{ij}' = \sum_{k,l} a_{ik} a_{jl} \Theta_{kl}. \tag{12.23}$$

This transformation relation is the reason for denoting $\widehat{\Theta}$ as a "tensor." A *tensor of rank* m is generally defined as any quantity which under orthogonal transformations behaves according to the logical extension of equation (12.23) (summation over m indices), e.g., a tensor of third rank

$$A_{ijk}' = \sum_{i',j',k'} a_{ii'} a_{jj'} a_{kk'} A_{i'j'k'}. \tag{12.24}$$

$\widehat{\Theta}$ is a tensor of second rank; a vector can because of (12.16) be considered as tensor of first rank, a scalar accordingly as as tensor of rank 0. One can easily memorize the transformation law of a tensor: Each component of the tensor transforms as a vector (see (12.16)).

For the tensor of inertia, the equation (12.23) can be more clearly represented in matrix notation:

$$\widehat{\Theta}' = \widehat{A}\widehat{\Theta}\widehat{A}^{-1}. \tag{12.25}$$

This is a *similarity transformation*.

The matrices $\widehat{A}$ ($\widehat{A}^{-1}$) reduce to row vectors (column vectors) if we want to determine only the moment of inertia Θ'_{ii} about a given axis $\mathbf{e}_i'$ from the tensor of inertia Θ_{kl} in the coordinate system $\mathbf{e}_k$. According to (12.23), the desired moment of inertia Θ'_{ii} is

$$\Theta'_{ii} = \sum_{j,l} a_{ij} a_{il} \Theta_{jl}.$$

Now, according to (12.17), $\mathbf{e}_i' = \{a_{i1}, a_{i2}, a_{i3}\}$ is the vector $\mathbf{e}_i'$ in the basis $\mathbf{e}_j$. Hence, the moment of inertia about the rotation axis $\mathbf{e}_i' = \mathbf{n} = (n_1, n_2, n_3)$ can obviously be written as follows:

$$\Theta_\mathbf{n} = \sum_{j,l} a_{ij} \Theta_{jl} a_{il} = \sum_{j,l} a_{ij} \Theta_{jl} (a^T)_{li} = \sum_{j,l} n_j \Theta_{jl} n_l$$

$$= (n_1, n_2, n_3) \widehat{\Theta} \begin{pmatrix} n_1 \\ n_2 \\ n_3 \end{pmatrix} = \mathbf{n}^T \cdot \widehat{\Theta} \cdot \mathbf{n} \tag{12.26}$$

$$= \sum_{i,j} \Theta_{ij} n_i n_j.$$

This relation will be derived more clearly in the context of the subsequent equation (12.33). It allows one to calculate the moment of inertia about an arbitrary rotation axis $\mathbf{n}$ rather quickly.

Tensor of inertia in the system of principal axes

If the three orientations of the principal axes $\mathbf{e}_i' = \boldsymbol{\omega}^{(i)}$ are selected as coordinate axes, then

$$\mathbf{e}_i' = \omega_1^{(i)} \mathbf{e}_1 + \omega_2^{(i)} \mathbf{e}_2 + \omega_3^{(i)} \mathbf{e}_3 = \sum_j \omega_j^{(i)} \mathbf{e}_j.$$

A comparison with equation (12.17) shows that in this case

$$a_{ij} = \omega_j^{(i)}.$$

Hence, according to equation (12.23) the tensor of inertia in the system of principal axes reads

$$\Theta'_{ij} = \sum_{k,l} a_{ik} a_{jl} \Theta_{kl} = \sum_{k,l} \omega_k^{(i)} \omega_l^{(j)} \Theta_{kl} = \sum_k \omega_k^{(i)} \left(\sum_l \Theta_{kl} \omega_l^{(j)} \right). \qquad (12.27)$$

Since $\omega^{(i)}$ is an eigenvector of the matrix $\widehat{\Theta}$ with the eigenvalue Θ_j, according to (12.5), we have

$$\widehat{\Theta} \omega^{(j)} = \Theta_j \omega^{(j)} \qquad (12.28)$$

or, explicitly,

$$\sum_l \Theta_{kl} \omega_l^{(j)} = \Theta_j \omega_k^{(j)}.$$

Therefore, equation (12.27) turns into

$$\Theta'_{ij} = \sum_k \omega_k^{(i)} \Theta_j \omega_k^{(j)} = \Theta_j \sum_k \omega_k^{(i)} \omega_k^{(j)} = \Theta_j \omega^{(i)} \cdot \omega^{(j)}$$
$$= \Theta_j \delta_{ij}. \qquad (12.29)$$

We thus used the orthonormality (12.16) of the principal axes vectors $\omega^{(i)}$. The $\omega^{(i)}$ were assumed to be normalized, which is possible because of the linearity of the eigenvalue equation (12.28) with respect to ω. Equation (12.29) expresses the interesting and important fact that the tensor of inertia in its eigenrepresentation (i.e., in the coordinate system with the principal axes $\omega^{(i)}$ as coordinate axes) is diagonal and exactly of the form (12.2). This was to be expected, but it is satisfactory to see how everything fits together consistently.

Ellipsoid of inertia

We define a rotation axis by the unit vector **n** with the direction cosines $\mathbf{n} = (\cos \alpha, \cos \beta, \cos \gamma)$.

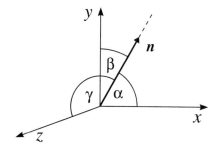

Figure 12.8. **n** characterizes the rotation axis.

According to (12.26), the moment of inertia Θ about this axis is

$$\Theta = \Theta_{\mathbf{n}} = (\cos\alpha, \cos\beta, \cos\gamma) \begin{pmatrix} \Theta_{xx} & \Theta_{xy} & \Theta_{xz} \\ \Theta_{xy} & \Theta_{yy} & \Theta_{yz} \\ \Theta_{xz} & \Theta_{yz} & \Theta_{zz} \end{pmatrix} \begin{pmatrix} \cos\alpha \\ \cos\beta \\ \cos\gamma \end{pmatrix}.$$

Multiplying out, we obtain

$$\Theta_{\mathbf{n}} = \Theta_{xx}\cos^2\alpha + \Theta_{yy}\cos^2\beta + \Theta_{zz}\cos^2\gamma$$
$$+ 2\Theta_{xy}\cos\alpha\cos\beta + 2\Theta_{xz}\cos\alpha\cos\gamma + 2\Theta_{yz}\cos\beta\cos\gamma. \tag{12.30}$$

By defining a vector $\varrho_{\mathbf{n}} = \mathbf{n}/\sqrt{\Theta_{\mathbf{n}}}$, we can rewrite the equation as

$$\Theta_{xx}\varrho_x^2 + \Theta_{yy}\varrho_y^2 + \Theta_{zz}\varrho_z^2 + 2\Theta_{xy}\varrho_x\varrho_y + 2\Theta_{xz}\varrho_x\varrho_z + 2\Theta_{yz}\varrho_y\varrho_z = 1. \tag{12.31}$$

This equation represents an ellipsoid in the coordinates $(\varrho_x, \varrho_y, \varrho_z)$, the so-called *ellipsoid of inertia*.

The distance ϱ from the center of rotation 0 along the direction $\mathbf{n}$ to the ellipsoid of inertia is $\varrho = 1/\sqrt{\Theta}$. This allows us to write down at once the moment of inertia if the ellipsoid of inertia is known. Each ellipsoid can now be brought to its normal form by a rotation of the coordinate system, i.e., the mixed terms can be made to vanish. We then obtain the form of the ellipsoid of inertia

$$\Theta_1\varrho_1^2 + \Theta_2\varrho_2^2 + \Theta_3\varrho_3^2 = 1. \tag{12.32}$$

This transformation of the ellipsoid of inertia obviously corresponds to the transformation of the tensor of inertia to principal axes. This becomes clear by comparing (12.31) and (12.32) with (12.1b) and (12.4a). The principal moments of inertia are given by the squares

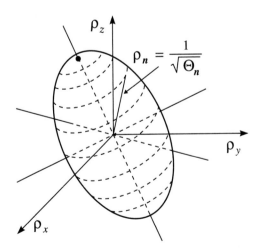

Figure 12.9. The ellipsoid of inertia.

ELLIPSOID OF INERTIA

of the reciprocal axis lengths of the ellipsoid. For two equal principal moments of inertia, the ellipsoid of inertia is a rotation ellipsoid, for three equal moments a sphere.

There is also a physical approach to the ellipsoid of inertia which will be presented now. Let $\mathbf{n} = \{\cos\alpha, \cos\beta, \cos\gamma\}$ be a unit vector pointing along the direction of the angular velocity $\boldsymbol{\omega}$, so that

$$\boldsymbol{\omega} = \omega\mathbf{n} = \omega\{\cos\alpha, \cos\beta, \cos\gamma\} = \omega\{n_1, n_2, n_3\} = \{\omega_1, \omega_2, \omega_3\}.$$

For the kinetic rotation energy, we then obtain according to (12.1a)

$$T_{\text{rot}} = \frac{1}{2}\sum_{ik}\Theta_{ik}\omega_i\omega_k$$

$$= \frac{1}{2}\omega^2(\Theta_{11}\cos^2\alpha + \Theta_{22}\cos^2\beta + \Theta_{33}\cos^2\gamma$$

$$+ 2\Theta_{12}\cos\alpha\cos\beta + 2\Theta_{13}\cos\alpha\cos\gamma + 2\Theta_{23}\cos\beta\cos\gamma)$$

$$= \frac{1}{2}\Theta_{\mathbf{n}}\omega^2.$$

$\Theta_{\mathbf{n}}$ denotes the moment of inertia about the axis $\mathbf{n}$. Hence, the moment of inertia about an axis with the orientation $\mathbf{n}$ is given by

$$\Theta_{\mathbf{n}} = \Theta_{11}\cos^2\alpha + \Theta_{22}\cos^2\beta + \Theta_{33}\cos^2\gamma$$
$$+ 2\Theta_{12}\cos\alpha\cos\beta + 2\Theta_{13}\cos\alpha\cos\gamma + 2\Theta_{23}\cos\beta\cos\gamma.$$

This agrees with the already known result (12.30). With the coordinates $\boldsymbol{\varrho} = \mathbf{n}/\sqrt{\Theta_{\mathbf{n}}} = (\varrho_1, \varrho_2, \varrho_3)$, we thus obtain the ellipsoid of inertia

$$\Theta_{11}\varrho_1^2 + \Theta_{22}\varrho_2^2 + \Theta_{33}\varrho_3^2 + 2\Theta_{12}\varrho_1\varrho_2 + 2\Theta_{13}\varrho_1\varrho_3 + 2\Theta_{23}\varrho_2\varrho_3 = 1. \quad (12.33)$$

The radius of the ellipsoid in the direction $\mathbf{n}$ is $\varrho_{\mathbf{n}} = 1/\sqrt{\Theta_{\mathbf{n}}}$.

Finally, there is still a third approach to the ellipsoid of inertia: According to Figure 12.10, the moment of inertia about the axis $\mathbf{n}$ is given by

$$\Theta_{\mathbf{n}} = \sum_{\nu} m_{\nu}d_{\nu}^2 = \sum_{\nu} m_{\nu}|\mathbf{r}_{\nu} \times \mathbf{n}|^2. \quad (12.34)$$

We check

$$\mathbf{r}_{\nu} \times \mathbf{n} = \begin{vmatrix} \mathbf{e}_1 & \mathbf{e}_2 & \mathbf{e}_3 \\ x_{\nu} & y_{\nu} & z_{\nu} \\ \cos\alpha & \cos\beta & \cos\gamma \end{vmatrix}$$

$$= (y_{\nu}\cos\gamma - z_{\nu}\cos\beta)\mathbf{e}_1 + (z_{\nu}\cos\alpha - x_{\nu}\cos\gamma)\mathbf{e}_2$$
$$+ (x_{\nu}\cos\beta - y_{\nu}\cos\alpha)\mathbf{e}_3$$

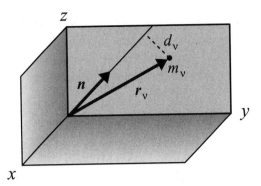

Figure 12.10. The rigid body rotates about the axis **n**: d_ν is the distance of the mass m_ν from the rotation axis.

and

$$\begin{aligned}
d_\nu^2 &= |\mathbf{r}_\nu \times \mathbf{n}|^2 \\
&= (y_\nu \cos\gamma - z_\nu \cos\beta)^2 + (z_\nu \cos\alpha - x_\nu \cos\gamma)^2 + (x_\nu \cos\beta - y_\nu \cos\alpha)^2 \\
&= (y_\nu^2 + z_\nu^2)\cos^2\alpha + (x_\nu^2 + z_\nu^2)\cos^2\beta + (x_\nu^2 + y_\nu^2)\cos^2\gamma \\
&\quad - 2x_\nu y_\nu \cos\alpha\cos\beta - 2x_\nu z_\nu \cos\alpha\cos\gamma - 2y_\nu z_\nu \cos\beta\cos\gamma.
\end{aligned} \quad (12.35)$$

Inserting this into equation (12.34) immediately yields

$$\Theta_\mathbf{n} = \sum_{i,j} \Theta_{ij} n_i n_j, \quad (12.36)$$

i.e., again the known ellipsoid of inertia.

One should realize at this point that the ellipsoid of inertia for a given tensor of inertia Θ_{ik} can immediately be written down and drawn according to equation (12.31). We use this method of evaluating moments of inertia in an arbitrary direction in Problem 12.4.

Example 12.2: Transformation of the tensor of inertia of a square covered with mass

The tensor of inertia of the square covered with mass in the x,y-plane was given by (compare Example 12.1)

$$\widehat{\Theta} = \begin{pmatrix} \frac{1}{3}\Theta_0 & -\frac{1}{4}\Theta_0 & 0 \\ -\frac{1}{4}\Theta_0 & \frac{1}{3}\Theta_0 & 0 \\ 0 & 0 & \frac{2}{3}\Theta_0 \end{pmatrix}.$$

ELLIPSOID OF INERTIA

The rotation of the coordinate system by $\varphi = \pi/4$ about the z-axis must bring $\widehat{\Theta}'$ to diagonal form, because the angle bisectors of the x,y-plane, as was shown (compare Problem 12.1), are principal axes. The corresponding rotation matrix reads

$$\widehat{A} = \begin{pmatrix} \cos\varphi & \sin\varphi & 0 \\ -\sin\varphi & \cos\varphi & 0 \\ 0 & 0 & 1 \end{pmatrix} = \begin{pmatrix} \frac{\sqrt{2}}{2} & \frac{\sqrt{2}}{2} & 0 \\ -\frac{\sqrt{2}}{2} & \frac{\sqrt{2}}{2} & 0 \\ 0 & 0 & 1 \end{pmatrix}.$$

Obviously,

$$\widehat{A}^{-1} = \widehat{A}^T.$$

Performing the matrix multiplication yields in accordance with the former result

$$\widehat{\Theta}' = \widehat{A}\widehat{\Theta}\widehat{A}^{-1} = \begin{pmatrix} \frac{1}{12}\Theta_0 & 0 & 0 \\ 0 & \frac{7}{12}\Theta_0 & 0 \\ 0 & 0 & \frac{2}{3}\Theta_0 \end{pmatrix}.$$

Example 12.3: Exercise: Rolling circular top

Find the kinetic energy of a homogeneous circular top (density ϱ, mass m, height h, vertex angle 2α),

(a) rolling on a plane, and

(b) whose base circle rolls on a plane while its longitudinal axis is parallel to the plane and the vertex is fixed at a point.

Solution For the calculation of the tensor of inertia, we choose the coordinate system so that the longitudinal axis coincides with the z-axis.

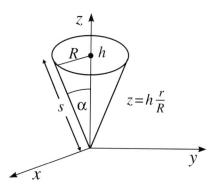

Figure 12.11. From the figure it is seen that $m = (1/3)\pi h R^2 \varrho$, $R = h\tan\alpha$, $s = R/\sin\alpha$.

Obviously,

$$\Theta_{xx} = \varrho \int_V (y^2 + x^2)\,dV = \varrho \int\int\int (r^2 \sin^2 \varphi + z^2) r\,dz\,dr\,d\varphi$$

$$= \varrho \int_0^{2\pi} d\varphi \int_0^R r\,dr \int_{h(r/R)}^h (r^2 \sin^2 \varphi + z^2)\,dz$$

$$= \varrho \frac{\pi}{20} h R^2 (R^2 + 4h^2),$$

$$\Theta_{xx} = \frac{3}{20} m h^2 (\tan^2 \alpha + 4). \tag{12.37}$$

For reasons of symmetry, we have

$$\Theta_{yy} = \Theta_{xx}.$$

Likewise,

$$\Theta_{zz} = \varrho \int_V (x^2 + y^2)\,dV = \varrho \int\int\int r^3\,dz\,dr\,d\varphi$$

$$= \varrho \int_0^{2\pi} d\varphi \int_0^R r^3\,dr \int_{h(r/R)}^h dz = \frac{\pi}{10} \varrho h R^4,$$

$$\Theta_{zz} = \frac{3}{10} m h^2 \tan^2 \alpha. \tag{12.38}$$

Since the integrals over φ of $xy = r^2 \cos\varphi \sin\varphi$, $xz = rz \cos\varphi$, $yz = rz \sin\varphi$ with the limits 0 and 2π vanish, follows $\Theta_{xy} = \Theta_{xz} = \Theta_{yz} = 0$. The adopted system is a system of principal axes. We therefore set $\Theta_1 = \Theta_{xx} = \Theta_2, \Theta_3 = \Theta_{zz}$.

(a) The kinetic energy in the representation of principal axes reads

$$T = \frac{1}{2}\Theta_1 \omega_1^2 + \frac{1}{2}\Theta_2 \omega_2^2 + \frac{1}{2}\Theta_3 \omega_3^2. \tag{12.39}$$

Since we already know the principal axes of inertia and moments of inertia, it remains only to express the motion of the cone by the corresponding angular velocities. The momentary rotation of the cone

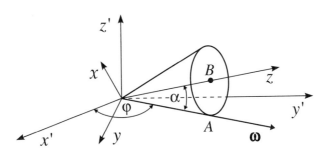

Figure 12.12. Rolling cone.

ELLIPSOID OF INERTIA

happens with the angular velocity $\boldsymbol{\omega}$ about a line of support. We can express $\boldsymbol{\omega}$ by $\dot{\varphi}$ by considering the velocity of the point A. On the one hand $v_A = \dot{\varphi} h \cos \alpha$, and on the other hand $v_A = \omega \cdot R \cos \alpha$. From this, we find

$$\omega = \dot{\varphi} \frac{h}{R}. \tag{12.40}$$

φ is the polar angle of the figure axis (or, equivalently, the tangential line) in the x',y'-plane; $\dot{\varphi}$ is the corresponding angular velocity.

A decomposition of $\boldsymbol{\omega}$ in the system of principal axes, where $\boldsymbol{\omega}$ lies in the x,z-plane, leads to $\omega_2 = 0$ and

$$\omega_3 = \omega \cos \alpha \quad \text{and} \quad \omega_1 = \sin \alpha. \tag{12.41}$$

For the kinetic energy, we thus obtain from (12.39)

$$\begin{aligned} T &= \frac{1}{2}\Theta_1 \omega_1^2 + \frac{1}{2}\Theta_2 \omega_2^2 + \frac{1}{2}\Theta_3 \omega_3^2 \\ &= \frac{1}{2}\frac{3}{20}mh^2(\tan^2 \alpha + 4)\omega_1^2 + \frac{1}{2}\frac{3}{10}mh^2 \tan^2 \alpha_3^2 \\ &= \frac{3}{40}mh^2\omega^2 \sin^2 \alpha(\tan^2 \alpha + 4) + \frac{3}{20}mh^2\omega^2 \sin^2 \alpha \\ &= \frac{3}{40}mh^2\omega^2 \left(\frac{\sin^4 \alpha}{\cos^2 \alpha} + 6 \sin^2 \alpha \right). \end{aligned} \tag{12.42}$$

If we replace ω by (12.40) and employ $R/h = \tan \alpha = \sin \alpha / \cos \alpha$, then

$$T = \frac{3}{40}mh^2\dot{\varphi}^2 \frac{h^2}{R^2} \left(\frac{R^2}{h^2} \sin^2 \alpha + 6 \cos^2 \alpha \frac{R^2}{h^2} \right) = \frac{3}{40}mh^2\dot{\varphi}^2 (1 + 5 \cos^2 \alpha). \tag{12.43}$$

(b) The momentary rotation axis $\boldsymbol{\omega}$ is again the connecting line between the fixed vertex and the point of support. The relation between ω and $\dot{\varphi}$ is likewise obtained by considering the velocity of point A.

We have $v_A = h \cdot \dot{\varphi} = \omega R \cos \alpha$, from which it follows that $\omega = \dot{\varphi} / \sin \alpha$. The projection of $\boldsymbol{\omega}$ onto the principal axes yields

$$\omega_1 = \omega \sin \alpha = \dot{\varphi},$$

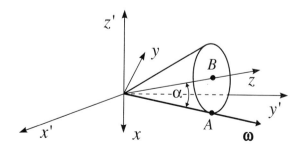

Figure 12.13. x', y', z' labels the laboratory system, x, y, z the system of principal axes.

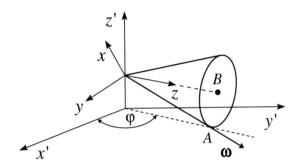

Figure 12.14. Cone rolling on the edge of its base.

$$\omega_2 = 0,$$
$$\omega_3 = \omega \cos\alpha = \dot\varphi \frac{h}{R}. \tag{12.44}$$

Hence, for the kinetic energy, it follows from (12.39) that

$$T = \frac{1}{2}\frac{3}{20}mh^2(\tan^2\alpha + 4)\omega_1^2 + \frac{1}{2}\frac{3}{10}mh^2\tan^2\alpha\,\omega_3^2$$
$$= \frac{3}{40}mh^2\dot\varphi^2\left(\frac{R^2}{h^2} + 4 + 2\frac{R^2}{h^2}\cdot\frac{h^2}{R^2}\right) = \frac{3}{40}mh^2\dot\varphi^2\left(6 + \frac{R^2}{h^2}\right). \tag{12.45}$$

Example 12.4:   Exercise: Ellipsoid of inertia of a quadratic disk

Determine the ellipsoid of inertia for the rotation of a quadratic disk about the origin, as described in problem 12.1. Find the moments of inertia of the disk for rotation about (a) the x-axis, (b) the y-axis, (c) the z-axis, (d) the three principal axes, and (e) the axis $\{\cos 45°, \cos 45°, \cos 45°\}$.

Solution   The ellipsoid of inertia reads

$$\frac{\Theta_0}{3}\varrho_x^2 - \frac{\Theta_0}{2}\varrho_x\varrho_y + \frac{\Theta_0}{3}\varrho_y^2 + \frac{2\Theta_0}{3}\varrho_z^2 = 1. \tag{12.46}$$

(a) For rotation about the x-axis $\mathbf{n} = \{1, 0, 0\}$, and thus $\varrho = \{1/\sqrt{\Theta_x}, 0, 0\}$. Insertion into (12.46) yields

$$\frac{\Theta_0}{3}\cdot\frac{1}{\Theta_x} = 1 \;\Rightarrow\; \Theta_x = \frac{\Theta_0}{3},$$

as expected.

(b) Here, $\mathbf{n} = \{0, 1, 0\}$, and following the procedure in (a), we find

$$\Theta_y = \frac{\Theta_0}{3}.$$

(c) Here, $\mathbf{n} = \{0, 0, 1\}$, and following the procedure in (a), we find

$$\Theta_z = \frac{2}{3}\Theta_0.$$

(d) The third principal axis is identical with the z-axis, which corresponds to (c). The first two principal axes are given by

$$\mathbf{n}_1 = \left\{\frac{1}{\sqrt{2}}, \frac{1}{\sqrt{2}}, 0\right\} \quad \text{and} \quad \mathbf{n}_2 = \left\{-\frac{1}{\sqrt{2}}, \frac{1}{\sqrt{2}}, 0\right\},$$

respectively. Therefore,

$$\varrho_1 = \frac{\mathbf{n}}{\sqrt{\Theta_1}} = \left\{\frac{1}{\sqrt{2\Theta_1}}, \frac{1}{\sqrt{2\Theta_1}}, 0\right\}$$

and

$$\varrho_2 = \frac{\mathbf{n}_2}{\sqrt{\Theta_2}} = \left\{-\frac{1}{\sqrt{2\Theta_2}}, \frac{1}{\sqrt{2\Theta_2}}, 0\right\}.$$

Insertion into (12.46) yields

$$\frac{\Theta_0}{3}\frac{1}{2\Theta_1} - \frac{\Theta_0}{2}\frac{1}{2\Theta_1} + \frac{\Theta_0}{3}\frac{1}{2\Theta_1} + 0 = 1 \quad \Rightarrow \quad \Theta_1 = \frac{\Theta_0}{12},$$

and

$$\frac{\Theta_0}{3}\frac{1}{2\Theta_2} + \frac{\Theta_0}{2}\frac{1}{2\Theta_2} + \frac{\Theta_0}{3}\frac{1}{2\Theta_2} + 0 = 1 \quad \Rightarrow \quad \Theta_2 = \frac{7}{12}\Theta_0.$$

These are the principal moments of inertia, as was expected.

(e) In this case, $\mathbf{n}$ is proportional to $\{\cos 45°, \cos 45°, \cos 45°\}$. Thus,

$$\mathbf{n} = \left\{\frac{1}{\sqrt{3}}, \frac{1}{\sqrt{3}}, \frac{1}{\sqrt{3}}\right\},$$

and therefore,

$$\varrho = \frac{\mathbf{n}}{\sqrt{\Theta}} = \left\{\frac{1}{\sqrt{3\Theta}}, \frac{1}{\sqrt{3\Theta}}, \frac{1}{\sqrt{3\Theta}}\right\}.$$

Insertion into (12.46) yields

$$\frac{\Theta_0}{3}\frac{1}{3\Theta} - \frac{\Theta_0}{2}\frac{1}{3\Theta} + \frac{\Theta_0}{3}\frac{1}{3\Theta} + \frac{2}{3}\Theta_0\frac{1}{3\Theta} = 1,$$

from which we find

$$\Theta = \frac{10}{36}\Theta_0.$$

This problem demonstrates the simple handling and the usefulness of the ellipsoid of inertia.

Example 12.5: Exercise: Symmetry axis as a principal axis

Demonstrate that an n-fold rotational symmetry axis is a principal axis of inertia, and that in the case $n \geq 3$, the two other principal axes can be freely chosen in the plane perpendicular to the first axis.

Solution If a body has an n-fold symmetry axis, then the tensor of inertia must be equal in two coordinate systems rotated from each other by $\varphi = 2\pi/n$:

$$\widehat{\Theta} = \widehat{\Theta}' = \widehat{A}\,\widehat{\Theta}\,\widehat{A}^{-1}.$$

If we select the z-axis as a rotation axis, the rotation matrix reads

$$\hat{A} = \begin{pmatrix} \cos\varphi & \sin\varphi & 0 \\ -\sin\varphi & \cos\varphi & 0 \\ 0 & 0 & 1 \end{pmatrix}.$$

Multiplying the matrices out, one obtains the components Θ'_{ij} of the new tensor of inertia which shall coincide with Θ_{ij}.

$$\Theta'_{11} = \Theta_{11} = \Theta_{11}\cos^2\varphi + \Theta_{22}\sin^2\varphi + 2\Theta_{12}\sin\varphi\cos\varphi,$$
$$\Theta'_{22} = \Theta_{22} = \Theta_{11}\sin^2\varphi + \Theta_{22}\cos^2\varphi - 2\Theta_{12}\sin\varphi\cos\varphi,$$
$$\Theta'_{12} = \Theta_{12} = -\Theta_{11}\cos\varphi\sin\varphi + \Theta_{22}\cos\varphi\sin\varphi + \Theta_{12}(1 - 2\sin^2\varphi),$$
$$\Theta'_{13} = \Theta_{13} = +\Theta_{13}\cos\varphi + \Theta_{23}\sin\varphi,$$
$$\Theta'_{23} = \Theta_{23} = -\Theta_{13}\sin\varphi + \Theta_{23}\cos\varphi.$$

The determinant of the system of the last two equations,

$$\begin{vmatrix} \cos\varphi - 1 & \sin\varphi \\ -\sin\varphi & \cos\varphi - 1 \end{vmatrix} = 2(1 - \cos\varphi),$$

vanishes only for $\varphi = 0, 2\pi, \ldots$. If there is symmetry ($n \geq 2$), then we must have $\Theta_{13} = \Theta_{23} = 0$, i.e., the z-axis must be a principal axis.

Two of the remaining three equations are identical, and there remains the system of equations

$$(\Theta_{22} - \Theta_{11})\sin^2\varphi + 2\Theta_{12}\sin\varphi\cos\varphi = 0,$$
$$(\Theta_{22} - \Theta_{11})\cos\varphi\sin\varphi - 2\Theta_{12}\sin^2\varphi = 0.$$

The determinant of coefficients has the value

$$D = -2\sin^4\varphi - 2\sin^2\varphi\cos^2\varphi = -2\sin^2\varphi.$$

There holds $D = 0$ for $\varphi = 0, \pi, 2\pi, \ldots$. Hence, $\Theta_{11} = \Theta_{22}$ and $\Theta_{12} = 0$, if $n > 2$. If the axis of rotational symmetry z is at least 3-fold, the tensor of inertia is diagonal for each orthogonal pair of axes in the x,y-plane.

Example 12.6: Exercise: Tensor of inertia and ellipsoid of inertia of a system of three masses

A rigid body consists of three mass points that are connected to the z-axis by rigid massless bars (see Figure 12.15).

(a) Find the elements of the tensor of inertia relative to the x,y,z-system.

(b) Calculate the ellipsoid of inertia with respect to the origin 0, and the moment of inertia of the entire body with respect to the axis $0a$.

Solution   (a) The elements of the tensor of inertia relative to the x,y,z-system are

$$\Theta_{xx} = \sum_i m_i(y_i^2 + z_i^2)$$

$$= m_1(y_1^2 + z_1^2) + m_2(y_2^2 + z_2^2) + m_3(y_3^2 + z_3^3),$$

and after inserting the numerical values from Figure 12.15, one has

$$\Theta_{xx} = 100(144 + 25) + 200(64 + 225) + 150(144 + 196)$$
$$= 125.7 (\text{kg cm}^2).$$

Likewise, one obtains

$$\Theta_{yy} = 117.5 (\text{kg cm}^2) \quad \text{and} \quad \Theta_{zz} = 104.75 (\text{kg cm}^2).$$

For the deviation moments of the tensor of inertia, it follows that

$$\Theta_{xy} = -\sum_i m_i(x_i y_i)$$
$$= 100(12 \cdot 10) - 200(10 \cdot 8) + 150(11 \cdot 14) = 19.1 (\text{kg cm}^2),$$

and likewise,

$$\Theta_{xz} = -44.8 (\text{kg cm}^2) \quad \text{and} \quad \Theta_{yz} = 4.800 (\text{kg cm}^2).$$

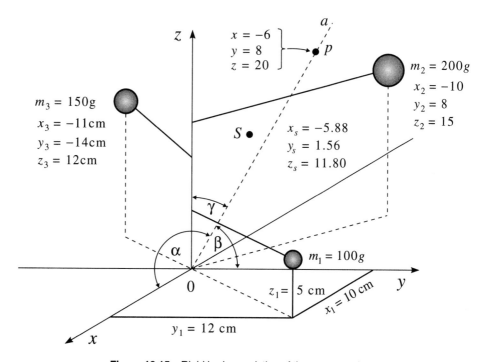

Figure 12.15. Rigid body consisting of three mass points.

(b) From (a) one now immediately obtains for the ellipsoid of inertia with respect to the origin 0 (see equation (12.30))

$$\Theta = \Theta_{xx} \cos^2 \alpha + \Theta_{yy} \cos^2 \beta + \Theta_{zz} \cos^2 \gamma$$
$$+ 2\Theta_{xy} \cos \alpha \cos \beta + 2\Theta_{xz} \cos \alpha \cos \gamma + 2\Theta_{yz} \cos \beta \cos \gamma. \quad (12.47)$$

To calculate the moment of inertia Θ_{0a}, we evaluate the direction cosines with the coordinates given in Figure 12.15,

$$\cos \alpha = \frac{-6}{\sqrt{6^2 + 8^2 + 20^2}} = -0.268,$$

$$\cos \beta = \frac{8}{\sqrt{6^2 + 8^2 + 20^2}} = 0.358,$$

and

$$\cos \gamma = \frac{20}{\sqrt{6^2 + 8^2 + 20^2}} = 0.895.$$

By inserting into equation (12.47) for the moment of inertia, we obtain

$$\Theta_{0a} = (0.268)^2 \cdot 125.7 + (0.358)^2 \cdot 117.25 + (0.895)^2 \cdot 104.75$$
$$- 2(0.268)(0.358) \cdot 19.1 + 2(0.268)(0.895) \cdot 44.8$$
$$- 2(0.358)(0.895) \cdot 4.800$$
$$= 128.87 (\text{kg cm}^2).$$

Example 12.7: Exercise: Friction forces and acceleration of a car

A car of mass M is driven by a motor that performs the torque $2D$ on the wheel axis. The radius of the wheels is R, and their moment of inertia is $\Theta = mR^2$ (m is the reduced mass of the wheels).

(a) Determine the friction force $\mathbf{f}$ which acts on each wheel and causes the acceleration of the car. The street is assumed to be planar.

(b) Calculate the acceleration of the car if the torque $2D = 10^3$ J, $M = 2 \cdot 10^3$ kg, $R = 0.5$ m and $m = 12.5$ kg.

Solution (a) Figure 12.16 shows one of the wheels and the force $\mathbf{f}$ acting on it. Since the linear acceleration of the wheel center is the same as that of the center of gravity of the car $\mathbf{a}_s$, one has

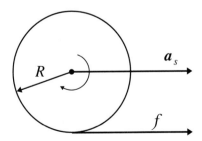

Figure 12.16. Wheel, acceleration $\mathbf{a}_s$, and friction force $\mathbf{f}$.

$$Ma_s = 2f - F. \tag{12.48}$$

The factor 2 accounts for the fact that a car in general is driven by two wheels. **F** is a possible external force which impedes the motion (air resistance), and $\mathbf{a}_s$ is the acceleration of the car. For the torque relative to the axis, one obtains

$$4\Theta\dot{\omega} = 2(D - fR), \tag{12.49}$$

where Θ is the moment of inertia of each of the *four* wheels, D is the accelerating torque, and $-fR$ is the torque performed by the friction force on each wheel. The moment of inertia is $\Theta = mR^2$. If the car does not glide, one has

$$\dot{\omega}R = a_s, \tag{12.50}$$

and with equation (12.49) and equation (12.48), it immediately follows that

$$\dot{\omega}R = a_s = \frac{1}{2}\left(\frac{DR - fR^2}{mR^2}\right) = \frac{2f - F}{M}. \tag{12.51}$$

Equation (12.51) finally yields for the friction force

$$f = \frac{1}{2}\frac{(2D/R)M + 4mF}{M + 4m} \tag{12.52}$$

and by neglecting the backdriving force F $(F = 0)$, we have

$$f = \frac{D/R}{1 + (4m/M)}. \tag{12.53}$$

(b) By replacing f in equation (12.49) by equation (12.53) and solving for a_s $(F = 0)$, one finds the acceleration of the car

$$a_s = \frac{2D/R}{M + 4m} = \frac{10^3/0.5}{2 \cdot 10^3 + 4 \cdot 12.4} = \frac{10^3}{1025} \approx 1\frac{\text{m}}{\text{s}^2}.$$

With the numerical values from (b), the friction force f is given by

$$f \approx \frac{D}{R} = 1000\,\text{N}.$$

13 Theory of the Top

The free top

A rigid, rotating body is called a top. A top is called symmetric if two of its principal moments of inertia are equal. If $\Theta_1 = \Theta_2$, we further distinguish

(a) $\Theta_3 > \Theta_1$ oblate top or flattened top, e.g., a disk;

(b) $\Theta_3 < \Theta_1$ prolate top or cigar top, e.g., an (extended) cylinder; and

(c) $\Theta_3 = \Theta_1$ spherical top, e.g., a cube.

The third principal axis of inertia which is related to Θ_3 is called the *figure axis*. It specifies the spatial orientation of the top. For rotationally symmetric bodies it coincides with their symmetry axis. Hence, the center of gravity of a rotational body always lies on the figure axis. Moreover, we must distinguish between the *free top* and the *heavy top*. For the free top one assumes that no external forces act on the body, so that the torque with respect to the fixed point vanishes. On the heavy top forces act, for example gravity. One can however imagine other forces (centrifugal forces, friction forces, etc.). For an experimental realization of a free top we only have to support an arbitrary body at the center of gravity. The body is then in an indifferent equilibrium, and there is no torque acting on it.

For a theoretical description of the top, we start from the basic equations

$$\mathbf{L} = \widehat{\Theta} \cdot \boldsymbol{\omega} = \mathbf{constant} \quad \text{(conservation of angular momentum)}, \tag{13.1}$$

$$T = \frac{1}{2} \boldsymbol{\omega} \cdot \mathbf{L} = \text{constant} \quad \text{(conservation of kinetic energy)}. \tag{13.2}$$

The angular momentum $\mathbf{L}$ and the kinetic energy T of the free top are constant in time. This is the content of the last two equations.

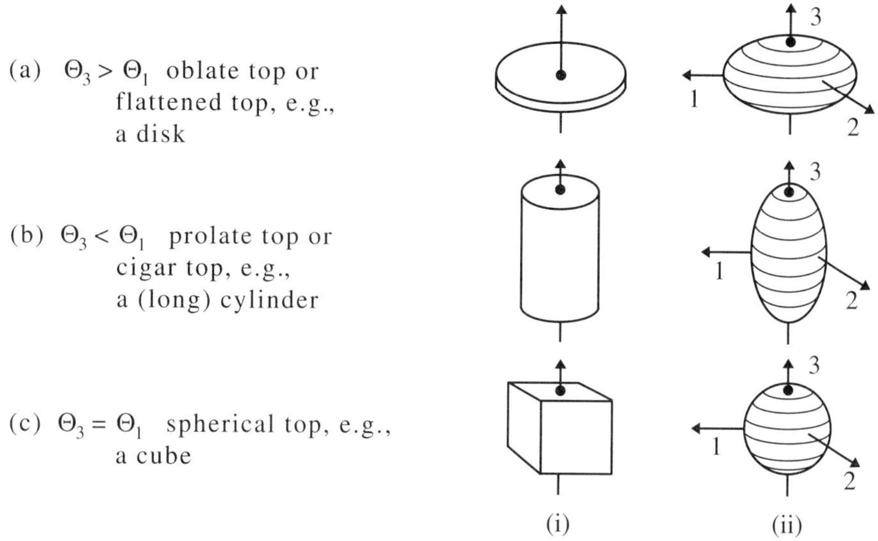

(a) $\Theta_3 > \Theta_1$ oblate top or flattened top, e.g., a disk

(b) $\Theta_3 < \Theta_1$ prolate top or cigar top, e.g., a (long) cylinder

(c) $\Theta_3 = \Theta_1$ spherical top, e.g., a cube

(i)   (ii)

Figure 13.1. (i) Possible real form of the top. (ii) Ellipsoid of inertia.

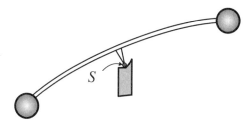

Figure 13.2. Model of a free top supported at the center of gravity S: The construction is so that S is also the supporting point.

Geometrical theory of the top

We first will derive the laws governing the free top from geometrical considerations. The geometrical theory of the top is based on *Poinsot's*[1] *ellipsoid* (also called the *energy ellipsoid*):

$$\Theta_{xx}\omega_x^2 + \Theta_{yy}\omega_y^2 + \Theta_{zz}\omega_z^2 + 2\Theta_{xy}\omega_x\omega_y + 2\Theta_{xz}\omega_x\omega_z + 2\Theta_{yz}\omega_y\omega_z$$
$$= 2T = \text{constant}. \quad \textbf{(13.3)}$$

---

[1] *Louis Poinsot*, French mathematician and physicist, b. Jan. 3, 1777, Paris–d. Dec. 5, 1859, Paris. Professor in Paris, introduced in his *Eléments de statique* (Paris, 1804) the concept of the couple to mechanics and used it to represent the motion of the top. Poinsot-motion means the motion of a free top.

This ellipsoid in the ω-space is immediately obtained from (13.2). It is similar to the ordinary ellipsoid of inertia and has the same body-fixed axes.

In the subsequent considerations, we shall utilize the property of (13.3) that the endpoint of the vector ω lies just on the surface of the ellipsoid.

Now follows Poinsot's construction of the motion of the free top. The angular momentum vector is constant and defines an orientation in space. The straight line determined by **L** is therefore called the *invariable straight line*. Moreover, the kinetic energy is constant, hence $2T = \omega \cdot \mathbf{L} =$ constant; from the definition of the scalar product immediately it follows that

$$\omega \cdot \cos(\omega, \mathbf{L}) = \text{constant}. \tag{13.4}$$

In other words, the projection of ω onto **L** is constant. If one now considers ω as the position vector for points in space, the parameter representation $\omega(t)$ fixes a plane which is called an *invariable plane*. The invariable straight line is then perpendicular to the invariable plane.

Now one can describe the motion of the top by the rolling of the Poinsot ellipsoid on the invariable plane. This is allowed since the endpoint of ω, as is evident from equation (13.4), lies on the surface of the ellipsoid and moves in the invariable plane. The invariable plane is also a tangent plane of Poinsot's ellipsoid, since there is only one common vector ω, and hence the ellipsoid and the plane have a common point. To prove this, we show that at the point ω the gradient of the ellipsoid is parallel to **L**. From vector analysis we know that the gradient of a surface is perpendicular to this plane. The surface of the ellipsoid F is described by (13.3).

Because[2]

$$\nabla_\omega F = \left(\frac{\partial F}{\partial \omega_x}, \frac{\partial F}{\partial \omega_y}, \frac{\partial F}{\partial \omega_z} \right),$$

we obtain

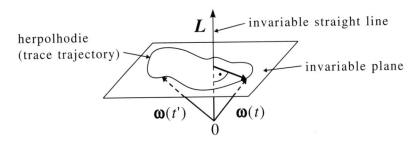

Figure 13.3. Invariable straight line and invariable plane.

---

[2]Since the surface (13.3) is defined in the ω-space, we mean by gradient the ω-gradient, i.e., $\nabla_\omega = \{\partial/\partial\omega_x, \partial/\partial\omega_y, \partial/\partial\omega_z\}$.

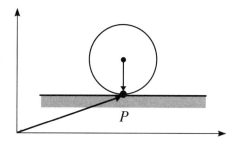

Figure 13.4. On the condition of rolling.

$$\frac{1}{2}\nabla_\omega F = \begin{pmatrix} \Theta_{xx}\omega_x + \Theta_{xy}\omega_y + \Theta_{xz}\omega_z \\ \Theta_{xy}\omega_x + \Theta_{yy}\omega_y + \Theta_{yz}\omega_z \\ \Theta_{xz}\omega_x + \Theta_{yz}\omega_y + \Theta_{zz}\omega_z \end{pmatrix} = \widehat{\Theta}\omega = \mathbf{L},$$

i.e., $\mathrm{grad}_\omega F$ is parallel to $\mathbf{L}$ or $F \perp \mathbf{L}$; therefore, the tangent plane of F at the point ω is parallel to the invariable plane.

Since the center of the ellipsoid is a constant distance from the invariable plane (see equation (13.4)), the motion of the top can be described as follows: The body-fixed Poinsot ellipsoid rolls without gliding on the invariable plane, where the center of the ellipsoid is fixed. The instantaneous value of the angular velocity is then given by the distance from the center to the contact point of the ellipsoid.

The ellipsoid rolls but does not glide. This follows from the fact that all points along the ω-axis are momentarily at rest; hence the contact point is also momentarily at rest. Rolling without gliding means that the changes of the rotation vector ω measured from the laboratory system and from the body-fixed system are equal. Actually,

$$\left.\frac{d\omega}{dt}\right|_L = \left.\frac{d\omega}{dt}\right|_K + \omega \times \omega, \quad \text{i.e.,} \quad \left.\frac{d\omega}{dt}\right|_L = \left.\frac{d\omega}{dt}\right|_K.$$

Concerning the difference between gliding and rolling, if a wheel rolls on a plane, the velocities of change of the contact point P in the body-fixed and in the laboratory system are equal. If the wheel glides, the contact point in the body-fixed system is fixed; in the laboratory system its position changes permanently.

The trajectory of ω on the invariable plane is called the *herpolhodie* or *trace trajectory*, the corresponding curve on the ellipsoid is called the *polhodie* or *pole trajectory*. See Figure 13.5.

The polhodie and the herpolhodie are in general complicated, not closed curves. For the special case of a symmetric top, the Poinsot ellipsoid turns into a rotation ellipsoid, and by rolling of the rotation ellipsoid there arise circles. ω has constant magnitude but permanently changes the direction, i.e., ω rotates on a cone about the angular momentum axis. This cone is called the *herpolhodie-* or *trace cone*. For a symmetric cone it is efficient to use the symmetry axis (figure axis) as third axis for describing the motion. The figure axis that is tightly fixed to the ellipsoid rotates just like ω rotates about $\mathbf{L}$. The cone resulting

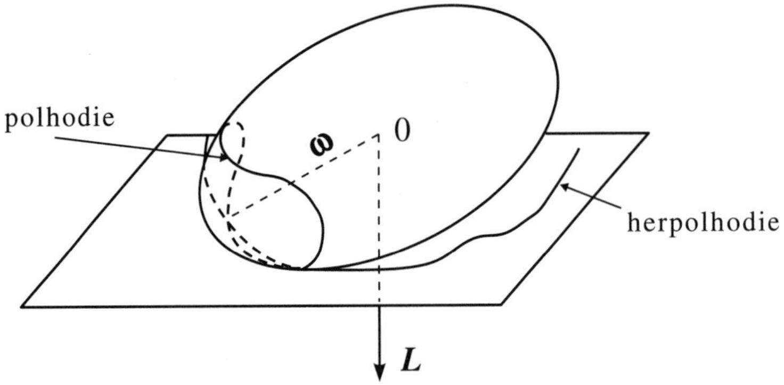

Figure 13.5. The Poinsot ellipsoid rolls on the invariable plane.

this way is called the *nutation cone*. The motion of the figure axis of the top in space is called *nutation*. (The term *precession* used in the American literature makes little sense, since the term means a motion of the heavy top that is of a completely different origin.)

An observer who is in the system of the top and considers the figure axis as fixed will find that $\boldsymbol{\omega}$ and $\mathbf{L}$ rotate about this axis. For the cone arising by the rotation of $\boldsymbol{\omega}$ the term *polhodie-* or *pole cone* is introduced. The precise orientation of the axes and cones depends essentially on the shape of the rotation ellipsoid. This is shown by the following two diagrams on the orientation of the axes. Note that a large principal momentum of inertia Θ_3 corresponds to a small radius of the Poinsot ellipsoid, namely, $\sqrt{2T/\Theta_3}$. The other axes of the Poinsot ellipsoid accordingly have the lengths $\sqrt{2T/\Theta_1}$ and $\sqrt{2T/\Theta_2}$, respectively.

This is immediately seen from the form of (13.3) in terms of the principal axes:

$$\frac{\omega_1^2}{1/\Theta_1} + \frac{\omega_2^2}{1/\Theta_2} + \frac{\omega_3^2}{1/\Theta_3} = 2T = \text{constant}.$$

Figure 13.6(a) shows the ellipsoid of a flattened (oblate) top; Figure 13.6(b) represents a prolate top. In the first case, the axes have the sequence $\boldsymbol{\omega} - \mathbf{L} -$ figure axis; in the second case, the sequence is $\mathbf{L} - \boldsymbol{\omega} -$ figure axis.

Likewise is the sequence of the cones introduced above. Figure 13.6(c) shows the case of an oblate top, and Figure 13.6(d) that of a prolate top. We note that the three axes lie in a plane.

Analytical theory of the free top

We consider the motion of the angular momentum and angular velocity vectors from a coordinate system that is tightly fixed to the top and moves with it. For the angular velocity, we have

$$\boldsymbol{\omega} = \omega_1 \mathbf{e}_1 + \omega_2 \mathbf{e}_2 + \omega_3 \mathbf{e}_3,$$

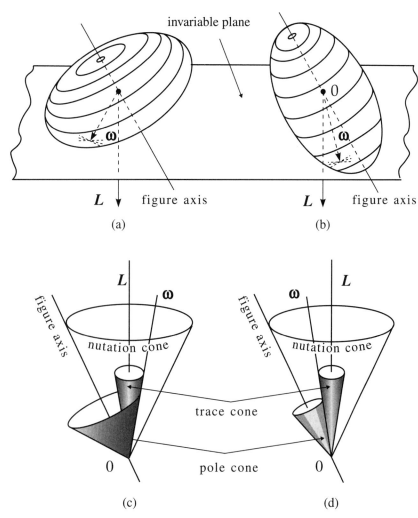

Figure 13.6. (a) Oblate symmetric top. (b) Prolate symmetric top. (c) Oblate symmetric top: The pole cone rolls inside of the trace cone. (d) Prolate symmetric top: The pole cone rolls outside of the trace cone.

where $\mathbf{e}_1$, $\mathbf{e}_2$, and $\mathbf{e}_3$ are *body-fixed principal axes* of the top. We now investigate the angular momentum of the top no longer in the moving coordinate system, i.e., in the system of the top that rotates with $\boldsymbol{\omega}$ in the laboratory system, but transformed into the laboratory system, using our knowledge of moving coordinate systems. We then obtain

$$\dot{\mathbf{L}}\big|_{\text{lab}} = \dot{\mathbf{L}}\big|_{\text{top}} + \boldsymbol{\omega} \times \mathbf{L}.$$

Because $\dot{\mathbf{L}}\big|_{\text{top}} = \widehat{\Theta}\dot{\boldsymbol{\omega}}$, for the component in the laboratory system we have

$$\dot{\mathbf{L}}\big|_{\text{lab}} = \Theta_1\dot{\omega}_1\mathbf{e}_1 + \Theta_2\dot{\omega}_2\mathbf{e}_2 + \Theta_3\dot{\omega}_3\mathbf{e}_3 + \begin{vmatrix} \mathbf{e}_1 & \mathbf{e}_2 & \mathbf{e}_3 \\ \omega_1 & \omega_2 & \omega_3 \\ \Theta_1\omega_1 & \Theta_2\omega_2 & \Theta_3\omega_3 \end{vmatrix}.$$

Solved for the components $\mathbf{e}_1$, $\mathbf{e}_2$, and $\mathbf{e}_3$ and combined, this reads

$$\begin{aligned}\dot{\mathbf{L}}\big|_{\text{lab}} &= (\Theta_1\dot{\omega}_1 + \Theta_3\omega_2\omega_3 - \Theta_2\omega_2\omega_3)\mathbf{e}_1 \\ &+ (\Theta_2\dot{\omega}_2 + \Theta_1\omega_1\omega_3 - \Theta_3\omega_1\omega_3)\mathbf{e}_2 \\ &+ (\Theta_3\dot{\omega}_3 + \Theta_2\omega_1\omega_2 - \Theta_1\omega_1\omega_2)\mathbf{e}_3.\end{aligned}$$

Since the laboratory system is an inertial frame of reference, we have the relation

$$\dot{\mathbf{L}} = \mathbf{D}.$$

The torque is again expressed by the body-fixed coordinates, and we obtain

$$\dot{\mathbf{L}}\big|_{\text{lab}} = D_1\mathbf{e}_1 + D_2\mathbf{e}_2 + D_3\mathbf{e}_3.$$

Thus, we find the *Euler equations*:

$$\begin{aligned} D_1 &= \Theta_1\dot{\omega}_1 + (\Theta_3 - \Theta_2)\omega_2\omega_3, \\ D_2 &= \Theta_2\dot{\omega}_2 + (\Theta_1 - \Theta_3)\omega_1\omega_3, \\ D_3 &= \Theta_3\dot{\omega}_3 + (\Theta_2 - \Theta_1)\omega_1\omega_2. \end{aligned} \tag{13.5}$$

These three coupled differential equations for $\omega_1(t)$, $\omega_2(t)$, and $\omega_3(t)$ are not linear. This suggests that in general the solutions $\omega_i(t)$ are rather complicated functions of time. Only in the case of free motion ($\mathbf{D} = 0$) can one obtain a transparent solution which will be discussed now. Later we shall deal with the heavy top for which $\mathbf{D} \neq 0$.

We choose the body-fixed coordinate system so that the $\mathbf{e}_3$-axis corresponds to the figure axis. Since we will restrict the analytical consideration of the theory of the top to a free symmetric top that shall be symmetric about the figure axis, the following conditions hold:

$$\dot{\mathbf{L}}\big|_{\text{lab}} = \mathbf{D} = 0, \quad \text{i.e.,} \quad D_1 = D_2 = D_3 = 0, \quad \text{and} \quad \Theta_1 = \Theta_2.$$

We show that for a symmetric top $\mathbf{e}_3$, $\boldsymbol{\omega}$, and $\mathbf{L}$ lie in a plane. For this we have to calculate the scalar triple product of the three vectors which must vanish:

$$\mathbf{e}_3 \cdot (\boldsymbol{\omega} \times \mathbf{L}) = \mathbf{e}_3 \cdot \begin{vmatrix} \mathbf{e}_1 & \mathbf{e}_2 & \mathbf{e}_3 \\ \omega_1 & \omega_2 & \omega_3 \\ \Theta_1\omega_1 & \Theta_2\omega_2 & \Theta_3\omega_3 \end{vmatrix}$$

$$= (\Theta_2 - \Theta_1)\omega_1\omega_2 = 0,$$

because $\Theta_1 = \Theta_2$.

ANALYTICAL THEORY OF THE FREE TOP

With the conditions for the free symmetric top, the Euler equations read

$$\Theta_3 \dot{\omega}_3 = 0 \quad \Rightarrow \quad \omega_3 = \text{constant},$$
$$\Theta_1 \dot{\omega}_1 + (\Theta_3 - \Theta_1)\omega_2 \omega_3 = 0,$$
$$\Theta_1 \dot{\omega}_2 + (\Theta_1 - \Theta_3)\omega_1 \omega_3 = 0.$$

Thus, the component of $\boldsymbol{\omega}$ along the figure axis is constant. To show this in the subsequent calculation, we set

$$\omega_3 = u.$$

To solve the two differential equations, we differentiate the second equation with respect to time:

$$\Theta_1 \ddot{\omega}_1 + (\Theta_3 - \Theta_1) u \dot{\omega}_2 = 0, \qquad \Theta_1 \dot{\omega}_2 + (\Theta_1 - \Theta_3) u \omega_1 = 0.$$

By solving the last equation for $\dot{\omega}_2$ and inserting into the first one, we obtain

$$\ddot{\omega}_1 + \frac{(\Theta_3 - \Theta_1)^2}{\Theta_1^2} u^2 \omega_1 = 0.$$

This form of the differential equation is already known: setting

$$\frac{|\Theta_3 - \Theta_1|}{\Theta_1} u = k,$$

we see that $\ddot{\omega}_1 + k^2 \omega_1 = 0$ is exactly the differential equation of the harmonic oscillator, which is solved by

$$\omega_1 = B \sin kt + C \cos kt.$$

Considering the initial condition $\omega_1(t=0) = 0$, it follows that $\omega_1 = B \sin kt$, or from the second equation, $\omega_2 = -B \cos kt$.

The result means that $\boldsymbol{\omega}$ moves on a circle about the figure axis, as seen from the system of the top:

$$\boldsymbol{\omega} = B(\sin kt\, \mathbf{e}_1 - \cos kt\, \mathbf{e}_2) + u \mathbf{e}_3.$$

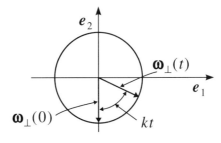

Figure 13.7. Motion of $\omega(t)$ in the $\mathbf{e}_1, \mathbf{e}_2$-plane.

The rotational frequency is thereby given by k; for $k > 0$ the rotation proceeds in the mathematically positive sense. The cone arising in the rotation is again called the *pole cone*. The angular momentum, which is given by $\mathbf{L} = \widehat{\Theta} \cdot \boldsymbol{\omega}$, also changes with time:

$$\mathbf{L} = \Theta_1 B \sin kt\, \mathbf{e}_1 - \Theta_1 B \cos kt\, \mathbf{e}_2 + \Theta_3 u \mathbf{e}_3,$$

i.e., the $\mathbf{L}$-axis rotates with the same frequency k but with a different amplitude about the figure axis (nutation). This is no contradiction to the statement $|\mathbf{L}|_{\text{lab}} = $ constant, since we measure the angular momentum from the system of the top.

Finally, we determine the angles between the three axes. We set

$$\sphericalangle(\mathbf{e}_3, \mathbf{L}) = \alpha, \qquad \sphericalangle(\mathbf{e}_3, \boldsymbol{\omega}) = \beta,$$

and scalar multiply $\mathbf{e}_3$ and $\mathbf{L}$; this yields

$$\mathbf{e}_3 \cdot \mathbf{L} = L \cos \alpha = \sqrt{(\Theta_1 B)^2 + (\Theta_3 u)^2}\, \cos \alpha$$

or

$$\mathbf{e}_3 \cdot \mathbf{L} = \mathbf{e}_3 \cdot (\Theta_1 \omega_1 \mathbf{e}_1 + \Theta_2 \omega_2 \mathbf{e}_2 + \Theta_3 \omega_3 \mathbf{e}_3) = \Theta_3 \omega_3 = \Theta_3 u.$$

Equating both equations leads to

$$\cos \alpha = \frac{\Theta_3 u}{\sqrt{(\Theta_1 B)^2 + (\Theta_3 u)^2}} = \frac{1}{\sqrt{(\Theta_1 B / \Theta_3 u)^2 + 1}}$$

or

$$\cos \alpha \sqrt{\left(\frac{\Theta_1 B}{\Theta_3 u}\right)^2 + 1} = 1.$$

Comparison of the coefficients with the trigonometric formula

$$\cos x \sqrt{\tan^2 x + 1} = 1$$

yields $\tan \alpha = \Theta_1 B / \Theta_3 u = $ constant.

Performing the same calculation for $\mathbf{e}_3 \cdot \boldsymbol{\omega}$, for β we find

$$\tan \beta = \frac{B}{u} = \text{constant}.$$

The comparison of the last two results shows the dependence of the orientation of the axes on Θ_1 and Θ_3: One has

$$\tan \alpha / \tan \beta = \Theta_1 / \Theta_3,$$

from which it follows that

(1) $\Theta_1 > \Theta_3$ (prolate top) $\Rightarrow \alpha > \beta$ for $\alpha, \beta < \pi/2$; sequence of axes: $\mathbf{e}_3 - \boldsymbol{\omega} - \mathbf{L}$;

(2) $\Theta_1 < \Theta_3$ (oblate top) $\Rightarrow \alpha < \beta$ for $\alpha, \beta < \pi/2$; sequence of axes: $\mathbf{e}_3 - \mathbf{L} - \boldsymbol{\omega}$;

ANALYTICAL THEORY OF THE FREE TOP

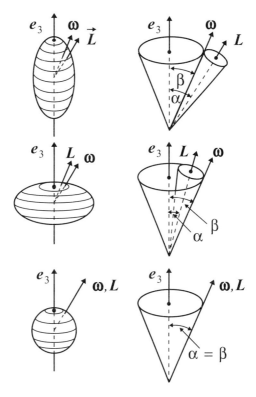

Figure 13.8.

(3) $\Theta_1 = \Theta_3$ (spherical top) $\Rightarrow \alpha = \beta$ for $\alpha, \beta < \pi$; $\boldsymbol{\omega}$ lies on the **L**-axis. Since the $\mathbf{e}_3$-axis of a spherical top can be chosen arbitrarily, there is no loss of generality if we set $\alpha = \beta = 0$.

For (3) we note that, for the spherical top, $k = u(\Theta_3 - \Theta_1)/\Theta_1 = 0$ because $\Theta_1 = \Theta_3$. For the spherical top, as was discussed above, we can set the figure axis (i.e., the $\mathbf{e}_3$-axis) arbitrarily, e.g., also along the **L**- or $\boldsymbol{\omega}$-axis. The result $\alpha = \beta$ would follow also from

$$\boldsymbol{\omega} \times \mathbf{L} = \boldsymbol{\omega} \times \Theta_1 \boldsymbol{\omega} = 0 \quad \text{because} \quad \widehat{\Theta} = \begin{pmatrix} \Theta_1 & 0 & 0 \\ 0 & \Theta_1 & 0 \\ 0 & 0 & \Theta_1 \end{pmatrix}.$$

Example 13.1: Nutation of the earth

The earth is not a spherical top but a flattened rotation ellipsoid. The half-axes are

$a = b = 6378$ km (equator) and $c = 6357$ km.

If the angular momentum axis and the figure axis do not coincide, the figure axis performs nutations about the angular momentum axis. The angular velocity of the nutations is

$$k = \frac{\Theta_3 - \Theta_1}{\Theta_1}\omega_3.$$

The third axis is the principal axis of inertia (pole axis). If we consider the earth as a homogeneous ellipsoid of mass M, we obtain the two moments of inertia:

$$\Theta_1 = \Theta_2 = \frac{M}{5}(b^2 + c^2), \qquad \Theta_3 = \frac{M}{5}(a^2 + b^2).$$

From this, we obtain

$$k = \frac{a^2 - c^2}{b^2 + c^2}\omega_3.$$

Since the half-axes differ only a little, we set $a = b \approx c$, and thus,

$$k = \frac{(a-c)(a+c)}{b^2 + c^2}\omega_3 \approx \frac{a-c}{a}\omega_3.$$

The rotation velocity of the earth is $\omega_3 = 2\pi/\text{day}$. Thus, we obtain for the period of nutation

$$T = \frac{2\pi}{k} = 304 \text{ days}.$$

The figure axis of the earth (geometrical north pole) and the rotation axis $\boldsymbol{\omega}$ of the earth (kinematical north pole) rotate about each other. The measured period (the so-called *Chandler period*)[3] is 433 days. The deviation is essentially caused by the fact that the earth is not rigid. The amplitude of this nutation is about $\pm 0.2''$. The kinematical north pole moves along a spiral trajectory within a circle of 10 m radius in the sense of the earth's rotation.

Example 13.2:   Ellipsoid of inertia of a regular polyhedron

The ellipsoid of inertia of any regular polyhedron is a sphere, which will be shown by the example of the tetrahedron; the reasoning for the octahedron, dodecahedron, and icosahedron is analogous. Suppose there were a principal axis of inertia with a moment which differs from those of the two other principal axes of inertia. In a rotation by 120° about the axis g (perpendicular from point C on the opposite plane; see Figure 13.9), this axis of inertia must turn into itself, since the tetrahedron is transferred into itself. It is easily seen that only the axis g has this property and therefore must be the distinguished axis of inertia. But since h is a symmetry axis too, a rotation by 120° about h must also transfer the axis of inertia g into itself, which however is not true. Assuming the existence of a distinct axis of inertia leads to a contradiction, and hence the ellipsoid of inertia of a tetrahedron must be a sphere.

---

[3] *Seth Carlo Chandler*, b. Sept. 17, 1846, Boston, Mass.–d. Dec. 31, 1913, Wellesley Hills, Mass. American astronomer, detected the Chandler period of 14 months in the pole height fluctuations. He observed variable stars, and for a long time he edited the *Astronomical Journal*.

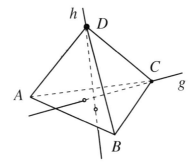

Figure 13.9. Regular tetrahedron: *g* and *h* are straight lines (axes) that are perpendicular to the planes opposite *C* and *D*.

Example 13.3: Exercise: Rotating ellipsoid

A homogeneous three-axial ellipsoid with the moments of inertia Θ_1, Θ_2, Θ_3 rotates with the angular velocity $\dot{\varphi}$ about the principal axis of inertia 3. The axis 3 rotates with $\dot{\vartheta}$ about the axis $\overline{AB}$. The axis $\overline{AB}$ passes through the center of gravity and is perpendicular to 3. Find the kinetic energy.

Solution We decompose the angular velocity $\boldsymbol{\omega}$ into its components along the principal axes of inertia:

$$\boldsymbol{\omega} = (\omega_1, \omega_2, \omega_3), \quad \text{where} \quad \omega_1 = \dot{\vartheta}\cos\varphi, \quad \omega_2 = -\dot{\vartheta}\sin\varphi, \quad \omega_3 = \dot{\varphi}.$$

The kinetic energy is then

$$T = \frac{1}{2}\sum_i \Theta_i \omega_i^2 = \frac{1}{2}(\Theta_1 \cos^2\varphi + \Theta_2 \sin^2\varphi)\dot{\vartheta}^2 + \frac{1}{2}\dot{\varphi}^2.$$

The ellipsoid shall now be symmetric, $\Theta_1 = \Theta_2$; the axis $\overline{AB}$ is tilted from the third axis by the angle α. For the total angular velocity, we have

$$\boldsymbol{\omega} = \dot{\varphi}\mathbf{e}_3 + \dot{\vartheta}\mathbf{e}_{AB}.$$

We decompose the unit vector $\mathbf{e}_{AB}$ along the axis $\overline{AB}$ with respect to the principal axes

$$\mathbf{e}_{AB} = \mathbf{e}_3 \cdot \cos\alpha + (\cos\varphi\,\mathbf{e}_1 - \sin\varphi\,\mathbf{e}_2)\sin\alpha.$$

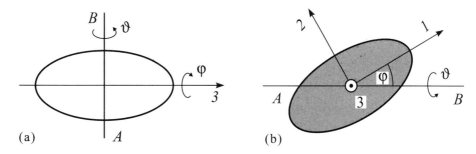

Figure 13.10. Homogeneous three-axial ellipsoid: (a) side view, and (b) top view.

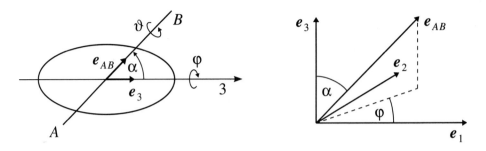

Figure 13.11. The axis $\overline{AB}$ tilted from the third axis by α: Positions of the axes are shown in perspective.

Thus, the components of $\boldsymbol{\omega}$ along the directions of the principal axes are

$$\omega_1 = \sin\alpha \cos\varphi\, \dot\vartheta,$$
$$\omega_2 = -\sin\alpha \sin\varphi\, \dot\vartheta,$$
$$\omega_3 = \dot\varphi + \cos\alpha\, \dot\vartheta.$$

Hence, the kinetic energy reads

$$T = \frac{1}{2}\Theta_1 \sin^2\alpha\, \dot\vartheta^2 + \frac{1}{2}\Theta_3(\dot\varphi + \dot\vartheta \cos\alpha)^2.$$

For $\alpha = 90°$, we obtain for the first case a rotation ellipsoid.

Example 13.4: Exercise: Torque of a rotating plate

Find the torque that is needed to rotate a rectangular plate (edges a and b) with constant angular velocity ω about a diagonal.

Solution   The principal moments of inertia of the rectangle are already known from Example 11.7:

$$I_1 = \frac{1}{12}Ma^2, \qquad I_2 = \frac{1}{12}Mb^2, \qquad I_3 = \frac{1}{12}M(a^2+b^2). \tag{13.6}$$

The angular velocity is

$$\boldsymbol{\omega} = (\boldsymbol{\omega}\cdot \mathbf{e}_x)\mathbf{e}_x + (\boldsymbol{\omega}\cdot \mathbf{e}_y)\mathbf{e}_y,$$

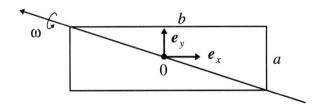

Figure 13.12. The rectangular plate rotates about the diagonal axis.

ANALYTICAL THEORY OF THE FREE TOP

i.e.,

$$\boldsymbol{\omega} = -\frac{\omega b}{\sqrt{a^2 + b^2}}\mathbf{e}_x + \frac{\omega a}{\sqrt{a^2 + b^2}}\mathbf{e}_y$$

$$\Rightarrow \quad \omega_1 = \frac{-\omega b}{\sqrt{a^2 + b^2}}, \quad \omega_2 = \frac{+\omega a}{\sqrt{a^2 + b^2}}, \quad \omega_3 = 0. \tag{13.7}$$

Inserting (13.6) and (13.7) into the Euler equations yields

$$I_1\dot{\omega}_1 + (I_3 - I_2)\omega_2\omega_3 = D_1,$$
$$I_2\dot{\omega}_2 + (I_1 - I_3)\omega_3\omega_1 = D_2,$$
$$I_3\dot{\omega}_3 + (I_2 - I_1)\omega_2\omega_1 = D_3,$$

and furthermore, $D_1 = 0$, $D_2 = 0$, and

$$D_3 = \frac{-M(b^2 - a^2)ab\omega^2}{12(a^2 + b^2)}.$$

Hence, the torque is

$$\mathbf{D} = \frac{-M(b^2 - a^2)ab\omega^2}{12(a^2 + b^2)}\mathbf{e}_z.$$

For $a = b$ (square), $\mathbf{D} = 0$!

Example 13.5: Exercise: Rotation of a vibrating neutron star

The surface of a neutron star (sphere) vibrates slowly, so that the principal moments of inertia are harmonic functions of time:

$$I_{zz} = \frac{2}{5}mr^2(1 + \varepsilon \cos \omega t),$$

$$I_{xx} = I_{yy} = \frac{2}{5}mr^2\left(1 - \varepsilon\frac{\cos \omega t}{2}\right), \quad \varepsilon \ll 1.$$

The star simultaneously rotates with the angular velocity $\boldsymbol{\Omega}(t)$.

(a) Show that the z-component of $\boldsymbol{\Omega}$ remains nearly constant!

(b) Show that $\boldsymbol{\Omega}(t)$ nutates about the z-axis and determine the nutation frequency for $\Omega_z \gg \omega$.

Solution (a) If the total angular momentum is given in an inertial system, then

$$\left(\frac{d\mathbf{L}}{dt}\right)_{\text{inertial}} = 0.$$

The principal moments of inertia are, however, given in a body-fixed system that rotates itself with the angular velocity $\boldsymbol{\Omega}$ with respect to the inertial system. Then

$$\left(\frac{d\mathbf{L}}{dt}\right)_{\text{inertial}} = \left(\frac{d\mathbf{L}}{dt}\right)_k + \boldsymbol{\Omega} \times \mathbf{L} = 0.$$

One therefore gets in the body-fixed system (Euler equations)

$$\frac{d}{dt}(I_{zz}\Omega_z) = 0, \tag{13.8}$$

$$\frac{d}{dt}(I_{xx}\Omega_x) + \frac{3}{2}I_0\Omega_y\Omega_z\varepsilon\cos\omega t = 0, \tag{13.9}$$

$$\frac{d}{dt}(I_{yy}\Omega_y) - \frac{3}{2}I_0\Omega_x\Omega_z\varepsilon\cos\omega t = 0, \tag{13.10}$$

where $I_0 = (2/5)mr^2$ is the moment of inertia of the sphere. (13.8) has the solution

$$\Omega_z = \frac{\Omega_{0z}}{1+\varepsilon\cos\omega t},$$

where Ω_{0z} follows from the initial conditions; this means that Ω_z is only very weakly time dependent.

(b) We suppose that $\omega \ll \Omega_z$, i.e.,

$$\frac{dI_{xx}}{dt} \approx 0 \quad \text{and} \quad \frac{dI_{yy}}{dt} \approx 0.$$

From this, we find

$$I_{xx}\dot\Omega_x + \frac{3}{2}I_0\Omega_z\varepsilon\cos\omega t\,\Omega_y = 0, \qquad I_{yy}\dot\Omega_y - \frac{3}{2}I_0\Omega_z\varepsilon\cos\omega t\,\Omega_x = 0. \tag{13.11}$$

Differentiating again and inserting (13.8), (13.9), and (13.10) yield

$$I_{xx}\ddot\Omega_x + \frac{1}{I_{yy}}\left(\frac{3}{2}I_0\Omega_z\varepsilon\cos\omega t\right)^2\Omega_x = 0,$$

$$I_{yy}\ddot\Omega_y + \frac{1}{I_{xx}}\left(\frac{3}{2}I_0\Omega_z\varepsilon\cos\omega t\right)^2\Omega_y = 0. \tag{13.12}$$

If $I_{xx} = I_{yy} \approx I_0$, then

$$\ddot\Omega_x + \left(\frac{3}{2}\varepsilon\Omega_z\cos\omega t\right)^2\Omega_x = 0, \qquad \ddot\Omega_y + \left(\frac{3}{2}\varepsilon\Omega_z\cos\omega t\right)^2\Omega_y = 0.$$

Since $\omega \ll \Omega_z$ (we further assume that $\omega \ll \varepsilon\Omega_z$), we find

$$\omega_n = \frac{3}{2}\varepsilon\Omega_z\cos\omega t \quad \text{(nutation frequency)},$$

i.e., Ω_x and Ω_y perform a nutation motion with ω_n.

Example 13.6: Exercise: Pivot forces of a rotating circular disk

A homogeneous circular disk (mass M, radius R) rotates with constant angular velocity ω about a body-fixed axis passing through the center. The axis is inclined by the angle α from the surface normal and is pivoted at both sides of the disk center with spacing d. Determine the forces acting on the pivots.

Solution The Euler equations read

$$I_1\dot\omega_1 - \omega_2\omega_3(I_2 - I_3) = D_1, \tag{13.13}$$

$$I_2\dot\omega_2 - \omega_1\omega_3(I_3 - I_1) = D_2, \tag{13.14}$$

$$I_3\dot\omega_3 - \omega_1\omega_2(I_1 - I_2) = D_3, \tag{13.15}$$

ANALYTICAL THEORY OF THE FREE TOP

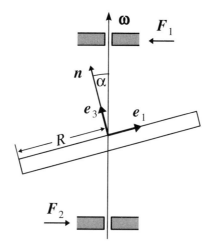

Figure 13.13. Geometry and pivoting of the rotating circular disk.

where $\mathbf{D} = \{D_1, D_2, D_3\}$ represents the torque in the body-fixed system. We choose the body-fixed coordinate system in such a way that $\mathbf{n} = \mathbf{e}_3$ and $\mathbf{e}_1$ lies in the plane spanned by $\mathbf{n}, \boldsymbol{\omega}$. For the principal momentum of inertia I_1, we have

$$I_1 = \sigma \int_0^{2\pi}\!\!\int_0^R y^2 r\, dr d\varphi = \sigma \int_0^{2\pi}\!\!\int_0^R r^3 \sin^2\varphi\, dr d\varphi = \frac{1}{4}\sigma R^4 \int_0^{2\pi} \sin^2\varphi\, d\varphi$$

$$= \frac{1}{4}\sigma R^4 \pi = \frac{1}{4}\left(\frac{M}{\pi R^2}\right) R^4 \pi = \frac{1}{4} M R^2, \tag{13.16}$$

since the surface density σ is given by $\sigma = M/F = M/\pi R^2$. And likewise for I_2 and I_3

$$I_1 = I_2 = \frac{1}{2} I_3 = \frac{1}{4} M R^2. \tag{13.17}$$

The components of the angular velocity vector are given by

$$\boldsymbol{\omega} = \{\omega_1 = \omega \sin\alpha, \omega_2 = 0, \omega_3 = \omega \cos\alpha\}. \tag{13.18}$$

Because $\dot{\boldsymbol{\omega}} = \mathbf{0}$, inserting (13.17) and (13.18) into (13.13) to (13.15) yields

$$D_1 = D_3 = 0 \quad \text{and} \quad D_2 = -\omega^2 \sin\alpha \cos\alpha \frac{1}{4} M R^2. \tag{13.19}$$

Because $\mathbf{D} = \mathbf{r} \times \mathbf{F}$, in the pivots act equal but oppositely directed forces of magnitude

$$|\mathbf{F}| = \frac{|D_2|}{2d} = M R^2 \omega^2 \frac{1}{4d}\left(\frac{1}{4}\sin 2\alpha\right) = M R^2 \omega^2 \frac{\sin 2\alpha}{16d} \tag{13.20}$$

(see Figure 13.13).

Example 13.7: Exercise: Torque on an elliptic disk

What torque is needed to rotate an elliptic disk with the half-axes a and b about the rotation axis $0A$ with constant angular velocity ω_0? The rotation axis is tilted from the large half-axis a by the angle α.

Solution

We choose the $\mathbf{e}_1$-axis orthogonal to the plane of the drawing, $\mathbf{e}_2$ along the small half-axis b, and $\mathbf{e}_3$ along the large half-axis. The principal moments of inertia are then ($M = \sigma\pi ab$, $dM = \sigma dF$)

$$I_2 = \sigma \int_{-a}^{+a} \int_{-\varphi(z)}^{\varphi(z)} z^2 \, dz \, dy \quad \text{with} \quad \varphi(z) = b\sqrt{1 - \frac{z^2}{a^2}}$$

from the ellipse equation $z^2/a^2 + y^2/b^2 = 1$.

$$I_2 = \sigma \int_{-a}^{+a} z^2 y \Big|_{-b\sqrt{1-z^2/a^2}}^{b\sqrt{1-z^2/a^2}} dz = 2\sigma b \int_{-a}^{+a} z^2 \sqrt{1 - \frac{z^2}{a^2}} \, dz$$

$$= 2\sigma \frac{b}{a} \left\{ \frac{z}{8}(2z^2 - a^2)\sqrt{a^2 - z^2} + \frac{a^4}{8} \arcsin \frac{z}{|a|} \right\} \Big|_{-a}^{+a}$$

$$= \frac{1}{4}\sigma ba^3 \pi = \frac{1}{4} M a^2; \tag{13.21}$$

accordingly,

$$I_3 = \sigma \int_{-b}^{+b} \int_{-\overline{\varphi}(y)}^{\overline{\varphi}(y)} y^2 \, dy \, dz \quad \text{with} \quad \overline{\varphi}(y) = a\sqrt{1 - \frac{y^2}{b^2}}$$

$$\Rightarrow \quad I_3 = \frac{1}{4} M b^2. \tag{13.22}$$

We can immediately write down I_1 (because $I_1 = I_2 + I_3$ for thin plates):

$$I_1 = \frac{1}{4} M(a^2 + b^2). \tag{13.23}$$

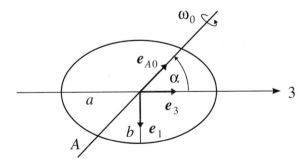

Figure 13.14.

For $\boldsymbol{\omega}$, we obtain

$$\boldsymbol{\omega} = 0 \cdot \mathbf{e}_1 - \omega_0 \sin\alpha \cdot \mathbf{e}_2 + \omega_0 \cos\alpha \cdot \mathbf{e}_3.$$

We insert into the Euler equations of the top:

$$\begin{aligned} D_1 &= I_1\dot{\omega}_1 + (I_3 - I_2)\omega_2\omega_3 \\ &= -\frac{1}{4}M(b^2 - a^2)\sin\alpha\cos\alpha \cdot \omega_0^2, \\ D_2 &= I_2\dot{\omega}_2 + (I_1 - I_3)\omega_1\omega_3 = 0, \\ D_3 &= I_3\dot{\omega}_3 + (I_2 - I_1)\omega_1\omega_2 = 0. \end{aligned} \qquad (13.24)$$

Thus, we obtain for the desired torque $\mathbf{D}$

$$\mathbf{D} = -\omega_0^2 \cdot \frac{M}{8}(b^2 - a^2)\sin 2\alpha \cdot \mathbf{e}_1. \qquad (13.25)$$

It is obvious that

(1) for $\alpha = 0, \pi/2, \pi, \ldots$, the torque vanishes, since the rotation is performed about a principal axis of inertia; and

(2) for $b^2 = a^2$, i.e., the case of a circular disk, the torque vanishes for all angles α.

We will consider once again the last conclusions: Given an elliptic disk with half-axes a and b, for $\alpha = 0°, 180°$, or for $\alpha = 90°, 270°$, the rotation axis coincides with one of the principal axes of inertia along the half-axes. In this case the orientation of the angular momentum is identical with the momentary rotation axis. Because $\omega_0 =$ constant, we also have $\mathbf{L} =$ constant, and therefore the resulting torque vanishes. This also results by insertion into the Euler equations of the top: $\dot{\omega} = (0, 0, 0)$, $\omega = (0, 0, \omega_0)$ or $\omega = (0, \omega_0, 0)$

$$\Rightarrow \quad \begin{aligned} D_1 &= I_1\dot{\omega}_1 + (I_3 - I_2)\omega_2\omega_3 = 0, \\ D_2 &= I_2\dot{\omega}_2 + (I_1 - I_3)\omega_1\omega_3 = 0, \\ D_3 &= I_3\dot{\omega}_3 + (I_2 - I_1)\omega_1\omega_2 = 0. \end{aligned} \qquad (13.26)$$

The heavy symmetric top: Elementary considerations

We now consider the motion of the top under the action of gravity. If the bearing point O of the top does not coincide with the center of gravity S, gravity performs a torque. To distinguish the top from the freely moving top, it is called the *heavy top*. First we restrict ourselves to the symmetric top which rotates with the angular velocity $\boldsymbol{\omega}$ about its figure axis. The origin of the space-fixed coordinate system is set into the bearing point O, the negative z-axis points along the gravity force. Let the distance $\overrightarrow{OS} = \mathbf{l}$; gravity acts on the top of mass m with the torque $\mathbf{D} = \mathbf{l} \times m\mathbf{g}$. Hence, the angular momentum vector is not constant in time:

$$\dot{\mathbf{L}} = \mathbf{D} \quad \text{or} \quad d\mathbf{L} = \mathbf{D} \cdot dt.$$

This differential form of the equation of motion expresses that the torque causes a change $d\mathbf{L}$ of the angular momentum which is parallel to the torque $\mathbf{D}$.

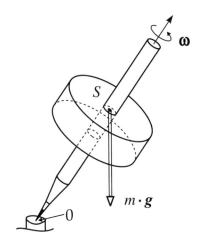

Figure 13.15. A heavy top: The center of gravity S and the bearing point O are sketched in.

Sommerfeld[4] and Klein[5] in their book *Theory of the Top* called this phenomenon—philosophizing—*"Die Tendenz zum gleichsinnigen Parallelismus."* The z-component of the

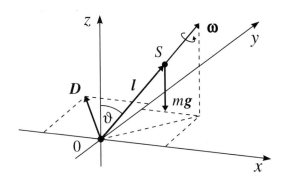

Figure 13.16. Position of the heavy top in the coordinate system.

---

[4]*Arnold Sommerfeld*, b. Dec. 5, 1868, Königsberg–d. April 26, 1951, Munich, physicist. From 1897, professor in Clausthal-Zellerfeld and from 1900 in Aachen, successfully tried for a mathematical backup of technique. In 1906, Sommerfeld became professor for theoretical physics in Munich, where he was an excellent academic teacher for generations of physicists (among others P. Debye, P.P. Ewald, W. Heisenberg, W. Pauli, and H.A. Bethe). He extended Bohr's ideas in 1915 to the "Bohr–Sommerfeld theory of atom" and discovered many of the laws for the number, wavelength, and intensity of spectral lines. His work *Atomic Structure and Spectral Lines* (Vol. 1, 1919; Vol. 2, 1929) was accepted for decades as a standard work of atomic physics. Further works: *Lectures on Theoretical Physics*, six volumes (1942–1962).

[5]*Felix Klein*, b. April 25, 1849, Düsseldorf–d. June 22, 1925, Göttingen. Klein studied from 1865 to 1870 in Bonn. During a stay in 1870 in Paris, he became familiar with the rapidly developing group theory. From 1871, Klein was a private lecturer in Göttingen, from 1872, professor in Erlangen, from 1875, in Munich, from 1880, in Leipzig and from 1886, in Göttingen. He contributed fundamental papers on function theory, geometry, and

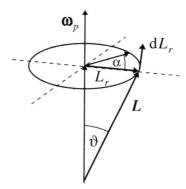

Figure 13.17. On the calculation of the precession frequency ω_p.

angular momentum is however conserved. This results from the following consideration: Because $\mathbf{g} = -g\mathbf{e}_z$, we have $\mathbf{D} = mg\mathbf{e}_z \times \mathbf{l}$, i.e., the torque has no component along the z-direction, and hence L_z is constant. Hence the torque $\mathbf{D}$ causes a motion of the angular momentum vector $\mathbf{L}$ on a cone about the z-axis; this motion of the heavy top is called *precession*. The precession frequency of the top is thereby constant by reason of symmetry; the relative orientation of the torque and angular momentum vectors is also constant.

We now calculate the *precession frequency*. For this purpose we start from the radial component L_r of the angular momentum:

$$L_r = L \sin \vartheta.$$

The angle covered by L_r in the time dt is

$$d\alpha = \frac{dL_r}{L_r} = \frac{dL}{L_r} = \frac{D\,dt}{L \sin \vartheta}.$$

For the *precession frequency* $\omega_p = d\alpha/dt$, we thus have

$$\omega_p = \frac{D}{L \sin \vartheta} = \frac{mgl}{L}$$

or in vector notation

$$\boldsymbol{\omega}_p \times \mathbf{L} = \mathbf{D}.$$

Hence, the precession frequency is independent of the inclination ϑ of the top if $\vartheta \neq 0$ is presupposed. In the general case nutation motions are superimposed on the precession, so that the tip of the figure axis F no longer moves on a circle but on a much more complicated trajectory about the z-axis. The angle ϑ varies between two extreme values

$$\vartheta_0 - \Delta\vartheta \leq \vartheta \leq \vartheta_0 + \Delta\vartheta$$

---

algebra. In particular, group theory and its applications attracted his interest. In 1872, he published the *Program of Erlangen*. In later life Klein became interested in pedagogical and historical problems.

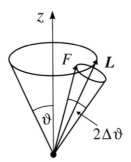

Figure 13.18. In general, a nutation is superimposed on the precession.

(see Figure 13.18). If $\mathbf{D} = 0$, there is only the nutation of the figure axis F about the then-invariable straight line $\mathbf{L}$. For $\mathbf{D} \neq 0$, the precession of $\mathbf{L}$ about the z-axis dominates. The nutation is superimposed on this precession.

Since in the special case considered here the vectors of angular momentum and angular velocity coincide with the figure axis of the top, we can write for the angular momentum

$$\mathbf{L} = \Theta_3 \boldsymbol{\omega},$$

where Θ_3 is the moment of inertia about the figure axis. For the torque we then have the relation

$$\mathbf{D} = \Theta_3 \boldsymbol{\omega}_p \times \boldsymbol{\omega}.$$

The inverse of this moment $\mathbf{D}' = -\mathbf{D} = \Theta_3 \boldsymbol{\omega} \times \boldsymbol{\omega}_p$ is called the *top moment*. It is that torque the top performs on its bearing if it is turned with the angular velocity $\boldsymbol{\omega}_p$. This top moment—sometimes also called the *deviation resistance*—can reach very large values for a quickly rotating top and a sudden turn of its rotation axis. We feel it for example if we suddenly turn the axis of a quickly rotating flywheel held in the hand (e.g., wheel of a bicycle). The top moment observed from a reference system rotating with the angular velocity $\boldsymbol{\omega}_p$ is identical with the moment of the Coriolis forces, which can be proved.

The earth also precesses under the action of the gravitational attraction of the sun and moon. The earth is a top with a fully free rotation axis, but it is not free of forces. As a result of its flattening and of the tilted ecliptic, the attraction of the sun and moon generates a torque. We imagine the earth as an ideal sphere with a bulge upon it, which is largest at the equator, and we first consider only the action of the sun. In the center of earth (center of gravity), the attraction exactly balances the centrifugal forces due to the orbit of the earth about the sun. We divide the bulge into the halves pointing to the sun and away from it, respectively. The attraction of the sun on the former half is larger than at the earth's center, because of the smaller distance. The centrifugal force, however, is by the same reason smaller. At the center of gravity S_1 of the half-bulge results a force $\mathbf{K}$ pointing toward the sun.

On the side away from the sun, the situation is reversed. Here the centrifugal force dominates over the attraction by the sun, and at the center of gravity S_2 of the half-bulge

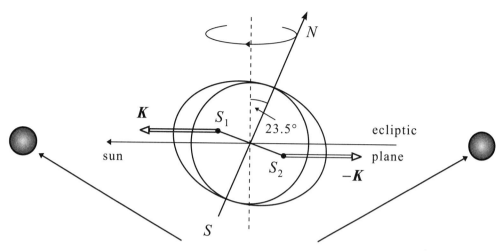

Figure 13.19. Mass ring caused by the orbiting sun as seen from the earth (See also *Classical Mechanics: Point Particles and Relativity*, Chapter 28).

there is a force $-\mathbf{K}$ pointing away from the sun, which – because of the inclination of the ecliptic – forms a couple with the former force. The couple tries to turn the earth axis and the axis which is perpendicular to the earth orbit radius and lies in the orbital plane. From this follows the precession motion about the axis perpendicular to the earth orbit. The moon acts in the same sense, but even more strongly than the sun, because of its small distance. The earth axis rotates in 25800 years ("Platonic year") once on a conical surface with the vertex angle of twice the inclination of the ecliptic, i.e., 47°; it therefore changes its orientation over the millennia. This precession motion must be distinguished from the nutations of earth (Chandler's nutations) discussed in Problem 13.1. The latter are superimposed on the precession motion.

Possibly the most important practical application of the top is the *gyrocompass*. The idea goes back to Foucault (1852). The gyrocompass consists in principle of a quickly rotating, semi-cardanic suspended top, with the rotation axis kept in the horizontal plane by the suspension.

The earth is not an inertial system; it rotates with the angular velocity $\boldsymbol{\omega}_E$. Since the top wants to preserve the orientation of its angular momentum, it is forced to precess with $\boldsymbol{\omega}_E$. Hence, there is a top moment $\mathbf{D}'$:

$$\mathbf{D}' = \Theta_3 \boldsymbol{\omega}_K \times \boldsymbol{\omega}_E,$$

where we set

$$\boldsymbol{\omega}_E = \omega_E \sin\varphi \, \mathbf{e}_z + \omega_E \cos\varphi \, \mathbf{e}_N \equiv \boldsymbol{\omega}_{E_z} + \boldsymbol{\omega}_{E_N}$$

with φ as the geographic latitude. $\mathbf{e}_N$ is a unit vector pointing along the meridian. By splitting $\boldsymbol{\omega}_E$ one obtains

$$\mathbf{D}' = \Theta_3(\boldsymbol{\omega}_K \times \boldsymbol{\omega}_{E_z} + \boldsymbol{\omega}_K \times \boldsymbol{\omega}_{E_N}).$$

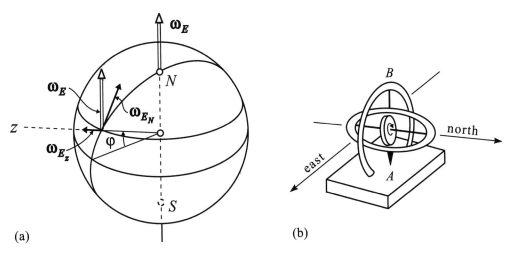

Figure 13.20. (a) Decomposition of the angular frequency ω_E into a vertical and a horizontal component. (b) Semi-cardanic suspension: This top can freely rotate about the *AB*-axis.

The first term is compensated by the bearing of the top. This part of the top moment tends to turn the AB-axis (see Figure 13.20(b)) The second term causes a rotation of the top about the z-axis. The splitting of $\boldsymbol{\omega}_K$ leads to the acting torque

$$\mathbf{D}' = \Theta_3 \omega_K \sin\alpha\, \omega_E \cos\varphi\, \mathbf{e}_z.$$

Hence, a torque arises which always tends to turn the top along the meridian ($\alpha = 0$).

If the suspension of the top is damped, the top adjusts along the north–south direction provided that it is not just at one of the two poles ($\varphi \pm 90°$). Otherwise, it performs damped pendulum vibrations about the north–south direction. One can therefore use the top as a direction indicator if one is not close to a pole.

Foucault's experiments with a "gyroscope" led only to indications of the described effect. Anschütz-Kaempfe succeeded in constructing the first useful gyrocompass (1908).

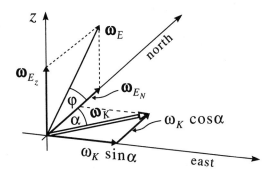

Figure 13.21. Decomposition of the angular velocity of the earth (ω_E) and of the top (ω_K).

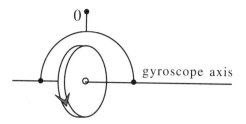

Figure 13.22. Principle of the gyroscope.

To reduce the friction, the gyroscope body—a three-phase current motor—hangs at a float that floats in a basin of mercury. The top axis is kept horizontal by placing the center of gravity of the top lower than the buoyancy center (corresponding to the suspension point). In this setup the gyroscope axis vibrates under the influence of the moment not only in the horizontal, but also in the vertical plane about the north–south direction.

By an appropriate damping of the latter of these coupled vibrations, one can also reach a damping of the vibrations in the horizontal plane, which is needed for the adjustment. The deviations arising from ship vibrations and from other effects could be removed in more recent construction (multiple gyrocompass), or accounted for by calculation.

Further applications of the top

In order to stabilize free motions of bodies, e.g., of a disk or a projectile, these are set into rapid eigenrotation (spin). A disk thereby maintains its tilted position almost unchanged, and therefore gets a buoyancy similarly to the wing of a plane and thus reaches a much larger range of flight than without rotation. A prolate projectile rotating about its longitudinal axis experiences a torque from the air resistance which tends to turn the projectile about a center-of-gravity axis perpendicular to the flight direction. The projectile responds with a kind of precession motion which is very intricate because of the variable air resistance. The vibrations of the projectile tip remain close to the tangent of the trajectory, but for a projectile with "right-hand spin" drift off to the right of the shot plane. The projectile therefore hits the target with the tip ahead, but on firing one has to account for a right-hand deviation.

The gyroscope torques acting on guided tops tend to turn up the axes of the wheels of a car passing a bend, which causes an additional pressure on the outer wheels and a relief of the inner ones. The same gyroscope effect provides an increase of the milling pressure in grinding mills and finds further application in the turn and bank indicator of airplanes. If the plane performs a turn, the gyroscope actions on the propeller must be taken into account.

The top can also serve to stabilize systems (cars) which by their nature are unstable, as in the one-rail track, or to reduce the vibrations of an internally stable system, as in *Schlick's ship gyroscope*. In the latter device a heavy top with a vertical axis, driven by a motor, is set into a frame that can rotate about the transverse horizontal axis. During ship vibrations about the longitudinal axis—these "rolling vibrations" shall be damped—the top performs vibrations because of the precession about the axis lying across the ship; ship vibrations and top vibrations represent coupled vibrations. The top vibrations are appropriately damped by a brake. By the coupling the released energy is pulled out of the ship vibrations; as a result these are considerably reduced. As is seen from the above example, for stabilization by a top it is generally essential that its rotation axis is not fixed relative to the body, but that all degrees of freedom are available. For this reason a bicycle with a tightly mounted front wheel would not be stable. Moreover, riding a bicycle without support is also partly based on the laws of top motion.

An indirect stabilization is used by the devices which control the straight motion of torpedoes. On deviation from the shot direction, the top activates a relay which causes the adjustment of the corresponding rudder.

An important problem is to stabilize a horizontal plane so that it remains horizontal on a moving ship or airplane. This so-called artificial horizon (gyroscope horizon, flight horizon) could, according to Schuler, be realized by a gravitation pendulum with a period of 84 minutes (pendulum length = earth radius), since such a pendulum, even when the suspension point is moved, points to the earth's center. Useful artificial horizons could be realized by tops with cardanic suspension ("top pendulum," center of gravity below the rotation point).

We finally note that an ordinary play top that moves with a rounded tip on a horizontal plane, and thus has five degrees of freedom, does not fit the definition of a top, since in general no point remains fixed during its motion; i.e., translation and rotation motion cannot be dynamically separated. The fact that a play top with an initially tilted axis straightens up under sufficiently fast rotation—which also happens, for example, with a cooked egg—can be explained by a torque created by the friction.

Example 13.8: Exercise: Gyrocompass

A simple gyrocompass consists of a gyroscope that rotates about its axis with the angular velocity ω. Let the moment of inertia about this axis be C and the moment of inertia about a perpendicular axis be A. The suspension of the gyroscope floats on mercury, hence the only acting torque forces the gyroscope axis to stay in the horizontal plane. The gyroscope is brought to the equator. Let the angular velocity of the earth be Ω. What is the response of the gyroscope?

Solution Since the earth rotates with angular velocity Ω, the angular momentum in the earth system satisfies

$$\frac{d\mathbf{L}}{dt} = \mathbf{D} - \mathbf{\Omega} \times \mathbf{L},$$

where $\mathbf{D}$ is the total torque. At the equator $\mathbf{\Omega}$ points along the y-axis, and the z-axis is perpendicular to it.

The components of the angular momentum are

$$L_x = C\omega \sin\varphi,$$
$$L_y = C\omega \cos\varphi, \quad \frac{\omega}{\Omega} \ll 1,$$
$$L_z = A\dot\varphi.$$

FURTHER APPLICATIONS OF THE TOP

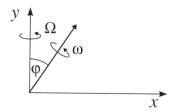

Figure 13.23. Ω is the angular velocity of the earth, ω that of the gyroscope.

We suppose that φ is small. Then

$$L_x \cong C\omega\varphi,$$
$$L_y \cong C\omega, \quad \frac{\omega}{\Omega} \ll 1,$$
$$L_z \cong -A\dot{\varphi}.$$

Since there are no forces acting in the x,y-plane, $D_z = 0$. Hence, the equation for L_z is

$$A\ddot{\varphi} = -C\omega\Omega\varphi$$

or

$$\ddot{\varphi} + \frac{C}{A}\omega\Omega\varphi = 0; \quad \text{i.e.,} \quad \ddot{\varphi} + \omega_r^2\varphi = 0, \quad \omega_r^2 = \frac{C}{A}\omega\Omega.$$

φ oscillates with the frequency

$$\omega_r = \left(\frac{C}{A}\omega\Omega\right)^{1/2}$$

in the north–south direction!

Example 13.9: Exercise: Tidal forces, and lunar and solar eclipses: The Saros cycle[6]

The ancient Chinese court astronomers were able to predict lunar and solar eclipses with great reliability. The fact that such eclipses arise only occasionally—while otherwise we have a full moon or a new moon—is caused by the inclination of the orbital plane of the earth-moon system from the ecliptic, i.e., the orbital plane of the motion of the common center of gravity about the sun. This inclination is about 5.15°. It is not fixed in space but processes because of the tidal forces exerted by the sun. This leads to the so-called Saros cycle, which is of great importance for the prediction of eclipses.

Consider the earth-moon system as a dumbbell-shaped top which rotates about its center of gravity S_p; the center of gravity orbits about the sun on a circular path. The gravitation force between the

---

[6]The name goes back to the *Chaldeans*, a Babylonian tribe. Thales presumably used Babylonian tables for predicting the solar eclipse in 585 B.C.. The knowledge of natural science of the Babylonians was highly developed. They had tables for square roots and powers, approximated the number π by 3 1/8, and could solve quadratic equations. The subdivision of the celestial circle into 12 zodiacal symbols and the 360° division of the circle are modern examples of Babylonian nomenclature.

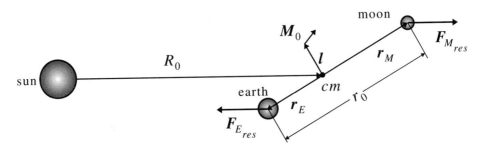

Figure 13.24.

earth and moon just balances the centrifugal force resulting from the eigenrotation of the system and thus fixes the almost rigid dumbbell length r_0. The gravitation of the sun and the centrifugal force due to orbiting about the sun don't compensate for each body independently but lead to resulting tidal forces. These forces create a torque M_0 on the top. Calculate M_0 for the sketched position where it just takes its maximum value. Realize that M_0 on the (monthly and annual) average has a quarter of this value. Calculate from this the precession period T_p. Can you find arguments for why the actual Saros cycle of 18.3 years is notably longer?

Hint: The only data you need for the calculation are the distances r_0, the length of year, and the length of the sidereal month.

Solution R_0 is defined as the vector pointing from the center of gravity of the sun to the center of gravity of the earth-moon system. The coordinate origin of the system is in the center of gravity of the sun.

Let **R** be given in cylindrical coordinates:

$$\mathbf{R} = \mathbf{R}_0 + \Delta\mathbf{R}$$
$$= R_0 \mathbf{e}_r + \Delta R_r \mathbf{e}_r + \Delta R_\varphi \mathbf{e}_\varphi + \Delta R_z \mathbf{e}_z$$

with

$$|\Delta\mathbf{R}| \ll |\mathbf{R}_0|.$$

We then have

$$|\mathbf{R}| \approx R_0 \left(1 + \frac{2\Delta R_r}{R_0}\right)^{1/2}.$$

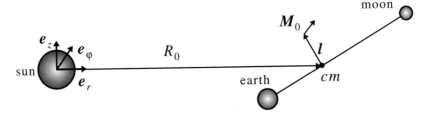

Figure 13.25. The adopted unit vectors.

FURTHER APPLICATIONS OF THE TOP

Hence, we can write for the gravitational force

$$\mathbf{F}_{\text{Gr}}(\mathbf{R}) = -\frac{\gamma m M}{|\mathbf{R}|^3}\mathbf{R}$$

$$\approx -\frac{\gamma m M}{R_0^3}\left(1 - \frac{3\Delta R_r}{R_0}\right)(R_0\mathbf{e}_r + \Delta R_r\mathbf{e}_r + \Delta R_\varphi\mathbf{e}_\varphi + \Delta R_z\mathbf{e}_z)$$

$$\approx \mathbf{e}_r\left(-\frac{\gamma m M}{R_0^3}(R_0 - 2\Delta R_r)\right) + \mathbf{e}_\varphi\left(-\frac{\gamma m M}{R_0^3}\Delta R_\varphi\right) + \mathbf{e}_z\left(-\frac{\gamma m M}{R_0^3}\Delta R_z\right). \quad (13.27)$$

The two masses m and M don't yet have a special meaning. We now consider the motion of the earth-moon system as a two-body problem with an external force:

$$\mathbf{R}_{\text{CM}} := \mathbf{R}_0 = \frac{m_E\mathbf{R}_E + m_M\cdot\mathbf{R}_M}{m_E + m_M} \quad \text{(CM means center of mass)},$$

$$\mathbf{V}_{\text{CM}} = \frac{m_E\mathbf{V}_E + m_M\mathbf{V}_M}{m_E + m_M},$$

$$\mathbf{B}_{\text{CM}} = \frac{m_E\mathbf{B}_E + m_M\mathbf{B}_M}{m_E + m_M},$$

$$= \frac{\mathbf{F}_{ES} + \mathbf{F}_{EM} + \mathbf{F}_{ME} + \mathbf{F}_{MS}}{m_E + m_M} \quad \text{(S means sun)}.$$

According to (13.27),

$$\Delta \mathbf{R}_E = \mathbf{r}_E,$$
$$\Delta \mathbf{R}_M = \mathbf{r}_M,$$
$$m_E\mathbf{r}_E = -m_M\mathbf{r}_M;$$

further from equation (13.27) we have

$$\mathbf{B}_{\text{CM}} = \frac{1}{m_E + m_M}\left(-\gamma\frac{(m_E + m_M)M_S}{R_0^2}\cdot\mathbf{e}_r\right)$$

$$= -\frac{\gamma M_{\text{CM}}}{R_0^2}\cdot\mathbf{e}_r = -\omega_{\text{CM}}^2 R_0\cdot\mathbf{e}_r \quad \text{(circular acceleration)}. \quad (13.28)$$

The last equality follows from the equilibrium condition for the center of gravity. The magnitude of the gravitational acceleration must be equal to the magnitude of the circular acceleration. From this, it follows that the center of mass CM rotates with the frequency ω_{CM} about the sun at the distance R_0.

We further know the following values:

$$T_{\text{CM}} = \frac{2\pi}{\omega_{\text{CM}}} = 365 \text{ days}; \quad R_0 = 149.6\cdot 10^6 \text{ km}. \quad (13.29)$$

We are mainly interested in the motion of the earth-moon system. To this end we consider the relative distance $\mathbf{r}_{\text{rel}}$ between the earth and moon.

$$\mathbf{r}_{\text{rel}} = \mathbf{R}_E - \mathbf{R}_M, \quad |\mathbf{r}_{\text{rel}}| = r_0,$$

$$\mathbf{P}_{\text{rel}} = \mu(\mathbf{V}_E - \mathbf{V}_M) \quad \text{with} \quad \mu = \frac{m_E\cdot m_M}{m_E + m_M},$$

$$\frac{d\mathbf{P}_{\text{rel}}}{dt} = \mu(\mathbf{B}_E - \mathbf{B}_M) = \mu\left(\frac{\mathbf{F}_{ES}}{m_E} + \frac{\mathbf{F}_{EM}}{m_E} - \frac{\mathbf{F}_{ME}}{m_M} - \frac{\mathbf{F}_{MS}}{m_M}\right),$$

$$\mathbf{F}_{EM} = -\gamma \frac{m_E m_M}{r_0^3} \mathbf{r}_{\rm rel} = -m\omega_M^2 \mathbf{r}_{\rm rel}.$$

The last equality holds because of the equilibrium condition, as for the center-of-mass acceleration.

The relative distance also performs a circle with the sidereal period of the moon at the distance r_0. One has

$$T_M = \frac{2\pi}{\omega_M} = 27 \text{ days} + 8 \text{ hours},$$
$$r_0 = r_E + r_M = 0.384 \cdot 10^6 \text{ km}. \tag{13.30}$$

We thus obtain the following combined motion:

$$\left.\begin{aligned} \mathbf{R}_E &= \mathbf{R}_0 + \frac{m_M}{m_M + m_E} \mathbf{r}_{\rm rel} \\ \mathbf{R}_M &= \mathbf{R}_0 - \frac{m_E}{m_M + m_E} \mathbf{r}_{\rm rel} \end{aligned}\right\} \quad \text{epicycle motion.} \tag{13.31}$$

The angular momentum with respect to the center of mass CM is given by

$$\begin{aligned}
\mathbf{L}_0 &= \mathbf{r}_E \times m_E (\mathbf{V}_E - \mathbf{V}_{\rm CM}) + \mathbf{r}_M \times m_M (\mathbf{V}_M - \mathbf{V}_{\rm CM}) \\
&= m_E \frac{m_M}{m_E + m_M} \mathbf{r}_{\rm rel} \times \left(\frac{m_M}{m_E + m_M}\right) \mathbf{v}_{\rm rel} \\
&\quad + m_M \frac{m_E}{m_E + m_M} \mathbf{r}_{\rm rel} \times \left(\frac{m_E}{m_E + m_M}\right) \mathbf{v}_{\rm rel} \\
&= \frac{m_E m_M}{m_E + m_M} \mathbf{r}_{\rm rel} \times \mathbf{v}_{\rm rel} \\
&= \mu \cdot \mathbf{r}_{\rm rel} \times \mathbf{v}_{\rm rel} \\
&\approx \frac{m_E m_M}{m_E + m_M} \omega_M \cdot r_0^2 \cdot \mathbf{1}_{EM},
\end{aligned} \tag{13.32}$$

where $\mathbf{1}_{EM}$ represents a normal vector to the orbital plane of the earth-moon system. The relation (13.32) does not hold exactly, since the motion of the earth-moon system is not perfectly circular, but can be well approximated by a circle.

The coordinate system at the center of gravity is oriented just as at the origin. The total angular momentum with respect to the sun is evaluated as

$$\begin{aligned}
\mathbf{L}_{\rm tot} &= m_E \mathbf{R}_E \times \mathbf{V}_E + m_M \cdot \mathbf{R}_M \times \mathbf{V}_M \\
&= m_E \left(\mathbf{R}_0 + \frac{m_M}{m_M + m_E} \mathbf{r}_E\right) \times \left(\mathbf{V}_{\rm CM} + \frac{m_M}{m_M + m_E} \mathbf{v}_{\rm rel}\right) \\
&\quad + m_M \left(\mathbf{R}_0 - \frac{m_E}{m_M + m_E} \mathbf{r}_{\rm rel}\right) \times \left(\mathbf{V}_{\rm CM} - \frac{m_E}{m_M + m_E} \mathbf{v}_{\rm rel}\right) \\
&= (m_E + m_M) \mathbf{R}_0 \times \mathbf{V}_{\rm CM} + \mu \cdot \mathbf{r}_{\rm rel} \times \mathbf{v}_{\rm rel} \\
&= (m_E + m_M) \omega_{\rm CM} \cdot R_0^2 \cdot \mathbf{1}_{S-CM} + \mathbf{L}_0 \\
&= \mathbf{L}_{\rm CM} + \mathbf{L}_0.
\end{aligned} \tag{13.33}$$

Similar as in equation (13.32), here it also holds that $\mathbf{1}_{S-CN}$ is a vector normal to the orbital plane defined by the sun and the center of gravity.

$$\frac{d\mathbf{L}_{\rm tot}}{dt} = (\mathbf{R}_E \times \mathbf{F}_{ES} + \mathbf{R}_M \times \mathbf{F}_{MS} + (\mathbf{R}_E - \mathbf{R}_M) \times \mathbf{F}_{EM}) = 0;$$

this means
$$\dot{\mathbf{L}}_{CM} = -\dot{\mathbf{L}}_o. \tag{13.34}$$

We now consider the resulting torque $\mathbf{M}_0$ with respect to the center of mass CM:
$$\mathbf{M}_0 = \mathbf{r}_E \times (\mathbf{F}_{ES} + \mathbf{F}_{EM} - m_E \mathbf{B}_{CM}) + \mathbf{r}_M \times (\mathbf{F}_{MS} + \mathbf{F}_{ME} - m_M \mathbf{B}_{CM})$$
$$= \mathbf{r}_E \times \mathbf{F}_{ES} + \mathbf{r}_M \times \mathbf{F}_{MS}.$$

The second terms in each bracket drop because the force and position vector have the same orientation. The third two terms cancel because
$$\mathbf{r}_E m_E = -\mathbf{r}_M m_M.$$

By inserting $\mathbf{r}_{\text{rel}}$, it follows that
$$\mathbf{M}_0 = \frac{m_M}{m_E + m_M} \mathbf{r}_{\text{rel}} \times \mathbf{F}_{ES} - \frac{m_E}{m_E + m_M} \mathbf{r}_{\text{rel}} \times \mathbf{F}_{MS}.$$

Using equation (13.27) for the two force vectors $\mathbf{F}_{ES}$ and $\mathbf{F}_{MS}$, one obtains by simplifying $\mathbf{M}_0$ with respect to cylindrical coordinates:
$$\mathbf{M}_0 = \frac{3\gamma m_E M}{R_0^3} (\Delta R_{rE} (\mathbf{r}_{\text{rel}} \times \mathbf{e}_r)). \tag{13.35}$$

Here, $\Delta R_{\varphi E}$ and ΔR_{zE} were set to zero, according to the definition of the problem. In order not to complicate the problem unnecessarily, we put the coordinate system at the center of gravity and thereby also that at the origin just so that the angular momentum on the average lies in the x, z-plane. This approach is justified since the precession frequency to be calculated is notably less than ω_{CM}. During one revolution about the sun the angular momentum has changed only insignificantly (by about 20°), so that the ecliptic of the earth-moon system has turned only slightly.

Ansatz:
$$\mathbf{r}'_{\text{rel}} = r_0 \begin{pmatrix} \cos \beta \\ \sin \beta \\ 0 \end{pmatrix}, \quad \text{where} \quad \beta \sim \omega_M t.$$

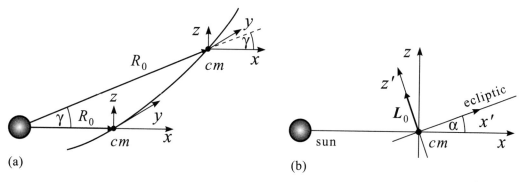

Figure 13.26. (a) The center-of-mass motion about the sun is parametrized by the angle γ. (b) The ecliptic plane is tilted by α from the plane spanned by the center-of-mass motion about the sun.

In the nonprimed system the vector has the components

$$\mathbf{r}_{\mathrm{rel}} = \begin{pmatrix} \cos\alpha & 0 & -\sin\alpha \\ 0 & 1 & 0 \\ \sin\alpha & 0 & \cos\alpha \end{pmatrix}, \qquad \mathbf{r}'_{\mathrm{rel}} = r_0 \begin{pmatrix} \cos\alpha \cos\beta \\ \sin\beta \\ \sin\alpha \sin\beta \end{pmatrix}.$$

Ansatz:

$$\mathbf{e}_r = \begin{pmatrix} \cos(\gamma + \varphi_0) \\ \sin(\gamma + \varphi_0) \\ 0 \end{pmatrix}, \quad \text{where} \quad \gamma \sim \omega_{CM} t.$$

Using the relation

$$\Delta R_{rE} = \frac{m_M}{m_E + m_M} (\mathbf{r}_{\mathrm{rel}} \cdot \mathbf{e}_r),$$

one obtains, by using the two approaches, the following new formulation of equation (13.35):

$$\mathbf{M}_0 = \frac{3\gamma m_E M}{R_0^3} r_0^2 \frac{m_M}{m_E + m_M} \mathbf{v}$$

with

$$\mathbf{v} = \begin{pmatrix} -\sin\alpha \cos\alpha \cos^2\beta \sin(\gamma+\varphi_0)\cos(\gamma+\varphi_0) - \sin\alpha \cos\beta \sin\beta \sin^2(\gamma+\varphi_0) \\ \sin\alpha \cos\alpha \cos^2\beta \cos^2(\gamma+\varphi_0) + \sin\alpha \cos\beta \sin\beta \sin(\gamma+\varphi_0)\cos(\gamma+\varphi_0) \\ \cos^2\alpha \cos^2\beta \cdot \sin(\gamma+\varphi_0)\cos(\gamma+\varphi_0) - \sin\beta \cos\beta \cos\alpha \cos^2(\gamma+\varphi_0) \\ +\cos\alpha \cos\beta \cdot \sin\beta \sin^2(\gamma+\varphi_0) - \sin^2\beta \sin(\gamma+\varphi_0)\cos(\gamma+\varphi_0) \end{pmatrix}.$$

This clumsy expression can be significantly simplified, assuming again that $\omega_P \gg \omega_{CM}$. This means that the angular momentum $\mathbf{L}_0$ changes only slightly during one revolution about the sun.

We first consider β. The revolution period of the moon about the earth is about 28 days. The moment $\mathbf{M}_0$ changes its orientation with varying β; for the "inert" angular momentum, however, only the average momentum $\langle \mathbf{M}_0 \rangle \beta$ counts; this is obtained by averaging over a full period of β. The moment impact (analogous to the force impact) of $\mathbf{M}_0$ and of $\langle \mathbf{M}_0 \rangle \beta$ has the same value, because of the linearity $\beta = \omega_M t$.

$$\langle \mathbf{M}_0 \rangle \beta = \frac{3\gamma M \mu}{R_0^3} r_0^2 \begin{pmatrix} -\sin\alpha \cos\alpha \frac{1}{2} \sin(\gamma+\varphi_0)\cos(\gamma+\varphi_0) \\ \sin\alpha \cos\alpha \frac{1}{2} \cos^2(\gamma+\varphi_0) \\ \cos^2\alpha \frac{1}{2} \sin(\gamma+\varphi_0)\cos(\gamma+\varphi_0) - \frac{1}{2}\sin(\gamma+\varphi_0)\cos(\gamma+\varphi_0) \end{pmatrix}.$$

The same consideration can be made for the "rotating" angle $\gamma \sim \omega_{CM} t$, since we assume that $\omega_p \gg \omega_{CM}$. We therefore average over $\langle \mathbf{M}_0 \rangle \beta$:

$$\langle \langle \mathbf{M}_0 \rangle \beta \rangle \gamma = \frac{3\gamma M \mu}{R_0^3} r_0^2 \begin{pmatrix} 0 \\ \frac{1}{4} \sin\alpha \cos\alpha \\ 0 \end{pmatrix}.$$

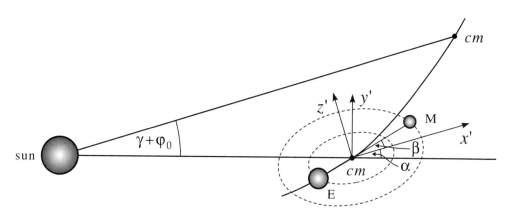

Figure 13.27. Meaning of the angles α, β, γ at one glance: β describes the rotation of the earth-moon axis in the primed coordinate system.

$$= \frac{3\gamma M \mu}{R_0^3} r_0^2 \cdot \frac{1}{4} \sin\alpha \cos\alpha \cdot \mathbf{e}_y := \langle \mathbf{M}_0 \rangle. \tag{13.36}$$

Not very much is left over from the extended expression; the resulting acting moment $\langle \mathbf{M}_0 \rangle$ is exactly perpendicular to $\mathbf{L}_0$ (see Figure 13.27) and points along the y-direction. If the angular momentum moved (slowly), we imagine the shifted angular momentum as being embedded in a fixed coordinate system and thus get, according to (13.36), the same result. $\langle \mathbf{M}_0 \rangle$ is constant and always perpendicular to $\mathbf{L}_0$. It therefore causes a precession.

But first we will illustrate equation (13.36) that has been obtained in a rather mathematical way: In this situation, we get a maximum moment. The moment $\mathbf{M}(\beta)$ now points in the y'-direction for all possible β. For $\beta = 90°$ or $\beta = 270°$, the vector $\mathbf{M}$ vanishes. On the average, we thus obtain

$$\langle \mathbf{M}(\beta) \rangle = 2 \langle \mathbf{M}_0 \rangle.$$

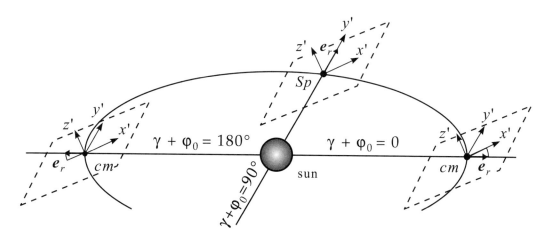

Figure 13.28. Orientation of $\mathbf{e}_r$ as a function of $\gamma+\varphi_0$: The averaged moment $\langle \mathbf{M} \rangle$ β always points along the y'-direction.

The diagram shows that there are also two "maximum" orientations with respect to $(\gamma + \varphi_0)$. Between these positions $\langle \mathbf{M}(\beta) \rangle$ must vanish. One thus obtains altogether

$$\langle \langle \mathbf{M}(\beta, \gamma) \rangle \beta \rangle \gamma = \langle \mathbf{M}_0 \rangle.$$

Thus, the result (13.36) is also clearly understood.

With equation (13.32) in our defined coordinate system, we have

$$\mathbf{L}_0 = \mu \cdot \omega_M \cdot r_0^2 \begin{pmatrix} -\sin\alpha \\ 0 \\ \cos\alpha \end{pmatrix}.$$

The angular momentum $\mathbf{L}_0$ now precesses:

$$\omega_p = \frac{M_S}{L \cdot \sin\alpha} = \frac{\langle \mathbf{M}_0 \rangle}{L_0 \cdot \sin\alpha}$$

$$= \frac{3}{4} \cdot \frac{\gamma M_S \mu r_0^2 \sin\alpha \cos\alpha}{R_0^3 \mu \omega_M r_0^2 \sin\alpha} = \frac{3}{4} \frac{\gamma M_S}{R_0^3} \cos\alpha \cdot \frac{1}{\omega_M}. \qquad (13.37)$$

It has been shown at the beginning (13.28) that

$$\frac{\gamma M_S}{R_0^2} = \omega_{CM}^2 R_0 \quad \Rightarrow \quad \frac{\gamma M_S}{R_0^3} = \omega_{CM}^2 = \left(\frac{2\pi}{T_{CM}}\right)^2,$$

and with

$$\omega_M = \frac{2\pi}{T_M},$$

one gets

$$\omega_p = \frac{3}{4} \cos\alpha \left(\frac{4\pi^2}{T_{CM}^2}\right) \cdot \frac{T_M}{2\pi} = \frac{3}{2} \cos\alpha \frac{T_M}{T_{CM}^2}, \qquad (13.38)$$

and for T_p

$$T_p = \frac{2\pi}{\omega_p} = \frac{4}{3} \cdot \frac{1}{\cos\alpha} \cdot \frac{T_{CM}^2}{T_M}. \qquad (13.39)$$

With $T_{CM} \approx 365.25$ days, $T_M \approx 27.3$ days, and $\alpha \approx 5.5°$, one obtains

$$T_p \approx 17.9 \text{ years}. \qquad (13.40)$$

The fact that the actual Saros cycle is larger by about 2% is partly due to the approximation when averaging γ (the angular momentum actually moves slightly), but possibly due to the elliptic path of the moon about the earth. In any case, the result is relatively accurate, considering the approximations made.

From (13.34), it further follows that

$$\dot{\mathbf{L}}_{CM} = -\dot{\mathbf{L}}_0,$$

i.e., the "large" angular momentum vector $\mathbf{L}_{CM}$ runs through an opposite precession cone.

The Euler angles

The motion of the heavy top suspended at a point can be described by specifying the orientation of a body-fixed coordinate system (x', y', z') relative to a space-fixed system (x, y, z). The two coordinate systems have a common origin at the suspension point of the top. To establish the relation between the two coordinate systems, one usually adopts the *Euler angles*. The coordinate system (x', y', z') is obtained from the system (x, y, z) by three subsequent rotations about defined axes. The corresponding rotation angles are called Euler angles. The sequence of the rotations is important, since rotations by finite angles are not commutative. In Figure 13.29 we see at once that a permutation of the sequence of two rotations about different axes leads to a different result.

A rotation by 90° about the x-axis, followed by a 90°-rotation about the y-axis (upper figure) leads to a different result than rotating first about the y-axis and then about the x-axis (lower figure) (noncommutativity of finite rotations).

The Euler angles are defined as follows: The *first rotation is performed about the z-axis by the angle α*. The x- and y-axis turn into the X- and Y-axis. The Z-axis coincides with the z-axis. The so defined X, Y, Z system is a *first intermediate system* which is only used to keep the calculation transparent. For the unit vectors, we have

$$\begin{aligned}
\mathbf{i} &= (\mathbf{i} \cdot \mathbf{I})\mathbf{I} + (\mathbf{i} \cdot \mathbf{J})\mathbf{J} + (\mathbf{i} \cdot \mathbf{K})\mathbf{K} = \cos\alpha \mathbf{I} - \sin\alpha \mathbf{J}, \\
\mathbf{j} &= (\mathbf{j} \cdot \mathbf{I})\mathbf{I} + (\mathbf{j} \cdot \mathbf{J})\mathbf{J} + (\mathbf{j} \cdot \mathbf{K})\mathbf{K} = \sin\alpha \mathbf{I} + \cos\alpha \mathbf{J}, \\
\mathbf{k} &= (\mathbf{k} \cdot \mathbf{I})\mathbf{I} + (\mathbf{k} \cdot \mathbf{J})\mathbf{J} + (\mathbf{k} \cdot \mathbf{K})\mathbf{K} = \mathbf{K}.
\end{aligned} \qquad (13.41a)$$

The second rotation is performed about the (new) X-axis by the angle β; the Y- and Z-axis turn into the Y'- and the Z'-axis. The X'-axis coincides with the X-axis. The X', Y', Z' system fixed in this way is a *second intermediate system*. It serves for mathematical clarity

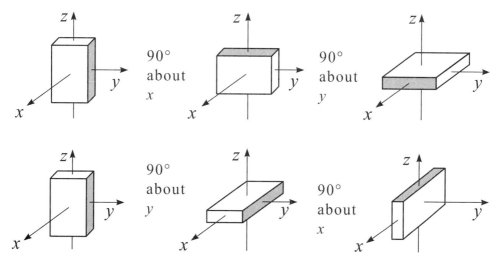

Figure 13.29. Demonstration of the noncommutativity of finite rotations.

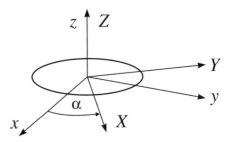

Figure 13.30. The first Euler angle.

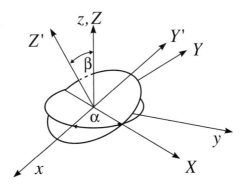

Figure 13.31. The first two Euler angles.

and transparency, just as the first intermediate system did. An analogous calculation yields for the unit vectors

$$\mathbf{I} = \mathbf{I}',$$
$$\mathbf{J} = \cos\beta \mathbf{J}' - \sin\beta \mathbf{K}', \quad (13.41b)$$
$$\mathbf{K} = \sin\beta \mathbf{J}' + \cos\beta \mathbf{K}'.$$

The third rotation is performed about the Z'-axis by the angle γ; the X'- and Y'-axis then turn into the x'- and y'-axis, respectively. The z'-axis is identical with the Z'-axis. The x', y', z' system constructed this way is the desired body-fixed coordinate system. For the unit vectors, one obtains

$$\mathbf{I}' = \cos\gamma\, \mathbf{i}' - \sin\gamma\, \mathbf{j}',$$
$$\mathbf{J}' = \sin\gamma\, \mathbf{i}' + \cos\gamma\, \mathbf{j}', \quad (13.41c)$$
$$\mathbf{K}' = \mathbf{k}'.$$

Using the relations between the unit vectors, we now determine the unit vectors $\mathbf{i}, \mathbf{j}, \mathbf{k}$ as functions of $\mathbf{i}', \mathbf{j}', \mathbf{k}'$. For this purpose, we insert

$$\mathbf{i} = \cos\alpha\, \mathbf{I} - \sin\alpha\, \mathbf{J}$$
$$= \cos\alpha\, \mathbf{I}' - \sin\alpha\cos\beta\, \mathbf{J}' + \sin\alpha\sin\beta\, \mathbf{K}'$$

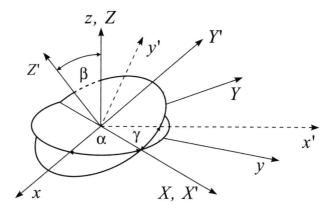

Figure 13.32. All three Euler angles.

$$= \cos\alpha\cos\gamma\,\mathbf{i}' - \cos\alpha\sin\gamma\,\mathbf{j}' - \sin\alpha\cos\beta\sin\gamma\,\mathbf{i}' \qquad (13.41\text{d})$$
$$- \sin\alpha\cos\beta\cos\gamma\,\mathbf{j}' + \sin\alpha\sin\beta\,\mathbf{k}'$$
$$= (\cos\alpha\cos\gamma - \sin\alpha\cos\beta\sin\gamma)\mathbf{i}'$$
$$+ (-\cos\alpha\sin\gamma - \sin\alpha\cos\beta\cos\gamma)\mathbf{j}' + \sin\alpha\sin\beta\,\mathbf{k}'.$$

An analogous calculation yields

$$\mathbf{j} = (\sin\alpha\cos\gamma + \cos\alpha\cos\beta\sin\gamma)\mathbf{i}'$$
$$+ (-\sin\alpha\sin\gamma + \cos\alpha\cos\beta\cos\gamma)\mathbf{j}' - \cos\alpha\sin\beta\,\mathbf{k}', \qquad (13.41\text{e})$$
$$\mathbf{k} = \sin\beta\sin\gamma\,\mathbf{i}' + \sin\beta\cos\gamma\,\mathbf{j}' + \cos\beta\,\mathbf{k}'.$$

The rotations can also be expressed by the corresponding rotation matrices. For the first rotation, we have

$$\mathbf{r} = \widehat{A}\mathbf{R},$$

where

$$\widehat{A} = \begin{pmatrix} \cos\alpha & -\sin\alpha & 0 \\ \sin\alpha & \cos\alpha & 0 \\ 0 & 0 & 1 \end{pmatrix} \quad \text{and} \quad \mathbf{R} = \begin{pmatrix} X \\ Y \\ Z \end{pmatrix}.$$

The matrices for the rotations by the angles β and γ accordingly read

$$\widehat{B} = \begin{pmatrix} 1 & 0 & 0 \\ 0 & \cos\beta & -\sin\beta \\ 0 & \sin\beta & \cos\beta \end{pmatrix},$$

$$\widehat{C} = \begin{pmatrix} \cos\gamma & -\sin\gamma & 0 \\ \sin\gamma & \cos\gamma & 0 \\ 0 & 0 & 1 \end{pmatrix}.$$

The matrix of the entire rotation $\widehat{D}$ is the product of the three matrices $\widehat{D} = \widehat{A}\widehat{B}\widehat{C}$. Hence,

$$\mathbf{r} = \widehat{D}\mathbf{r}' \quad \text{or} \quad \mathbf{r}' = \widehat{D}^{-1}\mathbf{r} = \widetilde{\widehat{D}}\mathbf{r}.$$

Since the rotation matrices are orthogonal, the inverse matrix equals the transposed one. By calculating the matrix product, one can easily show that the matrix $\widehat{D}$ agrees with the relations derived for the unit vectors. (This agrees with the general considerations from Chapter 30 of *Classical Mechanics: Point Particles and Relativity* of the lectures on theoretical physics.)

We first calculate the angular velocity $\boldsymbol{\omega}$ of the top as a function of the Euler angles. If $(\mathbf{i}, \mathbf{j}, \mathbf{k})$ define the laboratory system and $(\mathbf{i}', \mathbf{j}', \mathbf{k}')$ a body-fixed system of principal axes, for the angular velocity we have

$$\boldsymbol{\omega} = \omega_\alpha \mathbf{k} + \omega_\beta \mathbf{I} + \omega_\gamma \mathbf{K}' = \dot{\alpha}\mathbf{k} + \dot{\beta}\mathbf{I} + \dot{\gamma}\mathbf{K}',$$

where we presuppose that $\mathbf{k}$, $\mathbf{I}$, and $\mathbf{K}'$ are not coplanar. We utilize the derived relations between the unit vectors and obtain

$$\boldsymbol{\omega} = \dot{\alpha}\sin\beta\sin\gamma\,\mathbf{i}' + \dot{\alpha}\sin\beta\cos\gamma\,\mathbf{j}' + \dot{\alpha}\cos\beta\,\mathbf{k}' + \dot{\beta}\cos\gamma\,\mathbf{i}' - \dot{\beta}\sin\gamma\,\mathbf{j}' + \dot{\gamma}\mathbf{k}'$$
$$= (\dot{\alpha}\sin\beta\sin\gamma + \dot{\beta}\cos\gamma)\mathbf{i}' + (\dot{\alpha}\sin\beta\cos\gamma - \dot{\beta}\sin\gamma)\mathbf{j}' + (\dot{\alpha}\cos\beta + \dot{\gamma})\mathbf{k}'.$$

Setting $\boldsymbol{\omega} = \omega_{x'}\mathbf{i}' + \omega_{y'}\mathbf{j}' + \omega_{z'}\mathbf{k}'$, we get the components of the angular velocity relative to the body-fixed coordinate system:

$$\omega_{x'} \equiv \omega_1 = \dot{\alpha}\sin\beta\sin\gamma + \dot{\beta}\cos\gamma,$$
$$\omega_{y'} \equiv \omega_2 = \dot{\alpha}\sin\beta\cos\gamma - \dot{\beta}\sin\gamma, \tag{13.42}$$
$$\omega_{z'} \equiv \omega_3 = \dot{\alpha}\cos\beta + \dot{\gamma}. \tag{13.43}$$

The kinetic energy T of the top is then

$$T = \frac{1}{2}(\Theta_1\omega_1^2 + \Theta_2\omega_2^2 + \Theta_3\omega_3^2)$$
$$= \frac{1}{2}\Theta_1(\dot{\alpha}\sin\beta\sin\gamma + \dot{\beta}\sin\gamma)^2 \tag{13.44}$$
$$+ \frac{1}{2}\Theta_2(\dot{\alpha}\sin\beta\cos\gamma - \dot{\beta}\sin\gamma)^2$$
$$+ \frac{1}{2}\Theta_3(\dot{\alpha}\cos\beta + \dot{\gamma})^2.$$

If $\Theta_1 = \Theta_2$, i.e., the top is symmetric, the above expression simplifies to

$$T = \frac{1}{2}\Theta_1(\dot{\alpha}^2\sin^2\beta + \dot{\beta}^2) + \frac{1}{2}\Theta_3(\dot{\alpha}\cos\beta + \dot{\gamma})^2. \tag{13.45}$$

Motion of the heavy symmetric top

For the special case of the heavy symmetric top, we will determine the explicit equations of motion and the constants of motion, starting from the Euler equations. For simplification, we note that for the symmetric top the two orientations of the principal axes $\mathbf{e}_{x'}, \mathbf{e}_{y'}$ can be arbitrarily chosen in a plane perpendicular to $\mathbf{e}_{z'}$. *We therefore choose a coordinate system where the angle γ always vanishes. This system is then no longer body-fixed* (it does not rotate with the top about the $\mathbf{e}_{z'}$-axis). The axes $\mathbf{e}_{z'}, \mathbf{e}_z, \mathbf{e}_{y'}$ are then coplanar, as are $\mathbf{e}_x, \mathbf{e}_{x'}, \mathbf{e}_y$. This is illustrated in Figure 13.33.

Analytically, this follows from the fact that

$$\begin{aligned}
\mathbf{e}_x' &= \mathbf{I}' = \mathbf{I} = \cos\alpha\,\mathbf{i} + \sin\alpha\,\mathbf{j} = \cos\alpha\,\mathbf{e}_x + \sin\alpha\,\mathbf{e}_y, \\
\mathbf{e}_y' &= \mathbf{J}' = \cos\beta\,\mathbf{J} + \sin\beta\,\mathbf{K} = \cos\beta(-\sin\alpha\,\mathbf{i} + \cos\alpha\,\mathbf{j}) + \sin\beta\,\mathbf{K} \\
&= -\sin\alpha\cos\beta\,\mathbf{e}_x + \cos\alpha\cos\beta\,\mathbf{e}_y + \sin\beta\,\mathbf{e}_z, \\
\mathbf{e}_z' &= \mathbf{K}' = (-\sin\beta)\mathbf{J} + \cos\beta\,\mathbf{K} = -\sin\beta(-\sin\alpha\,\mathbf{i} + \cos\alpha\,\mathbf{j}) + \cos\beta\,\mathbf{K} \\
&= \sin\alpha\sin\beta\,\mathbf{e}_x - \cos\alpha\sin\beta\,\mathbf{e}_y + \cos\beta\,\mathbf{e}_z.
\end{aligned} \qquad (13.46)$$

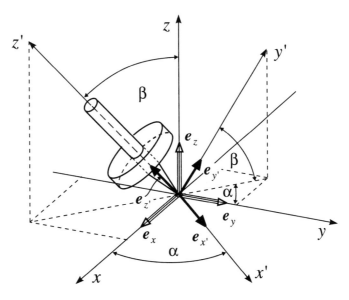

Figure 13.33. Heavy top in various coordinate systems.

We thereby have inverted the relations (13.41a) and (13.41b). By means of the expressions for $\mathbf{e}_x'$, $\mathbf{e}_y'$, $\mathbf{e}_z'$, one can now easily check the triple scalar products $\mathbf{e}_z' \cdot (\mathbf{e}_z \times \mathbf{e}_y')$ and $\mathbf{e}_x \cdot (\mathbf{e}_x' \times \mathbf{e}_y)$, e.g.,

$$\mathbf{e}_x \cdot (\mathbf{e}_x' \times \mathbf{e}_y) = \det \begin{pmatrix} 1 & 0 & 0 \\ \cos\alpha & \sin\alpha & 0 \\ 0 & 1 & 0 \end{pmatrix} = 0.$$

Similarly, one shows the vanishing of the other triple scalar product and thus confirms that the corresponding vectors are coplanar.

The coordinate system thus follows the precession (with $\dot\alpha$) and the nutation (with $\dot\beta$) of the top, but not its eigenrotation. To realize that $\dot\beta$ describes the nutation, we note that a nutation motion of the figure axis is superimposed onto the precession (compare the discussion in the section "Elementary considerations on the heavy top"). This manifests itself for β in an up-and-down motion (vibration) about a fixed value β_0 (see Figure 13.34).

For the angular velocities (13.43) of the $\mathbf{e}_{x'}$, $\mathbf{e}_{y'}$, $\mathbf{e}_{z'}$ system (which is only partly body-fixed) relative to the laboratory system $\mathbf{e}_x$, $\mathbf{e}_y$, $\mathbf{e}_z$ in this system ($\gamma = 0$) we have

$$\omega_1 = \omega_{x'} = \dot\beta,$$
$$\omega_2 = \omega_{y'} = \dot\alpha \sin\beta, \qquad (13.47a)$$
$$\omega_3 = \omega_{z'} = \dot\alpha \cos\beta,$$

or

$$\boldsymbol{\omega} = \dot\beta\,\mathbf{e}_{x'} + \dot\alpha \sin\beta\,\mathbf{e}_{y'} + \dot\alpha \cos\beta\,\mathbf{e}_{z'}. \qquad (13.47b)$$

The angular velocity of the top, on the other hand, is

$$\boldsymbol{\omega}_K = \omega_{x'}\mathbf{e}_{x'} + \omega_{y'}\mathbf{e}_{y'} + (\omega_{z'} + \omega_0)\mathbf{e}_{z'}$$
$$= \dot\beta\,\mathbf{e}_{x'} + \dot\alpha \sin\beta\,\mathbf{e}_{y'} + (\dot\alpha \cos\beta + \omega_0)\mathbf{e}_{z'}. \qquad (13.48)$$

Here, ω_0 is the additional angular velocity of the top relative to the $\mathbf{e}_{x'}$, $\mathbf{e}_{y'}$, $\mathbf{e}_{z'}$ system. The angular velocity $\omega_0(t)$ in general depends on the time. We must take care also when

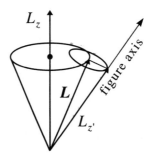

Figure 13.34. Precession and nutation of the angular momentum: The figure axis points along $L_{z'}$.

calculating the angular momentum, because in this particular $\mathbf{e}_{x'}, \mathbf{e}_{y'}, \mathbf{e}_{z'}$ system the rigid body still rotates with the angular velocity $\omega_0 \mathbf{e}_{z'}$. We can call this additional rotation *spin*. It is due to the particular choice of our (not exactly body-fixed) system $\mathbf{e}_{x'}, \mathbf{e}_{y'}, \mathbf{e}_{z'}$. The angular momentum is then

$$\begin{aligned}\mathbf{L} &= \widehat{\Theta}\omega_k = \Theta_1 \omega_1 \mathbf{e}_{x'} + \Theta_2 \omega_2 \mathbf{e}_{y'} + \Theta_3(\omega_3 + \omega_0)\mathbf{e}_{z'} \\ &= \Theta_1 \dot{\beta} \mathbf{e}_{x'} + \Theta_2 \dot{\alpha} \sin\beta \, \mathbf{e}_{y'} + \Theta_3(\dot{\alpha}\cos\beta + \omega_0)\mathbf{e}_{z'} \\ &= \{L_{x'}, L_{y'}, L_{z'}\},\end{aligned} \quad (13.49)$$

and the Euler equations read

$$\dot{\mathbf{L}}\big|_{\text{lab}} = \dot{\mathbf{L}}\big|_{e'} + \boldsymbol{\omega} \times \mathbf{L} = \mathbf{D}. \quad (13.50)$$

Note that the spin rotation ω_0 appears in $\mathbf{L}$ but not in $\boldsymbol{\omega}$!

The torque about the origin of the space-fixed system is

$$\mathbf{D} = (l \cdot \mathbf{e}_{z'}) \times (-mg\mathbf{e}_z) = mgl \sin\beta \, \mathbf{e}_{x'},$$

because $\mathbf{e}_{z'} \times \mathbf{e}_z = -\sin\beta \, \mathbf{e}_{x'}$, as is easily seen from the equations (13.46). By inserting this into the Euler equations (13.5) and (13.50) and noting that $\Theta_1 = \Theta_2$, we obtain

$$\begin{aligned}mgl\sin\beta &= \Theta_1 \ddot{\beta} + (\Theta_3 - \Theta_1)\dot{\alpha}^2 \sin\beta\cos\beta + \Theta_3 \omega_0 \dot{\alpha} \sin\beta, \\ 0 &= \Theta_1(\ddot{\alpha}\sin\beta + \dot{\alpha}\dot{\beta}\cos\beta) + (\Theta_1 - \Theta_3)\dot{\alpha}\dot{\beta}\cos\beta - \Theta_3 \omega_0 \dot{\beta}, \\ 0 &= \Theta_3(\ddot{\alpha}\cos\beta - \dot{\alpha}\dot{\beta}\sin\beta + \dot{\omega}_0).\end{aligned} \quad (13.51)$$

From the above system of equations, $\alpha(t)$, $\beta(t)$, and $\omega_0(t)$ can be determined. From the third equation, we have, because $\Theta_3 \neq 0$,

$$\ddot{\alpha}\cos\beta - \dot{\alpha}\dot{\beta}\sin\beta + \dot{\omega}_0 = \frac{d}{dt}(\dot{\alpha}\cos\beta + \omega_0) = 0 \quad (13.52a)$$

or

$$\dot{\alpha}\cos\beta + \omega_0 = A = \text{constant}, \quad (13.52b)$$

i.e., the angular momentum component $L_{z'} = \Theta_3 A$ (see (13.49)!) about the figure axis is constant.

We therefore set $\dot{\alpha}\cos\beta + \omega_0 = A$, calculate ω_0 from this and insert it into the first two equations. We then obtain two coupled differential equations for precession (α) and nutation (β), respectively:

$$mgl\sin\beta = \Theta_1 \ddot{\beta} - \Theta_1 \dot{\alpha}^2 \sin\beta\cos\beta + \Theta_3 A \sin\beta \cdot \dot{\alpha}, \quad (13.52c)$$

$$0 = \Theta_1(\ddot{\alpha}\sin\beta + 2\dot{\alpha}\dot{\beta}\cos\beta) - \Theta_3 A \dot{\beta}. \quad (13.52d)$$

We first investigate this system for the case that the top performs no nutation. Then $\ddot{\beta} = \dot{\beta} = 0$, and $\beta > 0$. By insertion, we obtain

$$mgl = -\Theta_1 \dot{\alpha}^2 \cos\beta + \Theta_3 A \dot{\alpha}, \qquad \ddot{\alpha} = 0. \quad (13.53)$$

The second equation means that the *precession is stationary*.

From the first equation, we determine the precession velocity $\dot{\alpha}$:

$$\dot{\alpha} = \frac{\Theta_3 A}{2\Theta_1 \cos\beta}\left(1 \pm \sqrt{1 - \frac{4mgl\Theta_1 \cos\beta}{\Theta_3^2 A^2}}\right). \tag{13.54}$$

For a top rotating quickly about the $\mathbf{e}_{z'}$-axis, A becomes very large, and the fraction in the radicand becomes very small. We terminate the expansion of the root after the second term and get as solutions to first order for $\dot{\alpha}_{\text{small}}$:

$$\dot{\alpha}_{\text{small}} = \frac{mgl}{\Theta_3 A}; \tag{13.55a}$$

to zeroth order for $\dot{\alpha}_{\text{large}}$, we have

$$\dot{\alpha} = \frac{\Theta_3}{\Theta_1 \cos\beta} A. \tag{13.55b}$$

A stationary precession without nutation (regular precession) occurs only if the heavy symmetric top gets a certain precession velocity ($\dot{\alpha}_{\text{small}}$ or $\dot{\alpha}_{\text{large}}$) by an impact. In the general case the precession is always coupled with a nutation. The heavy top will always begin its motion with an deviation toward the direction of the gravitational force, i.e., with a nutation. We still note that $\dot{\alpha}_{\text{small}}$ agrees with the precession frequency

$$\omega_p = \frac{mgl}{L} = \frac{mgl}{\Theta_3 \omega_0} \approx \frac{mgl}{\Theta_3 A}$$

obtained in the section "The heavy symmetric top: Elementary considerations."

Before we continue the discussion on the general motion of the top, we determine additional constants of motion. We have already seen that from the last equation of the system (13.51) we have

$$\dot{\alpha}\cos\beta + \omega_0 = A = \text{constant}, \tag{13.52b}$$

hence, the corresponding part of the kinetic energy is

$$T_3 = \frac{1}{2}\Theta_3(\dot{\alpha}\cos\beta + \omega_0)^2 = \frac{1}{2}\Theta_3 A^2 = \text{constant}. \tag{13.56}$$

Multiplying the first of the Euler equations (13.51) by $\dot{\beta}$ and the second one by $\dot{\alpha}\sin\beta$, after addition one gets the total differential

$$mgl\sin\beta \cdot \dot{\beta} = \Theta_1 \ddot{\beta}\dot{\beta} + \Theta_1(\ddot{\alpha}\dot{\alpha}\sin^2\beta + \dot{\alpha}^2\dot{\beta}\sin\beta\cos\beta)$$

or

$$\frac{d}{dt}(-mgl\cos\beta) = \frac{d}{dt}\left(\frac{1}{2}\Theta_1\dot{\beta}^2 + \frac{1}{2}\Theta_1\dot{\alpha}^2\sin^2\beta\right). \tag{13.57}$$

This means that the energy (more precisely, the sum of the kinetic parts $T_1 + T_2$ plus the potential energy)

$$E' = \frac{1}{2}\Theta_1(\dot{\beta}^2 + \dot{\alpha}^2\sin^2\beta) + mgl\cos\beta \tag{13.58}$$

MOTION OF THE HEAVY SYMMETRIC TOP

is also a constant of motion. This must be so, of course, since the total energy of the top must be constant.

The total energy of the top is then

$$E = E' + T_3$$
$$= \frac{1}{2}\Theta_1(\dot{\beta}^2 + \dot{\alpha}^2 \sin^2\beta) + \frac{1}{2}\Theta_3(\dot{\alpha}\cos\beta + \omega_0)^2 + mgl\cos\beta. \quad (13.59)$$

The last term obviously describes the potential energy of the top in the gravitational field. The energy law (13.59) must of course hold in general. We could have written it down immediately and skipped the derivation (13.57) from the Euler equations. Nevertheless, it is of interest to see how the equations of motion succeed too.

In the second Euler equation (13.52d), we insert

$$L_{z'} = \Theta_3 A = \text{constant} \quad (13.60)$$

and multiply by $\sin\beta$. This yields

$$\Theta_1(\ddot{\alpha}\sin^2\beta + 2\dot{\alpha}\dot{\beta}\sin\beta\cos\beta) - L_{z'}\sin\beta \cdot \dot{\beta} = 0. \quad (13.61)$$

Since $L_{z'}$ is constant, this is a total differential, and then it follows that

$$\Theta_1(\dot{\alpha}\sin^2\beta) + L_{z'}\cos\beta = \text{constant}. \quad (13.62)$$

This constant is the z-component of the angular momentum in the space-fixed system. This is seen immediately if we multiply the angular momentum

$$\mathbf{L} = \Theta_1(\omega_{x'}\mathbf{e}_{x'} + \omega_{y'}\mathbf{e}_{y'}) + L_{z'}\mathbf{e}_{z'} \quad (13.63)$$

by $\mathbf{e}_z$. From the equations (13.46) or from Figure 13.33, we see that

$$\mathbf{e}_{x'} \cdot \mathbf{e}_z = 0, \quad \mathbf{e}_{y'} \cdot \mathbf{e}_z = \sin\beta, \quad \mathbf{e}_{z'} \cdot \mathbf{e}_z = \cos\beta. \quad (13.64)$$

By noting that $\omega_{y'} = \dot{\alpha}\sin\beta$, we see that

$$\mathbf{L} \cdot \mathbf{e}_z = L_z = \Theta_1\dot{\alpha}\sin^2\beta + L_{z'}\cos\beta = \text{constant}. \quad (13.65)$$

The two angular components L_z and $L_{z'}$ are constant, because the moment of the gravitational force acts only in $\mathbf{e}_{x'}$-direction, i.e., perpendicular both to the z- as well as to the z'-axis. The conditions $L_z = $ constant and $L_{z'} = $ constant' can be realized by a precession of $\mathbf{L}$ about the z-axis, and an additional rotation of the z'-axis about the $\mathbf{L}$-axis. The latter motion is the *nutation*. This obviously means that the angular momentum $\mathbf{L}$ precesses about the laboratory axis $\mathbf{e}_z$, and the figure axis $\mathbf{e}_{z'}$ simultaneously performs a nutation about the angular momentum $\mathbf{L}$.

With the constants of motion we will now further discuss the motion of the top. From the equation of the angular momentum component L_z in the laboratory system,

$$\Theta_1\dot{\alpha}\sin^2\beta + L_{z'}\cos\beta = L_z, \quad (13.66)$$

we determine $\dot{\alpha}$:

$$\dot{\alpha} = \frac{L_z - L_{z'}\cos\beta}{\Theta_1\sin^2\beta}, \quad (13.67)$$

and insert this into equation (13.59):

$$\frac{1}{2}\Theta_1\dot\beta^2 + \frac{(L_z - L_{z'}\cos\beta)^2}{2\Theta_1 \sin^2\beta} + T_3 + mgl\cos\beta = E.$$

Since $L_z, L_{z'}, T_3$, and E are constants of motion, this is a differential equation for the nutation $\beta(t)$. We now substitute

$$u = \cos\beta, \tag{13.68}$$

then $\dot u = -\sin\beta \cdot \dot\beta$ and $\sin^2\beta = 1 - u^2$. From this, we get

$$\frac{1}{2}\Theta_1\frac{\dot u^2}{1-u^2} + \frac{(L_z - L_{z'}u)^2}{2\Theta_1(1-u^2)} + mglu = E - T_3 \tag{13.69}$$

or

$$\dot u^2 + \frac{(L_z - L_{z'}u)^2}{\Theta_1^2} + \frac{2mglu(1-u^2)}{\Theta_1} = \frac{2(1-u^2)}{\Theta_1}(E - T_3). \tag{13.70}$$

With the abbreviations

$$\varepsilon = 2\frac{E - T_3}{\Theta_1}, \qquad \xi = \frac{2mgl}{\Theta_1}, \qquad \gamma = \frac{L_z}{\Theta_1}, \qquad \delta = \frac{L_{z'}}{\Theta_1}, \tag{13.71}$$

this equation can be written as follows:

$$\dot u^2 = (\varepsilon - \xi u)(1 - u^2) - (\gamma - \delta u)^2. \tag{13.72}$$

It cannot be solved by elementary methods. We therefore give the graphical representation of the functional dependence. In the following we use the abbreviation $\dot u^2 = f(u)$. For large u the leading term is u^3, i.e., the curve approaches $f(u) = \xi u^3$. For $f(1)$ and $f(-1)$, we have

$$f(1) = -(\gamma - \delta)^2 \le 0, \qquad f(-1) = -(\gamma + \delta)^2 < 0. \tag{13.73}$$

From this, we obtain Figure 13.35.

In general, the function $f(u)$ has three zeros. Because of its asymptotic behavior for large, positive u and because $f(1) < 0$, for one zero we have $u_3 > 1$.

For the motion of the top, we must have $\dot u^2 \ge 0$. Since $0 \le \beta \le \pi/2$, we have $0 \le u \le 1$. To ensure that the top moves at all in the physically relevant region $0 \le u \le 1$, in a certain interval of this region we must have $\dot u^2 = f(u) > 0$. Hence, *for physical reasons two physically interesting zeros u_1, u_2 must exist between zero and unity.* Therefore in the general case there are two corresponding angles β_1 and β_2 with

$$\cos\beta_1 = u_1 \quad \text{and} \quad \cos\beta_2 = u_2. \tag{13.74}$$

In special cases, we can have (1) $u_1 = u_2$ and (2) $u_1 = u_2 = 1$. We first consider these special cases:

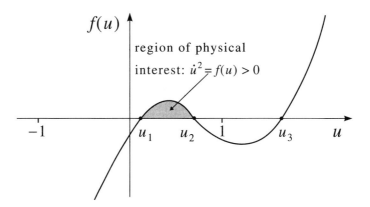

Figure 13.35. Qualitative trend of the function f(u).

(1) $u_1 = u_2 \neq 1$: The tip of the figure axis orbits on a circle (this is called *stationary precession*); no nutation occurs (the angle β has a fixed value). According to (13.67) the precession velocity reads

$$\dot{\alpha} = \frac{\gamma - \delta u}{1 - u^2} \tag{13.75}$$

and is constant.

(2) $u_1 = u_2 = 1$: In this case, the figure axis points vertically upward. The top performs neither nutation nor precession motion (*sleeping top*). This is obviously a special case of the stationary precession (compare Example 13.8).

In the general case ($u_1 \neq u_2$), a nutation of the top is superimposed on the precession between the angles β_1 and β_2. According to the angular momentum law (13.66), (13.67) for the precession velocity, we have

$$\dot{\alpha} = \frac{L_z - L_{z'} \cos \beta}{\Theta_1 \sin^2 \beta} = \frac{\gamma - \delta u}{1 - u^2}. \tag{13.76}$$

The zeros of this equation, i.e., the solution of $\dot{\alpha}(u) = 0$, specify those angles β at which the precession velocity $\dot{\alpha}$ momentarily vanishes. In order to illustrate the gyroscope motion, we give the curve described by the intersection point of the figure axis on a sphere centered about the bearing point.

There are three different types of motion, as illustrated in Figures 13.36, 13.37, and 13.38.

(1) $\gamma/\delta = u_2$: The precession velocity just vanishes at β_2; hence, the peaks appear.

(2) $u_1 < \gamma/\delta < u_2$: The upper peaks at β_2 extended to loops. The precession velocity vanishes between β_2 and β_1.

(3) $\gamma/\delta > u_2$: The precession velocity would vanish beyond β_2 (as indicated in Figure 13.38). A peak cannot arise.

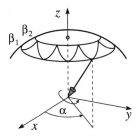
Figure 13.36. $\gamma/\delta = u_2$.

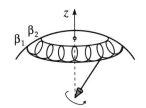
Figure 13.37. $u_1 < \gamma/\delta < u_2$.

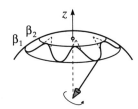
Figure 13.38. $\gamma/\delta > u_2$.

Example 13.10: The sleeping top

In the case of the so-called "sleeping top," the figure axis points up vertically, so that neither nutation nor precession occurs. For this special case, we must have $\beta = 0$ and $\dot\beta = 0$.

From energy conservation, we obtain

$$\frac{1}{2}\Theta_1(\dot\beta^2 + \dot\alpha^2 \sin^2\beta) + \frac{1}{2}\Theta_3 A^2 + mgl\cos\beta = E, \tag{13.77}$$

and because

$$\beta = 0, \quad \dot\beta = 0,$$

it follows that

$$\Theta_3 A^2 = 2(E - mgl). \tag{13.78}$$

Constancy of the z-component of the angular momentum yields

$$\Theta_1 \dot\alpha \sin^2\beta + \Theta_3 A \cos\beta = \text{constant} = K, \tag{13.79}$$

from which follows ($A = \omega_3 + \omega_0 = \text{constant}$)

$$\Theta_3 A = K. \tag{13.80}$$

For the quantities ε, ξ, γ, and δ in the differential equation for the nutation motion in $u = \cos\beta$, we have

$$\varepsilon = \frac{2(E - (1/2)\Theta_3 A^2)}{\Theta_1} = \frac{2mgl}{\Theta_1},$$

$$\xi = \frac{2mgl}{\Theta_1},$$

$$\gamma = \frac{K}{\Theta_1} = \frac{\Theta_3 A}{\Theta_1}, \tag{13.81}$$

$$\delta = \frac{\Theta_3 A}{\Theta_1},$$

$$\Rightarrow \quad \varepsilon = \xi \quad \text{and} \quad \gamma = \delta.$$

Inserting this into the differential equation (13.72) for u, one obtains

$$\dot u^2 = f(u) = \varepsilon(1-u)(1-u)(1+u) - \gamma^2(1-u)^2 \tag{13.82}$$

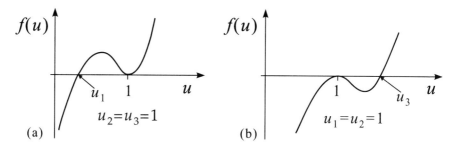

Figure 13.39.

$$\Rightarrow \quad f(u) = (1-u)^2[\varepsilon(1+u) - \gamma^2]. \tag{13.83}$$

Equation (13.83) has a twofold zero, which will be denoted by $u_2 = u_3 = 1$ or $u_1 = u_2 = 1$ (compare Figure 13.39). The third zero is at

$$\bar{u} = \frac{\gamma^2}{\varepsilon} - 1 = \frac{\Theta_3^2 A^2}{\Theta_1 2mgl} - 1. \tag{13.84}$$

Accordingly, $f(u)$ has one of the two courses (see Figure 13.39).

For Figure 13.39(a), $\dot{\beta}$ actually vanishes since $f(u)$ has a zero. Thus, we can have $\dot{\beta}$ = constant $\neq$ 0, i.e., the case of stationary precession. But since we also require $\dot{\beta} = 0$ for the "sleeping top," only Figure 13.39(b) is left over where $\dot{\beta} \neq 0$ does not exist as a solution ($u_1 \geq 1$).

Hence, from (13.84) we obtain as condition equations for the "sleeping top":

$$\frac{\Theta_3^2 A^2}{\Theta_1 2mgl} - 1 \geq 1 \quad \Leftrightarrow \quad A^2 \geq \frac{4mgl\Theta_1}{\Theta_3^2}. \tag{13.85}$$

Equation (13.85) will be satisfied only in the initial phase of the gyroscope motion. Because of friction, $A^2 = (\omega_3 + \omega_0)^2$ decreases, so that

$$A^2 < \frac{4mgl\Theta_1}{\Theta_3^2}$$

and therefore one observes precession with overlaid nutation. Further energy loss inevitably causes the top to tilt down.

Example 13.11: The heavy symmetric top

(a) Write the total energy of the top as a function of the Euler angles.

(b) Determine the constants of motion, and use them to eliminate the Euler angles α and γ from the energy law. Propose approaches for solving the resulting one-dimensional differential equation:

$$E = \frac{1}{2}\Theta_1 \dot{\beta}(t)^2 + V_{\text{eff}}(\beta). \tag{13.86}$$

(c) Discuss the effective potential $V_{\text{eff}}(\beta)$, and solve the differential equation of the heavy top for infinitesimal displacements from the stable position in the minimum of the potential:

$$\beta(t) = \beta_0 + \eta(t).$$

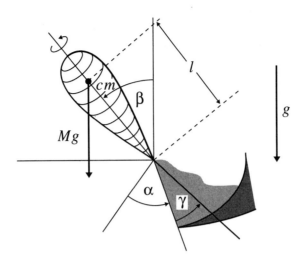

Figure 13.40.

We consider the symmetric top bound to a fixed point in the gravitation field. The energy law reads

$$E = T + V, \tag{13.87}$$

where

$$T = \frac{1}{2} \sum_i \Theta_i \omega_i^2 \quad \text{and} \quad V = M \cdot g \cdot h. \tag{13.88}$$

M is the mass of the top, and $h = l \cos \beta$ is the distance between the center of gravity and the bearing plane.

In order to express the angular velocity $\boldsymbol{\omega} = (\omega_{x'}, \omega_{y'}, \omega_{z'})$ by the Euler angles and their time derivatives, we note that $\dot{\alpha}, \dot{\beta}, \dot{\gamma}$ are rotation velocities by themselves. The rotation velocity of the body is obtained as the vector sum

$$\boldsymbol{\omega} = \boldsymbol{\omega}_\alpha + \boldsymbol{\omega}_\beta + \boldsymbol{\omega}_\gamma = \dot{\alpha} \mathbf{e}_\alpha + \dot{\beta} \mathbf{e}_\beta + \dot{\gamma} \mathbf{e}_\gamma. \tag{13.89}$$

The vectors $\mathbf{e}_\gamma, \mathbf{e}_\alpha, \mathbf{e}_\beta$ follow from the definition of the Euler angles:

γ: Rotation about the new (body-fixed) z'-axis:

$$\mathbf{e}_\gamma = \mathbf{e}_{z'} = (0, 0, 1). \tag{13.90}$$

α: Rotation about the space-fixed z-axis:

$$\mathbf{e}_\alpha = \mathbf{e}_z = (\sin \beta \sin \gamma, \sin \beta \cos \gamma, \cos \beta). \tag{13.91}$$

β: Rotation about the nodal line, x-axis:

$$\mathbf{e}_\beta = \mathbf{e}_x = (\cos \gamma, -\sin \gamma, 0). \tag{13.92}$$

One thus obtains the components of the rotation velocity in the body-fixed coordinate system

$$\omega_{x'} = \omega_1 = \dot{\alpha} \sin \beta \sin \gamma + \dot{\beta} \cos \gamma,$$
$$\omega_{y'} = \omega_2 = \dot{\alpha} \sin \beta \cos \gamma - \dot{\beta} \sin \gamma,$$

MOTION OF THE HEAVY SYMMETRIC TOP

$$\omega_{z'} = \omega_3 = \dot{\alpha}\cos\beta + \dot{\gamma}. \tag{13.93}$$

By inserting this into (13.88) and using $\Theta_1 = \Theta_2$, we obtain for the kinetic energy

$$T = \frac{1}{2}\Theta_1(\dot{\alpha}^2\sin^2\beta + \dot{\beta}^2) + \frac{1}{2}\Theta_3(\dot{\alpha}\cos\beta + \dot{\gamma})^2. \tag{13.94}$$

Since the gravitational force acts only along the z-direction, the torque acts only along the nodal line $\mathbf{e}_\beta$.

$$\begin{aligned}\mathbf{D} = \mathbf{r}\times\mathbf{F} &= -Mgl\mathbf{e}_{z'}\times\mathbf{e}_z \\ &= -Mgl\sin\beta\,\mathbf{e}_\beta.\end{aligned}$$

Thus, the angular momentum components in the $\mathbf{e}_z, \mathbf{e}_{z'}$-plane remain unchanged. L_z and $L_{z'}$ are constants of motion:

$$\begin{aligned}L_{z'} &= \Theta_3\omega_3 = \Theta_3(\dot{\alpha}\cos\beta + \dot{\gamma}) = \text{constant}, \\ L_z &= \mathbf{L}\cdot\mathbf{e}_z = \text{constant}.\end{aligned} \tag{13.95}$$

We evaluate the scalar product in the body-fixed coordinates:

$$\begin{aligned}L_z &= \mathbf{L}\cdot\mathbf{e}_z \\ &= \Theta_1(\dot{\alpha}\sin\beta\sin\gamma + \dot{\beta}\cos\gamma)(\sin\gamma\sin\beta) \\ &\quad + \Theta_1(\dot{\alpha}\sin\beta\cos\gamma - \dot{\beta}\sin\gamma)(\cos\gamma\sin\beta) \\ &\quad + \Theta_3(\dot{\alpha}\cos\beta + \dot{\gamma})(\cos\beta) \\ &= \Theta_1(\dot{\alpha}\sin^2\beta) + \Theta_3\cos\beta(\dot{\alpha}\cos\beta + \dot{\gamma}).\end{aligned} \tag{13.96}$$

Here, we utilized (13.93). The equations (13.95) and (13.96) can be inverted, i.e., solved for $\dot{\gamma}$ and $\dot{\alpha}$.

$$\dot{\alpha} = \frac{L_z - L_{z'}\cos\beta}{\Theta_1\sin^2\beta}, \tag{13.97}$$

$$\dot{\gamma} = L_{z'}\left(\frac{1}{\Theta_3} + \frac{\cot^2\beta}{\Theta_1}\right) - \frac{L_z\cos\beta}{\Theta_1\sin^2\beta}. \tag{13.98}$$

We insert the relations (13.97) and (13.98) obtained this way into the expression for the kinetic energy (13.94) and obtain

$$\begin{aligned}E &= \frac{1}{2}\Theta_1\dot{\beta}^2 + \frac{\Theta_1}{2}\left(\frac{L_z - L_{z'}\cos\beta}{\Theta_1\sin^2\beta}\right)^2\sin^2\beta \\ &\quad + \frac{1}{2}\Theta_3(\dot{\alpha}^2\cos^2\beta + \dot{\gamma}^2 + 2\dot{\alpha}\dot{\gamma}\cos\beta) + Mgl\cos\beta \\ &= \frac{1}{2}\Theta_1\dot{\beta}^2 + \frac{1}{2\Theta_1}\frac{(L_z - L_{z'}\cos\beta)^2}{\sin^2\beta} + \frac{1}{2}\frac{L_{z'}}{\Theta_3} + Mgl\cos\beta \\ &= \frac{1}{2}\Theta_1\dot{\beta}^2 + V_{\text{eff}}(\beta).\end{aligned} \tag{13.99}$$

We used the constancy of L_z and $L_{z'}$ to eliminate the two Euler angles α and γ. This simplified the problem greatly. From the energy law (13.99) we can in principle determine $\beta(t)$ and then obtain $\alpha(t)$ and $\gamma(t)$ via (13.97) and (13.98).

To proceed further, various possibilities offer themselves:

(a) We can establish the equation of motion for $\beta(t)$:

$$\frac{dE}{dt} = 0 = \Theta_1 \dot{\beta}\ddot{\beta} + \frac{\partial V_{\text{eff}}(\beta)}{\partial \beta} \dot{\beta}. \tag{13.100}$$

Hence, energy conservation leads to the equation of motion:

$$\Theta_1 \ddot{\beta} = -\frac{\partial}{\partial \beta} V_{\text{eff}}(\beta) \tag{13.101}$$

$$= \frac{1}{\Theta_1}(L_z - L_{z'}\cos\beta)^2 \frac{\cos\beta}{\sin^3\beta} - \frac{L_{z'}}{\Theta_1 \sin\beta}(L_z - L_{z'}\cos\beta) + Mgl\sin\beta$$

$$= \frac{\cos\beta}{\Theta_1 \sin^3\beta}(L_z^2 - 2L_z L_{z'}\cos\beta + L_{z'}^2) - \frac{L_z L_{z'}}{\Theta_1 \sin\beta} + Mgl\sin\beta. \tag{13.102}$$

This is a one-dimensional differential equation for β, although highly nonlinear. For a given solution $\beta(t)$, $\alpha(t)$, and $\gamma(t)$ can be found by integration of (13.97) and (13.98).

(b) Another principal approach is to solve the differential equation (13.99) by separation of variables and integration.

$$\frac{d\beta}{dt} = \sqrt{\frac{2}{\Theta_1}(E - V_{\text{eff}}(\beta))},$$

$$t - t_0 = \int_{\beta_0}^{\beta} d\beta' \frac{1}{\sqrt{2/\Theta_1(E - V_{\text{eff}}(\beta'))}}. \tag{13.103}$$

Thus, the time dependence of β can be determined by integration.

Since (a) and (b) are likewise complicated, we restrict ourselves to a discussion of the effective potential, in order to understand the essentials of gyroscopic motion.

Discussion of the effective potential

$$V_{\text{eff}} = \frac{1}{2\Theta_1}\frac{(L_z - L_{z'}\cos\beta)^2}{\sin^2\beta} + \frac{1}{2}\frac{L_{z'}}{\Theta_3} + Mgl\cos\beta. \tag{13.104}$$

The effective potential is composed of three terms which we will discuss separately.

Spin term

$$\frac{1}{2}\frac{L_{z'}}{\Theta_3}. \tag{13.105}$$

The second term is constant and is due to the energy of the eigenrotation of the top about its figure axis. It shifts the zero point of the energy scale and is independent of β.

Angular momentum barrier

$$\frac{1}{2\Theta_1}\frac{(L_z - L_{z'}\cos\beta)^2}{\sin^2\beta}. \tag{13.106}$$

The first term is understood by analogy to the $l^2/2mr^2$ angular momentum term, which appeared in the effective potential when treating the central force problem. It is positive and vanishes for $L_z/L_{z'} = \cos\beta$. Then β is only a physically meaningful angle if $L_z < L_{z'}$. This is in general

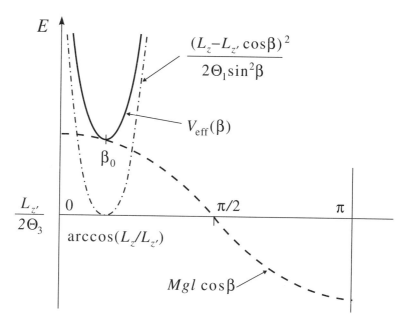

Figure 13.41. The effective potential is composed of three terms.

fulfilled for a top with fast eigenrotation. For $\beta = 0$, $\beta = \pi$, the first term diverges because of the factor $\sin^2 \beta$ in the denominator. As a consequence, the term then has a minimum which lies at $\beta = \arccos(L_z/L_{z'}) < \pi/2$. For $\beta \to 0$ and $\beta \to \pi$ the potential rises steeply.

Gravitation term

$$Mgl \cos \beta. \tag{13.107}$$

The last contribution is caused by the gravitational potential. It is antisymmetric about the center $\pi/2$ and shifts the minimum of the effective potential to the right side of $\arccos(L_z/L_{z'})$ without changing its qualitative form.

For a given energy E ($> L_z/2\Theta_3$), the motion is restricted to the region $E > V_{\text{eff}}(\beta)$ with reversal points $\beta_\pm$; these points are defined by $E = V_{\text{eff}}(\beta_\pm)$.

For a more precise analysis of the motion, we determine the stationary solution for which $\beta = \beta_0 = $ constant. It is located exactly in the minimum of the effective potential, so that the reversal points $\beta_\pm$ coincide, and $E = V_{\text{eff}}(\beta_0)$. β_0 is then determined from the minimum property:

$$\left(\frac{\partial V_{\text{eff}}}{\partial \beta} \right)_{\beta = \beta_0} = 0. \tag{13.108}$$

Hence, (13.102) leads to

$$L_z^2 - 2 L_z L_{z'} \cos \beta_0 + L_{z'}^2 = \frac{L_z L_{z'} \sin^2 \beta_0}{\cos \beta_0} - \frac{\Theta_1 Mgl \sin \beta_0}{\cos \beta_0}. \tag{13.109}$$

This equation fixes β_0 for given values of L_z and $L_{z'}$.

The equations (13.97), (13.98) imply for $\dot{\alpha}$ and $\dot{\gamma}$ constant values $\dot{\alpha}_0$ and $\dot{\gamma}_0$. Hence, in dynamic equilibrium the top performs a constant rotation about its own axis $\gamma(t) = \dot{\gamma}_0 t$ at fixed angle β_0, as well as a precession motion $\alpha(t) = \dot{\alpha}_0 t$ with constant precession frequency $\dot{\alpha}_0$.

Small oscillations about the dynamic equilibrium position

In order to investigate the motion in the vicinity of β_0, we consider small displacements from the equilibrium. Instead of explicitly solving equation (13.102), we write

$$\beta(t) = \beta_0 + \eta(t) \tag{13.110}$$

with an infinitesimal displacement $\eta(t)$. We expand the potential into a Taylor series

$$V_{\text{eff}}(\beta) = V_{\text{eff}}(\beta_0) + \eta \left(\frac{\partial V_{\text{eff}}}{\partial \beta} \right)_{\beta=\beta_0} + \frac{1}{2} \eta^2 \left(\frac{\partial^2 V_{\text{eff}}}{\partial t^2} \right)_{\beta=\beta_0} + \cdots . \tag{13.111}$$

The linear term vanishes by construction, and the quadratic term follows by differentiation of the negative right side of equation (13.102):

$$\frac{\partial^2 V_{\text{eff}}}{\partial \beta^2} = -Mgl\cos\beta - \frac{3L_z L_{z'} \cos\beta}{\Theta_1 \sin^2\beta}$$
$$+ (L_z^2 - 2L_z L_{z'} \cos\beta + L_{z'}^2) \frac{3 - 2\sin^2\beta}{\Theta_1 \sin^4\beta}. \tag{13.112}$$

By inserting equation (13.109) for the last term, one obtains

$$\left. \frac{\partial^2 V_{\text{eff}}}{\partial \beta^2} \right|_{\beta=\beta_0} = \frac{L_z L_{z'} - \Theta_1 Mgl(4 - 3\sin^2\beta_0)}{\Theta_1 \cos\beta_0}. \tag{13.113}$$

For the total energy, one then obtains likewise

$$E = \frac{1}{2}\Theta_1 \dot{\eta}^2 + \frac{1}{2}\eta^2 \left(\frac{\partial^2 V_{\text{eff}}}{\partial \beta^2} \right)\bigg|_{\beta=\beta_0} + V_{\text{eff}}(\beta_0). \tag{13.114}$$

Differentiation with respect to the time finally leads to the differential equation of the harmonic oscillator:

$$\ddot{\eta} + \Omega^2 \eta = 0 \tag{13.115}$$

with

$$\Omega^2 = \frac{1}{\Theta_1} \left(\frac{\partial^2 V_{\text{eff}}}{\partial \beta^2} \right)_{\beta=\beta_0} = \frac{L_z L_{z'} - \Theta_1 Mgl(4 - 3\sin^2\beta_0)}{\Theta_1^2 \cos\beta_0}. \tag{13.116}$$

The corresponding motion

$$\beta(t) = \beta_0 + \eta_0 \cos(\Omega t + \Phi_0) \tag{13.117}$$

is stable if $\Omega^2 > 0$. Obviously the product $L_z L_{z'}$ must be sufficiently large to ensure stable vibrations.

Precession and nutation

We insert the explicit solution of (13.115) into (13.97) and (13.98), and expand with respect to $\eta(t)$:

$$\dot{\alpha}(t) \approx \frac{L_z - L_{z'} \cos\beta_0}{\Theta_1 \sin^2\beta_0} + \eta(t) \frac{\partial}{\partial \beta} \left(\frac{L_z - L_{z'} \cos\beta}{\Theta_1 \sin^2\beta} \right)_{\beta=\beta_0} + \cdots \tag{13.118}$$

MOTION OF THE HEAVY SYMMETRIC TOP

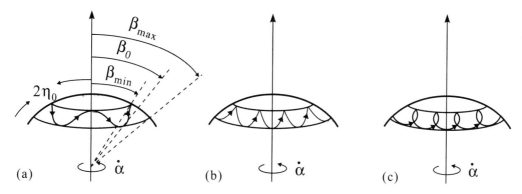

Figure 13.42.

$$\equiv \dot{\alpha}_0 + \eta(t)\dot{\alpha}_1,$$
$$\dot{\gamma}(t) \simeq \dot{\gamma}_0 + \eta(t)\dot{\gamma}_1, \tag{13.119}$$

where $\dot{\alpha}_0, \dot{\alpha}_1, \dot{\gamma}_0, \dot{\gamma}_1$ are constants which depend on $L_z, L_{z'}$, and E (through β_0). For a qualitative investigation of the superposition of nutation ($\beta(t)$) and precession ($\alpha(t)$), we start from equation (13.118):

$$\dot{\alpha}(t) \simeq \dot{\alpha}_0 + \eta_0 \dot{\alpha}_1 \cos(\Omega t + \varphi_0)$$
$$= \dot{\alpha}_1 \eta_0 \left(\frac{\dot{\alpha}_0}{\dot{\alpha}_1 \eta_0} + \cos(\Omega t + \varphi_0) \right). \tag{13.120}$$

For $\dot{\alpha}_0/\dot{\alpha}_1 \eta_0 > 1$, $\dot{\alpha}$ always remains larger than zero (Figure 13.42(a)). For $\dot{\alpha}_0/\dot{\alpha}_1 \eta_0 = 1$, the precession frequency may become equal to zero (Figure 13.42(b)). For $\dot{\alpha}_0/\dot{\alpha}_1 \eta_0 < 1$, we have a backward motion in parts (Figure 13.42(c)).

Example 13.12:   Exercise: Stable and unstable rotations of the asymmetric top

Use the Euler equations to show that for an asymmetric top the rotations about the axes of the largest and smallest moment of inertia are stable; however, the rotation about the axis of the intermediate moment of inertia is unstable.

Solution   We start from the Euler equations for the free top:

$$\dot{\omega}_1 = \frac{\Theta_2 - \Theta_3}{\Theta_1} \omega_2 \omega_3, \tag{13.121}$$

$$\dot{\omega}_2 = \frac{\Theta_3 - \Theta_1}{\Theta_2} \omega_1 \omega_3, \tag{13.122}$$

$$\dot{\omega}_3 = \frac{\Theta_1 - \Theta_2}{\Theta_3} \omega_1 \omega_2. \tag{13.123}$$

Let the top rotate about the body-fixed z-axis, i.e., $\omega_3 = \omega_0 =$ constant and $\omega_1 = \omega_2 = 0$. To investigate the stability of the rotation about this principal axis, we tilt the rotation axis by a small amount, so that new components $\delta\omega_1, \delta\omega_2$ and an additional $\delta\omega_3$ arise. For $\delta\dot{\omega}_3$, we have from the Euler equation

$$\delta\dot{\omega}_3 = \frac{\Theta_1 - \Theta_2}{\Theta_3}\delta\omega_1\delta\omega_2 \simeq 0. \tag{13.124}$$

Neglecting quadratic small terms, we can set $\omega_3 = \omega_0$. From the other two Euler equations, we then obtain

$$\delta\dot{\omega}_1 + \frac{\Theta_3 - \Theta_2}{\Theta_1}\delta\omega_2\omega_0 = 0, \tag{13.125}$$

$$\delta\dot{\omega}_2 + \frac{\Theta_1 - \Theta_3}{\Theta_2}\delta\omega_1\omega_0 = 0. \tag{13.126}$$

To solve this coupled system, we use the *ansatz*

$$\delta\omega_1 = Ae^{\lambda t},$$
$$\delta\omega_2 = Be^{\lambda t}. \tag{13.127}$$

This leads to a linear set of equations in A and B, where the determinant must vanish for nontrivial solutions:

$$\begin{vmatrix} \lambda & \frac{\Theta_3 - \Theta_2}{\Theta_1}\omega_0 \\ \frac{\Theta_1 - \Theta_3}{\Theta_2}\omega_0 & \lambda \end{vmatrix} = 0. \tag{13.128}$$

From this, we find the characteristic equation

$$\lambda^2 = \omega_0^2 \frac{(\Theta_3 - \Theta_2)(\Theta_1 - \Theta_3)}{\Theta_1\Theta_2}. \tag{13.129}$$

For the rotation about the axis of the smallest moment of inertia $\Theta_3 < \Theta_1, \Theta_2$, and for the rotation about the axis of the largest moment of inertia $\Theta_3 > \Theta_1, \Theta_2$, equation (13.129) leads to a purely imaginary λ:

$$\lambda^2 < 0, \tag{13.130}$$

and therefore to vibration solutions for $\delta\omega_1$ and $\delta\omega_2$. The rotation about the axis of the largest and smallest moment of inertia, respectively, is therefore stable.

The rotation about the axis of the intermediate moment of inertia

$$\Theta_1 > \Theta_3 > \Theta_2 \quad \text{or} \quad \Theta_2 > \Theta_3 > \Theta_1 \tag{13.131}$$

leads to a real λ and thus to a time evolution of $\delta\omega_1$ and $\delta\omega_2$ according to

$$\delta\omega_{1/2} = C_{1/2}\cosh\lambda t + D_{1/2}\sinh\lambda t. \tag{13.132}$$

The rotation axis turns away exponentially from the initial position. The rotation about the axis of the intermediate moment of inertia is not stable!

PART V

LAGRANGE EQUATIONS

14 Generalized Coordinates

In many cases, the motion of bodies considered in mechanics is not free but is restricted by certain constraint conditions. The constraints can take different forms. For instance, a mass point can be bound to a space curve or to a surface. The constraints for a rigid body state that the distances between the individual points are constant. If one considers gas molecules in a vessel, the constraints specify that the molecules cannot penetrate the wall of the vessel. Since the constraints are important for solving a mechanical problem, mechanical systems are classified according to the type of constraints. A system is called *holonomic* if the constraints can be represented by equations of the form

$$f_k(\mathbf{r}_1, \mathbf{r}_2, \ldots, t) = 0, \qquad k = 1, 2, \ldots, s. \tag{14.1}$$

This form of the constraints is important since it can be used for eliminating dependent coordinates. For a pendulum of length l the equation (14.1) reads $x^2 + y^2 - l^2 = 0$ if we put the coordinate origin at the suspension point. The coordinates x and y can be expressed by this equation.

We already met another simple example of holonomic constraints in the context of the rigid body, i.e., the constancy of the distances between two points: $(\mathbf{r}_i - \mathbf{r}_j)^2 - C_{ij}^2 = 0$. In this case the constraints served to reduce the $3N$ degrees of freedom of a system of N mass points to the 6 degrees of freedom of the rigid body.

All constraints that cannot be represented in the form (14.1) are called *nonholonomic*. These are conditions that cannot be described by a closed form or by inequalities. An example of this type of constraint is the system of gas molecules enclosed in a sphere of radius R. Their coordinates must satisfy the conditions $r_i \leq R$.

A further classification of the constraint conditions is made based on their time dependence. If the constraint is an explicit function of the time, then it is called *rheonomic*. If the time does not enter explicitly, the constraint is called *scleronomic*. A rheonomic constraint appears if a mass point moves along a moving space curve, or if gas molecules are enclosed in a sphere with a time-dependent radius.

In certain cases the constraints may also be given in differential form, for example if there is a condition on velocities, e.g., for the rolling of a wheel. The constraints then have the form

$$\sum_{k}^{N} a_k(x_1, x_2, \ldots, x_N) dx_k = 0, \tag{14.2}$$

where the x_k represent the various coordinates, and the a_k are functions of these coordinates. We now have to distinguish between two cases.

If the equation (14.2) represents the total differential of a function U, we can integrate it immediately and obtain an equation of the form of equation (14.1). In this case the constraints are holonomic. If equation (14.2) is not a total differential, we can integrate it only after having solved the full problem. Then equation (14.2) is not suitable for eliminating dependent coordinates; it is nonholonomic.

From the requirement that equation (14.2) be a total differential, one can derive a criterion for the holonomity of differential constraints. One must have

$$\sum_k a_k \, dx_k = dU \quad \text{with} \quad a_k = \frac{\partial U}{\partial x_k}.$$

This leads to

$$\frac{\partial a_k}{\partial x_i} = \frac{\partial^2 U}{\partial x_i \partial x_k} = \frac{\partial a_i}{\partial x_k}.$$

Thus, equation (14.2) represents a holonomic constraint if the coefficients obey the *integrability conditions*

$$\frac{\partial a_k}{\partial x_i} = \frac{\partial a_i}{\partial x_k}.$$

These only mean that the "vector" $\mathbf{a} = \{a_1, a_2, \ldots, a_N\}$ must be rotation-free (irrotational). In N-dimensional space, the situation is analogous.

To classify a mechanical system, we additionally specify whether the system is conservative or not.

Example 14.1: Small sphere rolls on a large sphere

A sphere in the gravitational field rolls without friction from the upper pole of a larger sphere. The system is conservative. The constraints change completely after getting away from the sphere and

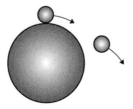

Figure 14.1. A small sphere rolls on a large sphere.

GENERALIZED COORDINATES

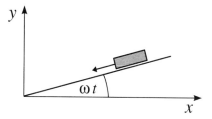

Figure 14.2.

cannot be represented in the closed form of equation (14.1), and therefore the system is nonholonomic. Since the time does not enter explicitly, the system is scleronomic.

Example 14.2: Body glides on an inclined plane

A body glides with friction down on an inclined plane (see Figure 14.2). The inclination angle of the plane varies with time. The coordinates and the inclination angle are related by

$$\frac{y}{x} - \tan \omega t = 0.$$

Thus, the time occurs explicitly in the constraint. The system is holonomic and rheonomic. Since friction occurs, the system is furthermore not conservative.

Example 14.3: Wheel rolls on a plane

An example of a system with differential constraints is a wheel that rolls on a plane without gliding. The wheel cannot fall over. The radius of the wheel is a.

For the calculation, we use the coordinates x_M, y_M of the center, the angle φ that describes the rotation, and the angle ψ that characterizes the orientation of the wheel plane relative to the y-axis.

The velocity v of the wheel center and the rotation velocity are related by the rolling condition

$$v = a\dot{\varphi}.$$

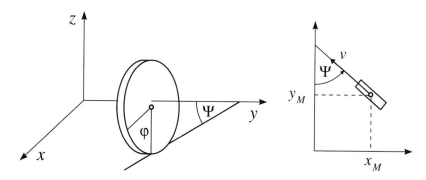

Figure 14.3.

The components of the velocity are

$$\dot{x}_M = -v \sin \psi,$$
$$\dot{y}_M = v \cos \psi.$$

By inserting v, we obtain

$$dx_M + a \sin \psi \cdot d\varphi = 0,$$
$$dy_M - a \cos \psi \cdot d\varphi = 0,$$

i.e., a constraint of the type of equation (14.2).

Since the angle ψ is known only after solving the problem, the equations are not integrable. Hence, the problem is nonholonomic, scleronomic, and conservative.

If a body moves along a trajectory specified (or restricted) by constraints, there appear constraint reactions that keep it on this trajectory. Such constraint reactions are support forces, bearing forces (-moments), string tensions, etc. If one is not especially interested in the load of a string or a bearing, one tries to formulate the problem in such a way that the constraint (and thus the constraint reaction) no longer appears in the equations to be solved. We have tacitly used this approach in the problems treated so far. A simple example is the plane pendulum. Instead of the formulation in Cartesian coordinates, where the constraint $x^2 + y^2 = l^2$ must be considered explicitly, we use polar coordinates (r, φ).

The constancy of the pendulum length means that the r-coordinate remains constant and that the motion of the pendulum can be completely described by the angle coordinate alone. This procedure—the transformation to coordinates adapted to the problem—shall now be formulated more generally.

If we consider a system of n mass points, then it is described by $3n$ coordinates $\mathbf{r}_1, \mathbf{r}_2, \ldots, \mathbf{r}_n$. The number of degrees of freedom also equals $3n$. If there are s *constraints*, the number of degrees of freedom reduces to $3n - s$. The set of originally $3n$ independent coordinates now involves s dependent coordinates. Now the meaning of the holonomic constraints becomes transparent. If the constraints are expressed by equations of the form (14.1), the dependent coordinates can be eliminated. We can transform to $3n - s$ coordinates $q_1, q_2, \ldots, q_{3n-s}$ that implicitly incorporate the constraints and that are independent of each other. The old coordinates $\mathbf{r}_i$ are expressed by the new coordinates q_j by means of equations of the form

$$\mathbf{r}_1 = \mathbf{r}_1(q_1, q_2, \ldots, q_{3n-s}, t),$$
$$\mathbf{r}_2 = \mathbf{r}_2(q_1, q_2, \ldots, q_{3n-s}, t),$$
$$\vdots$$
$$\mathbf{r}_n = \mathbf{r}_n(q_1, q_2, \ldots, q_{3n-s}, t).$$

(14.3)

These coordinates q_i, which now can be considered free, are called *generalized coordinates*. In the practical cases considered here, the choice of the generalized coordinates is already suggested by the formulation of the problem, and the transformation equations (14.3) need not be established explicitly. Using generalized coordinates is also helpful for problems

GENERALIZED COORDINATES

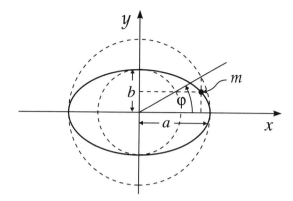

Figure 14.4. Ellipse: $y = b\sin\varphi$, $x = a\cos\varphi$.

without constraint conditions. For instance, a central force problem can be described more simply and completely by the coordinates (r, ϑ, φ) instead of the (x, y, z).

As a rule, lengths and angles serve as generalized coordinates. As will be seen below, moments and energies etc. can also be used as generalized coordinates.

Example 14.4: Generalized coordinates

An ellipse is given in the x,y-plane. A particle moving on the ellipse has the coordinates (x, y). The Cartesian coordinates can be expressed by the parameter φ:

$$y = b\sin\varphi, \qquad x = a\cos\varphi.$$

Thus, the motion of the particle can be completely described by the angle φ (the generalized coordinate φ).

Example 14.5: Cylinder rolls on an inclined plane

The position of a cylinder on an inclined plane is completely specified by the distance l from the origin to the center of mass and by the rotational angle φ of the cylinder about its axis.

If the cylinder glides on the plane, both generalized coordinates are significant.

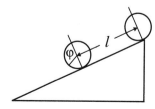

Figure 14.5. A cylinder rolls on an inclined plane.

If the cylinder does not glide, l depends on φ through a rolling condition. Only one of the two generalized coordinates will then be needed for a complete description of the motion of the cylinder.

Example 14.6: Exercise: Classification of constraints

Classify the following systems according to whether or not they are scleronomic or rheonomic, holonomic or nonholonomic, and conservative or nonconservative:

(a) a sphere rolling downward without friction on a fixed sphere;

(b) a cylinder rolling down on a rough inclined plane (inclination angle α);

(c) a particle gliding on the rough inner surface of a rotation paraboloid; and

(d) a particle moving without friction along a very long bar. The bar rotates with the angular velocity ω in the vertical plane about a horizontal axis.

Solution

(a) Scleronomic, since the constraint is not an explicit function of time. Nonholonomic, since the rolling sphere leaves the fixed sphere. Conservative, since the gravitational force can be derived from a potential.

(b) Scleronomic, holonomic, nonconservative: The equation of the constraint represents either a line or a surface. Since the surface is rough, friction occurs. Therefore this system is not conservative.

(c) Scleronomic, holonomic, but not conservative, since the friction force does not result from a potential!

(d) Rheonomic: The constraint is an explicit function of time. Holonomic: The equation of the constraint is a straight line that contains the time explicitly; conservative.

Quantities of mechanics in generalized coordinates

The velocity of the mass point i can according to the transformation equation

$$\mathbf{r}_i = \mathbf{r}_i(q_1, \ldots, q_\nu, t)$$

be represented as

$$\dot{\mathbf{r}}_i = \frac{\partial \mathbf{r}_i}{\partial q_1}\frac{dq_1}{dt} + \cdots + \frac{\partial \mathbf{r}_i}{\partial q_\nu}\frac{dq_\nu}{dt} + \frac{\partial \mathbf{r}_i}{\partial t}.$$

In the scleronomic case, the last term drops. The velocity can also be written in the form

$$\dot{\mathbf{r}}_i = \sum_\alpha^f \frac{\partial \mathbf{r}_i}{\partial q_\alpha}\dot{q}_\alpha + \frac{\partial \mathbf{r}_i}{\partial t}, \quad \text{where} \quad \dot{q}_\alpha = \frac{dq_\alpha}{dt} \tag{14.4}$$

and $\dot{q}_\alpha$ denotes the *generalized velocity*. In the following, we restrict ourselves to the x-component. Moreover, we consider only the scleronomic case and write for the x-component of equation (14.4)

$$\dot{x}_i = \sum_\alpha \frac{\partial x_i}{\partial q_\alpha} \dot{q}_\alpha. \tag{14.5}$$

By differentiating (14.5) once again with respect to time, we obtain for the Cartesian components of the acceleration

$$\ddot{x}_i = \sum_\alpha \frac{d}{dt}\left(\frac{\partial x_i}{\partial q_\alpha}\right) \dot{q}_\alpha + \frac{\partial x_i}{\partial q_\alpha} \ddot{q}_\alpha.$$

The total derivative in the first term is written as usual:

$$\frac{d}{dt}\left(\frac{\partial x_i}{\partial q_\alpha}\right) = \sum_\beta \frac{\partial^2 x_i}{\partial q_\beta \partial q_\alpha} \dot{q}_\beta.$$

The index to be summed over is denoted here by the letter β, to avoid confusion with the summation index α. Then we have

$$\ddot{x}_i = \sum_{\alpha,\beta} \frac{\partial^2 x_i}{\partial q_\beta \partial q_\alpha} \dot{q}_\beta \dot{q}_\alpha + \sum_\alpha \frac{\partial x_i}{\partial q_\alpha} \ddot{q}_\alpha.$$

The first term involves double summation over α and β.

Let a system have the generalized coordinates $q_1, \ldots, q_f$ that now shall be increased by $dq_1, \ldots, dq_f$. We will determine the work performed by this infinitesimal displacement. For an infinitesimal displacement of the particle i, we have

$$d\mathbf{r}_i = \sum_{\alpha=1}^f \frac{\partial \mathbf{r}_i}{\partial q_\alpha} dq_\alpha. \tag{14.6}$$

From this, we obtain the work performed:

$$dW = \sum_{i=1}^n \mathbf{F}_i \cdot d\mathbf{r}_i = \sum_{i=1}^n \left(\sum_{\alpha=1}^f \mathbf{F}_i \cdot \frac{\partial \mathbf{r}_i}{\partial q_\alpha} \right) dq_\alpha = \sum_\alpha Q_\alpha dq_\alpha,$$

where

$$Q_\alpha = \sum_i \mathbf{F}_i \cdot \frac{\partial \mathbf{r}_i}{\partial q_\alpha}. \tag{14.7}$$

Q_α is called the *generalized force*. Since the generalized coordinate must not have the dimension of a length, Q_α must not have the dimension of a force. The product $Q_\alpha q_\alpha$, however, always has the dimension of work.

In conservative systems, i.e., if W does not depend on time, one has

$$dW = \sum_\alpha \frac{\partial W}{\partial q_\alpha} dq_\alpha \quad \text{and} \quad dW = \sum_\alpha Q_\alpha dq_\alpha.$$

Then we must have

$$dW - dW = 0 = \sum_\alpha \left(Q_\alpha - \frac{\partial W}{\partial q_\alpha} \right) dq_\alpha = 0.$$

Since the q_α are generalized coordinates, they are independent of each other, and therefore it follows that $(Q_\alpha - \partial W/\partial q_\alpha) = 0$ in order to satisfy the equation

$$\sum_\alpha \left(Q_\alpha - \frac{\partial W}{\partial q_\alpha} \right) dq_\alpha = 0.$$

But this holds only if

$$Q_\alpha = \frac{\partial W}{\partial q_\alpha}.$$

The components of the generalized force are thus obtained as the derivative of the work with respect to the corresponding generalized coordinate.

15 D'Alembert[1] Principle and Derivation of the Lagrange Equations

Virtual displacements

A virtual displacement $\delta \mathbf{r}$ is an infinitesimal displacement of the system that is compatible with the constraints. Contrary to the case of a real infinitesimal displacement $d\mathbf{r}$, in a virtual displacement the forces and constraints acting on the system do not change. A virtual displacement will be characterized by the symbol δ, a real displacement by d. Mathematically we operate with the element δ just as with a differential. For example,

$$\delta \sin x = \frac{\delta \sin x}{\delta x} \delta x = (\cos x) \delta x, \quad \text{etc.}$$

We consider a system of mass points in equilibrium. Then the total force $\mathbf{F}_i$ acting on each individual mass point vanishes; hence, $\mathbf{F}_i = 0$. The product of force and virtual

---

[1] *Jean le Rond d'Alembert*, b. Nov. 16 or 17, 1717, Paris, as the son of a general–d. Oct. 29, 1783, Paris. D'Alembert, who was abandoned by his mother, was found near the church Jean le Rond and was brought up by the family of a glazier. Later he was educated according to his social status, supported by grants. He studied at the Collège des Quatre Nations, and in 1741, he became a member of the Académie des sciences. In mechanics, the d'Alembert principle is named after him; moreover, he worked on the theory of analytic functions (1746), on partial differential equations (1747), and on the foundations of algebra. D'Alembert is the author of the mathematical articles of the *Encyclopédie*.

displacement $\mathbf{F}_i \cdot \delta \mathbf{r}_i$ is called the *virtual* work. Since the force for each individual mass point vanishes, the sum over the virtual work performed on the individual mass points also equals zero:

$$\sum_i \mathbf{F}_i \cdot \delta \mathbf{r}_i = 0. \tag{15.1}$$

The force $\mathbf{F}_i$ will now be subdivided into the constraint reaction $\mathbf{F}_i^z$ and the acting (imposed) force $\mathbf{F}_i^a$:

$$\sum_i (\mathbf{F}_i^a + \mathbf{F}_i^z) \cdot \delta \mathbf{r}_i = 0. \tag{15.2}$$

We now restrict ourselves to such systems where the work performed by the constraint reactions vanishes. In many cases (except, e.g., for those with friction) the constraint reaction is perpendicular to the direction of motion, and the product $\mathbf{F}^z \cdot \delta \mathbf{r}$ vanishes. For instance, if a mass point is forced to move along a given spatial curve, its direction of motion is always tangential to the curve; the constraint reaction points perpendicular to the curve. There are, however, examples where the individual constraint reactions perform work, while the sum of the works of all constraint forces vanishes; thus,

$$\sum_i \mathbf{F}_i^z \cdot \delta \mathbf{r}_i = 0.$$

The string tensions of two masses hanging on a roller represent such a case. We refer to example 15.1. This is the proper, true meaning of the d'Alembert principle: The constraint reactions *in total* do not perform work. We always have

$$\sum_i \mathbf{F}_i^z \cdot \delta \mathbf{r}_i = 0.$$

This is *the* fundamental characteristic of the constraint reactions. One can, of course, trace this presupposition back to Newton's axiom "action equals reaction," as we just have seen in the example of the string tensions between two masses. But in general it does not follow from Newton's axioms alone. The assumption that the *total virtual work of the constraint reactions vanishes* can be considered to be a new postulate. It accounts for systems of not freely movable mass points and can be expressed by the forces imposed on the system, as we shall see below (see equation (15.5)). Then the constraint position drops out from equation (14.2), and one has

$$\sum_i \mathbf{F}_i^a \cdot \delta \mathbf{r}_i = 0. \tag{15.3}$$

While in equation (15.1) each term vanishes individually, now only the sum in total vanishes. The statement of equation (15.3) is called the *principle of virtual work*. It says that a system is only in equilibrium if the entire virtual work of the *imposed* (external) forces vanishes. In the next chapter (equations (16.8) and (16.9)) the principle of virtual work (the *total* virtual work vanishes) will be established by the Lagrangian formalism.

For holonomic constraints, the effect of the constraint reactions can be elucidated by the following: If we consider the ith constraint in the form

$$g_i(\mathbf{r}_1, \mathbf{r}_2, \ldots, \mathbf{r}_N, t) = 0,$$

then the change of g_i with respect to a change of the position vector $\mathbf{r}_j$ must be a measure of the constraint reaction $\mathbf{F}_{ji}^z$ on the jth particle due to the constraint $g_i(\mathbf{r}_1, \mathbf{r}_2, \ldots, \mathbf{r}_N, t) = 0$. We thus can write

$$\mathbf{F}_{ji}^z = \lambda_i \frac{\partial g_i(\mathbf{r}_1, \mathbf{r}_2, \ldots, \mathbf{r}_N, t)}{\partial \mathbf{r}_j} = \lambda_i \nabla_j g_i(\mathbf{r}_1, \ldots, t).$$

Here λ_i is an unknown factor, since the constraints $g_i(\mathbf{r}_1, \mathbf{r}_2, \ldots, \mathbf{r}_N, t) = 0$ are known up to a nonvanishing factor. The total constraint reaction on the jth particle is then the sum over all constraint reactions originating from the individual k constraints; hence,

$$\mathbf{F}_j^z = \sum_{i=1}^{k} \mathbf{F}_{ji}^z = \sum_{i=1}^{k} \lambda_i \frac{\partial g_i(\mathbf{r}_1, \ldots, \mathbf{r}_N, t)}{\partial \mathbf{r}_j}.$$

The virtual work performed by all constraints is then

$$\delta W = \sum_{j=1}^{N} \mathbf{F}_j^z \cdot \delta \mathbf{r}_j = \sum_{i=1}^{k} \sum_{j=1}^{N} \lambda_i \frac{\partial g_i}{\partial \mathbf{r}_j}(\mathbf{r}_1, \ldots, \mathbf{r}_N, t) \cdot \delta \mathbf{r}_j$$

$$= \sum_{i=1}^{k} \lambda_i \delta g_i(\mathbf{r}_1, \ldots, \mathbf{r}_N, t),$$

where

$$\delta g_i(\mathbf{r}_1, \ldots, \mathbf{r}_N, t) = \sum_{j=1}^{N} \frac{\partial g_i}{\partial \mathbf{r}_j} \cdot \delta \mathbf{r}_j.$$

This is just the change of g_i caused by the virtual displacements $\delta \mathbf{r}_j$. Since the virtual displacements are by assumption compatible with the constraints, i.e., the $\delta \mathbf{r}_j$ satisfy the constraints, we must have

$$\delta g_i(\mathbf{r}_1, \ldots, \mathbf{r}_N, t) = 0.$$

From this, we see immediately that

$$\mathbf{F}_i^z \cdot \mathbf{r}_i = 0 \tag{15.4a}$$

and therefore also

$$\delta W = \sum_{j=1}^{N} \mathbf{F}_j^z \cdot \delta \mathbf{r}_j = 0. \tag{15.4b}$$

Hence, for holonomic constraints the constraint reactions are perpendicular to the displacements that are compatible with the constraints, and the virtual work of the *individual* constraint reactions vanishes. In Chapter 16, equations (16.8) and (16.9), we shall understand from a very general point of view that in the general case (hence including the case of nonholonomic constraints), the sum of the virtual work of all constraint reactions must vanish. Therefore, $\sum_i \mathbf{F}_i^z \cdot \delta \mathbf{r}_i = 0$ always holds, while $\mathbf{F}_i^z \cdot \delta \mathbf{r}_i = 0$ holds only in special (holonomic) cases.

The principle of virtual work at first only allows us to treat problems of statics. By introducing the inertial force according to Newton's axiom

$$\mathbf{F}_i = \dot{\mathbf{p}}_i, \tag{15.5}$$

D'Alembert succeeded in applying the principle of virtual work to problems of dynamics as well. We proceed in an analogous way to derive the principle of virtual work. Because of equations (15.4a) and (15.4b) in the sum

$$\sum_i (\mathbf{F}_i - \dot{\mathbf{p}}_i) \cdot \delta \mathbf{r}_i = 0, \tag{15.6a}$$

every individual term vanishes. If we again subdivide the total force $\mathbf{F}_i$ into the imposed force $\mathbf{F}_i^a$ and the constraint reaction $\mathbf{F}_i^z$, with the same restriction as above we find the equation

$$\sum_i (\mathbf{F}_i^a - \dot{\mathbf{p}}_i) \cdot \delta \mathbf{r}_i = 0, \tag{15.6b}$$

where the individual terms can differ from zero; only the sum in (15.6b) vanishes. This equation expresses the *d'Alembert principle*.

Example 15.1: Two masses on concentric rollers

Two masses m_1 and m_2 hang on two concentrically fixed rollers with the radii R_1 and R_2. The mass of the rollers can be neglected. The equilibrium condition shall be determined by means of the principle of virtual work.

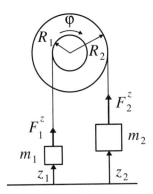

Figure 15.1. Two masses on concentric rollers: The string tensions $\mathbf{F}_1^z$ and $\mathbf{F}_2^z$ are parallel but have different magnitudes.

For the conservative system under consideration (where no friction appears), the total work performed by the constraint reactions vanishes, i.e.,

$$\sum_i \mathbf{F}_i^z \cdot \delta \mathbf{r}_i = 0.$$

In the present example, the constraint forces are the string tensions F_1^z and F_2^z.

The vanishing of $\sum_i \mathbf{F}_i^z \cdot \delta \mathbf{r}_i$ *in the equilibrium state* is equivalent to the equality of the *torques* imposed by the string tensions F_1^z, F_2^z through the radii R_1, R_2:

$$D_1 = R_1 F_1^z = D_2 = R_2 F_2^z.$$

By means of the constraint, it follows with $\delta z_1 = R_1 \delta \varphi$, $\delta z_2 = -R_2 \delta \varphi$, that

$$F_1^z \delta z_1 + F_2^z \delta z_2 = (F_1^z R_1 - F_2^z R_2) \delta \varphi = (D_1 - D_2) \delta \varphi = 0.$$

In the case of equal radii ($R_1 = R_2$), the string tensions are equal.
From

$$\sum_i \mathbf{F}_i^a \cdot \delta \mathbf{r}_i = 0,$$

it follows that

$$m_1 g \delta z_1 + m_2 g \delta z_2 = 0.$$

The displacements are correlated by the constraint condition; we have

$$\delta z_1 = R_1 \delta \varphi, \qquad \delta z_2 = -R_2 \delta \varphi.$$

Hence, we obtain

$$(m_1 R_1 - m_2 R_2) \delta \varphi = 0$$

or

$$m_1 R_1 = m_2 R_2$$

as the equilibrium condition.

Example 15.2: Two masses connected by a rope on an inclined plane

In the setup shown in Figure 15.2, two masses connected by a rope move without friction. The

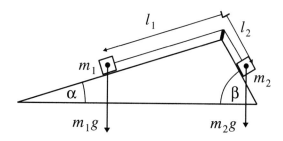

Figure 15.2. Two masses on an inclined plane connected by a rope.

equation of motion shall be established by means of the d'Alembert principle. For the two masses, this principle reads

$$(\mathbf{F}_1^a - \dot{\mathbf{p}}_1) \cdot \delta\mathbf{l}_1 + (\mathbf{F}_2^a - \dot{\mathbf{p}}_2) \cdot \delta\mathbf{l}_2 = 0. \tag{15.7}$$

The length of the rope is constant (constraint):

$$l_1 + l_2 = l.$$

This leads to

$$\delta l_1 = -\delta l_2 \quad \text{and} \quad \ddot{l}_1 = -\ddot{l}_2.$$

The inertial forces are

$$\dot{\mathbf{p}}_1 = m_1 \ddot{\mathbf{l}}_1 \quad \text{and} \quad \dot{\mathbf{p}}_2 = m_2 \ddot{\mathbf{l}}_2.$$

By inserting this into equation (15.7) and taking into account that the accelerations are parallel to the displacements, we have

$$(m_1 g \sin\alpha - m_1 \ddot{l}_1)\delta l_1 + (m_2 g \sin\beta - m_2 \ddot{l}_2)\delta l_2 = 0,$$
$$(m_1 g \sin\alpha - m_1 \ddot{l}_1 - m_2 g \sin\beta - m_2 \ddot{l}_2)\delta l_1 = 0,$$

or

$$\ddot{l}_1 = \frac{m_1 \sin\alpha - m_2 \sin\beta}{m_1 + m_2} g.$$

Example 15.3: Exercise: Equilibrium condition of a bascule bridge

Find by means of the d'Alembert principle the equilibrium condition for

(a) a lever of length l_1, with a mass m at a distance l_2 from the bearing point, and with a force F_1 acting vertically upward at its end; and

(b) the bascule bridge in Figure 15.4, with the forces G and Q acting.

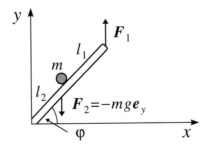

Figure 15.3. Lever with mass m and force $\mathbf{F}_l$.

VIRTUAL DISPLACEMENTS

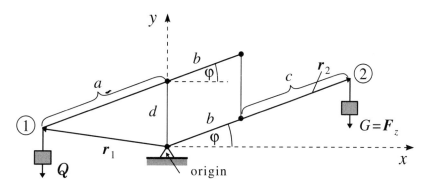

Figure 15.4. Geometry of the bascule bridge described in the problem.

Solution (a) The d'Alembert principle yields

$$\sum_{\nu} \mathbf{F}_{\nu} \cdot \delta \mathbf{r}_{\nu} = 0.$$

We have

$$\mathbf{F}_1 = F_1 \mathbf{e}_y, \qquad \mathbf{F}_2 = -mg \mathbf{e}_y$$

and

$$\mathbf{r}_1 = l_1 \cos \varphi \mathbf{e}_x + l_1 \sin \varphi \mathbf{e}_y,$$
$$\mathbf{r}_2 = l_2 \cos \varphi \mathbf{e}_x + l_2 \sin \varphi \mathbf{e}_y = \frac{l_2}{l_1} \mathbf{r}_1.$$

Furthermore,

$$\delta \mathbf{r}_1 = (-l_1 \sin \varphi \mathbf{e}_x + l_1 \cos \varphi \mathbf{e}_y) \delta \varphi \quad \text{and} \quad \delta \mathbf{r}_2 = \frac{l_2}{l_1} \mathbf{r}_1.$$

This leads to

$$\sum_{\nu=1}^{2} \mathbf{F}_{\nu} \cdot \delta \mathbf{r}_{\nu} = (F_1 l_1 \cos \varphi - mg l_2 \cos \varphi) \delta \varphi = 0,$$

i.e., the equilibrium condition reads

$$F_1 = \frac{l_2}{l_1} mg \quad \text{for} \quad \varphi \neq \frac{\pi}{2}, \frac{3\pi}{2}, \ldots.$$

(b) The forces acting at the points 1 and 2 are

$$\mathbf{F}_2 = -G \mathbf{e}_y, \qquad \mathbf{F}_1 = -Q \mathbf{e}_y.$$

Furthermore,

$$\mathbf{r}_1 = -a \cos \varphi \mathbf{e}_x + (d - a \sin \varphi) \mathbf{e}_y$$

and

$$\mathbf{r}_2 = (b+c)\cos\varphi \mathbf{e}_x + (b+c)\sin\varphi \mathbf{e}_y;$$

i.e.,

$$\delta\mathbf{r}_1 = (a\sin\varphi \mathbf{e}_x - a\cos\varphi \mathbf{e}_y)\delta\varphi$$

and

$$\delta\mathbf{r}_2 = (-(b+c)\sin\varphi \mathbf{e}_x + (b+c)\cos\varphi \mathbf{e}_y)\delta\varphi.$$

The d'Alembert principle reads

$$0 = \sum_{\nu=1}^{2} \mathbf{F}_\nu \cdot \delta\mathbf{r}_\nu = (Qa\cos\varphi - G(b+c)\cos\varphi)\delta\varphi = [Qa - G(b+c)]\cos\varphi\delta\varphi.$$

The equilibrium condition

$$Q = G\frac{b+c}{a}$$

is independent of the angle φ!

As is seen in Examples 15.1 and 15.2, the drawback of the principle of virtual displacements is that one still must eliminate displacements that are dependent through constraints before one can find an equation of motion. We therefore introduce generalized coordinates q_i. If we transform the $\delta\mathbf{r}_i$ in equation (15.6a) to δq_i, the coefficients of the δq_i can immediately be set to zero.

Starting from equation (15.6a), we introduce in the first sum according to equations (14.6) and (14.7), the generalized forces

$$\sum_{i=1}^{n} \mathbf{F}_i \cdot \delta\mathbf{r}_i = \sum_{i=1}^{n} \mathbf{F}_i \cdot \sum_{\alpha=1}^{f} \frac{\partial \mathbf{r}_i}{\partial q_\alpha}\delta q_\alpha = \sum_{\alpha=1}^{f} Q_\alpha \delta q_\alpha. \tag{15.8}$$

We now consider the other term in equation (15.6a):

$$\sum_i \dot{\mathbf{p}}_i \cdot \delta\mathbf{r}_i = \sum_i m_i \ddot{\mathbf{r}}_i \cdot \delta\mathbf{r}_i.$$

If we express $\delta\mathbf{r}_i$ according to (14.6) by the δq_i, we obtain

$$\sum_i \dot{\mathbf{p}}_i \cdot \delta\mathbf{r}_i = \sum_{i,\nu} m_i \ddot{\mathbf{r}}_i \cdot \frac{\partial \mathbf{r}_i}{\partial q_\nu}\delta q_\nu. \tag{15.9}$$

By adding and simultaneously subtracting equal terms, we rewrite the right-hand side of the equation:

$$\sum_i m_i \ddot{\mathbf{r}}_i \cdot \frac{\partial \mathbf{r}_i}{\partial q_\nu} = \sum_i \left(\frac{d}{dt}(m_i \dot{\mathbf{r}}_i) \cdot \frac{\partial \mathbf{r}_i}{\partial q_\nu}\right) + \sum_i \left(m_i \dot{\mathbf{r}}_i \cdot \frac{d}{dt}\left(\frac{\partial \mathbf{r}_i}{\partial q_\nu}\right)\right)$$

VIRTUAL DISPLACEMENTS

$$-\sum_i \left(m_i \dot{\mathbf{r}}_i \cdot \frac{d}{dt}\left(\frac{\partial \mathbf{r}_i}{\partial q_v}\right)\right)$$
$$= \sum_i \left(\frac{d}{dt}\left(m_i \dot{\mathbf{r}}_i \cdot \frac{\partial \mathbf{r}_i}{\partial q_v}\right) - m_i \dot{\mathbf{r}}_i \frac{d}{dt}\left(\frac{\partial \mathbf{r}_i}{\partial q_v}\right)\right). \quad (15.10)$$

To derive the expression for the kinetic energy, we change the order of differentiation with respect to t and q_v in the last term of equation (15.10):

$$\frac{d}{dt}\left(\frac{\partial \mathbf{r}_i}{\partial q_v}\right) = \frac{\partial}{\partial q_v}\left(\frac{d}{dt}\mathbf{r}_i\right) = \frac{\partial}{\partial q_v}\mathbf{v}_i. \quad (15.11)$$

Insertion in equation (15.10) yields

$$\sum_i \left(m_i \ddot{\mathbf{r}}_i \cdot \frac{\partial \mathbf{r}_i}{\partial q_v}\right) = \sum_i \left(\frac{d}{dt}\left(m_i \dot{\mathbf{r}}_i \cdot \frac{\partial \mathbf{r}_i}{\partial q_v}\right) - m_i \mathbf{v}_i \cdot \frac{\partial}{\partial q_v}\mathbf{v}_i\right). \quad (15.12)$$

We can rewrite the expression $\partial \mathbf{r}_i / \partial q_v$ in the first term of the right side of equation (15.12) by partially differentiating equation (14.4) with respect to $\dot{q}_v$:

$$\frac{\partial \mathbf{v}_i}{\partial \dot{q}_v} = \frac{\partial \mathbf{r}_i}{\partial q_v},$$

since $(\partial/\partial \dot{q}_v)(\partial \mathbf{r}_i/\partial t) = 0$ and from the sum remains only the factor at $\dot{q}_v$. By inserting this relation into (15.12), we obtain

$$\sum_i \left(m_i \ddot{\mathbf{r}}_i \cdot \frac{\partial \mathbf{r}_i}{\partial q_v}\right) = \sum_i \left(\frac{d}{dt}\left(m_i \mathbf{v}_i \cdot \frac{\partial \mathbf{v}_i}{\partial \dot{q}_v}\right)\right) - \sum_i \left(m_i \mathbf{v}_i \cdot \frac{\partial \mathbf{v}_i}{\partial q_v}\right)$$
$$= \frac{d}{dt}\left(\frac{\partial}{\partial \dot{q}_v}\left(\sum_i \frac{1}{2}m_i \mathbf{v}_i^2\right)\right) - \frac{\partial}{\partial q_v}\left(\sum_i \frac{1}{2}m_i \mathbf{v}_i^2\right).$$

Here, $\sum_i (1/2) m_i \mathbf{v}_i^2$ is the kinetic energy T:

$$\sum_i \left(m_i \ddot{\mathbf{r}}_i \cdot \frac{\partial \mathbf{r}_i}{\partial q_v}\right) = \frac{d}{dt}\left(\frac{\partial T}{\partial \dot{q}_v}\right) - \frac{\partial T}{\partial q_v}.$$

Insertion into equation (15.9) leads to

$$\sum_i \dot{\mathbf{p}}_i \cdot \delta \mathbf{r}_i = \sum_v \left(\frac{d}{dt}\left(\frac{\partial T}{\partial \dot{q}_v}\right) - \frac{\partial T}{\partial q_v}\right) \delta q_v. \quad (15.13)$$

Using equations (15.8) and (15.13), we can express the d'Alembert principle by generalized coordinates. Insertion of

$$\sum_i \mathbf{F}_i \cdot \delta \mathbf{r}_i = \sum_v Q_v \delta q_v \quad \text{(compare (15.8))}$$

into equation (15.6a) yields

$$\sum_v \left(\frac{d}{dt}\left(\frac{\partial T}{\partial \dot{q}_v}\right) - \frac{\partial T}{\partial q_v} - Q_v\right) \delta q_v = 0. \quad (15.14)$$

The q_ν are generalized coordinates; thus, the q_ν and the related δq_ν are independent of each other. Therefore, equation (15.14) is satisfied only if the individual coefficients vanish, i.e., for any coordinate q_ν we must have

$$\frac{d}{dt}\left(\frac{\partial T}{\partial \dot{q}_\nu}\right) - \frac{\partial T}{\partial q_\nu} - Q_\nu = 0, \qquad \nu = 1, \ldots, f. \tag{15.15}$$

As a further simplification, we assume that all forces $\mathbf{F}_i$ can be derived from a potential V (conservative force field):

$$\mathbf{F}_i = -\text{grad}_i(V) = -\nabla_i(V).$$

In this case, the generalized forces Q_ν can be written as

$$Q_\nu = \sum_i \mathbf{F}_i \cdot \frac{\partial \mathbf{r}_i}{\partial q_\nu} = -\sum_i \nabla_i V \cdot \frac{\partial \mathbf{r}_i}{\partial q_\nu} = -\frac{\partial V}{\partial q_\nu},$$

because

$$\sum_i \left(\frac{\partial V}{\partial x_i}\mathbf{e}_x + \frac{\partial V}{\partial y_i}\mathbf{e}_y + \frac{\partial V}{\partial z_i}\mathbf{e}_z\right) \cdot \left(\frac{\partial x_i}{\partial q_\nu}\mathbf{e}_x + \frac{\partial y_i}{\partial q_\nu}\mathbf{e}_y + \frac{\partial z_i}{\partial q_\nu}\mathbf{e}_z\right)$$

$$= \sum_i \left(\frac{\partial V}{\partial x_i}\frac{\partial x_i}{\partial q_\nu} + \frac{\partial V}{\partial y_i}\frac{\partial y_i}{\partial q_\nu} + \frac{\partial V}{\partial z_i}\frac{\partial z_i}{\partial q_\nu}\right)$$

$$= \frac{\partial V}{\partial q_\nu}.$$

By inserting $Q_\nu = -\partial V/\partial q_\nu$ into equation (15.15), we obtain

$$\frac{d}{dt}\left(\frac{\partial T}{\partial \dot{q}_\nu}\right) - \frac{\partial T}{\partial q_\nu} + \frac{\partial V}{\partial q_\nu} = 0$$

and

$$\frac{d}{dt}\left(\frac{\partial T}{\partial \dot{q}_\nu}\right) - \frac{\partial T - V}{\partial q_\nu} = 0.$$

V is independent of the generalized velocity; i.e., V is only a function of the position:

$$\frac{\partial V}{\partial \dot{q}_\nu} = 0.$$

Therefore, we can write

$$\frac{d}{dt}\frac{\partial}{\partial \dot{q}_\nu}(T - V) - \frac{\partial}{\partial q_\nu}(T - V) = 0, \tag{15.16}$$

or, by defining a new function, the *Lagrangian*[2]

$$L = T - V, \tag{15.17}$$

$$\frac{d}{dt}\frac{\partial L}{\partial \dot{q}_\nu} - \frac{\partial L}{\partial q_\nu} = 0, \quad \nu = 1, \ldots, f. \tag{15.18}$$

These equations are called *Lagrange equations*, and the quantities $\partial L/\partial \dot{q}_\nu$ are called *generalized momenta*. In Newton's formulation of mechanics, the equations of motions are established directly. The forces are thus put in the foreground; they must be specified for a given problem and inserted into the basic dynamic equations

$$\dot{\mathbf{p}}_i = \mathbf{F}_i, \quad i = 1, \ldots, N.$$

In the Lagrangian formulation the Lagrangian is the central quantity, and L includes both the kinetic energy T and the potential energy V. The latter one implicitly involves the forces. After L is established, the Lagrange equations can be established and solved. Both methods are equivalent to each other, as can be seen by stepwise inversion of the steps leading from (15.6a) to (15.18).

Example 15.4: Two blocks connected by a bar

Two blocks of equal mass that are connected by a rigid bar of length l move without friction along a given path (compare Figure 15.5). The attraction of the earth acts along the negative y-axis. The generalized coordinate is the angle α (corresponding to the single degree of freedom of the system).

For the relative distances x and y of the two blocks, we have

$$x = l\cos\alpha, \quad y = l\sin\alpha.$$

The constraint is holonomic and scleronomic. We will determine the Lagrangian

$$L = T - V.$$

The kinetic energy of the system is

$$T = \frac{1}{2}m(\dot{x}^2 + \dot{y}^2).$$

For this purpose, we form $\dot{x}$ and $\dot{y}$:

$$\dot{x} = -l(\sin\alpha)\dot{\alpha}, \quad \dot{y} = l(\cos\alpha)\dot{\alpha}.$$

---

[2] *Joseph Louis Lagrange*, b. Jan. 25, 1736, Torino–d. April 10, 1813, Paris. Lagrange came from a French-Italian family and in 1755 became professor in Torino. In 1766, he went to Berlin as director of the mathematical-physical class of the academy. In 1786, after the death of Friedrich II, he went to Paris. There he essentially supported the reformation of the system of measures and was a professor at various universities. His very extensive work includes a new foundation of variational calculus (1760) and its application to dynamics, contributions to the three-body problem (1772), application of the theory of continued fractions to the solution of equations (1767), number-theoretical problems, and an unsuccessful reduction of infinitesimal calculus to algebra. With his *Mécanique Analytique* (1788), Lagrange became the founder of analytical mechanics.

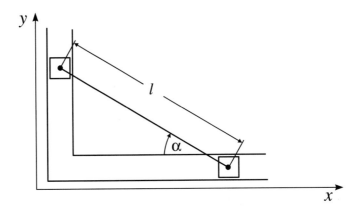

Figure 15.5. Two blocks are connected by a bar.

Thus, we get for T

$$T = \frac{1}{2}m\left(l^2(\sin^2\alpha)\dot{\alpha}^2 + l^2(\cos^2\alpha)\dot{\alpha}^2\right) = \frac{1}{2}ml^2\dot{\alpha}^2.$$

For the potential, we have (conservative system),

$$V = mgy = mgl\sin\alpha.$$

The Lagrangian therefore reads

$$L = T - V = \frac{1}{2}ml^2\dot{\alpha}^2 - mgl\sin\alpha.$$

We insert L into the Lagrange equation (15.16):

$$\frac{d}{dt}\frac{\partial L}{\partial \dot{\alpha}} - \frac{\partial L}{\partial \alpha} = \frac{d}{dt}(ml^2\dot{\alpha}^2) + mgl\cos\alpha = 0$$

and

$$ml^2\ddot{\alpha} + mgl\cos\alpha = 0, \qquad \ddot{\alpha} + \frac{g}{l}\cos\alpha = 0.$$

Multiplication by $\dot{\alpha}$ yields

$$\ddot{\alpha}\dot{\alpha} + \frac{g}{l}\dot{\alpha}\cos\alpha = 0.$$

These equations can be integrated directly. One obtains

$$\frac{1}{2}\dot{\alpha}^2 + \frac{g}{l}\sin\alpha = \text{constant} = c$$

or

$$\dot{\alpha} = \sqrt{2\left(c - \frac{g}{l}\sin\alpha\right)}.$$

Separation of the variables α and t leads to the equation

$$dt = \frac{d\alpha}{\sqrt{2(c - (g/l)\sin\alpha)}}, \qquad t - t_0 = \int_{\alpha_0}^{\alpha} \frac{d\alpha}{\sqrt{2(c - (g/l)\sin\alpha)}}.$$

The constants c and t_0 are determined from the given initial conditions.

Example 15.5: Ignorable coordinate

We will use the following example for the Lagrangian formalism to explain the concept of the *ignorable coordinate*. The arrangement is shown in Figure 15.6.

Two masses m and M are connected by a string of constant total length $l = r + s$. The string mass is negligibly small compared to $m + M$. The mass m can rotate with the string (with varying partial length r) on the plane. The string leads from m through a hole in the plane to M, where the mass M hangs from the tightly stretched string (with the also variable partial length $s = l - r$). Depending on the values ω of the rotation of m on the plane, the arrangement can glide upward or downward. Thus, the mass M moves only along the z-axis. The constraints characterizing the system are holonomic and scleronomic. This arrangement has two degrees of freedom. The two corresponding generalized coordinates φ and s uniquely describe the state of motion of this conservative system.

We have

$$x = r\cos\varphi = (l - s)\cos\varphi,$$
$$y = r\sin\varphi = (l - s)\sin\varphi.$$

For the kinetic energy T of the system, we obtain

$$T = \frac{1}{2}m\left(\frac{d}{dt}(l-s)\right)^2 + \frac{1}{2}(l-s)^2 m\dot\varphi^2 + \frac{1}{2}M\dot s^2$$
$$= \frac{1}{2}(m+M)\dot s^2 + \frac{1}{2}(l-s)^2 m\dot\varphi^2.$$

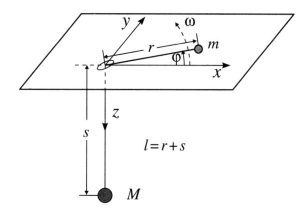

Figure 15.6. Two masses m and M are connected by a string.

The potential V reads

$$V = -Mgs.$$

For the Lagrangian L, we get

$$L = T - V = \frac{1}{2}(m+M)\dot{s}^2 + \frac{1}{2}(l-s)^2 m\dot{\varphi}^2 + Mgs.$$

We now form

$$\frac{d}{dt}\frac{\partial L}{\partial \dot{s}} = (m+M)\ddot{s}, \qquad \frac{\partial L}{\partial s} = -(l-s)m\dot{\varphi}^2 + Mg,$$

$$\frac{d}{dt}\frac{\partial L}{\partial \dot{\varphi}} = \frac{d}{dt}((l-s)^2 m\dot{\varphi}), \qquad \frac{\partial L}{\partial \varphi} = 0.$$

Because $\partial L/\partial \varphi = 0$, φ is called an *ignorable* or *cyclic coordinate*. The Lagrange equation for φ then reduces to

$$\frac{d}{dt}\frac{\partial L}{\partial \dot{\varphi}} = \frac{d}{dt}((l-s)^2 m\dot{\varphi}) = 0$$

or

$$(l-s)^2 \dot{\varphi} m = \tilde{L} = \text{constant}.$$

Here, $\tilde{L}$ is the angular momentum of the rotating mass m.

This first integral of motion is the angular momentum conservation law. Generally speaking, the Lagrangian equation of motion

$$\frac{d}{dt}\frac{\partial L}{\partial \dot{q}_j} - \frac{\partial L}{\partial q_j} = 0$$

for an ignorable (cyclic) variable reduces to

$$\frac{d}{dt}\frac{\partial L}{\partial \dot{q}_j} = 0 \quad \text{or} \quad \frac{dp_j}{dt} = 0.$$

Here, $p_j = \partial L/\partial \dot{q}_j$ is the generalized momentum. The generalized momentum related to the cyclic coordinate is thus constant in time. Therefore, the general conservation law holds:

The generalized momentum related to a cyclic coordinate is conserved.

The Lagrange equation for s reads

$$(m+M)\ddot{s} + (l-s)m\dot{\varphi}^2 - Mg = 0$$

or, after multiplication by $\dot{s}$,

$$(m+M)\ddot{s}\dot{s} + \frac{\tilde{L}^2 \dot{s}}{(l-s)^3 m} - Mg\dot{s} = 0, \quad \text{with} \quad \tilde{L} = (l-s)^2 m\dot{\varphi}.$$

The last equation can be integrated immediately, and we obtain as a second integral of motion

$$\frac{1}{2}(m+M)\dot{s}^2 + \frac{\tilde{L}^2}{2(l-s)^2 m} - Mgs = \text{constant} = T + V = E;$$

i.e., the total energy of the system is conserved. The given system is in a state of equilibrium (gravitation force = centrifugal force) for vanishing acceleration, $d^2s/dt = 0$:

$$0 = \ddot{s} = \frac{1}{m+M}\left[Mg - (l-s)m\left(\frac{\tilde{L}}{(l-s)^2 m}\right)^2\right]$$

$$= \frac{1}{m+M}\left[Mg - \frac{\tilde{L}^2}{(l-s)^3 m}\right].$$

The result states that s must be constant. For a fixed distance s_0, equilibrium therefore appears for a definite angular momentum $\tilde{L} = \tilde{L}_0$, which corresponds to a definite angular velocity $\omega = \dot{\varphi}$:

$$\tilde{L}_0 = \sqrt{Mmg(l-s_0)^3}.$$

For $\tilde{L} > \tilde{L}_0$, the entire arrangement glides upward; for $\tilde{L} < \tilde{L}_0$, the string with the two masses m and M glides downward. For $\tilde{L} = \tilde{L}_0$, the system is in an equilibrium state. For the special case $\tilde{L} = 0$ (i.e., $\dot{\varphi} = 0$, no rotation on the plane), one simply has the retarded free fall of the mass M.

Example 15.6: Sphere in a rotating tube

As a further example of the Lagrangian formalism, we discuss a problem with a holonomic rheonomic constraint. A sphere moves in a tube that rotates in the x,y-plane about the z-axis with constant angular velocity ω.

This arrangement has one degree of freedom. Accordingly, we need only one generalized coordinate for a complete description of the state of motion of the system: the radial distance r of the sphere from the rotation center.

One has

$$x = r\cos\omega t,$$
$$y = r\sin\omega t.$$

The Lagrangian $L = T - V$ then reads

$$L = \frac{1}{2}m(\dot{x}^2 + \dot{y}^2) = \frac{1}{2}m(\dot{r}^2 + \omega^2 r^2),$$

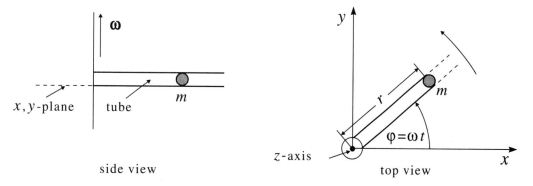

Figure 15.7. Sphere in a rotating tube.

if we take into account that for this arrangement the potential $V = 0$.

We now form

$$\frac{d}{dt}\frac{\partial L}{\partial \dot r} = m\ddot r, \qquad \frac{\partial L}{\partial r} = m\omega^2 r.$$

Then we obtain the Lagrange equation

$$m\ddot r - m\omega^2 r = 0,$$

or

$$\ddot r - \omega^2 r = 0.$$

This differential equation corresponds–up to the minus sign–to the equation for the nondamped harmonic oscillator. It has a general solution of the type

$$r(t) = A e^{\omega t} + B e^{-\omega t}.$$

With increasing time t, this expression for $r(t)$ also increases; i.e.,

$$\lim_{t \to \infty} r(t) = \infty \quad \text{for} \quad A > 0.$$

From the physical point of view, this means that the sphere is hurled outward by the centrigugal force that results from the rotation of the arrangement.

The energy of the sphere increases. The reason is that the constraint reaction performs work on the sphere. Although the constraint force is perpendicular to the tube wall, it is not perpendicular to the trajectory of the sphere. Hence, the product $\mathbf{F}^z \cdot \delta \mathbf{s}$ does not vanish.

Example 15.7: Exercise: Upright pendulum

Determine the Lagrangian and the equation of motion of the following system: Let m be a point mass on a massless bar of length l which in turn is fixed to a hinge. The hinge oscillates in the vertical direction according to $h(t) = h_0 \cos \omega t$. The only degree of freedom is the angle ϑ between the bar and the vertical (upright pendulum).

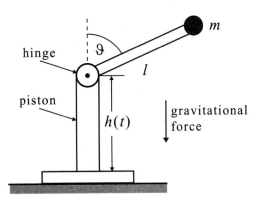

Figure 15.8. A mass m is fixed to one end of a bar; the other end of the bar is fixed to an oscillating hinge.

Solution The position of the point mass m is (x, y):

$$x = l \sin \vartheta, \qquad y = h(t) + l \cos \vartheta = h_0 \cos \omega t + l \cos \vartheta.$$

Differentiation of this equation yields

$$\dot{x} = \dot\vartheta l \cos \vartheta, \qquad \dot{y} = -(\omega h_0 \sin \omega t + \dot\vartheta l \sin \vartheta).$$

Hence, the kinetic energy T becomes

$$T = \frac{1}{2} m(\dot{x}^2 + \dot{y}^2)$$
$$= \frac{1}{2} m(\dot\vartheta^2 l^2 + \omega^2 h_0^2 \sin^2 \omega t + 2\omega h_0 \dot\vartheta l \sin \vartheta \sin \omega t),$$

and the potential energy reads

$$V = mgy = mg(h_0 \cos \omega t + l \cos \vartheta).$$

Then the Lagrangian becomes

$$L = T - V$$
$$= \frac{m}{2} \left[\dot\vartheta^2 l^2 + \omega^2 h_0^2 \sin^2 \omega t + 2\omega h_0 \dot\vartheta \sin \vartheta \sin \omega t - 2g(h_0 \cos \omega t + l \cos \vartheta) \right].$$

The Lagrange equation reads

$$\frac{d}{dt} \left(\frac{\partial L}{\partial \dot\vartheta} \right) - \frac{\partial L}{\partial \vartheta} = 0,$$

$$\frac{\partial L}{\partial \dot\vartheta} = ml^2 \dot\vartheta + m\omega h_0 l \sin \vartheta \sin \omega t,$$

$$\frac{\partial L}{\partial \vartheta} = m\omega h_0 \dot\vartheta l \cos \vartheta \sin \omega t + mgl \sin \vartheta,$$

$$\frac{d}{dt} \frac{\partial L}{\partial \dot\vartheta} = ml^2 \ddot\vartheta + m\omega h_0 l \dot\vartheta \cos \vartheta \sin \omega t + m\omega^2 h_0 l \sin \vartheta \cos \omega t,$$

$$l^2 \ddot\vartheta + \omega h_0 l \dot\vartheta \cos \vartheta \sin \omega t + \omega^2 h_0 l \sin \vartheta \cos \omega t - \omega h_0 l \dot\vartheta \cos \vartheta \sin \omega t - gl \sin \vartheta = 0,$$

or

$$l\ddot\vartheta + \omega^2 h_0 \sin \vartheta \cos \omega t - g \sin \vartheta = 0.$$

The substitution $\vartheta' = \vartheta - \pi \Rightarrow \sin \vartheta = -\sin \vartheta'$; for small displacements, $-\sin \vartheta' \approx -\vartheta'$, i.e.,

$$l\ddot\vartheta' + (g - \omega^2 h_0 \cos \omega t)\vartheta' = 0.$$

This is the desired equation of motion. If the piston is at rest, i.e., $h(t) = h_0 = 0$, we get

$$\ddot\vartheta' + \frac{g}{l} \vartheta' = 0.$$

This is the equation of motion of the ordinary pendulum!

Example 15.8: Exercise: Stable equilibrium position of an upright pendulum

Find the position of stable equilibrium of the pendulum of Example 15.7 if the hinge oscillates with the frequency $\omega \gg \sqrt{g/l}$.

Solution We first rewrite the Lagrangian of the pendulum of Example 15.7 as follows: The terms

$$\frac{m\omega^2}{2} h_0^2 \sin^2 \omega t \quad \text{and} \quad -mgh_0 \cos \omega t$$

can be written as total differentials with respect to time:

$$\frac{m\omega^2}{2} h_0^2 \sin^2 \omega t = \frac{d}{dt}\left(-\frac{1}{4} m\omega h_0^2 \sin \omega t \cos \omega t\right) + C,$$

$$-mgh_0 \cos \omega t = \frac{d}{dt}\left(-\frac{mgh_0}{\omega} \sin \omega t\right).$$

We can omit these terms, since Lagrangians that differ only by a total derivative with respect to time, according to the Hamilton principle $\delta \int_{t_1}^{t_2} L \, dt = 0$, are equivalent. Hence,

$$L = \frac{m}{2}[\dot\vartheta^2 l^2 + \omega^2 h_0^2 \sin^2 \omega t + 2\omega h_0 \dot\vartheta l \sin \vartheta \sin \omega t - 2g(h_0 \cos \omega t + l \cos \vartheta)] \quad (15.19)$$

$$= \frac{m}{2}[\dot\vartheta^2 l^2 + 2\omega h_0 \dot\vartheta l \sin \vartheta \sin \omega t - 2gl \cos \omega t].$$

Another transformation yields

$$m\omega h_0 \dot\vartheta l \sin \vartheta \sin \omega t = -\frac{d}{dt}(m\omega h_0 l \cos \vartheta \sin \omega t) + m\omega^2 h_0 l \cos \vartheta \cos \omega t,$$

so that the Lagrangian finally reads

$$L = \frac{m}{2}[\dot\vartheta^2 l^2 + 2\omega^2 h_0 l \cos \vartheta \cos \omega t - 2gl \cos \vartheta]. \quad (15.20)$$

From this, one obtains of course the equation of motion as in Example 15.7.

We consider ϑ as a generalized coordinate with the appropriate mass coefficient ml^2. The equation of motion then reads

$$ml^2 \ddot\vartheta = mgl \sin \vartheta - m\omega^2 h_0 l \sin \vartheta \cos \omega t$$

$$= -\frac{du}{d\vartheta} + f \quad (15.21)$$

with $u = mgl \cos \vartheta$ and $f = -m\omega^2 h_0 l \sin \vartheta \cos \omega t$. The additional force f is due to the motion of the hinge. For very fast oscillations of the hinge, we assume that the motion of the pendulum in the potential u is superposed by quick oscillations ξ:

$$\vartheta(t) = \widetilde\vartheta(t) + \xi(t).$$

The average value of the oscillations over a period $2\pi/\omega$ equals zero, while $\widetilde\vartheta$ changes only slowly; therefore,

$$\widetilde\vartheta(t) = \frac{\omega}{2\pi} \int_0^{2\pi/\omega} \vartheta(t) \, dt = \widetilde\vartheta(t). \quad (15.22)$$

Equations (15.21) with (15.22) can then be written as

$$ml^2 \ddot{\widetilde\vartheta}(t) + ml^2 \ddot\xi(t) = -\frac{du}{d\vartheta} + f(\vartheta).$$

Because $f(\vartheta) = f(\widetilde{\vartheta} + \xi) = f(\widetilde{\vartheta}) + \xi df/d\vartheta$, an expansion up to first order in ξ yields

$$ml^2\ddot{\widetilde{\vartheta}} + ml^2\ddot{\xi} = -\frac{dU}{d\widetilde{\vartheta}} - \xi\frac{d^2U}{d\widetilde{\vartheta}^2} + f(\widetilde{\vartheta}) + \xi\frac{df}{d\widetilde{\vartheta}}. \tag{15.23}$$

The dominant terms for the oscillations are $ml^2\ddot{\xi}$ and $f(\widetilde{\vartheta})$:

$$ml^2\ddot{\xi} = f(\widetilde{\vartheta})$$
$$\Rightarrow \quad \ddot{\xi} = -\frac{\omega^2 h_0}{l}\sin\widetilde{\vartheta}\cos\omega t,$$

and from this, we obtain

$$\xi = \frac{h_0}{l}\sin\widetilde{\vartheta}\cos\omega t = -\frac{f}{m\omega^2 l^2}. \tag{15.24}$$

We now calculate an *effective potential* created by the oscillations, and for this purpose we average (15.23) over a period $2\pi/\omega$ (the mean values over ξ and f vanish):

$$ml^2\ddot{\widetilde{\vartheta}} = -\frac{dU}{d\widetilde{\vartheta}} + \overline{\xi\frac{df}{d\widetilde{\vartheta}}} = -\frac{dU}{d\widetilde{\vartheta}} - \frac{1}{m\omega^2 l^2}\overline{f\frac{df}{d\widetilde{\vartheta}}}.$$

This can be written as

$$ml^2\ddot{\widetilde{\vartheta}} = -\frac{dU_{\text{eff}}}{d\widetilde{\vartheta}} \quad \text{with} \quad U_{\text{eff}} = U + \frac{1}{2m\omega^2 - l^2}\overline{f^2}. \tag{15.25}$$

Because $\overline{\cos^2\omega t} = 1/2$, we get

$$U_{\text{eff}} = U + \frac{m\omega^2 h_0^2}{4}\sin^2\vartheta$$
$$= mgl\cos\vartheta + \frac{m\omega^2 h_0^2}{4}\sin^2\vartheta. \tag{15.26}$$

The minima of U_{eff} give the stable equilibrium positions:

$$\frac{dU_{\text{eff}}}{d\vartheta} = -mgl\sin\vartheta + \frac{m\omega^2 h_0^2}{4}\sin\vartheta\cos\vartheta \stackrel{!}{=} 0$$

$$\Rightarrow \quad \sin\vartheta = 0 \quad \text{or} \quad \cos\vartheta = \frac{2gl}{\omega^2 h_0^2}. \tag{15.27}$$

From this, it follows that for any ω the position vertically downwards ($\vartheta = \pi$) is stable. $\vartheta = 0$ is excluded because $U_{\text{eff}}(\vartheta = 0) = mgl$. Additional stable equilibrium positions arise for $\omega^2 > 2gl/h_0^2$ with the angle given above.

Example 15.9: Exercise: Vibration frequencies of a three-atom symmetric molecule

Find the vibration frequencies of a linear three-atom symmetric molecule ABA. We assume that the

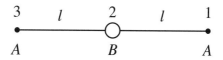

Figure 15.9. Linear three-atom symmetric molecule.

potential energy of the molecule depends only on the distances AB and BA and the angle ABA. Write the Lagrangian of the molecule in appropriate coordinates (normal coordinates) where the Lagrangian has the form

$$L = \sum_\alpha \frac{m_\alpha}{2}(\dot{\Theta}_\alpha^2 - \omega_\alpha^2 \Theta_\alpha^2).$$

The ω_α are the desired vibration frequencies of the normal modes. If one cannot find the normal coordinates of the system, one can proceed as follows: If a system has s degrees of freedom and does not vibrate, then the Lagrangian generally reads

$$L = \frac{1}{2}\sum_{i,k}(m_{ik}\dot{x}_i\dot{x}_k - k_{ik}x_ix_k).$$

The eigenfrequencies of the system are then determined by the so-called characteristic equation

$$\det|k_{ik} - \omega^2 m_{ik}| = 0.$$

Solution We describe the geometry of the molecule in the x,y-plane. Let the displacement of the atom α from the rest position $\mathbf{r}_{\alpha 0}$ be denoted by $\mathbf{x}_\alpha = (x_\alpha, y_\alpha)$, i.e., $\mathbf{r}_\alpha = \mathbf{r}_{\alpha 0} + \mathbf{x}_\alpha$. The forces that keep the atoms together are assumed to be to first order linear in the displacement from the rest position, i.e.,

$$L = \frac{m_A}{2}(\dot{x}_1^2 + \dot{x}_3^2) + \frac{m_B}{2}\dot{x}_2^2 - \frac{K_L}{2}[(x_1 - x_2)^2 + (x_3 - x_2)^2],$$

if we consider longitudinal vibrations. For these modes the conservation of the center of gravity can be written as follows:

$$m_A(x_1 + x_3)m_Bx_2 = 0, \qquad \left[\sum_\alpha m_\alpha \mathbf{r}_\alpha = \sum_\alpha m_\alpha \mathbf{r}_{\alpha 0}\right],$$

and we can eliminate x_2 from L:

$$L = \frac{m_A}{2}(\dot{x}_1^2 + \dot{x}_3^2) + \frac{m_A^2}{2m_B}(\dot{x}_1 + \dot{x}_3)^2$$
$$- \frac{K_L}{2}\left[x_1^2 + x_3^2 + 2\frac{m_A}{m_B}(x_1 + x_3)^2 + \frac{2m_A^2}{m_B^2}(x_1 + x_3)^2\right].$$

Hence, only two normal coordinates for the longitudinal motion can exist, because of the conservation of the center of gravity.

Let $\Theta_1 = x_1 + x_3$, $\Theta_2 = x_1 - x_3$. L can then be written as

$$L = \frac{m_A}{4}\dot{\Theta}_2^2 + \frac{m_A\mu}{4m_B}\dot{\Theta}_1^2 - \frac{K_L}{4}\Theta_2^2 - \frac{K_L\mu}{4m_B^2}\Theta_1^2, \qquad \mu \equiv 2m_A + m_B,$$

i.e., Θ_1 and Θ_2 are the two normal coordinates of the longitudinal vibration (μ represents the total mass of the molecule).

(a) For $x_1 = x_3$, Θ_2 vanishes; i.e., Θ_1 describes antisymmetric longitudinal vibrations (Figure 15.10).

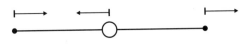

Figure 15.10.

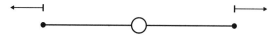

Figure 15.11.

(b) For $x_1 = -x_3$, Θ_1 vanishes; i.e., Θ_2 describes symmetric longitudinal vibrations (Figure 15.11). A comparison of kinetic and potential energy yields the normal frequencies

$$\omega_a = \sqrt{\frac{K_L \mu}{m_A m_B}}, \quad \text{antisymmetric vibration,}$$

$$\omega_s = \sqrt{\frac{K_L}{m_A}}, \quad \text{symmetric vibration.}$$

For transverse vibrations of the form in Figure 15.12, we set

$$L = \frac{m_A}{2}(\dot{y}_1^2 + \dot{y}_3^2) + \frac{m_B}{2}\dot{y}_2^2 - \frac{K_T}{2}(l\delta)^2,$$

where δ is the deviation of the angle $\sphericalangle (ABA)$ from π. For small values of δ, we can set

$$\begin{aligned}
\delta &= \left(\frac{\pi}{2} - \alpha_1\right) + \left(\frac{\pi}{2} - \alpha_2\right) \\
&= \sin\left(\frac{\pi}{2} - \alpha_1\right) + \sin\left(\frac{\pi}{2} - \alpha_2\right) \\
&= \cos\alpha_1 + \cos\alpha_2 \\
&= \frac{y_2 - y_1}{l} + \frac{y_2 - y_3}{l}.
\end{aligned}$$

We utilize the conservation of the center of gravity and angular momentum conservation to eliminate y_2 and y_3 from L.

$$m_A(y_1 + y_3) + m_B y_2 = 0 \quad \text{(conservation of the center of gravity).}$$

To exclude rotation of the molecule, the total angular momentum must vanish:

$$D = \sum_\alpha m_\alpha [\mathbf{r}_\alpha \times \mathbf{v}_\alpha] \simeq \sum_\alpha m_\alpha [\mathbf{r}_{\alpha 0} \times \dot{\mathbf{x}}_\alpha] = \frac{d}{dt} \sum_\alpha m_\alpha [\dot{\mathbf{r}}_{\alpha 0} \times \mathbf{x}_\alpha],$$

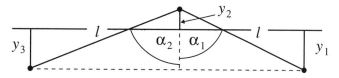

Figure 15.12.

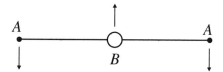

Figure 15.13.

which can be achieved by

$$\sum_\alpha m_\alpha [\mathbf{r}_{\alpha 0} \times \mathbf{x}_\alpha] = 0.$$

For our case, it thus follows that $y_1 = y_3$. Then we get

$$(l\dot\delta)^2 = \frac{4\mu^2 \dot y_1^2}{m_B^2} \quad \text{and} \quad L = \frac{m_A m_B}{4\mu} l^2 \dot\delta^2 - \frac{K_T l^2}{2} \delta^2.$$

We thus obtain the eigenfrequency of the transverse vibration:

$$\omega_T = \sqrt{\frac{2 K_T \mu}{m_A m_B}}.$$

Example 15.10: Exercise: Normal frequencies of a triangular molecule

Calculate the normal frequencies of a symmetric molecule ABA of triangular shape:

Solution   Conservation of the center of gravity here reads

$$m_A(x_1 + x_3) + m_B x_2 = 0, \qquad m_A(y_1 + y_3) + m_B y_2 = 0.$$

For angular momentum conservation, we go to the rest position of atom B, and because $m_1 = m_3 = m_A$, it follows that

$$\mathbf{r}_{10} \times \mathbf{x}_1 + \mathbf{r}_{30} \times \mathbf{x}_3 = 0.$$

We have

$$\mathbf{r}_{10} \times \mathbf{x}_1 = |\mathbf{r}_{10}|(-x_1 \cos\alpha + y_1 \sin\alpha)\mathbf{e},$$
$$\mathbf{r}_{30} \times \mathbf{x}_3 = |\mathbf{r}_{30}|(-x_3 \cos\alpha - y_3 \sin\alpha)\mathbf{e}.$$

Because $|\mathbf{r}_{10}| = |\mathbf{r}_{30}|$, the angular momentum conservation law follows:

$$\sin\alpha(y_1 - y_3) = \cos\alpha(x_1 + x_3) \quad \text{or} \quad y_1 - y_3 = \cot\alpha(x_1 + x_3).$$

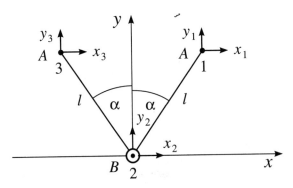

Figure 15.14. Triangular molecule.

VIRTUAL DISPLACEMENTS

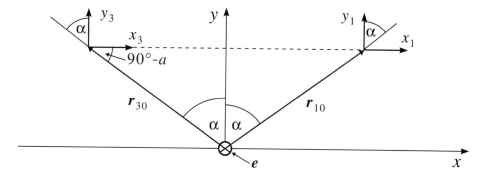

Figure 15.15. The various coordinates for Example 15.10.

The changes δl_1 and δl_2 of the distances AB and BA result by projection of the vectors $\mathbf{x}_1 - \mathbf{x}_2$ and $\mathbf{x}_3 - \mathbf{x}_2$ onto the directions of the lines AB and BA:

$$\delta l_1 = (x_1 - x_2)\sin\alpha + (y_1 - y_2)\cos\alpha,$$
$$\delta l_2 = -(x_3 - x_2)\sin\alpha + (y_3 - y_2)\cos\alpha.$$

The change of the angle $2\alpha = \angle(ABA)$ is found by projection of the vectors $\mathbf{x}_1 - \mathbf{x}_2$ and $\mathbf{x}_3 - \mathbf{x}_2$ onto the directions orthogonal to the line segments AB and BA:

$$\delta = \frac{1}{l}[(x_1 - x_2)\cos\alpha - (y_1 - y_2)\sin\alpha] + \frac{1}{l}[-(x_3 - x_2)\cos\alpha - (y_3 - y_2)\sin\alpha].$$

We write the Lagrangian of the molecule as

$$L = \frac{m_A}{2}(\dot{\mathbf{x}}_1^2 + \dot{\mathbf{x}}_3^2) + \frac{m_B}{2}\dot{\mathbf{x}}_2^2 - \frac{K_1}{2}[(\delta l_1)^2 + (\delta l_2)^2] - \frac{K_2}{2}(l\delta)^2.$$

Here, $(K_1/2)[(\delta l_1)^2 + (\delta l_2)^2]$ is the potential energy of the rotation, and $K_2(l\delta)^2/2$ is the potential energy of the bending of the molecule. We adopt as new coordinates

$$Q_\alpha = x_1 + x_3, \qquad q_{s1} = x_1 - x_3, \qquad q_{s2} = y_1 + y_3,$$

and then have

$$x_1 = \frac{1}{2}(Q_\alpha + q_{s1}), \qquad x_2 = -\frac{m_A}{m_B}Q_\alpha, \qquad x_3 = \frac{1}{2}(Q_\alpha - q_{s1}),$$
$$y_1 = \frac{1}{2}(q_{s2} + Q_\alpha \cot\alpha), \qquad y_2 = -\frac{m_A}{m_B}q_{s2}, \qquad y_3 = \frac{1}{2}(q_{s2} - Q_\alpha \cot\alpha).$$

Because $y_1 - y_3 = Q_\alpha \cot\alpha$, we find for L

$$L = \frac{m_A}{4}\left(\frac{2m_A}{m_B} + \frac{1}{\sin^2\alpha}\right)\dot{Q}_\alpha^2 + \frac{m_A}{4}\dot{q}_{s1}^2 + \frac{m_A\mu}{4m_B}\dot{q}_{s2}^2$$
$$- Q_\alpha^2 \frac{K_1}{4}\left(\frac{2m_A}{m_B} + \frac{1}{\sin^2\alpha}\right)\left(1 + \frac{2m_A}{m_B}\sin^2\alpha\right)$$

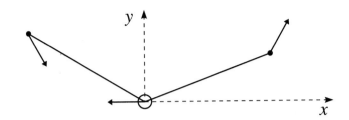

Figure 15.16.

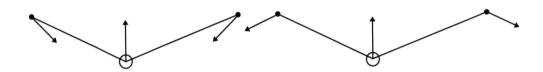

Figure 15.17.

$$-\frac{q_{s1}^2}{4}(K_1 \sin^2 \alpha + 2K_2 \cos^2 \alpha) - q_{s2}^2 \frac{\mu^2}{4m_B^2}(K_1 \cos^2 \alpha + 2K_2 \sin^2 \alpha)$$
$$+ q_{s1} q_{s2} \frac{\mu}{2m_B}(2K_2 - K_1) \sin \alpha \cos \alpha.$$

Obviously, Q_α is a normal coordinate, with the vibration frequency

$$\omega_\alpha^2 = \frac{K_1}{m_A}\left(1 + \frac{2m_A}{m_B} \sin^2 \alpha\right).$$

Pure Q_α-vibrations occur for $x_1 = x_3$, $y_1 = -y_3$; i.e., Q_α describes antisymmetric vibrations with respect to the y-axis in Figure 15.16.

The eigenfrequencies ω_{s1} and ω_{s2} of the normal vibrations for q_{s1} and q_{s2} must be determined by the characteristic equation

$$\omega^4 - \omega^2 \left[\frac{K_1}{m_A}\left(1 + \frac{2m_A}{m_B}\cos^2 \alpha\right) + \frac{2K_2}{m_A}\left(1 + \frac{2m_A}{m_B}\sin^2 \alpha\right)\right] + \frac{2\mu K_1 K_2}{m_B m_A^2} = 0.$$

The coordinates q_{s1} and q_{s2} correspond to vibrations that are symmetric about the y-axis (Figure 15.17):

$$(x_1 = -x_3, \quad Q_\alpha = 0 \quad \Rightarrow \quad y_1 = y_3).$$

Example 15.11: **Exercise: Normal frequencies of an asymmetric linear molecule**

Find the normal frequencies for a linear, asymmetric molecule with the shape in Figure 15.18.

VIRTUAL DISPLACEMENTS

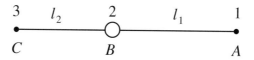

Figure 15.18.

Solution   Conservation of the center of gravity and of angular momentum now read

$$m_A x_1 + m_B x_2 + m_C x_3 = 0, \quad \text{x-center of gravity,}$$
$$m_A y_1 + m_B y_2 + m_C y_3 = 0, \quad \text{y-center of gravity,}$$
$$m_A l_1 y_1 = m_C l_2 y_3, \quad \text{angular momentum conservation.}$$

For the potential energy of bending, we write

$$V = \frac{K_2}{2}(l\delta)^2, \quad (2l = l_1 + l_2);$$

for that of rotation,

$$V = \frac{K_2}{2}(x_1 - x_2)^2 + \frac{K_1'}{2}(x_2 - x_3)^2.$$

The analogous calculation as for Example 15.9 after some effort yields

$$\omega_T^2 = \frac{K_2 l^2}{l_1^2 l_2^2}\left(\frac{l_1^2}{m_C} + \frac{l_2^2}{m_A} + \frac{4l^2}{m_B}\right)$$

for the frequency of the transverse vibration, and also the equation quadratic in ω^2

$$\omega^4 - \omega^2 \left[K_1\left(\frac{1}{m_A} + \frac{1}{m_B}\right) + K_1'\left(\frac{1}{m_B} + \frac{1}{m_C}\right)\right] + \frac{\mu K_1 K_1'}{m_A m_B m_C} = 0$$

for the frequencies $\omega_{L_1}, \omega_{L_2}$ of the two longitudinal vibrations.

Example 15.12: Exercise: Double pendulum

Determine

(a) the generalized coordinates of the double pendulum;

(b) the Lagrangian of the system;

(c) the equations of motion;

(d) for $m_1 = m_2 = m$ and $l_1 = l_2 = l$;

(e) as (d) for small amplitudes; and

(f) for the case (e) the normal vibrations and frequencies.

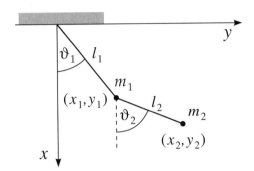

Figure 15.19. Coordinates of the double pendulum.

Solution (a) The appropriate generalized coordinates are the two angles ϑ_1 and ϑ_2 that are related to the Cartesian coordinates by

$$x_1 = l_1 \cos \vartheta_1, \qquad y_1 = l_1 \sin \vartheta_1, \tag{15.28}$$
$$x_2 = l_1 \cos \vartheta_1 + l_2 \cos \vartheta_2, \qquad y_2 = l_1 \sin \vartheta_1 + l_2 \sin \vartheta_2.$$

(b) From (15.28), it follows by differentiation that

$$\dot{x}_1 = -l_1 \dot{\vartheta}_1 \sin \vartheta_1, \qquad \dot{y}_1 = l_1 \dot{\vartheta}_1 \cos \vartheta_1,$$
$$\dot{x}_2 = -l_1 \dot{\vartheta}_1 \sin \vartheta_1 - l_2 \dot{\vartheta}_2 \sin \vartheta_2, \qquad \dot{y}_2 = l_1 \dot{\vartheta}_1 \cos \vartheta_1 + l_2 \dot{\vartheta}_2 \cos \vartheta_2.$$

The kinetic energy of the system is

$$T = \frac{1}{2} m_1 (\dot{x}_1^2 + \dot{y}_1^2) + \frac{1}{2} m_2 (\dot{x}_2^2 + \dot{y}_2^2)$$
$$= \frac{1}{2} m_1 l_1^2 \dot{\vartheta}_1^2 + \frac{1}{2} m_2 (l_1^2 \dot{\vartheta}_1^2 + l_2^2 \dot{\vartheta}_2^2 + 2 l_1 l_2 \dot{\vartheta}_1 \dot{\vartheta}_2 \cos(\vartheta_1 - \vartheta_2)).$$

(Addition theorem!)
To get the potential energy, we adopt a plane as a reference height, at the distance $l_1 + l_2$ below the suspension point:

$$V = m_1 g [l_1 + l_2 - l_1 \cos \vartheta_1] + m_2 g [l_1 + l_2 - (l_1 \cos \vartheta_1 + l_2 \cos \vartheta_2)].$$

The Lagrangian then becomes

$$L = T - V$$
$$= \frac{1}{2} m_1 l_1^2 \dot{\vartheta}_1^2 + \frac{1}{2} m_2 \left[l_1^2 \dot{\vartheta}_1^2 + l_2^2 \dot{\vartheta}_2^2 + 2 l_1 l_2 \dot{\vartheta}_1 \dot{\vartheta}_2 \cos(\vartheta_1 - \vartheta_2) \right] \tag{15.29}$$
$$- m_1 g [l_1 + l_2 - l_1 \cos \vartheta_1] - m_2 g [l_1 + l_2 - (l_1 \cos \vartheta_1 + l_2 \cos \vartheta_2)].$$

(c) The Lagrange equations with ϑ_1 and ϑ_2 read

$$\frac{d}{dt} \left(\frac{\partial L}{\partial \dot{\vartheta}_1} \right) - \frac{\partial L}{\partial \vartheta_1} = 0, \qquad \frac{d}{dt} \left(\frac{\partial L}{\partial \dot{\vartheta}_2} \right) - \frac{\partial L}{\partial \vartheta_2} = 0.$$

VIRTUAL DISPLACEMENTS

One has

$$\frac{\partial L}{\partial \vartheta_1} = -m_2 l_1 l_2 \dot{\vartheta}_1 \dot{\vartheta}_2 \sin(\vartheta_1 - \vartheta_2) - m_1 g l_1 \sin \vartheta_1 - m_2 g l_1 \sin \vartheta_1,$$

$$\frac{\partial L}{\partial \dot{\vartheta}_1} = m_1 l_1^2 \dot{\vartheta}_1 + m_2 l_1^2 \dot{\vartheta}_1 + m_2 l_1 l_2 \dot{\vartheta}_2 \cos(\vartheta_1 - \vartheta_2),$$

$$\frac{\partial L}{\partial \vartheta_2} = m_2 l_1 l_2 \dot{\vartheta}_1 \dot{\vartheta}_2 \sin(\vartheta_1 - \vartheta_2) - m_2 g l_2 \sin \vartheta_2,$$

$$\frac{\partial L}{\partial \dot{\vartheta}_2} = m_2 l_2^2 \dot{\vartheta}_2 + m_2 l_1 l_2 \dot{\vartheta}_1 \cos(\vartheta_1 - \vartheta_2).$$

Thus, the Lagrange equations read

$$m_1 l_1^2 \ddot{\vartheta}_1 + m_2 l_1^2 \ddot{\vartheta}_1 + m_2 l_1 l_2 \ddot{\vartheta}_2 \cos(\vartheta_1 - \vartheta_2) - m_2 l_1 l_2 \dot{\vartheta}_2 (\dot{\vartheta}_1 - \dot{\vartheta}_2) \sin(\vartheta_1 - \vartheta_2)$$
$$= -m_2 l_1 l_2 \dot{\vartheta}_1 \dot{\vartheta}_2 \sin(\vartheta_1 - \vartheta_2) - m_1 g l_1 \sin \vartheta_1 - m_2 g l_1 \sin \vartheta_1$$

and

$$m_2 l_2^2 \ddot{\vartheta}_2 + m_2 l_1 l_2 \ddot{\vartheta}_1 \cos(\vartheta_1 - \vartheta_2) - m_2 l_1 l_2 \dot{\vartheta}_1 (\dot{\vartheta}_1 - \dot{\vartheta}_2) \sin(\vartheta_1 - \vartheta_2)$$
$$= m_2 l_1 l_2 \dot{\vartheta}_1 \dot{\vartheta}_2 \sin(\vartheta_1 - \vartheta_2) - m_2 g l_2 \sin \vartheta_2,$$

or

$$(m_1 + m_2) l_1^2 \ddot{\vartheta}_1 + m_2 l_1 l_2 \ddot{\vartheta}_2 \cos(\vartheta_1 - \vartheta_2) + m_2 l_1 l_2 \dot{\vartheta}_2^2 \sin(\vartheta_1 - \vartheta_2)$$
$$= -(m_1 + m_2) g l_1 \sin \vartheta_1 \tag{15.30}$$

and

$$m_2 l_2^2 \ddot{\vartheta}_2 + m_2 l_1 l_2 \ddot{\vartheta}_1 \cos(\vartheta_1 - \vartheta_2) - m_2 l_1 l_2 \dot{\vartheta}_1^2 \sin(\vartheta_1 - \vartheta_2)$$
$$= -m_2 g l_2 \sin \vartheta_2.$$

These are the desired equations of motion.

(d) For the case

$$m_1 = m_2 = m \quad \text{and} \quad l_1 = l_2 = l,$$

the equations (15.30) reduce to

$$2l\ddot{\vartheta}_1 + l\ddot{\vartheta}_2 \cos(\vartheta_1 - \vartheta_2) + l\dot{\vartheta}_2^2 \sin(\vartheta_1 - \vartheta_2) = -2g \sin \vartheta_1,$$
$$l\ddot{\vartheta}_1 \cos(\vartheta_1 - \vartheta_2) + l\ddot{\vartheta}_2 - l\dot{\vartheta}_1^2 \sin(\vartheta_1 - \vartheta_2) = -g \sin \vartheta_2. \tag{15.31}$$

(e) If moreover the oscillations are small, then $\sin \vartheta = \vartheta$, $\cos \vartheta = 1$, and terms proportional to $\dot{\vartheta}^2$ are negligible, which leads to

$$2l\ddot{\vartheta}_1 + l\ddot{\vartheta}_2 = -2g\vartheta_1, \qquad l\ddot{\vartheta}_1 + l\ddot{\vartheta}_2 = -g\vartheta_2. \tag{15.32}$$

(f) With the *ansatz*

$$\vartheta_1 = A_1 e^{i\omega t}, \qquad \vartheta_2 = A_2 e^{i\omega t},$$

we then obtain

$$2(g - l\omega^2) A_1 - l\omega^2 A_2 = 0, \qquad -l\omega^2 A_1 + (g - l\omega^2) A_2 = 0. \tag{15.33}$$

To ensure that A_1 and A_2 do not vanish simultaneously, the determinant of the coefficients must vanish:

$$\begin{vmatrix} 2(g-l\omega^2) & -l\omega^2 \\ -l\omega^2 & g-l\omega^2 \end{vmatrix} = 0,$$

and therefore,

$$l^2\omega^4 - 4lg\omega^2 + 2g^2 = 0$$

with the solutions

$$\omega^2 = \frac{4lg \pm \sqrt{16l^2g^2 - 8l^2g^2}}{2l^2} = (2 \pm \sqrt{2})\frac{g}{l};$$

i.e.,

$$\omega_1^2 = (2+\sqrt{2})\frac{g}{l}, \qquad \omega_2^2 = (2-\sqrt{2})\frac{g}{l}. \tag{15.34}$$

By inserting (15.34) into (15.33), we obtain

$\omega_1^2 : A_2 = -\sqrt{2}A_1,$ i.e., the pendulums oscillate out of phase,

$\omega_2^2 : A_2 = \sqrt{2}A_1,$ i.e., the pendulums oscillate in phase.

Example 15.13: **Exercise: Mass point on a cycloid trajectory**

A mass point glides without friction on a cycloid, which is given by $x = a(\vartheta - \sin\vartheta)$ and $y = a(1 + \cos\vartheta)$ (with $0 \leq \vartheta \leq 2\pi$). Determine

(a) the Lagrangian, and

(b) the equation of motion.

(c) Solve the equation of motion.

Solution   The cycloid is represented by

$$x = a(\vartheta - \sin\vartheta), \qquad y = a(1 + \cos\vartheta),$$

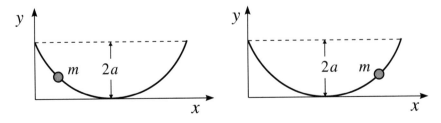

Figure 15.20. Mass point glides on a cycloid.

VIRTUAL DISPLACEMENTS

where $0 \le \vartheta \le 2\pi$. The kinetic energy is

$$T = \frac{1}{2}m(\dot{x}^2 + \dot{y}^2) = \frac{1}{2}ma^2\{[(1-\cos\vartheta)\dot{\vartheta}]^2 + [-(\sin\vartheta)\dot{\vartheta}]^2\},$$

and the potential energy is

$$V = mgy = mga(1+\cos\vartheta).$$

The Lagrangian is given by

$$L = T - V = ma^2(1-\cos\vartheta)\dot{\vartheta}^2 - mga(1+\cos\vartheta).$$

The equation of motion then reads

$$\frac{d}{dt}\left(\frac{\partial L}{\partial \dot{\vartheta}}\right) - \frac{\partial L}{\partial \vartheta} = 0,$$

i.e.,

$$\frac{d}{dt}[2ma^2(1-\cos\vartheta)\dot{\vartheta}] - [ma^2(\sin\vartheta)\dot{\vartheta}^2 + mga\sin\vartheta] = 0$$

or

$$\frac{d}{dt}[(1-\cos\vartheta)\dot{\vartheta}] - \frac{1}{2}(\sin\vartheta)\dot{\vartheta}^2 - \frac{g}{2a}\sin\vartheta = 0,$$

i.e.,

$$(1-\cos\vartheta)\ddot{\vartheta} + \frac{1}{2}(\sin\vartheta)\dot{\vartheta}^2 - \frac{g}{2a}\sin\vartheta = 0. \qquad (15.35)$$

By setting $u = \cos(\vartheta/2)$, one has

$$\frac{du}{dt} = -\frac{1}{2}\sin\left(\frac{\vartheta}{2}\right)\dot{\vartheta}$$

and

$$\frac{d^2u}{dt^2} = -\frac{1}{2}\sin\left(\frac{\vartheta}{2}\right)\ddot{\vartheta} - \frac{1}{4}\cos\left(\frac{\vartheta}{2}\right)\dot{\vartheta}^2.$$

Since $\cot(\vartheta/2) = \sin\vartheta/(1-\cos\vartheta)$, we can write (15.35) as

$$\ddot{\vartheta} + \frac{1}{2}\cot\left(\frac{\vartheta}{2}\right)\dot{\vartheta}^2 - \frac{g}{2a}\cot\left(\frac{\vartheta}{2}\right) = 0,$$

and therefore,

$$\frac{d^2u}{dt^2} + \frac{g}{4a}u = 0. \qquad (15.36)$$

The solution of this differential equation is

$$u = \cos\left(\frac{\vartheta}{2}\right) = C_1\cos\sqrt{\frac{g}{4a}}t + C_2\sin\sqrt{\frac{g}{4a}}t.$$

The motion is just like the vibration of an ordinary pendulum of length $l = 4a$. The arrangement is therefore called a "cycloid pendulum."

Example 15.14: Exercise: String pendulum

A mass m is suspended by a spring with spring constant k in the gravitational field. Besides the longitudinal spring vibration, the spring performs a plane pendulum motion (Figure 15.21). Find the Lagrangian, derive the equations of motion, and discuss the resulting terms.

Solution

We introduce plane polar coordinates for solving the problem and adopt the radius r and the polar angle φ as generalized coordinates.

$$y = r \cos\varphi, \qquad \dot{y} = \dot{r}\cos\varphi + r\dot{\varphi}\sin\varphi,$$
$$x = r \sin\varphi, \qquad \dot{x} = \dot{r}\sin\varphi - r\dot{\varphi}\cos\varphi. \tag{15.37}$$

The kinetic energy is given by

$$T = \frac{1}{2}m(\dot{x}^2 + \dot{y}^2) = \frac{1}{2}m(\dot{r}^2 + r^2\dot{\varphi}^2). \tag{15.38}$$

The length of the spring in its rest position, i.e., without the displacement caused by the mass m, is denoted by r_0. The potential energy then reads

$$V = -mgy + \frac{k}{2}(r - r_0)^2 = -mgr\cos\varphi + \frac{k}{2}(r - r_0)^2. \tag{15.39}$$

The Lagrangian is then

$$L = T - V = \frac{1}{2}m(\dot{r}^2 + r^2\dot{\varphi}^2) + mgr\cos\varphi - \frac{k}{2}(r - r_0)^2. \tag{15.40}$$

The equations of motion of the system are obtained immediately via the Lagrange equations:

$$\frac{d}{dt}(mr^2\dot{\varphi}) = -mgr\sin\varphi. \tag{15.41}$$

This is just the angular momentum law with reference to the coordinate origin. If we take the time dependence of r into account, then we have

$$mr\ddot{\varphi} = -mg\sin\varphi - 2m\dot{r}\dot{\varphi}. \tag{15.42}$$

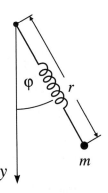

Figure 15.21.

The last term on the right-hand side is the Coriolis force caused by the time variation of the pendulum length r.

For the coordinate r, one obtains

$$m\ddot{r} = mr\dot{\varphi}^2 + mg\cos\varphi - k(r - r_0). \tag{15.43}$$

The first term on the right side represents the radial acceleration, the second term follows from the radial component of the weight force, and the last term represents Hooke's law. For small amplitudes φ the motion appears as a superposition of harmonic vibrations in the r, φ-plane.

Example 15.15: Exercise: Coupled mass points on a circle

Four mass points of mass m move on a circle of radius R. Each mass point is coupled to its two neighboring points by a spring with spring constant k (Figure 15.22). Find the Lagrangian of the system, and derive the equations of motion of the system. Calculate the eigenfrequencies of the system, and discuss the related eigenvibrations.

Solution The kinetic energy of the system is given by

$$T = \frac{1}{2}m \sum_{\nu=1}^{4} \dot{s}_\nu^2. \tag{15.44}$$

For small displacements from the equilibrium position, the potential reads

$$V = \frac{1}{2}k \sum_{\nu=1}^{4} (s_{\nu+1} - s_\nu)^2, \qquad s_{4+1} = s_1. \tag{15.45}$$

We set $s_\nu = R\varphi_\nu$, and take the angles φ_ν as generalized coordinates. Then the Lagrangian is

$$L = T - V = \frac{1}{2}mR^2 \sum_{\nu=1}^{4} \dot{\varphi}_\nu^2 - \frac{1}{2}kR^2 \sum_{\nu=1}^{4} (\varphi_{\nu+1} - \varphi_\nu)^2. \tag{15.46}$$

From the Lagrange equations

$$\frac{d}{dt}\frac{\partial L}{\partial \dot{\varphi}_\nu} = \frac{\partial L}{\partial \varphi_\nu}, \tag{15.47}$$

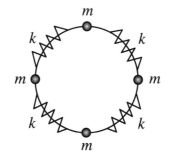

Figure 15.22.

we find the equations of motion:

$$\frac{d}{dt}\frac{\partial L}{\partial \dot{\varphi}_\nu} = mR\ddot{\varphi}_\nu$$
$$= -\frac{1}{2}kR^2[2(\varphi_\nu - \varphi_{\nu+1}) + 2(\varphi_\nu - \varphi_{\nu-1})] = \frac{\partial L}{\partial \varphi_\nu}. \tag{15.48}$$

For the case of four mass points, we then obtain

$$\ddot{\varphi}_1 = \frac{k}{m}(\varphi_2 - 2\varphi_1 + \varphi_4),$$
$$\ddot{\varphi}_2 = \frac{k}{m}(\varphi_3 - 2\varphi_2 + \varphi_1),$$
$$\ddot{\varphi}_3 = \frac{k}{m}(\varphi_4 - 2\varphi_3 + \varphi_2),$$
$$\ddot{\varphi}_4 = \frac{k}{m}(\varphi_1 - 2\varphi_4 + \varphi_3). \tag{15.49}$$

With the *ansatz* $\varphi_\nu = A_\nu \cos \omega t$, $\ddot{\varphi}_\nu = -A_\nu \omega^2 \cos \omega t$, we are led to the following linear system of equations:

$$\begin{pmatrix} 2\frac{k}{m} - \omega^2 & -\frac{k}{m} & 0 & -\frac{k}{m} \\ -\frac{k}{m} & 2\frac{k}{m} - \omega^2 & -\frac{k}{m} & 0 \\ 0 & -\frac{k}{m} & 2\frac{k}{m} - \omega^2 & -\frac{k}{m} \\ -\frac{k}{m} & 0 & -\frac{k}{m} & 2\frac{k}{m} - \omega^2 \end{pmatrix} \begin{pmatrix} A_1 \\ A_2 \\ A_3 \\ A_4 \end{pmatrix} = 0. \tag{15.50}$$

For the nontrivial solutions, the determinant of the coefficient matrix must vanish. This condition leads to the determining equation for the eigenfrequencies:

$$\left(2\frac{k}{m} - \omega^2\right)^2 \left(4\frac{k}{m} - \omega^2\right)(-\omega^2) = 0. \tag{15.51}$$

The frequencies are

$$\omega_1^2 = 0, \qquad \omega_2^2 = 4\frac{k}{m}, \qquad \omega_3^2 = \omega_4^2 = 2\frac{k}{m}. \tag{15.52}$$

To calculate the related eigenvibrations, we insert these frequencies into the system of equations (15.50).

(1) $\omega_1^2 = 0$: $A_1 = A_2 = A_3 = A_4$: The system does not vibrate but performs a uniform rotation (Figure 15.23(a)).

(2) $\omega_2^2 = 4k/m$: $A_1 = A_3 = -A_2 = -A_4$: Two neighboring mass points perform an out-of-phase vibration (Figure 15.23(b)).

(3) $\omega_3^2 = \omega_4^2 = 2k/m$: $A_1 = A_2 = -A_3 = -A_4$ or $A_1 = A_4 = -A_2 = -A_3$: Two neighboring mass points vibrate in phase (Figure 15.24(a,b)).

VIRTUAL DISPLACEMENTS

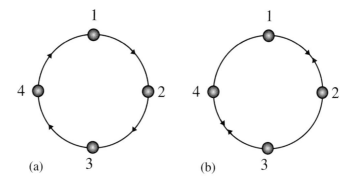

Figure 15.23. Uniform rotation and out-of-phase vibration.

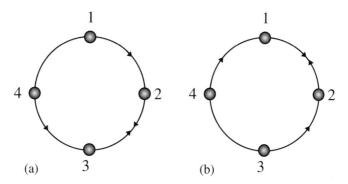

Figure 15.24. In-phase vibration.

Example 15.16: Exercise: Lagrangian of the asymmetric top

Write down the Lagrangian of the heavy asymmetric top. Use the Euler angles as generalized coordinates and determine the related generalized momenta. Which coordinate is cyclic? Which further cyclic coordinate appears for a symmetric top?

Solution In the system of principal axes, the kinetic energy of motion is given by

$$T = \frac{1}{2}\Theta_1\omega_1^2 + \frac{1}{2}\Theta_2\omega_2^2 + \frac{1}{2}\Theta_3\omega_3^2.$$

The potential energy is

$$V = mgh = mgl\cos\beta.$$

We take the Euler angles (α, β, γ) as generalized coordinates. The angular velocities expressed by these coordinates read (see (13.43)),

$$\omega_1 = \dot\alpha \sin\beta \sin\gamma + \dot\beta \cos\gamma,$$
$$\omega_2 = \dot\alpha \sin\beta \cos\gamma - \dot\beta \sin\gamma,$$
$$\omega_3 = \dot\alpha \cos\beta + \dot\gamma.$$

By inserting this into the Lagrangian $L = T - V$, we get

$$L = \frac{1}{2}\Theta_1(\dot{\alpha}^2 \sin^2\beta \sin^2\gamma + \dot{\beta}^2 \cos^2\gamma + 2\dot{\alpha}\dot{\beta} \sin\beta \sin\gamma \cos\beta)$$
$$+ \frac{1}{2}\Theta_2(\dot{\alpha}^2 \sin^2\beta \cos^2\gamma + \dot{\beta}^2 \sin^2\gamma - 2\dot{\alpha}\dot{\beta} \sin\beta \sin\gamma \cos\beta)$$
$$+ \frac{1}{2}\Theta_3(\dot{\alpha} \cos\beta + \dot{\gamma})^2 - Mgl \cos\beta.$$

The Euler angles as generalized coordinates obey the Euler–Lagrange equations

$$\frac{d}{dt}\frac{\partial L}{\partial \dot{\alpha}} = \frac{\partial L}{\partial \alpha}$$

and the analogous equations for β and γ. The Lagrangian does not depend on the angle α; hence, this coordinate is cyclic, and the related generalized momentum is conserved.

We determine the generalized momenta:

$$p_\alpha = \frac{\partial L}{\partial \dot{\alpha}} = \Theta_1 \omega_1 \frac{\partial \omega_1}{\partial \dot{\alpha}} + \Theta_2 \omega_2 \frac{\partial \omega_2}{\partial \dot{\alpha}} + \Theta_3 \omega_3 \frac{\partial \omega_3}{\partial \dot{\alpha}}$$
$$= \Theta_1(\dot{\alpha} \sin\beta \sin\gamma + \dot{\beta} \cos\gamma) \sin\beta \sin\gamma$$
$$+ \Theta_2(\dot{\alpha} \sin\beta \cos\gamma - \dot{\beta} \sin\gamma) \sin\beta \cos\gamma$$
$$+ \Theta_3(\dot{\alpha} \cos\beta + \dot{\gamma}) \cos\beta$$
$$= \dot{\alpha} \sin^2\beta(\Theta_1 \sin^2\gamma + \Theta_2 \cos^2\gamma) + (\Theta_1 - \Theta_2)\dot{\beta} \cos\gamma \sin\beta \sin\gamma$$
$$+ \Theta_3(\dot{\alpha} \cos\beta + \dot{\gamma}) \cos\beta.$$

In an analogous way, we obtain

$$p_\beta = \frac{\partial L}{\partial \dot{\beta}} = \Theta_1 \dot{\beta} \cos^2\gamma + \Theta_2 \dot{\beta} \sin^2\gamma$$
$$+ (\Theta_1 - \Theta_2)\dot{\alpha} \cos\gamma \sin\beta \sin\gamma + \Theta_3(\dot{\alpha} \cos\beta + \dot{\gamma}),$$
$$p_\gamma = \frac{\partial L}{\partial \dot{\gamma}} = \Theta_3(\dot{\alpha} \cos\beta + \dot{\gamma}).$$

For the symmetric top, $\Theta_1 = \Theta_2$, and thus, the Lagrangian simplifies considerably:

$$L = \frac{1}{2}\Theta_1(\dot{\alpha}^2 \sin^2\beta + \dot{\beta}^2) + \frac{1}{2}\Theta_3(\dot{\alpha} \cos\beta + \dot{\gamma})^2 - Mgl \cos\beta.$$

The Lagrangian of the symmetric top no longer depends on the angle γ; therefore, the angle γ becomes cyclic too. Hence, the momentum p_γ is also conserved.

The generalized momenta then read

$$p_\alpha = \dot{\alpha} \sin^2\beta \Theta_1 + \Theta_3(\dot{\alpha} \cos\beta + \dot{\gamma}) \cos\beta = \text{constant},$$
$$p_\beta = \dot{\beta}\Theta_1 + \Theta_3(\dot{\alpha} \cos\beta + \dot{\gamma}),$$
$$p_\gamma = \Theta_3(\dot{\alpha} \cos\beta + \dot{\gamma}) = \text{constant}.$$

The generalized momenta, being the projection of the total angular momentum onto the rotational axis related to the particular Euler angle, have a direct physical meaning.

p_α is the projection of the total angular momentum onto the space-fixed z-axis (see Example 13.12):

$$p_\alpha = \mathbf{L} \cdot \mathbf{e}_\alpha = \mathbf{L} \cdot \mathbf{e}_z.$$

This projection is a conserved quantity for the asymmetric and the symmetric top. Since the gravitational force acts only along the z-direction, the angular momentum about this axis remains unchanged.

p_β is the projection of the total angular momentum onto the nodal line, i.e., the axis about which the second Euler rotation is being performed:

$$p_\beta = \mathbf{L} \cdot \mathbf{e}_\beta = \mathbf{L} \cdot \mathbf{e}_x.$$

This momentum is not conserved.

p_γ can be interpreted as the projection of the total angular momentum onto the body-fixed $\mathbf{e}_{z'}$-axis:

$$p_\gamma = \mathbf{L} \cdot \mathbf{e}_\gamma = \mathbf{L} \cdot \mathbf{e}_{z'}.$$

For a symmetric top, the body-fixed z'-axis is a symmetry axis, and the angular momentum projection $\mathbf{L} \cdot \mathbf{e}_{z'}$ is conserved.

16 Lagrange Equation for Nonholonomic Constraints

For systems with holonomic constraints, the dependent coordinates can be eliminated by introducing generalized coordinates. If the constraints are nonholonomic, this approach does not work. There is no general method for treating nonholonomic problems. Only for those special nonholonomic constraints that can be given in differential form can one eliminate the dependent equations by the method of Lagrange multipliers. We therefore consider a system with constraints given in the form

$$\sum_{\nu=1}^{n} a_{l\nu} dq_\nu + a_{lt} dt = 0 \qquad (16.1)$$

($\nu = 1, 2, \ldots, n$ = number of coordinates; $n > s$; $l = 1, 2, \ldots, s$ = number of constraints).

The following considerations do not depend on whether the equations (16.1) are integrable or not; i.e., they hold both for holonomic as well as for nonholonomic constraints.

Therefore, the *method of Lagrange multipliers* derived below can be used also for holonomic constraints, if it is inconvenient to reduce all q_ν to independent coordinates or if one wants to keep the constraint reactions. Equation (16.1) is not the most general type of a nonholonomic constraint, e.g., constraints in the form of inequalities are not covered.

In our considerations, we start again—as in deriving the Lagrange equations—from the d'Alembert principle. According to (15.13), it reads in generalized coordinates as follows:

$$\sum_{\nu=1}^{n} \left(\frac{d}{dt} \frac{\partial T}{\partial \dot{q}_\nu} - \frac{\partial T}{\partial q_\nu} - Q_\nu \right) \delta q_\nu = 0. \qquad (16.2)$$

This equation holds for constraints of any kind.

The q_ν shall now depend on each other. Therefore, the virtual displacements δq_ν cannot be freely chosen as earlier (compare (15.13)). To reduce the number of virtual displacements to the number of independent displacements, we introduce the—for the present—freely chosen *Lagrange multipliers* λ_l. In the general case, the Lagrange multipliers λ_l with $l = 1, 2, \ldots, s$ are functions of the time and of the q_ν and $\dot{q}_\nu$. Virtual displacements δq_ν are performed at fixed time, i.e., with $\delta t = 0$. Then (16.1) changes to

$$\sum_{\nu=1}^{n} a_{l\nu} \delta q_\nu = 0.$$

These are also called *instantaneous* (belonging to a fixed time) constraints. This in turn leads to

$$\sum_{l=1}^{s} \lambda_l \sum_{\nu=1}^{n} a_{l\nu} \delta q_\nu = 0$$

or

$$\sum_{\nu=1}^{n} \left(\sum_{l=1}^{s} \lambda_l a_{l\nu} \right) \delta q_\nu = 0. \tag{16.3}$$

Equation (16.3) is now subtracted from (16.2):

$$\sum_{\nu=1}^{n} \left(\frac{d}{dt} \frac{\partial T}{\partial \dot{q}_\nu} - \frac{\partial T}{\partial q_\nu} - Q_\nu - \sum_{l=1}^{s} \lambda_l a_{l\nu} \right) \delta q_\nu = 0 \quad \text{for } \nu = 1, \ldots, s, \ldots n. \tag{16.4}$$

These equations involve in total n of the variables q_ν; s of them are dependent q_ν which are connected with the independent ones through the constraints, and $n - s$ are independent q_ν. For the dependent q_ν, the index ν shall run from $\nu = 1$ to $\nu = s$, for the independent q_ν from $\nu = s + 1$ to $\nu = n$. The coefficients of the δq_ν in equation (16.4) are such that through the s Lagrange multipliers λ_l ($l = 1, \ldots, s$) they can be chosen as freely as allowed by the s equations for the constraints. Since the λ_l can take any value, we can choose them in such a way that

$$\sum_{l=1}^{s} \lambda_l a_{l\nu} = \frac{d}{dt} \frac{\partial T}{\partial \dot{q}_\nu} - \frac{\partial T}{\partial q_\nu} - Q_\nu \quad (\nu = 1, \ldots, s),$$

i.e., the first s coefficients in (16.4) that correspond to the dependent q_ν are set to zero:

$$\frac{d}{dt} \frac{\partial T}{\partial \dot{q}_\nu} - \frac{\partial T}{\partial q_\nu} - Q_\nu - \sum_{l} \lambda_l a_{l\nu} = 0 \quad \text{for } \nu = 1, \ldots, s.$$

From equations (16.4) then remains

$$\sum_{\nu=s+1}^{n} \left(\frac{d}{dt} \frac{\partial T}{\partial \dot{q}_\nu} - \frac{\partial T}{\partial q_\nu} - Q_\nu - \sum_{l} \lambda_l a_{l\nu} \right) \delta q_\nu = 0.$$

These δq_ν (for $\nu = s + 1, \ldots, n$) are no longer subject to constraints. This means that these δq_ν are independent of each other. One then must set the coefficients of the δq_ν

($\nu = s+1, \ldots, n$) equal to zero, just as in the derivation of the Lagrange equation for holonomic systems.

This leads, together with the s equations for the dependent q_ν, to n equations in total:

$$\frac{d}{dt}\frac{\partial T}{\partial \dot{q}_\nu} - \frac{\partial T}{\partial q_\nu} - Q_\nu - \sum_{l=1}^{s}\lambda_l a_{l\nu} = 0 \quad \text{for } \nu = 1, \ldots, s, s+1, \ldots, n. \tag{16.5}$$

For conservative systems, the Q_ν can be derived from a potential:

$$Q_\nu = -\frac{\partial V}{\partial q_\nu}.$$

As in the derivation of the Lagrange equation for holonomic systems, we can reformulate the equation (16.5) with the Lagrangian $L = T - V$ as follows:

$$\frac{d}{dt}\frac{\partial L}{\partial \dot{q}_\nu} - \frac{\partial L}{\partial q_\nu} - \sum_{l=1}^{s}\lambda_l a_{l\nu} = 0, \quad \nu = 1, \ldots, n. \tag{16.6}$$

These n equations involve $n + s$ unknown quantities, namely the n coordinates q_ν and the s Lagrange multipliers λ_l. The additionally needed equations are just the s constraints (equation (16.1)) which couple the q_ν; however, these are now to be considered as differential equations:

$$\sum_\nu a_{l\nu}\dot{q}_\nu + a_{lt} = 0, \quad l = 1, 2, \ldots, s.$$

Thus, we have in total $n + s$ equations for $n + s$ unknowns. We thereby obtain both the q_ν we were looking for, and also the s quantities λ_l.

To understand the physical meaning of the λ_l, we assume that the constraints of the system are removed, but are replaced by external forces Q_ν^* which act in such a way that the motion of the system remains unchanged. The equations of motion would then also remain the same. These additional forces must be equal to the constraint reactions, since they act on the system in such a way that the constraint conditions are being fulfilled. With regard to these forces Q_ν^* the equations of motion read

$$\frac{d}{dt}\frac{\partial L}{\partial \dot{q}_\nu} - \frac{\partial L}{\partial q_\nu} = Q_\nu^*, \tag{16.7}$$

where the Q_ν^* enter in addition to the Q_ν. The equations (16.6) and (16.7) must be identical. This leads to

$$Q_\nu^* = \sum_l \lambda_l a_{l\nu}; \tag{16.8}$$

i.e., the Lagrange multipliers λ_l determine the *generalized constraint reactions* Q_ν^*; they will not be eliminated but are part of the solution of the problem (see also the statements in Chapter 17 on this topic). The relation (16.3) thus changes to

$$\sum_\nu Q_\nu^* \delta q_\nu = 0, \tag{16.9}$$

implying that the total virtual work performed by all constraint reactions vanishes. This can be considered as the general proof of the thesis introduced in equation (15.3), that constraint reactions do not perform work.

Example 16.1: Cylinder rolls down an inclined plane

As an example of the method of Lagrange multipliers, we consider a solid cylinder that rolls down without gliding on an inclined plane with height h and inclination angle α. This rolling condition is a holonomic constraint, but this is immaterial for the demonstration of the method.

The two generalized coordinates are s, φ. The constraint reads

$$R\dot{\varphi} = \dot{s} \quad \text{or} \quad Rd\varphi - ds = 0.$$

These equations can, of course, be integrated immediately, and with $R\varphi = s+$constant, the constraint is holonomic. But we stick to the differential form of the constraints and demonstrate the method of Lagrange multipliers. In this way we even find the constraint reactions.

The coefficients occurring in the constraint are

$$a_s = -1, \quad a_\varphi = R,$$

as is seen by comparison of coefficients with equation (16.1):

$$\sum_\nu a_{l\nu} \delta q_\nu = 0,$$

where $l = 1$ is the number of constraints, and $\delta t = 0$. The kinetic energy T can be represented as the sum of the kinetic energy of the center-of-mass motion and of the kinetic energy of the motion about the center of mass:

$$T = \frac{1}{2}m\dot{s}^2 + \frac{1}{2}\Theta\dot{\varphi}^2 = \frac{m}{2}\left(\dot{s}^2 + \frac{R^2}{2}\dot{\varphi}^2\right),$$

with the mass moment of inertia of the solid cylinder

$$\Theta_{\text{solcyl}} = \frac{1}{2}mR^2.$$

The potential energy V is

$$V = mgh - mgs\sin\alpha.$$

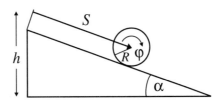

Figure 16.1. A cylinder rolls without gliding on an inclined plane.

The Lagrangian reads

$$L = T - V = \frac{m}{2}\left(\dot{s}^2 + \frac{R^2}{2}\dot{\varphi}^2\right) - mg(h - s\sin\alpha).$$

One should note that this Lagrangian cannot be used directly to derive the equation of motion according to equation (15.17). The reason is that the two coordinates s and φ are not independent of each other. Thus, φ is *not* an ignorable coordinate, although it does not explicitly appear in the Lagrangian.

Since there is only one constraint, only one Lagrange multiplier λ is needed. With the coefficients

$$a_s = -1, \qquad a_\varphi = R,$$

we obtain for the Lagrange equations

$$m\ddot{s} - mg\sin\alpha + \lambda = 0, \tag{16.10}$$

$$\frac{m}{2}R^2\ddot{\varphi} - \lambda R = 0, \tag{16.11}$$

which together with the constraint

$$R\dot{\varphi} = \dot{s} \tag{16.12}$$

represent three equations for three unknown quantities φ, s, λ. Differentiation of (16.12) with respect to the time yields

$$R\ddot{\varphi} = \ddot{s}.$$

From this, it follows, together with (16.11), that

$$m\ddot{s} = 2\lambda.$$

Hence, equation (16.10) changes to

$$mg\sin\alpha = 3\lambda.$$

From this equation, we obtain for the Lagrange multiplier

$$\lambda = \frac{1}{3}mg\sin\alpha.$$

The generalized constraint reactions are

$$a_s\lambda = -\frac{1}{3}mg\sin\alpha, \qquad a_\varphi\lambda = \frac{1}{3}Rmg\sin\alpha.$$

Here, $a_s\lambda$ is the constraint reaction caused by the friction; $a_\varphi\lambda$ is the torque generated by this force, which causes the rolling of the cylinder. One should clearly understand that the constraint "rolling" demands a particular constraint reaction (friction force). We have evaluated it here. We further note that the gravity is reduced exactly by the amount of the constraint reaction $a_s\lambda$.

Inserting the Lagrange multiplier λ into equation (16.10), we obtain the differential equation for s:

$$\ddot{s} = \frac{2}{3}g\sin\alpha.$$

The differential equation for φ is obtained from this by inserting

$$R\ddot{\varphi} = \ddot{s},$$

LAGRANGE EQUATION FOR NONHOLONOMIC CONSTRAINTS

$$\ddot{\varphi} = \frac{2}{3}\frac{g}{R}\sin\alpha.$$

We have seen by this example that the method of Lagrange multipliers yields not only the desired equations of motion but also the constraint reactions, which otherwise do not appear in the Lagrangian.

Example 16.2: Exercise: Particle moves in a paraboloid

A particle of mass m moves without friction under the action of gravitation on the inner surface of a paraboloid, which is given by

$$x^2 + y^2 = ax.$$

(a) Determine the Lagrangian and the equation of motion.

(b) Show that the particle moves on a horizontal circle in the plane $z = h$, provided that it gets an initial angular velocity. Find this angular velocity.

(c) Show that the particle oscillates about the circular orbit if it is displaced only weakly. Determine the oscillation frequency.

Solution (a) The appropriate coordinates are the cylindrical coordinates r, φ, z. The kinetic energy expressed in cylindrical coordinates reads

$$T = \frac{1}{2}m(\dot{r}^2 + r^2\dot{\varphi}^2 + \dot{z}^2).$$

Hence, the Lagrangian is

$$L = \frac{1}{2}m(\dot{r}^2 + r^2\dot{\varphi}^2 + \dot{z}^2) - mgz. \tag{16.13}$$

The constraint is $x^2 + y^2 = ax$. Since $x^2 + y^2 = r^2$, we have $r^2 - az = 0$, or in differential form, $2r\delta r - a\delta z = 0$.

Adopting the notation $r = q_1, \varphi = q_2, z = q_3$, from

$$\sum_\alpha A_\alpha q_\alpha = 0$$

we find that $A_1 = 2r, A_2 = 0, A_3 = -a$.

The Lagrange equations read

$$\frac{d}{dt}\left(\frac{\partial L}{\partial \dot{q}_\alpha}\right) - \frac{\partial L}{\partial q_\alpha} = \lambda_1 A_\alpha, \quad \alpha = 1, 2, 3;$$

i.e.,

$$\frac{d}{dt}\left(\frac{\partial L}{\partial \dot{r}}\right) - \frac{\partial L}{\partial r} = 2\lambda_1 r,$$

$$\frac{d}{dt}\left(\frac{\partial L}{\partial \dot{\varphi}}\right) - \frac{\partial L}{\partial \varphi} = 0,$$

and

$$\frac{d}{dt}\left(\frac{\partial L}{\partial \dot{z}}\right) - \frac{\partial L}{\partial z} = -\lambda_1 a. \tag{16.14}$$

With (16.13), we obtain

$$m(\ddot{r} - r\dot{\varphi}^2) = 2\lambda_1 r, \qquad m\frac{d}{dt}(r^2\dot{\varphi}), \qquad m\ddot{z} = -mg - \lambda_1 a, \tag{16.15}$$

and the constraint $2r\dot{r} - a\dot{z} = 0$. From this system, we can determine r, φ, z, λ_1.

(b) The radius of the circle arising by intersection of the plane $z = h$ with the paraboloid is $r^2 = az$,

$$r_0 = \sqrt{ah}.$$

From $m\ddot{z} = -mg - \lambda_1 a$, it follows with $z = h$ that

$$\lambda_1 = -\frac{mg}{a}.$$

From $m(\ddot{r} - r\dot{\varphi}^2) = 2\lambda_1 r$, it follows with $\dot{\varphi} = \omega, r = r_0$ that

$$m(-r_0\omega^2) = 2\left(-\frac{mg}{a}\right)r_0 \quad \text{or} \quad \omega^2 = \frac{2g}{a};$$

i.e.,

$$\omega = \sqrt{\frac{2g}{a}}$$

is the desired initial angular velocity.

(c) From $md(r^2\dot{\varphi})/dt = 0$, it follows that $r^2\dot{\varphi} = \text{constant} = A$. We suppose that the particle has the initial angular velocity ω; i.e.,

$$A = ah\omega, \quad \text{and therefore} \quad \dot{\varphi} = \frac{ah\omega}{r^2}.$$

Since the particle oscillates about $z = h$ with only small amplitude, we use $\lambda_1 = -mg/a$, which holds for $z = h$, and we obtain

$$m(\ddot{r} - r\dot{\varphi}^2) = -\frac{2mg}{a}r \quad \Rightarrow \quad \ddot{r} - a^2h^2\frac{\omega^2}{r^3} = -\frac{2gr}{a}.$$

Since the oscillation is small, we have $r = r_0 + u$; i.e.,

$$\ddot{u} - \frac{a^2h^2\omega^2}{(r_0+u)^3} = -\frac{2g}{a}(r_0+u). \tag{16.16}$$

We have

$$\frac{1}{(r_0+u)^3} = \frac{1}{r_0^3(1+u/r_0)^3} = \frac{1}{r_0^3}\left(1+\frac{u}{r_0}\right)^{-3} \approx \left(1-\frac{3u}{r_0}\right)\frac{1}{r_0^3},$$

since $u/r_0 \ll 1$ (power series expansion)!

Thus, from (16.16), we obtain with $r_0 = \sqrt{ah}$, $\omega = \sqrt{2g/a}$, the differential equation

$$\ddot{u} + \frac{8g}{a}u = 0 \tag{16.17}$$

with the solution

$$u = \varepsilon_1 \cos\sqrt{\frac{8g}{a}}t + \varepsilon_2 \sin\sqrt{\frac{8g}{a}}t,$$

and thus,

$$r = r_0 + u = \sqrt{ah} + \varepsilon_1 \cos\sqrt{\frac{8g}{a}}t + \varepsilon_2 \sin\sqrt{\frac{8g}{a}}t;$$

i.e., r oscillates with $\omega^2 = 8g/a$ about the equilibrium value $r_0 = \sqrt{ah}$. The oscillation period is

$$T_0 = \pi\sqrt{\frac{a}{2g}},$$

while the orbital period is

$$T_u = 2\pi\sqrt{\frac{a}{2g}} = 2T_0.$$

Example 16.3: Exercise: Three masses coupled by rods glide in a circular tire

Three mass points m_1, m_2, m_3 are fixed to the ends of two massless rods and glide without friction in a circular tire of radius R, which stands vertically in the gravitational field of the earth. Find the equations of motion by means of Lagrange multipliers, and determine the equilibrium position. Find the frequency of small oscillations about the equilibrium position.

Solution We use the angles φ_1, φ_2, and φ_3 as generalized coordinates. The angles are not independent of each other, but are coupled by the rigid rods connecting the mass points, via the constraints

$$\varphi_3 - \varphi_2 = \alpha = \text{constant},$$
$$\varphi_2 - \varphi_1 = \beta = \text{constant}. \tag{16.18}$$

In differential form, $\sum_\nu a_{l\nu}\delta q_\nu = 0$, they read

$$\delta\varphi_3 - \delta\varphi_2 = 0,$$
$$\delta\varphi_2 - \delta\varphi_1 = 0. \tag{16.19}$$

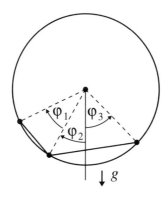

Figure 16.2.

The Lagrangian of the system can be immediately given in these coordinates:

$$L = \sum_\nu T_\nu - V_\nu$$
$$= \frac{1}{2} \sum_\nu m_\nu R^2 \dot{\varphi}^2 - \sum_\nu m_\nu g R (1 - \cos \varphi_\nu). \tag{16.20}$$

The Euler–Lagrange equations, generalized to nonintegrable constraints, i.e., constraints that are given *only* in the form (16.19), can be formulated by means of the Lagrange multipliers λ_l (equation (16.6)):

$$\frac{d}{dt} \frac{\partial L}{\partial \dot{\varphi}_\nu} - \frac{\partial L}{\partial \varphi_\nu} = \sum_{l=1}^{s} \lambda_l a_{l\nu}. \tag{16.21}$$

The number of constraints in the case considered here is $s = 2$. We thus obtain the three equations of motion:

$$m_\nu R^2 \ddot{\varphi}_\nu + m_\nu g R \sin \varphi_\nu = \lambda_1 a_{1\nu} + \lambda_2 a_{2\nu}, \qquad \nu = 1, 2, 3. \tag{16.22}$$

From (16.19), we obtain by comparison the coefficients $a_{l\nu}$:

$$\begin{aligned} a_{11} &= 0, & a_{12} &= -1, & a_{13} &= 1, \\ a_{21} &= -1, & a_{22} &= 1, & a_{23} &= 0. \end{aligned} \tag{16.23}$$

Then equation (16.18) implies that

$$\ddot{\varphi}_3 = \ddot{\varphi}_2 = \ddot{\varphi}_1. \tag{16.24}$$

By inserting this into (16.22), we get

$$\frac{g}{R} \sin \varphi_1 + \frac{\lambda_2}{m_1 R^2} = \frac{g}{R} \sin \varphi_2 + \frac{\lambda_1}{m_2 R^2} - \frac{\lambda_2}{m_2 R^2} = \frac{g}{R} \sin \varphi_3 - \frac{\lambda_1}{m_3 R^2}. \tag{16.25}$$

Solving for λ_1 and λ_2 leads to

$$\begin{aligned} \frac{\lambda_1}{R^2} &= m_3 \frac{\omega_0^2}{M} [m_1(\sin \varphi_3 - \sin \varphi_1) + m_2(\sin \varphi_3 - \sin \varphi_2)], \\ \frac{\lambda_2}{R^2} &= m_1 \frac{\omega_0^2}{M} [m_2(\sin \varphi_2 - \sin \varphi_1) + m_3(\sin \varphi_3 - \sin \varphi_1)]. \end{aligned} \tag{16.26}$$

Next, we set $M = m_1 + m_2 + m_3$ and $\omega_0^2 = g/R$. The angles φ_3 and φ_2 can be expressed by φ_1 via the constraint (16.18), so that one differential equation in the variable φ_1 describes the entire system. Hence, from (16.22) and (16.26) we obtain

$$\begin{aligned} \ddot{\varphi}_1 &= -\frac{\omega_0^2}{M} [m_1 \sin \varphi_1 + m_2 \sin \varphi_2 + m_3 \sin \varphi_3] \\ &= -\frac{\omega_0^2}{M} [m_1 \sin \varphi_1 + m_2 \sin(\varphi_1 + \beta) + m_3 \sin(\varphi_1 + \alpha + \beta)]. \end{aligned} \tag{16.27}$$

The equilibrium position is at the point of vanishing acceleration $\ddot{\varphi}_1 = 0$:

$$m_1 \sin \varphi_1 + m_2(\sin \varphi_1 \cos \beta + \cos \varphi_1 \sin \beta) + m_3[\sin \varphi_1 \cos(\alpha + \beta) + \cos \varphi_1 \sin(\alpha + \beta)] = 0. \tag{16.28}$$

Solving for φ_1 yields

$$\varphi_1|_{\ddot{\varphi}_1=0} = \varphi_1^0 = \arctan\left(-\frac{m_2 \sin\beta + m_3 \sin(\alpha+\beta)}{m_1 + m_2 \cos\beta + m_3 \cos(\alpha+\beta)}\right). \tag{16.29}$$

We now consider small vibrations ϑ about the equilibrium position determined by (16.29):

$$\varphi_1 = \varphi_1^0 + \vartheta \quad \text{with} \quad |\vartheta| \ll 1. \tag{16.30}$$

By means of the addition theorem $\sin(\Theta+\vartheta) \simeq \sin\Theta + \vartheta\cos\Theta$ and $\ddot{\varphi}_1 = \ddot{\vartheta}$, we obtain from (16.27) the desired frequency:

$$\ddot{\vartheta} = -\frac{\omega_0^2}{M}[m_1\cos\varphi_1^0 + m_2\cos(\varphi_1^0+\alpha) + m_3\cos(\varphi_1^0+\alpha+\beta)]\vartheta,$$
$$\ddot{\vartheta} \equiv -\Omega^2\vartheta. \tag{16.31}$$

For small amplitudes, this differential equation describes the vibrations of a physical pendulum. For $\alpha = \beta = 0$, and, hence, $\varphi_1^0 = 0$, it turns into the equation of the mathematical pendulum.

17 Special Problems

Velocity-dependent potentials

So far, we defined conservative forces **F** by the condition that they can be derived from a potential $V(\mathbf{r})$ by forming gradients, i.e.,

$$\mathbf{F}(\mathbf{r}, t) = -\nabla V(\mathbf{r}, t). \tag{17.1}$$

The potential $V(\mathbf{r}, t)$ is a function of the position and in general also of the time. This is possible as long as the forces do not depend on velocities or accelerations. There are, however, such cases: for instance, the *Lorentz force* which acts on a charged particle in the electromagnetic field is velocity-dependent:

$$\mathbf{F}^{(e)} = e\left(\mathbf{E} + \frac{\mathbf{v}}{c} \times \mathbf{B}\right). \tag{17.2}$$

Here, e is the charge of the particle, and **E** and **B** are the electric and magnetic field strength, respectively. $\mathbf{F}^{(e)}$ indicates that this shall be an external force.

If external forces depend on the velocity or the acceleration, we shall call them *conservative* as well if they can be expressed by a potential V that depends on the generalized coordinates q_j, the generalized velocities $\dot{q}_j$ and the time t, according to

$$Q_j = -\frac{\partial V}{\partial q_j} + \frac{d}{dt}\frac{\partial V}{\partial \dot{q}_j} \tag{17.3}$$

with $V = V(q_j, \dot{q}_j, t)$.

In some cases, such a representation can be possible also for the ordinary coordinates $\mathbf{r}_i$ and the velocity $\mathbf{v}_i$. The relation for $V = V(\mathbf{r}_i, \mathbf{v}_i, t)$ analogous to (17.3) then reads

$$\mathbf{F}_i = -\nabla_i V + \frac{d}{dt}\nabla_{\mathbf{v}_i} V = -\frac{\partial V}{\partial \mathbf{r}_i} + \frac{d}{dt}\frac{\partial V}{\partial \mathbf{v}_i}.$$

Here,

$$\nabla_{\mathbf{v}_i} = \left\{\frac{\partial}{\partial v_{ix}}, \frac{\partial}{\partial v_{iy}}, \frac{\partial}{\partial v_{iz}}\right\}$$

means the gradient vector with respect to the components of the velocity of the ith particle.

VELOCITY-DEPENDENT POTENTIALS

The *velocity-dependent potential*

$$V = V(q_j, \dot{q}_j, t) \tag{17.4}$$

is sometimes called the *generalized potential*. We know from (15.14) that the kinetic energy T and the generalized forces Q_j are related by

$$\frac{d}{dt}\frac{\partial T}{\partial \dot{q}_j} - \frac{\partial T}{\partial q_j} = Q_j. \tag{17.5}$$

Now using equation (17.3), we obtain

$$\frac{d}{dt}\frac{\partial T}{\partial \dot{q}_j} - \frac{\partial T}{\partial q_j} = -\frac{\partial V}{\partial q_j} + \frac{d}{dt}\frac{\partial V}{\partial \dot{q}_j}$$

or

$$\frac{d}{dt}\frac{\partial L}{\partial \dot{q}_j} - \frac{\partial L}{\partial q_j} = 0,$$

if we define the *generalized Lagrangian L* by

$$L = T - V$$

with the *generalized potential* $V(q_j, \dot{q}_j, t)$.

Sometimes, it is desirable to use another set of coordinates $\widetilde{q}_j$ instead of the set of generalized coordinates q_j. We now will show that this potential represents a generalized potential also in the new coordinates $\widetilde{q}_j$ and the related velocities $\dot{\widetilde{q}}_j$. This property is therefore independent of the selected special coordinates.

As in equation (14.7), the generalized forces $\widetilde{Q}_j$ belonging to $\widetilde{q}_j, \dot{\widetilde{q}}_j$ and the forces Q_j from (17.3) are related by

$$\widetilde{Q}_j = \sum_{\nu=1}^{3N} Q_\nu \frac{\partial q_\nu}{\partial \widetilde{q}_\nu}. \tag{17.6}$$

We have to show that

$$\widetilde{Q}_j = -\frac{\partial V}{\partial \widetilde{q}_j} + \frac{d}{dt}\left(\frac{\partial V}{\partial \dot{\widetilde{q}}_j}\right). \tag{17.7}$$

For the proof, we need the relation

$$\frac{\partial \dot{q}_k}{\partial \dot{\widetilde{q}}_j} = \frac{\partial q_k}{\partial \widetilde{q}_j}, \tag{17.8}$$

which immediately follows from

$$\dot{q}_k = \frac{d}{dt}(q_k) = \sum_{j=1}^{3N} \frac{\partial q_k}{\partial \widetilde{q}_j}\dot{\widetilde{q}}_j + \frac{\partial q_k}{\partial t}.$$

Then

$$\frac{\partial V}{\partial \widetilde{q}_j} = \sum_{\nu=1}^{3N} \frac{\partial V}{\partial q_\nu} \frac{\partial q_\nu}{\partial \widetilde{q}_j} + \sum_{\nu=1}^{3N} \frac{\partial V}{\partial \dot{q}_\nu} \frac{\partial \dot{q}_\nu}{\partial \widetilde{q}_j} + \frac{\partial V}{\partial t} \underbrace{\frac{\partial t}{\partial \widetilde{q}_j}}_{=0}$$

$$= \sum_{\nu=1}^{3N} \frac{\partial V}{\partial q_\nu} \frac{\partial q_\nu}{\partial \widetilde{q}_j} + \sum_{\nu=1}^{3N} \frac{\partial V}{\partial \dot{q}_\nu} \left(\frac{\partial}{\partial \widetilde{q}_j} \left(\sum_{\alpha=1}^{3N} \frac{\partial q_\nu}{\partial \widetilde{q}_\alpha} \dot{\widetilde{q}}_\alpha + \frac{\partial}{\partial t} q_\nu \right) \right). \quad (17.9)$$

Because

$$Q_\nu = -\frac{\partial V}{\partial q_\nu} + \frac{d}{dt}\left(\frac{\partial V}{\partial \dot{q}_\nu}\right),$$

we write the generalized force $\widetilde{Q}_j$ (17.6) as follows:

$$\widetilde{Q}_j = -\sum_{\nu=1}^{3N} \frac{\partial V}{\partial q_\nu} \frac{\partial q_\nu}{\partial \widetilde{q}_j} + \sum_{\nu=1}^{3N} \frac{d}{dt}\left(\frac{\partial V}{\partial \dot{q}_\nu}\right) \frac{\partial q_\nu}{\partial \widetilde{q}_j}$$

$$= -\sum_{\nu=1}^{3N} \frac{\partial V}{\partial q_\nu} \frac{\partial q_\nu}{\partial \widetilde{q}_j} + \sum_{\nu=1}^{3N} \frac{d}{dt}\left(\frac{\partial V}{\partial \dot{q}_\nu} \cdot \frac{\partial q_\nu}{\partial \widetilde{q}_j}\right) - \sum_{\nu=1}^{3N} \frac{\partial V}{\partial \dot{q}_\nu} \frac{d}{dt}\left(\frac{\partial q_\nu}{\partial \widetilde{q}_j}\right).$$

By inserting this expression into equation (17.9), we get

$$\widetilde{Q}_j = -\frac{\partial V}{\partial \widetilde{q}_j} + \sum_{\nu=1}^{3N} \frac{\partial V}{\partial \dot{q}_\nu} \left(\frac{\partial}{\partial \widetilde{q}_j}\left(\sum_{\alpha=1}^{3N} \frac{\partial q_\nu}{\partial \widetilde{q}_\alpha}\dot{\widetilde{q}}_\alpha + \frac{\partial}{\partial t} q_\nu\right)\right)$$

$$+ \sum_{\nu=1}^{3N} \frac{d}{dt}\left(\frac{\partial V}{\partial \dot{q}_\nu}\frac{\partial q_\nu}{\partial \widetilde{q}_j}\right) - \sum_{\nu=1}^{3N} \frac{\partial V}{\partial \dot{q}_\nu}\frac{d}{dt}\left(\frac{\partial q_\nu}{\partial \widetilde{q}_j}\right).$$

The third term yields with (17.8)

$$\sum_{\nu=1}^{3N} \frac{d}{dt}\left(\frac{\partial V}{\partial \dot{q}_\nu} \cdot \frac{\partial \dot{q}_\nu}{\partial \dot{\widetilde{q}}_j}\right) = \frac{d}{dt}\left(\frac{\partial V}{\partial \dot{\widetilde{q}}_j}\right).$$

Thus, $\widetilde{Q}_j$ becomes

$$\widetilde{Q}_j = -\frac{\partial V}{\partial \widetilde{q}_j} + \frac{d}{dt}\left(\frac{\partial V}{\partial \dot{\widetilde{q}}_j}\right)$$

$$+ \sum_{\nu=1}^{3N} \frac{\partial V}{\partial \dot{q}_\nu}\left(\frac{\partial}{\partial \widetilde{q}_\nu}\left(\sum_{\alpha=1}^{3N} \frac{\partial q_\nu}{\partial \widetilde{q}_\alpha}\dot{\widetilde{q}}_\alpha + \frac{\partial}{\partial t}q_\nu\right)\right)$$

$$- \sum_{\nu=1}^{3N} \frac{\partial V}{\partial \dot{q}_\nu}\left(\sum_{\alpha=1}^{3N} \frac{\partial}{\partial \widetilde{q}_\alpha}\left(\frac{\partial q_\nu}{\partial \widetilde{q}_j}\right)\dot{\widetilde{q}}_\alpha + \frac{\partial}{\partial t}\frac{\partial q_\nu}{\partial \widetilde{q}_j}\right).$$

VELOCITY-DEPENDENT POTENTIALS

Since $\dot{\tilde{q}}_\alpha$ does not depend on $\tilde{q}_j$, the last two terms cancel each other, and we find that (17.7) is valid. Thus, it has been shown that

$$V(q_j, \dot{q}_j, t) = V(q_j(\tilde{q}_j, t), \dot{q}_j(\dot{\tilde{q}}_j, t), t) \equiv V(\tilde{q}_j, \dot{\tilde{q}}_j, t)$$

also represents a generalized potential in the new coordinates $\tilde{q}_j$.

Example 17.1: Charged particle in an electromagnetic field

In the lectures of electrodynamics, we shall show that the electric field strength $\mathbf{E}$ and the magnetic field strength $\mathbf{B}$ can be derived from the *scalar potential* $\Phi(\mathbf{r}, t)$ and the *vector potential* $\mathbf{A}(\mathbf{r}, t)$, namely,

$$\mathbf{E} = -\nabla\Phi - \frac{1}{c}\frac{\partial \mathbf{A}}{\partial t}, \qquad \mathbf{B} = \nabla \times \mathbf{A}. \tag{17.10}$$

In other words, the electromagnetic phenomena can be described by $\Phi, \mathbf{A}$ instead of $\mathbf{E}, \mathbf{B}$. Now we show that in the frame of the Lagrangian formalism the Lorentz force (17.2) can be described by the velocity-dependent potential

$$V = e\Phi - \frac{e}{c}\mathbf{A} \cdot \mathbf{v}. \tag{17.11}$$

The Lagrangian then reads

$$L = T - V = \frac{1}{2}m\mathbf{v}^2 - e\Phi + \frac{e}{c}\mathbf{A} \cdot \mathbf{v}. \tag{17.12}$$

We restrict ourselves to the Lagrange equation for the x-component:

$$\frac{d}{dt}\frac{\partial L}{\partial v_x} - \frac{\partial L}{\partial x} = 0. \tag{17.13}$$

The other components follow likewise. We calculate

$$\frac{\partial L}{\partial x} = -e\frac{\partial \Phi}{\partial x} + \frac{e}{c}\frac{\partial \mathbf{A}}{\partial x} \cdot \mathbf{v}, \qquad \frac{\partial L}{\partial v_x} = mv_x + \frac{e}{c}A_x,$$

and furthermore according to (17.13),

$$\frac{d}{dt}mv_x = -e\frac{\partial \Phi}{\partial x} + \frac{e}{c}\frac{\partial \mathbf{A}}{\partial x} \cdot \mathbf{v} - \frac{e}{c}\frac{dA_x}{dt}. \tag{17.14}$$

For the last term, we obtain

$$\frac{dA_x}{dt} = \frac{\partial A_x}{\partial t} + \frac{\partial A_x}{\partial x}\frac{dx}{dt} + \frac{\partial A_x}{\partial y}\frac{dy}{dt} + \frac{\partial A_x}{\partial z}\frac{dz}{dt}$$

$$= \frac{\partial A_x}{\partial t} + \frac{\partial A_x}{\partial x}v_x + \frac{\partial A_x}{\partial y}v_y + \frac{\partial A_x}{\partial z}v_z, \tag{17.15}$$

and for the intermediate term,

$$\frac{\partial \mathbf{A}}{\partial x} \cdot \mathbf{v} = \frac{\partial A_x}{\partial x}v_x + \frac{\partial A_y}{\partial x}v_y + \frac{\partial A_z}{\partial x}v_z. \tag{17.16}$$

(17.15) and (17.16) are now inserted into (17.14) and yield

$$\frac{dmv_x}{dt} = e\left(-\frac{\partial \Phi}{\partial x} - \frac{1}{c}\frac{\partial A_x}{\partial t}\right) + \frac{e}{c}\left(\frac{\partial A_y}{\partial x} - \frac{\partial A_x}{\partial y}\right)v_y - \frac{e}{c}\left(\frac{\partial A_x}{\partial z} - \frac{\partial A_z}{\partial x}\right)v_z$$

$$= eE_x + \frac{e}{c}(B_x v_y - B_y v_z)$$

$$= e\left(\mathbf{E} + \frac{1}{c}\mathbf{v} \times \mathbf{B}\right)_x.$$

Corresponding expressions are obtained for the y- and z-components, so that we get in total

$$\frac{d}{dt}(m\mathbf{v}) = \left(\mathbf{E} + \frac{1}{c}\mathbf{v} \times \mathbf{B}\right), \tag{17.17}$$

i.e., Newton's equation of motion with the Lorentz force.

Nonconservative forces and dissipation function (friction function)

So far, the discussion was restricted to conservative forces only. We now consider systems with conservative and nonconservative forces. Such systems are first of all systems with friction. They play an important role in classical physics and recently also in heavy-ion physics. If two atomic nuclei collide, many internal degrees of freedom are excited; one can say that the nuclei are being heated up. Energy of relative motion is lost. This is a signature for friction forces, which are generally considered as being responsible for the energy loss.

We begin our discussion of nonconservative (e.g., friction) forces with the Lagrange equations in the form

$$\frac{d}{dt}\frac{\partial T}{\partial \dot{q}_j} - \frac{\partial T}{\partial q_j} = Q_j, \qquad j = 1, 2, \ldots, n, \tag{17.18}$$

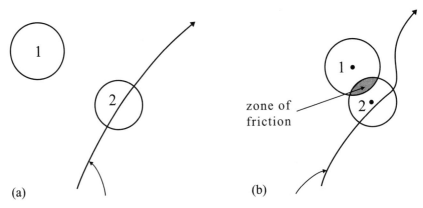

Figure 17.1. (a) Trajectory of nucleus 2 in the Coulomb field of nucleus 1. (b) Trajectory of nucleus 2 in the Coulomb plus nuclear field of nucleus 1.

and split the generalized forces Q_j in a conservative part $Q_j^{(c)}$ and a nonconservative part $Q_j^{(f)}$ (f for friction):

$$Q_j = Q_j^{(c)} + Q_j^{(f)}. \tag{17.19}$$

Since $Q_j^{(c)}$ can be derived by definition from a potential according to (17.3), we can introduce $L = T - V$ and bring (17.18) into the form

$$\frac{d}{dt}\frac{\partial L}{\partial \dot{q}_j} - \frac{\partial L}{\partial q_j} = Q_l^{(f)}, \qquad j = 1, 2, \ldots, n. \tag{17.20}$$

If the nonconservative forces are friction forces, on the right-hand side appear only these friction forces $Q_j^{(f)}$. For these, we make the *ansatz*

$$Q_j^{(f)} = -\sum_{k=1}^{n} f_{jk}\dot{q}_k, \tag{17.21}$$

where the f_{jk} are the *friction coefficients*. If the *friction tensor* f_{jk} is symmetric, i.e., $f_{jk} = f_{kj}$, the friction forces $Q_j^{(f)}$ can be obtained by partial derivation with respect to the generalized velocities $\dot{q}_j$ from the function

$$D = \frac{1}{2}\sum_{k,l=1}^{n} f_{kl}\dot{q}_k\dot{q}_l \tag{17.22}$$

according to

$$Q_j^{(f)} = -\frac{\partial D}{\partial \dot{q}_j}.$$

D is called the *dissipation function* (friction function). The Lagrange equations (17.20) can now be written as

$$\frac{d}{dt}\frac{\partial L}{\partial \dot{q}_j} + \frac{\partial D}{\partial \dot{q}_j} - \frac{\partial L}{\partial q_j} = 0. \tag{17.23}$$

In order to understand the physical meaning of the dissipation function, we calculate the work performed by the friction force $Q_j^{(f)}$ per unit time,

$$\frac{dW^{(r)}}{dt} = \sum_j Q_j^{(f)}\dot{q}_j = -\sum_{j,k} f_{jk}\dot{q}_j\dot{q}_k = -2D, \tag{17.24}$$

i.e., the energy consumed by the friction force per unit time is twice the dissipation function:

$$\frac{dE}{dt} = \frac{d}{dt}(T + V) = -2D. \tag{17.25}$$

This can also be directly derived from the Lagrange equations:

$$\frac{d}{dt}(T + V) = \sum_i \frac{\partial T}{\partial q_i}\dot{q}_i + \sum_i \frac{\partial T}{\partial \dot{q}_i}\ddot{q}_i + \frac{d}{dt}V. \tag{17.26}$$

With (17.23) we find

$$\sum_i \frac{\partial T}{\partial \dot{q}_i} \ddot{q}_i = \frac{d}{dt}\left(\sum_i \frac{\partial T}{\partial \dot{q}_i} \dot{q}_i\right) - \sum_i \dot{q}_i \frac{d}{dt}\left(\frac{\partial T}{\partial \dot{q}_i}\right)$$

$$= \frac{d}{dt}(2T) + \sum_i \dot{q}_i \frac{\partial D}{\partial \dot{q}_i} - \sum_i \frac{\partial T}{\partial \dot{q}_i} \dot{q}_i + \sum_i \frac{\partial V}{\partial q_i} \dot{q}_i$$

$$= \frac{d}{dt}(2T) + 2D - \sum_i \frac{\partial T}{\partial q_i} \dot{q}_i + \frac{d}{dt}V. \tag{17.27}$$

By inserting this into (17.26), we obtain

$$\frac{d(T+V)}{dt} = \frac{d(2T+2V)}{dt} + 2D$$

or

$$\frac{dE}{dt} = -2D,$$

i.e., the result (17.25).

Example 17.2: Motion of a projectile in air

The particle shall be under the action of the conservative gravitational force with the potential

$$V = mgz$$

and the nonconservative friction resistance of air. The air resistance depends on the projectile velocity. We suppose the friction force to be proportional to the velocity. It can then be derived from the *dissipation function*

$$D = \frac{1}{2}\alpha(\dot{x}^2 + \dot{y}^2 + \dot{z}^2),$$

and the Lagrange equations follow with $L = (1/2)m(\dot{x}^2 + \dot{y}^2 + \dot{z}^2) - mgz$ according to (17.23) as

$$m\ddot{x} + \alpha\dot{x} = 0, \qquad m\ddot{y} + \alpha\dot{y} = 0, \qquad m\ddot{z} + \alpha\dot{z} + mg = 0.$$

These equations of motion are known from *Classical Mechanics: Point Particles and Relativity* (Chapter 20).

Nonholonomic systems and Lagrange multipliers

In the preceding text, we have already discussed holonomic and nonholonomic systems. A brief recapitulation seems appropriate: For *holonomic* systems, the supplementary conditions can be expressed in the closed form

$$g_i(\mathbf{r}_\nu, t) = 0, \qquad i = 1, 2, \ldots, s, \qquad \nu = 1, 2, \ldots, N. \tag{17.28}$$

N = number of particles. We therefore can eliminate s coordinates and express the $\mathbf{r}_\nu$ as functions of $n = 3N - s$ independent generalized coordinates q_i. For *nonholonomic systems*, this is not possible, since the supplementary conditions appear in the differential form

$$\sum_{l=1}^{N} \mathbf{g}_{il}(\mathbf{r}_\nu, t) \cdot d\mathbf{r}_l + g_{it}(\mathbf{r}_\nu, t) dt = 0, \qquad i = 1, 2, \ldots, s. \tag{17.29}$$

Since these equations shall be nonintegrable, one cannot eliminate s dependent coordinates from them in the form (17.29). One therefore simply expresses the $\mathbf{r}_i$ as functions of $3N$ generalized coordinates q_i. The q_i are of course not all independent, but are subject to supplementary conditions which are obtained by rewriting (17.29) in terms of the q_i:

$$\sum_{l=1}^{3N} a_{il}(q, t) dq_l + a_{it}(q, t) dt = 0, \qquad i = 1, 2, \ldots, s. \tag{17.30}$$

For virtual displacements δq_l, i.e., $\delta t = 0$, these supplementary conditions change to

$$\sum_{l=1}^{3N} a_{il}(q, t) \delta q_l = 0, \qquad i = 1, 2, \ldots, s. \tag{17.31}$$

In this form, the supplementary conditions can be combined with the Lagrange equations in the same form, namely,

$$\sum_{j=1}^{3N} \left(Q_j^{(r)} + \frac{\partial L}{\partial q_j} - \frac{d}{dt}\frac{\partial L}{\partial \dot{q}_j} \right) \delta q_j = 0. \tag{17.32}$$

The conservative forces were taken into account in the Lagrangian L. Because of the conditions (17.31), not all δq_i in (17.32) are independent. To take this fact into account, one multiplies in (17.31) by the—at the moment still unknown—factors λ_i and sums up over i,

$$\sum_{i=1}^{s} \sum_{l=1}^{3N} \lambda_i a_{il}(q, t) \delta q_l = 0. \tag{17.33}$$

Addition of (17.32) and (17.33) then yields

$$\sum_{j=1}^{3N} \left(Q_j^{(r)} + \frac{\partial L}{\partial q_j} - \frac{d}{dt}\frac{\partial L}{\partial \dot{q}_j} + \sum_{i=1}^{s} \lambda_i a_{ij}(q, t) \right) \delta q_j = 0. \tag{17.34}$$

The factors λ_i are called *Lagrange multipliers*. They can be chosen arbitrarily in (17.34). Among the $3N$ quantities δq_j, however, only $3N - s$ can be chosen arbitrarily, since the s supplementary conditions (17.31) still must be satisfied. We number the δq_j so that the first s of them are just the dependent ones; the last $(3N - s)$ of the δq_j can be freely chosen.

Now we utilize the free choice of the s Lagrange parameters λ_i, which are determined in such a way that the coefficients of the first s variations δq_j in (17.34) vanish. This obviously leads to the s equations

$$Q_j^{(r)} + \frac{\partial L}{\partial q_j} - \frac{d}{dt}\frac{\partial L}{\partial \dot{q}_j} + \sum_{i=1}^{s} \lambda_i a_{ij}(q, t) = 0, \qquad j = 1, 2, \ldots, s, \tag{17.35}$$

and equation (17.34) reduces to

$$\sum_{j=s+1}^{3N} \left(Q_j^{(r)} + \frac{\partial L}{\partial q_j} - \frac{d}{dt}\frac{\partial L}{\partial \dot{q}_j} + \sum_{i=1}^{s} \lambda_i a_{ij}(q,t) \right) \delta q_j = 0. \qquad (17.36)$$

In equation (17.36), all of the δq_j can now be freely chosen. Therefore, the expression in the round bracket must vanish for every individual j; i.e., it follows that

$$Q_j^{(r)} + \frac{\partial L}{\partial q_j} - \frac{d}{dt}\frac{\partial L}{\partial \dot{q}_j} + \sum_{i=1}^{s} \lambda_i a_{ij}(q,t) = 0, \qquad j = s+1, s+2, \ldots, 3N. \qquad (17.37)$$

Now we see that the two sets of equations (17.35) and (17.37) have the same form and can be simply combined to

$$\frac{\partial L}{\partial q_j} - \frac{d}{dt}\frac{\partial L}{\partial \dot{q}_j} + Q_j^{(r)} + \sum_{i=1}^{s} \lambda_i a_{ij}(q,t) = 0, \qquad j = 1, 2, \ldots, 3N. \qquad (17.38)$$

These are $3N$ equations which, together with the s supplementary conditions in the form

$$\sum_{l=1}^{3N} a_{il}(q,t)\dot{q}_l + a_{it}(q,t) = 0, \qquad i = 1, 2, \ldots, s, \qquad (17.39)$$

determine the $3N + s$ unknown quantities, namely the $3N$ coordinates q_j and s Lagrange multipliers λ_i. Hence, the total number of desired quantities (q_j, λ_l) is $3N + s$. This is also the number of equations (17.38) and (17.39) which determine these quantities.

The meaning of the Lagrange multipliers can be understood even more precisely if we interpret the last term in (17.38) as an additional force $Q_j^{(z)}$, namely,

$$Q_j^{(z)} = \sum_{i=1}^{s} \lambda_i a_{ij}(q,t). \qquad (17.40)$$

These forces $Q_j^{(z)}$ are *constraint reactions* which appear since the motion of the system is restricted by supplementary conditions. Indeed, if the supplementary conditions disappear ($a_{ij} = 0$), the constraint reactions also vanish; $Q_j^{(z)} = 0$. The former equation (17.33) can now be written as

$$\sum_{i=1}^{3N} Q_i^{(z)} \delta q_i = 0 \qquad (17.41)$$

and can be interpreted as the *vanishing of the virtual work of the constraint reactions*.

It is clear that the method of Lagrange multipliers developed here for nonholonomic systems can be applied to holonomic systems, too. The holonomic constraints (17.28)

$$g_i(\mathbf{r}_\nu, t) = 0, \qquad i = 1, 2, \ldots, s, \qquad \nu = 1, 2, \ldots, N,$$

can immediately be written in differential form:

$$\sum_{l=1}^{N} \frac{\partial g_i}{\partial \mathbf{r}_l} \cdot d\mathbf{r}_l + \frac{\partial g_i}{\partial t} dt = 0, \qquad i = 1, 2, \ldots, s. \qquad (17.42)$$

NONHOLONOMIC SYSTEMS AND LAGRANGE MULTIPLIERS

This is exactly the form (17.29) for nonholonomic systems. From now on, the approach with Lagrange multipliers can run on as explained above. We then obtain $(3N + s)$ coupled equations, while the former solution method for holonomic systems (based on the elimination of s coordinates from (17.28)) leads only to $(3N - s)$ coupled equations. By the additional $2s$ equations the procedure now became much more complicated. However, this complication has also a great advantage: We now can determine the constraint reactions $Q_j^{(z)}$ according to (17.40) without difficulty (by solving the $3N + s$ equations).

Example 17.3: Exercise: Circular disk rolls on a plane

Determine the equations of motion and the constraint reactions of a circular disk of mass M and radius R that rolls *without gliding* on the x,y-plane (see Figure 17.2). The disk shall always stand perpendicular to the x,y-plane.

Solution We first consider how to mathematically formulate the constraints "without gliding" and "always stands perpendicular to the x,y-plane." This actually means that the center of the disk is exactly above the contact point (x,y) (the disk stands perpendicular), and the velocity of the circumference $R\dot{\Phi}$ of the disk edge equals the velocity of the contact point in the x,y-plane (Φ is the rotation angle of the disk around its axis). The latter means that there is no gliding. If we introduce the angle Θ between the disk axis and the x-axis (see Figure 17.3), the condition "no gliding" mathematically reads

$$\dot{x} = R\dot{\Phi}\sin\Theta, \qquad \dot{y} = -R\dot{\Phi}\cos\Theta. \tag{17.43}$$

In another formulation, these differential supplementary conditions read

$$dx - R\sin\Theta\, d\Phi = 0,$$
$$dy + R\cos\Theta\, d\Phi = 0. \tag{17.44}$$

In the form (17.30), these conditions thus read

$$a_{11}dx + a_{12}dy + a_{13}d\Phi + a_{14}d\Theta = 0,$$
$$a_{21}dx + a_{22}dy + a_{23}d\Phi + a_{24}d\Theta = 0,$$

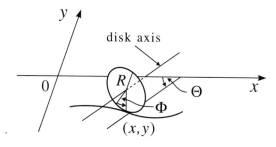

Figure 17.2. A circular disk rolls on the x,y-plane.

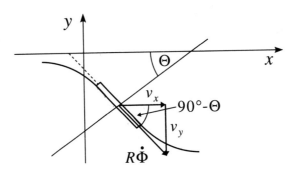

Figure 17.3. Projection onto the *x,y*-plane.

where

$$a_{11} = 1, \quad a_{12} = 0, \quad a_{13} = -R\sin\Theta, \quad a_{14} = 0,$$
$$a_{21} = 0, \quad a_{22} = 1, \quad a_{23} = R\cos\Theta, \quad a_{24} = 0.$$

Thus, according to equation (17.40) the constraint reactions are

$$\begin{aligned}
Q_x^{(z)} &= \lambda_1, \\
Q_y^{(z)} &= \lambda_2, \\
Q_\Phi^{(z)} &= -\lambda_1 R\sin\Theta + \lambda_2 R\cos\Theta, \\
Q_\Theta^{(z)} &= 0.
\end{aligned} \quad (17.45)$$

The kinetic energy of the disk is

$$T = \frac{1}{2}I_1\dot\Phi^2 + \frac{1}{2}I_2\dot\Theta^2 + \frac{1}{2}M\dot x^2 + \frac{1}{2}M\dot y^2, \quad (17.46)$$

where I_1 is the moment of inertia of the disk about the axis perpendicular to the disk through the center, and I_2 is the moment about the axis through the center and the contact point (x, y).

The Lagrange equations (17.38) now read explicitly

$$\begin{aligned}
M\ddot x &= Q_x + \lambda_1, \\
M\ddot y &= Q_y + \lambda_2, \\
I_1\ddot\Phi &= Q_\Phi - \lambda_1 R\sin\Theta + \lambda_2 R\cos\Theta, \\
I_2\ddot\Theta &= Q_\Theta.
\end{aligned} \quad (17.47)$$

$Q_x, Q_y, Q_\Phi, Q_\Theta$ are possible external forces. We study the case without such forces and therefore set them equal to zero. This transforms the equations (17.47) into

$$\begin{aligned}
M\ddot x &= \lambda_1, \\
M\ddot y &= \lambda_2, \\
I_1\ddot\Phi &= -\lambda_1 R\sin\Theta + \lambda_2 R\cos\Theta, \\
I_2\ddot\Theta &= 0,
\end{aligned} \quad (17.48)$$

which must be replaced by the equations (17.43) according to (17.39):

$$\dot{x} = R\dot{\Phi}\sin\Theta,$$
$$\dot{y} = -R\dot{\Phi}\cos\Theta. \qquad (17.49)$$

The last equation (17.48) can be immediately integrated, leading to

$$\Theta = \omega t + \Theta_0.$$

By inserting this into (17.49), one can calculate $\ddot{x}$ and $\ddot{y}$, which determine λ_1 and λ_2 through the first two equations (17.48):

$$\lambda_1 = M\ddot{x} = M(R\ddot{\Phi}\sin(\omega t + \Theta_0) + \omega R\dot{\Phi}\cos(\omega t + \Theta_0)),$$
$$\lambda_2 = M\ddot{y} = -M(R\ddot{\Phi}\cos(\omega t + \Theta_0) - \omega R\dot{\Phi}\sin(\omega t + \Theta_0)). \qquad (17.50)$$

This in turn is now inserted into the third equation (17.48), which then reads

$$\begin{aligned}I_1\ddot{\Phi} &= -MR(R\ddot{\Phi}\sin(\omega t + \Theta_0) + \omega R\dot{\Phi}\cos(\omega t + \Theta_0))\sin\omega t \\ &\quad - MR(R\ddot{\Phi}\cos(\omega t + \Theta_0) - \omega R\dot{\Phi}\sin(\omega t + \Theta_0))\cos\omega t \\ &= -MR^2\ddot{\Phi};\end{aligned}$$

i.e.,

$$(I_1 + MR^2)\ddot{\Phi} = 0.$$

This leads to $\ddot{\Phi} = 0$ and hence $\dot{\Phi} =$ constant. Therefore, we can explicitly write down the constraint reactions (17.45):

$$\begin{aligned}Q_x^{(z)} &= M\omega R\dot{\Phi}\cos(\omega t + \Theta_0), \\ Q_y^{(z)} &= M\omega R\dot{\Phi}\sin(\omega t + \Theta_0), \\ Q_\Phi^{(z)} &= 0, \qquad Q_\Theta^{(z)} = 0.\end{aligned} \qquad (17.51)$$

These constraint reactions must act to keep the disk vertical on the x,y-plane. If the disk rolls along a straight line ($\omega = 0$), the constraint reactions disappear.

Example 17.4: Exercise: Centrifugal force governor

Consider the degrees of freedom, and determine the equation of motion of the centrifugal force governor (Figure 17.4) through the Lagrangian.

Solution   The principle of the central force governor is applied, e.g., in automobiles. The distributor drive shaft is tightly fixed to the carrier plate of a central force governor which is attached below the interruptor plate. At higher speeds the centrifugal masses press on their carrier plate against a "cog." Thus, the distributor shaft set into the driving shaft is moved additionally in the rotation direction by a cam. This mechanism causes a preignition needed at higher speeds. For more advanced motors with "transistor ignition," this mechanism is dropped.

The system has two degrees of freedom, which can be described by the angles θ and φ. The motion of m, M is restricted by the constraints represented by the four rigid rods and the rotation axis.

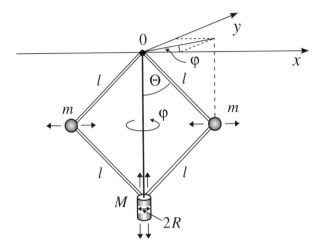

Figure 17.4.

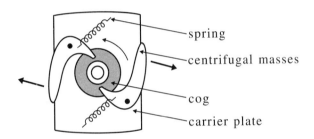

Figure 17.5. Centrifugal governor in automobiles.

Hence, θ and φ offer themselves as generalized coordinates. We first determine the kinetic energy. The moment of inertia of the cylinder is

$$\Theta_{ZZ} = \frac{1}{2}MR^2,$$

and therefore,

$$T_{\text{rot}} = \frac{1}{2}\left(\frac{1}{2}MR^2 + 2ml^2 \sin^2\theta\right)\dot{\varphi}^2. \tag{17.52}$$

The kinetic energy due to the motion in the x,y-plane is

$$T_{\text{plane}} = 2\frac{m}{2}v_m^2 + \frac{1}{2}Mv_M^2 \tag{17.53}$$

$$v_m = l\dot{\theta}, \qquad v_M = \frac{d}{dt}(-2l\cos\theta) = \dot{\theta}2l\sin\theta. \tag{17.54}$$

From this, it follows that

$$T_{\text{plane}} = (m + 2M\sin^2\theta)l^2\dot{\theta}^2. \tag{17.55}$$

With the potential energy $V = -2gl/(m + M)\cos\theta$, we can write down the Lagrangian:
$$\begin{aligned}L &= T_{\text{rot}} + T_{\text{plane}} - V \\ &= \frac{1}{2}(\Theta_{ZZ} + 2ml^2\sin^2\theta)\dot{\varphi}^2 + (m + 2M\sin^2\theta)l^2\dot{\theta}^2 + 2gl(m + M)\cos\theta.\end{aligned} \quad (17.56)$$

The Lagrange equations
$$\frac{d}{dt}\frac{\partial L}{\partial \dot{q}_\nu} - \frac{\partial L}{\partial q_\nu} = 0$$
immediately yield the equations of motion:
$$\begin{aligned}\frac{\partial L}{\partial \varphi} &= 0, \\ \frac{\partial L}{\partial \dot{\varphi}} &= (\Theta_{ZZ} + 2ml^2\sin^2\theta)\dot{\varphi}, \\ \frac{\partial L}{\partial \theta} &= 2ml^2\sin\theta\cos\theta\dot{\varphi}^2 + 4M\sin\theta\cos\theta l^2\dot{\theta}^2 - 2gl(m + M)\sin\theta, \\ \frac{\partial L}{\partial \dot{\theta}} &= 2(m + 2M\sin^2\theta)l^2\dot{\theta}.\end{aligned} \quad (17.57)$$
$$\quad (17.58)$$

From equation (17.57), we get
$$\frac{d}{dt}\left(\frac{\partial L}{\partial \dot{\varphi}}\right) = \frac{d}{dt}[(\Theta_{ZZ} + 2ml^2\sin^2\theta)\dot{\varphi}] = 0. \quad (17.59)$$

For the Lagrange equation in θ, we need
$$\frac{d}{dt}\left(\frac{\partial L}{\partial \dot{\theta}}\right) = (2m + 4M\sin^2\theta)l^2\ddot{\theta} + 8M\sin\theta\cos\theta l^2\dot{\theta}^2.$$

Then
$$(2m + 4M\sin^2\theta)l^2\ddot{\theta} + 4Ml^2\dot{\theta}^2\sin\theta\cos\theta \\ - 2ml^2\dot{\varphi}^2\sin\theta\cos\theta + 2gl(m + M)\sin\theta = 0, \quad (17.60)$$

and hence, we obtain the following equations of motion:
$$(2m + 4M\sin^2\theta)l^2\ddot{\theta} + 2l^2(2M\dot{\theta}^2 - m\dot{\varphi}^2)\sin\theta\cos\theta + 2gl(m + M)\sin\theta = 0,$$
$$\frac{d}{dt}\left[\left(\frac{1}{2}MR^2 + 2ml^2\sin^2\theta\right)\dot{\varphi}\right] = 0,$$
$$\Leftrightarrow \quad \dot{\varphi} = \frac{C}{(1/2)MR^2 + 2ml^2\sin^2\theta}. \quad (17.61)$$

From these equations of motion, the advantage of the Lagrangian formalism becomes evident. To account for the complicated constraint reactions in Newton's formulation would be much more laborious.

PART VI

HAMILTONIAN THEORY

18 Hamilton's Equations

The variables of the Lagrangian are the *generalized coordinates* q_α and the accompanying *generalized velocities* $\dot{q}_\alpha$. In Hamilton's theory, the generalized coordinates and the corresponding momenta are used as independent variables. In this theory the position coordinates and the "momentum coordinates" are treated on an equal basis. Hamiltonian theory leads to an essential understanding of the formal structure of mechanics and is of basic importance for the transition from classical mechanics to quantum mechanics.

We now look for a transition from the Lagrangian $L(q_i, \dot{q}_i, t)$ to the Hamiltonian $H(q_i, p_i, t)$ and remember that the *generalized momenta* are given by

$$p_i = \frac{\partial L}{\partial \dot{q}_i}.$$

We look for a transformation

$$L(q_i, \dot{q}_i, t) \implies H\left(q_i, \frac{\partial L}{\partial \dot{q}_i}, t\right) = H(q_i, p_i, t). \tag{18.1}$$

The question is, how to construct H? The recipe is simple and will be formulated in the following equation (18.2). The mathematical background of such a transformation (*Legendre*[1] *transformation*) can be easily demonstrated by a two-dimensional example. We change from the function $f(x, y)$ to the function $g(x, u) = g(x, \partial f/\partial y)$:

$$f(x, y) \implies g(x, u) \quad \text{with} \quad u = \frac{\partial f}{\partial y},$$

---

[1] *Adrien Marie Legendre*, b. Sept. 18, 1752–d. Jan. 10, 1833, Paris. Legendre made essential contributions to the foundation and development of number theory and geodesy. He also found important results on elliptic integrals, on foundations and methods of Euclidean geometry, on variational calculus, and on theoretical astronomy. For instance, he first applied the method of least squares and calculated voluminous tables. Legendre dealt with many problems that Gauss was also interested in, but he never reached his perfection. Beginning in 1775, Legendre served as professor at various universities at Paris and published excellent textbooks which had a long-lasting influence.

where $g(x, u)$ is defined by

$$g(x, u) = uy - f(x, y).$$

By forming the total differential, we realize that the function g formed this way no longer contains y as an independent variable:

$$\begin{aligned} dg &= y\,du + u\,dy - df \\ &= y\,du + u\,dy - \frac{\partial f}{\partial x}dx - \frac{\partial f}{\partial y}dy \\ &= y\,du - \frac{\partial f}{\partial y}dx, \end{aligned}$$

where now $y = \partial g/\partial u$ and $\partial g/\partial x = -\partial f/\partial x$.

According to this short insertion, we now construct the Hamiltonian from the Lagrangian. We write for the *Hamiltonian*

$$H(q_i, p_i, t) = \sum_i p_i \dot{q}_i - L(q_i, \dot{q}_i, t). \tag{18.2}$$

We look for those equations of motion which are equivalent to the Lagrange equations based on the Lagrangian L. To this end, we form the total differential:

$$dH = \sum p_i\,d\dot{q}_i + \sum \dot{q}_i\,dp_i - dL. \tag{18.3}$$

The total differential of the Lagrangian reads

$$dL = \sum \frac{\partial L}{\partial q_i} dq_i + \sum \frac{\partial L}{\partial \dot{q}_i} d\dot{q}_i + \frac{\partial L}{\partial t} dt. \tag{18.4}$$

We now utilize the definition of the generalized momentum, $p_i = \partial L/\partial \dot{q}_i$, and the Lagrange equation in the form

$$\frac{d}{dt}p_i - \frac{\partial L}{\partial q_i} = 0.$$

Inserting both into equation (18.4) yields

$$dL = \sum \dot{p}_i\,dq_i + \sum p_i\,d\dot{q}_i + \frac{\partial L}{\partial t} dt.$$

By insertion of dL into equation (18.3), it follows that

$$dH = \sum p_i\,d\dot{q}_i + \sum \dot{q}_i\,dp_i - \sum \dot{p}_i\,dq_i - \sum p_i\,d\dot{q}_i - \frac{\partial L}{\partial t} dt.$$

Since the first and fourth term mutually cancel, there remains

$$dH = \sum_i \dot{q}_i\,dp_i - \sum_i \dot{p}_i\,dq_i - \frac{\partial L}{\partial t} dt.$$

HAMILTON'S EQUATIONS

Therefore, H depends only on p_i, q_i, and t; thus, $H = H(q_i, p_i, t)$, and we have

$$dH = \sum \frac{\partial H}{\partial q_i} dq_i + \sum \frac{\partial H}{\partial p_i} dp_i + \frac{\partial H}{\partial t} dt = \sum \dot{q}_i dp_i - \sum \dot{p}_i dq_i - \frac{\partial L}{\partial t} dt.$$

From this immediately follow the *Hamilton equations:*[2]

$$\dot{q}_i = \frac{\partial H}{\partial p_i}, \qquad \dot{p}_i = -\frac{\partial H}{\partial q_i}, \qquad \frac{\partial H}{\partial t} = -\frac{\partial L}{\partial t}. \tag{18.5}$$

They are now the fundamental equations of motion in this formulation of mechanics. The Hamiltonian H here plays the central role, similar to the Lagrangian L in Lagrange's formulation of mechanics. This Hamiltonian H is constructed according to equation (18.2); but with the prescription that all velocities $\dot{q}_i$ are expressed by the generalized momenta p_i and the generalized coordinates q_i through equation (18.1). In other words, the equations (18.1) for the definition of the generalized momenta

$$p_i = \frac{\partial L(q_i, \dot{q}_i, t)}{\partial \dot{q}_i}$$

are solved for the generalized velocities $\dot{q}_i$, so that

$$\dot{q}_i = \dot{q}_i(q_i, p_i).$$

The $\dot{q}_i$ obtained this way are inserted into the definition of H (see equation (18.2)), so that the Hamiltonian H finally depends only on q_i, p_i, and the time t; hence, $H = H(q_i, p_i, t)$. From this, the Hamilton equations (18.5) are established and solved.

The Lagrange equations provide a set of n differential equations of second order in the time for the position coordinates. The Hamiltonian formalism yields $2n$ coupled differential equations of first order for the momentum and position coordinates. In any case, there are $2n$ integration constants when solving the system of equations.

From the equations (18.5), it is seen that for a coordinate that does not enter the Hamiltonian, the corresponding change of the momentum with time vanishes:

$$\frac{\partial H}{\partial q_i} = 0 \quad \Longrightarrow \quad p_i = \text{constant}.$$

If the Hamiltonian (the Lagrangian) is not explicitly time dependent, then H is a constant of motion since

$$\frac{dH}{dt} = \sum \frac{\partial H}{\partial q_i} \dot{q}_i + \sum \frac{\partial H}{\partial p_i} \dot{p}_i + \frac{\partial H}{\partial t},$$

---

[2] *Sir William Rowan Hamilton*, b. Aug. 4, 1805, Dublin–d. Sept. 2, 1865, Dunsik. Hamilton began his studies in 1824 in Dublin. In 1827, before finishing his studies, he became professor of astronomy and King's astronomer of Ireland. Hamilton contributed important papers on algebra and invented the quaternion calculus. His contributions to geometrical optics and classical mechanics, e.g., the canonical equations and the Hamilton principle, are of extraordinary importance.

and with the equations (18.5) this leads to

$$\frac{dH}{dt} = \frac{\partial H}{\partial t}.$$

Now it is clear that with $\partial H/\partial t = 0$ (since H shall not be explicitly time dependent) it follows that $dH/dt = 0$, and thus, $H =$ constant.

What is the meaning of the Hamiltonian; how can it be interpreted physically? To see that, we consider a special case: For a system with holonomic, scleronomic constraints and conservative internal forces, the Hamiltonian H represents the energy of the system.

To clarify this, we first consider the kinetic energy:

$$T = \frac{1}{2} \sum_\nu m_\nu \dot{\mathbf{r}}_\nu^2, \qquad \nu = 1, 2, \ldots, N \qquad (N = \text{number of particles}).$$

If the constraints are holonomic and not time-dependent, there exist transformation equations $\mathbf{r}_\nu = \mathbf{r}_\nu(q_i)$, and therefore,

$$\dot{\mathbf{r}}_\nu = \sum_i \frac{\partial \mathbf{r}_\nu}{\partial q_i} \dot{q}_i.$$

Insertion into the kinetic energy yields

$$T = \sum_\nu m_\nu \sum_{i,k} \left(\frac{\partial \mathbf{r}_\nu}{\partial q_i} \dot{q}_i \right) \cdot \left(\frac{\partial \mathbf{r}_\nu}{\partial q_k} \dot{q}_k \right)$$

$$= \sum_{i,k} \left(\frac{1}{2} \sum_\nu \frac{\partial \mathbf{r}_\nu}{\partial q_i} \cdot \frac{\partial \mathbf{r}_\nu}{\partial q_k} \right) \dot{q}_i \dot{q}_k$$

$$= \sum_{i,k} a_{ik} \dot{q}_i \dot{q}_k.$$

Thus, the kinetic energy is a *homogeneous quadratic function* of the generalized velocities. The arising *mass coefficients*

$$a_{ik} = \frac{1}{2} \sum_\nu m_\nu \frac{\partial \mathbf{r}_\nu}{\partial q_i} \cdot \frac{\partial \mathbf{r}_\nu}{\partial q_k}$$

are symmetric; i.e., $a_{ik} = a_{ki}$.

Now we can apply Euler's theorem on homogeneous functions. If f is a homogeneous function of rank n, i.e., if

$$f(\lambda x_1, \lambda x_2, \ldots, \lambda x_k) = \lambda^n f(x_1, x_2, \ldots, x_k),$$

then also

$$\sum_{i=1}^{k} x_i \frac{\partial f}{\partial x_i} = nf.$$

HAMILTON'S EQUATIONS

This can be shown by forming the derivative of the upper equation with respect to λ; thus,

$$\frac{\partial f}{\partial (\lambda x_1)} x_1 + \cdots + \frac{\partial f}{\partial (\lambda x_k)} x_k = n\lambda^{n-1} f.$$

By setting $\lambda = 1$, the assertion follows. Euler's theorem, applied to the kinetic energy ($n = 2$), means

$$\sum \frac{\partial T}{\partial \dot{q}_i} \cdot \dot{q}_i = 2T. \tag{18.6}$$

Since the forces are presupposed to be conservative, there exists a velocity-independent potential $V(q_i)$, so that

$$\frac{\partial L}{\partial \dot{q}_i} = \frac{\partial T}{\partial \dot{q}_i} = p_i,$$

and therefore,

$$H = \sum p_i \dot{q}_i - L = \sum \frac{\partial T}{\partial \dot{q}_i} \dot{q}_i - L.$$

By using the relation (18.6) and the definition of the Lagrangian, we see that

$$H = 2T - (T - V) = T + V = E.$$

Thus, under the given conditions the Hamiltonian represents the total energy. The energy $T - V$ represented by the Lagrangian is sometimes called the *free energy*.

One should note that H does not include a possible work performed by the constraint reactions.

The Hamiltonian formulation of mechanics emerges via the Lagrange equations from Newton's equations. This became evident in deriving the equations (18.5), where we explicitly used the Lagrange equations. The latter ones are however equivalent to Newton's formulation of mechanics (see d'Alembert's principle and following text). Conversely, one can easily derive Newton's equations from Hamilton's equations and thus show the equivalence of both formulations. It is sufficient to consider a single particle in a conservative force field and to use the Cartesian coordinates as generalized coordinates. Then

$$p_i = m\dot{x}_i, \qquad H = \frac{1}{2} \sum_i \dot{x}_i^2 + V(x_i) \qquad (i = 1, 2, 3),$$

or

$$H = \frac{1}{2} \sum_i \frac{p_i^2}{m} + V(q_i).$$

This leads to the Hamilton equations ($q_i = x_i$):

$$\dot{q}_i = \frac{\partial H}{\partial p_i} = \frac{p_i}{m} \quad \text{and} \quad \dot{p}_i = -\frac{\partial H}{\partial q_i} = -\frac{\partial V}{\partial q_i},$$

or in vector notation

$$\dot{\mathbf{p}} = -\text{grad } V.$$

These are Newton's equations of motion.

Example 18.1: Central motion

Let a particle perform a planar motion under the action of a potential that depends only on the distance from the origin. It is obvious that we should use plane polar coordinates (r, φ) as generalized coordinates.

$$L = T - V = \frac{1}{2}mv^2 - V = \frac{1}{2}m(\dot{r}^2 + r^2\dot{\varphi}^2) - V(r).$$

With $p_\alpha = \partial L/\partial \dot{q}_\alpha$, we get the momenta

$$p_r = \frac{\partial L}{\partial \dot{r}} = m\dot{r} \quad \text{or} \quad \dot{r} = \frac{p_r}{m},$$

$$p_\varphi = \frac{\partial L}{\partial \dot{\varphi}} = mr^2\dot{\varphi} \quad \text{or} \quad \dot{\varphi} = \frac{p_\varphi}{mr^2}.$$

Thus, the Hamiltonian reads

$$H = p_r \dot{r} + p_\varphi \dot{\varphi} - L = \frac{p_r^2}{2m} + \frac{p_\varphi^2}{2mr^2} + V(r).$$

The Hamilton equations then yield

$$\dot{r} = \frac{\partial H}{\partial p_r} = \frac{p_r}{m}, \quad \dot{\varphi} = \frac{\partial H}{\partial p_\varphi} = \frac{p_\varphi}{mr^2},$$

and

$$\dot{p}_r = -\frac{\partial H}{\partial r} = \frac{p_\varphi^2}{mr^3} - \frac{\partial V}{\partial r}, \quad \dot{p}_\varphi = -\frac{\partial H}{\partial \varphi} = 0.$$

φ is a cyclic coordinate. From this follows the conservation of the angular momentum in the central potential.

Example 18.2: The pendulum in the Newtonian, Lagrangian, and Hamiltonian theories

The equation of motion of the pendulum shall be derived within the frames of Newton's, Lagrange's, and Hamilton's theory.

Newtonian theory: We begin with Newton's axiom

$$\dot{\mathbf{p}} = \mathbf{K}.$$

The arclength of the displacement is denoted by s, and the tangent unit vector by $\mathbf{T}$. Then (see Figure 18.1)

$$\mathbf{K} = -mg \sin \Theta \mathbf{T},$$

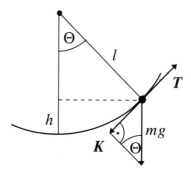

Figure 18.1. On pendulum motion.

and thus,

$$m\ddot{s}\mathbf{T} = -mg \sin \Theta \mathbf{T}.$$

With $s = l\Theta$, we have $\ddot{s} = l\ddot{\Theta}$. We therefore get the equation of motion

$$\ddot{\Theta} + \frac{g}{l} \sin \Theta = 0.$$

For small displacements ($\sin \Theta \sim \Theta + \ldots$), this becomes

$$\ddot{\Theta} + \frac{g}{l} \Theta = 0.$$

This differential equation has the general solution

$$\Theta = A \cos \sqrt{\frac{g}{l}} t + B \sin \sqrt{\frac{g}{l}} t,$$

where the constants A and B are to be determined from the initial conditions.

Lagrangian theory:

$$T = \frac{1}{2} m v^2 = \frac{1}{2} m (l\dot{\Theta})^2 = \frac{1}{2} m l^2 \dot{\Theta}^2,$$
$$V = mgh = mg(l - l\cos\Theta) = mgl(1 - \cos\Theta).$$

Hence, the Lagrangian for this conservative system reads

$$L = T - V = \frac{1}{2} m l^2 \dot{\Theta}^2 - mgl(1 - \cos\Theta).$$

Now we use the Lagrange equation

$$\frac{d}{dt}\left(\frac{\partial L}{\partial \dot{\Theta}}\right) - \frac{\partial L}{\partial \Theta} = 0.$$

With

$$\frac{\partial L}{\partial \Theta} = -mgl \sin \Theta \quad \text{and} \quad \frac{\partial L}{\partial \dot{\Theta}} = ml^2 \dot{\Theta},$$

we have

$$ml^2\ddot{\Theta} + mgl\sin\Theta = 0 \quad \text{or} \quad \ddot{\Theta} + \frac{g}{l}\sin\Theta = 0.$$

Hamiltonian theory: Using the generalized momentum

$$p_\Theta = \frac{\partial L}{\partial \dot\Theta} = ml^2\dot\Theta,$$

the kinetic energy can be written as

$$T = \frac{1}{2}\frac{p_\Theta^2}{ml^2}.$$

Since the total energy of the system is constant, the Hamiltonian reads

$$H = T + V = \frac{1}{2}\frac{p_\Theta^2}{ml^2} + mgl(1 - \cos\Theta).$$

The Hamilton equations yield

$$\dot p_\Theta = -\frac{\partial H}{\partial \Theta} = -mgl\sin\Theta \quad \text{and} \quad \dot\Theta = \frac{\partial H}{\partial p_\Theta} = \frac{p_\Theta}{ml^2}.$$

The last equation gives

$$p_\Theta = ml^2\dot\Theta.$$

Differentiation yields

$$\dot p_\Theta = ml^2\ddot\Theta.$$

By comparing this with the above expression for $\dot p_\Theta$, we finally get again

$$\ddot\Theta + \frac{g}{l}\sin\Theta = 0.$$

Example 18.3: Exercise: Hamiltonian and canonical equations of motion

A mass point m shall move in a cylindrically symmetric potential $V(\varrho, z)$. Determine the Hamiltonian and the canonical equations of motion with respect to a coordinate system that rotates with constant angular velocity ω about the symmetry axis,

(a) in Cartesian coordinates, and

(b) in cylindrical coordinates.

Solution (a) The coordinates of the inertial system (x, y, z) and those of the rotating reference system (x', y', z') are related by

$$x = \cos(\omega t)x' - \sin(\omega t)y',$$
$$y = \sin(\omega t)x' + \cos(\omega t)y', \quad (18.7)$$
$$z = z'.$$

Derivation of the coordinates yields

$$\dot x = \cos(\omega t)\dot x' - \sin(\omega t)\dot y' - \omega(\sin(\omega t)x' + \cos(\omega t)y'),$$
$$\dot y = \sin(\omega t)\dot x' + \cos(\omega t)\dot y' + \omega(\cos(\omega t)x' - \sin(\omega t)y'), \quad (18.8)$$

HAMILTON'S EQUATIONS

$$\dot{z} = \dot{z}'.$$

In the primed coordinate system, the Lagrangian takes the form

$$L = \frac{1}{2}m[\dot{x}'^2 + \dot{y}'^2 + \dot{z}'^2 + \omega^2(x'^2 + y'^2) + 2\omega(\dot{y}'x' - \dot{x}'y')] - V(x', y', z'). \tag{18.9}$$

From (18.9), we calculate the generalized momenta as

$$p'_x = \frac{\partial L}{\partial \dot{x}'} = m(\dot{x}' - \omega y'), \qquad p'_y = \frac{\partial L}{\partial \dot{y}'} = m(\dot{y}' + \omega x'), \qquad p'_z = \frac{\partial L}{\partial \dot{z}} = m\dot{z}'. \tag{18.10}$$

Now we solve (18.10) for the velocity components $\dot{x}'$, $\dot{y}'$, $\dot{z}'$:

$$\dot{x}' = \frac{p'_x}{m} + \omega y', \qquad \dot{y}' = \frac{p'_y}{m} - \omega x', \qquad \dot{z}' = \frac{p'_z}{m} \tag{18.11}$$

and calculate the Hamiltonian according to

$$H = \sum_i \dot{q}_i p_i - L. \tag{18.12}$$

This yields

$$\begin{aligned}
H &= m(\dot{x}'^2 - \omega \dot{x}' y') + m(\dot{y}'^2 + \omega \dot{y}' x') + m\dot{z}'^2 - L \\
&= \frac{1}{2}\left[2\dot{x}'^2 - 2\omega \dot{x}' y' + 2\dot{y}'^2 + 2\omega \dot{y}' x' + 2\dot{z}'^2 \right. \\
&\quad \left. - (\dot{x}'^2 + \dot{y}'^2 + \dot{z}'^2 + \omega^2(x'^2 + y'^2) + 2\omega(\dot{y}' x' - \dot{x}' y'))\right] + V \\
&= \frac{1}{2}m\left[\dot{x}'^2 + \dot{y}'^2 + \dot{z}'^2 - \omega^2(x'^2 + y'^2)\right] + V \\
&= \frac{1}{2}m\left[\frac{p_x'^2}{m^2} + 2\frac{\omega}{m}y' p'_x + \omega^2 y'^2 + \frac{p_y'^2}{m^2} - 2\frac{\omega}{m}x' p'_y + \omega^2 x'^2 \right. \\
&\quad \left. + \frac{p_z'^2}{m^2} - \omega^2 x'^2 - \omega^2 y'^2\right] + V \\
&= \frac{1}{2m}[p_x'^2 + p_y'^2 + p_z'^2] - \omega[x' p'_y - y' p'_x] + V\left(\sqrt{x'^2 + y'^2}, z'\right).
\end{aligned} \tag{18.13}$$

H is explicitly time-independent and is therefore a constant of motion. The canonical equations of motion read

$$\begin{aligned}
\dot{x}' &= \frac{\partial H}{\partial p'_x} = \frac{1}{m}p'_x + \omega y', \\
\dot{y}' &= \frac{1}{m}p'_y - \omega x', \\
\dot{z}' &= \frac{1}{m}p'_z,
\end{aligned} \tag{18.14}$$

$$\begin{aligned}
\dot{p}'_x &= -\frac{\partial H}{\partial x'} = \omega p'_y - \frac{\partial V}{\partial x'}, \\
\dot{p}'_y &= -\omega p'_x - \frac{\partial V}{\partial y'}, \\
\dot{p}'_z &= -\frac{\partial V}{\partial z'}.
\end{aligned} \tag{18.15}$$

(b) For the transition to cylindrical coordinates, we differentiate the transformation equations

$$x' = \varrho' \cos\varphi', \qquad y' = \varrho' \sin\varphi' \tag{18.16}$$

with respect to the time:

$$\dot{x}' = \dot{\varrho}' \cos\varphi' - \varrho'\dot{\varphi}' \sin\varphi',$$
$$\dot{y}' = \dot{\varrho}' \sin\varphi' + \varrho'\dot{\varphi}' \cos\varphi'. \tag{18.17}$$

From (18.9) and (18.17), we calculate the generalized momenta:

$$p'_\varrho = \frac{\partial L}{\partial \dot{\varrho}'} = \frac{\partial L}{\partial \dot{x}'}\frac{\partial \dot{x}'}{\partial \dot{\varrho}'} + \frac{\partial L}{\partial \dot{y}'}\frac{\partial \dot{y}'}{\partial \dot{\varrho}'}$$
$$= p'_x \cos\varphi' + p'_y \sin\varphi', \tag{18.18}$$

$$p'_\varphi = \frac{\partial L}{\partial \dot{\varphi}'} = \frac{\partial L}{\partial \dot{x}'}\frac{\partial \dot{x}'}{\partial \dot{\varphi}'} + \frac{\partial L}{\partial \dot{y}'}\frac{\partial \dot{y}'}{\partial \dot{\varphi}'}$$
$$= -p'_x \varrho' \sin\varphi' + p'_y \varrho' \cos\varphi'. \tag{18.19}$$

Now we solve for p'_x and p'_y. From (18.18), it follows that

$$p'_x = \frac{p'_\varrho - p'_y \sin\varphi'}{\cos\varphi'}, \tag{18.20}$$

and from (18.19) (with (18.20)),

$$p'_y = \frac{p'_\varphi \cos\varphi' + (p'_\varrho - p'_y \sin\varphi')\varrho' \sin\varphi'}{\varrho' \cos^2\varphi'}$$

$$\Rightarrow \frac{p'_y(\varrho' \cos^2\varphi' + \varrho' \sin^2\varphi')}{\varrho' \cos^2\varphi'} = \frac{p'_\varphi \cos\varphi' + p'_\varrho \varrho \sin\varphi'}{\varrho' \cos^2\varphi'}$$

$$\Rightarrow p'_y = \frac{1}{\varrho'} p'_\varphi \cos\varphi' + p'_\varrho \sin\varphi'. \tag{18.21}$$

Analogously, we obtain

$$p'_x = p'_\varrho \cos\varphi' - \frac{1}{\varrho'} p'_\varphi \sin\varphi'. \tag{18.22}$$

Now we insert (18.21) and (18.22) into (18.13) and obtain

$$H = \left[p'^2_\varrho \cos^2\varphi' - \frac{2}{\varrho'} p'_\varrho \cos\varphi' p'_\varphi \sin\varphi' + \frac{1}{\varrho'^2} \sin^2\varphi' p'^2_\varphi \right.$$
$$\left. + p'^2_\varrho \sin^2\varphi' + \frac{2}{\varrho'} p'_\varrho \sin\varphi' p'_\varphi \cos\varphi' + \frac{1}{\varrho'^2} p'^2_\varphi \cos^2\varphi' \right] \cdot \frac{1}{2m}$$
$$- \omega(x' p'_y - y' p'_x) + V(\varrho', z')$$

$$= \frac{1}{2m}\left[p'^2_\varrho + \frac{1}{\varrho'^2} p'^2_\varphi + p'^2_z \right] - \omega\left[\varrho' \cos\varphi' p'_\varrho \sin\varphi' + \varrho' \cos\varphi' \frac{1}{\varrho'} p'_\varphi \cos\varphi' \right.$$
$$\left. - \varrho' \sin\varphi' p'_\varrho \cos\varphi' + \varrho' \sin\varphi' \frac{1}{\varrho'} p'_\varphi \sin\varphi' \right] + V(\varrho', z')$$

$$= \frac{1}{2m}\left[p'^2_\varrho + \frac{1}{\varrho'^2} p'^2_\varphi + p'^2_z \right] - \omega p'_\varphi + V(\varrho', z). \tag{18.23}$$

A comparison of (18.13) and (18.23) shows that the Hamiltonian becomes especially simple if it is represented in coordinates adapted to the symmetry of the problem. From equation (18.23) we see that H does not depend on the angle φ' (φ' is a cyclic coordinate), hence the angular momentum component p'_φ is a constant of the motion.

The canonical equations of motion read

$$\dot{\varrho}' = \frac{1}{m} p'_\varrho, \qquad \dot{\varphi}' = \frac{1}{m\varrho'^2} p'_\varphi - \omega, \qquad \dot{z}' = \frac{1}{m} p'_z,$$
$$\dot{p}'_\varrho = \frac{1}{m\varrho'^3} p'^2_\varphi - \frac{\partial V}{\partial \varrho'}, \qquad \dot{p}'_\varphi = 0, \qquad \dot{p}'_z = -\frac{\partial V}{\partial z'}. \tag{18.24}$$

The Hamilton principle

The laws of mechanics can be expressed in two ways by variational principles that are independent of the coordinate system. The first of these are the *differential principles*. In this approach, one compares an arbitrarily selected momentary state of the system with (virtual) infinitesimal neighbor states. One example of this method is the d'Alembert principle. Another possibility is to vary a finite path element of the system. Such principles are called *integral principles*. The "path" is not understood as the trajectory of a point of the system in the three-dimensional position space, but rather as the path in a multidimensional space where the motion of the entire system is completely fixed. For a system with f degrees of freedom, this space is f-dimensional. In all integral principles the quantity to be varied has the dimension of an action (= energy · time); therefore, they are also called *principles of minimum action*. As an example we will consider the *Hamilton principle*. The Hamilton principle requires that a system moves in such a way that the time integral over the Lagrangian takes an extreme value:

$$I = \int_{t_1}^{t_2} L \, dt$$

shall have an extremum, which can also be expressed as follows:

$$\delta \int_{t_1}^{t_2} L \, dt = 0. \tag{18.25}$$

The path equation of the system can be determined by applying this principle.

Before considering equation (18.25) in more detail, we will briefly deal in general with the variational problem.

Example 18.4: A variational problem

As an example for substituting a description in terms of coordinates by a description independent of coordinates, one can consider the definition of a straight line in the plane. The straight line is uniquely determined by fixing two of its points, and it can also be described by a linear equation between the coordinates x and y. It can further be described by the differential equation

$$\frac{d^2 y}{dx^2} = 0 \tag{18.26}$$

with the further prescription that the values of the desired function $y(x)$ for $x = x_1$ and $x = x_2$ are given numbers. These are descriptions using rectangular coordinates. The straight line can however also be described as the shortest connection between two points, i.e., by

$$\int ds = \text{minimum}. \tag{18.27}$$

One may imagine the two given points as being connected by all possible curves, and among these curves that curve be selected which yields the minimum value for the given integral. This description of the straight line is independent of the choice of particular coordinates.

As a preparation for the following, we show how the search for the shortest connection between two points of the plane can be reduced mathematically to the equation (18.26). After introducing rectangular coordinates x and y, the problem is to look for a function $y(x)$ for which $y(x_1)$ and $y(x_2)$ have given values and the integral

$$I = \int_{x_1}^{x_2} \sqrt{1 + y'(x)^2}\, dx \tag{18.28}$$

takes a minimum value. Similar problems do not need to have a solution. So one could put the problem (18.27) or (18.28) and prescribe not only the start point and the endpoint, but also the direction of the curve at the start and endpoint, respectively. One easily recognizes that under these conditions there is no shortest connection, unless both of the given directions incidentally coincide with the straight connection.

The problem (18.28) has some similarity with the search for the minimum of a given function $f(x)$. There one considers a small change of x and forms

$$df(x) = f'(x)\, dx.$$

If $f'(x) \neq 0$, $f(x)$ can increase or decrease for small changes of x, and thus, there is no minimum at the point x. A necessary condition for a minimum is therefore $f'(x) = 0$. This condition is not sufficient; it is also fulfilled for a maximum.

In the problem (18.28), we do not have to change a variable but a function $y(x)$. We replace $y(x)$ by a "neighboring" function $y_0(x) + \varepsilon \eta(x)$ of the desired function y_0, where we will afterward assume the number ε is arbitrarily small. We must have $\eta(x_1) = \eta(x_2) = 0$. y' is then replaced by $y_0' + \varepsilon \eta'$, and instead of the integrand $\sqrt{1 + y'^2}$ we obtain the Taylor series expansion into powers of ε:

$$\sqrt{1 + (y_0' + \varepsilon \eta')^2} = \sqrt{1 + y_0'^2} + \varepsilon \frac{y_0'}{\sqrt{1 + y_0'^2}} \eta' + \varepsilon^2(\ldots),$$

where the term indicated by $\varepsilon^2(\ldots)$ can be neglected for sufficiently small $|\varepsilon|$. Therefore, we have

$$I(\varepsilon) = \int_{x_1}^{x_2} \sqrt{1 + (y_0' + \varepsilon \eta')^2}\, dx \approx \int_{x_1}^{x_2} \sqrt{1 + y_0'^2}\, dx + \varepsilon \int_{x_1}^{x_2} \frac{y_0'}{\sqrt{1 + y_0'^2}} \eta'\, dx,$$

which shall take a minimum for $\varepsilon = 0$. If the integral in the second term does not vanish, the integral

$$\int_{x_1}^{x_2} \sqrt{1 + y_0'^2}\, dx$$

can increase or decrease by changing the function $y_0(x)$, depending on the sign of ε. Hence, $y_0(x)$ does not provide a minimum of this integral. For a minimum rather exists the necessary condition

$$\int_{x_1}^{x_2} \frac{y_0'}{\sqrt{1 + y_0'^2}}\, \eta'\, dx = 0 \tag{18.29}$$

for any function $\eta(x)$ that vanishes at x_1 and x_2. To be able to exploit the far-reaching arbitrariness of the function $\eta(x)$, we transform (18.29) by integration by parts:

$$\left[\frac{y'}{\sqrt{1 + y_0'^2}} \eta\right]_{x_1}^{x_2} - \int_{x_1}^{x_2} \eta \frac{d}{dx} \frac{y_0'}{\sqrt{1 + y_0'^2}}\, dx = 0.$$

Because $\eta(x_1) = \eta(x_2) = 0$, the first term drops. The second term

$$\int_{x_1}^{x_2} \eta \cdot \frac{d}{dx} \frac{y_0'}{\sqrt{1 + y_0'^2}}\, dx \tag{18.30}$$

then and only then becomes zero for all allowed functions $\eta(x)$ if everywhere between x_1 and x_2 we have

$$\frac{d}{dx} \frac{y_0'}{\sqrt{1 + y_0'^2}} = 0. \tag{18.31}$$

If this equation were not satisfied everywhere, we could choose $\eta(x)$ so that it is always positive where

$$\frac{d}{dx} \frac{y_0'}{\sqrt{1 + y_0'^2}}$$

is positive, and choose it as negative where this expression is negative, and in this way establish a contradiction. We can also conclude this way: If (18.31) were not fulfilled anywhere, one should set $\eta(x)$ equal to zero everywhere, except for a certain interval about this place. But then the integral (18.30) does not vanish. We could not choose the quantity η' in (18.29) in this way; thus we could not draw the corresponding conclusion for (18.29). From (18.31) now follows $y_0' = $ constant or $y_0'' = 0$; that means the former description (18.26). Thus, our calculation has replaced the requirement that a definite integral be minimized by a function, by a differential equation for this function.

The equation (18.31) allows yet another interpretation. We have

$$\frac{d}{dx} \frac{y'}{\sqrt{1 + y'^2}} = \frac{y''}{(\sqrt{1 + y'^2})^3}.$$

As is shown in the theory of curves, this is an expression for the curvature of a curve. Equation (18.31) thus states that the desired curve everywhere has the curvature 0.

We just have treated a simple problem of the "variational calculus." Problems of the type (18.27) or (18.28) are called variational problems. In problems 18.5 and 18.6 we shall meet further, less trivial variational problems.

General discussion of variational principles

Given the integrable function $F = F(y(x), y'(x))$, we look for a function $y = y(x)$, so that the integral

$$I = \int_{x_1}^{x_2} F(y(x), y'(x))\, dx$$

takes an extremum value.

This problem is transformed into an elementary extremum value problem by covering the ensemble of all physically meaningful paths by a parametric representation:

$$y(x, \varepsilon) = y_0(x) + \varepsilon \eta(x),$$

where ε means a parameter that is constant for every path, $\eta(x)$ is an arbitrary differentiable function that vanishes at the endpoints:

$$\eta(x_1) = \eta(x_2) = 0.$$

The desired curve is given by $y_0(x) = y(x, 0)$.

The condition for an extremum value of the integral I is then

$$\left.\frac{dI}{d\varepsilon}\right|_{\varepsilon=0} = 0.$$

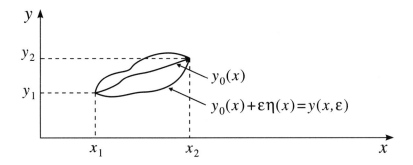

Figure 18.2. Possible paths from (x_1, y_1) to (x_2, y_2).

GENERAL DISCUSSION OF VARIATIONAL PRINCIPLES

The differentiation under the integral symbol (allowed if F is continuously differentiable with respect to ε) yields

$$\frac{dI}{d\varepsilon} = \int_{x_1}^{x_2} \left(\frac{\partial F}{\partial y} \frac{\partial y}{\partial \varepsilon} + \frac{\partial F}{\partial y'} \frac{\partial y'}{\partial \varepsilon} \right) dx = \int_{x_1}^{x_2} \left(\frac{\partial F}{\partial y} \eta + \frac{\partial F}{\partial y'} \eta' \right) dx.$$

The second integrand can be integrated by parts:

$$\int_{x_1}^{x_2} \frac{\partial F}{\partial y'} \frac{\partial \eta}{\partial x} dx = \left[\frac{\partial F}{\partial y'} \eta \right]_{x_1}^{x_2} - \int_{x_1}^{x_2} \left(\frac{d}{dx} \frac{\partial F}{\partial y'} \right) \eta \, dx.$$

Since the endpoints shall be fixed, the term integrated out vanishes, and the extremum condition reads

$$\int_{x_1}^{x_2} \left(\frac{\partial F}{\partial y} - \frac{d}{dx} \frac{\partial F}{\partial y'} \right) \eta \, dx = 0.$$

Since $\eta(x)$ can be an arbitrary function, this equation is generally satisfied only then if

$$\frac{d}{dx} \frac{\partial F(y(x), y'(x))}{\partial y'} - \frac{\partial F(y(x), y'(x))}{\partial y} = 0. \tag{18.32}$$

This relation (18.32) is called the *Euler–Lagrange equation*. It is a necessary condition for an extremum value of the integral I. The solution of the Euler–Lagrange equation, a differential equation of second order, together with the boundary conditions yields the wanted path. To simplify notation, we define the *variation of a function* $y(x, \varepsilon)$ as the difference between $y(x, \varepsilon)$ and $y(x, 0)$

$$\delta y = y(x, \varepsilon) - y(x, 0) = \left. \frac{\partial y}{\partial \varepsilon} \right|_{\varepsilon=0} \cdot \varepsilon$$

for very small ε. Then the variational problem can be formulated as

$$\delta \int_{x_1}^{x_2} F(y(x), y'(x)) \, dx = 0.$$

F can also include constraints by means of Lagrange multipliers (compare Chapter 16).

Example 18.5: Exercise: Catenary

This is an example with a constraint. A chain of constant density σ (mass per unit length: $\sigma = dm/ds$) and length l hangs in the gravitational field between two points $P_1(x_1, y_1)$ and $P_2(x_2, y_2)$. We look for the form of the curve, assuming that the potential energy of the chain takes a minimum.

Solution The potential energy of a chain element is

$$dV = g\sigma y \, ds.$$

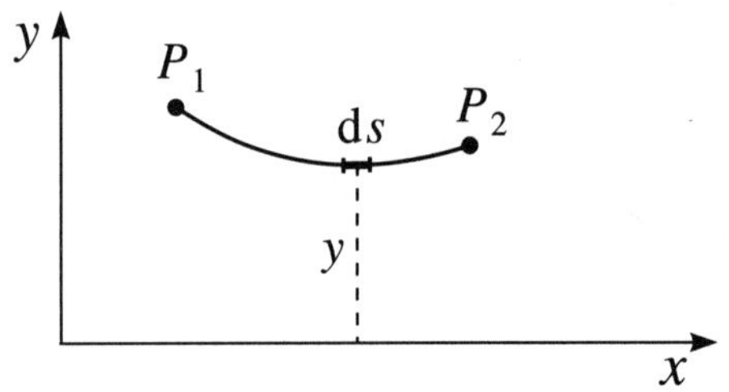

Figure 18.3. A chain hangs in the gravitational field.

The total potential energy is then

$$V = g\sigma \int_{x_1}^{x_2} y \, ds,$$

where the line element is given by

$$ds = \sqrt{1 + y'^2} \, dx, \qquad y' = \frac{dy}{dx}.$$

The constraint of given length l is represented by

$$0 = \int_{x_1}^{x_2} ds - l = \int_{x_1}^{x_2} \sqrt{1 + y'^2} \, dx - l.$$

With the Lagrange multiplier λ, the variational problem reads

$$g\sigma\delta \int_{x_1}^{x_2} y\sqrt{1 + y'^2} \, dx - \lambda\delta \left(\int_{x_1}^{x_2} \sqrt{1 + y'^2} \, dx - l \right) = 0.$$

Since $\delta l = 0$, we can introduce the function

$$F(y, y') = (y - \mu)\sqrt{1 + y'^2}$$

in the Euler equation (18.32), where we chose $\mu = \lambda/g\sigma$. From

$$\frac{\partial F}{\partial y} - \frac{d}{dx}\frac{\partial F}{\partial y'} = 0,$$

it follows that

$$(y - \mu)y'' - y'^2 - 1 = 0.$$

We rewrite the last equation. With

$$y'' = \frac{dy'}{dx} = \frac{dy'}{dy}\frac{dy}{dx} = y'\frac{dy'}{dy},$$

GENERAL DISCUSSION OF VARIATIONAL PRINCIPLES

we obtain

$$(y-\mu)y'\frac{dy'}{dy} = y'^2 + 1, \qquad \frac{dy}{y-\mu} = \frac{y'dy'}{1+y'^2}.$$

Integration yields

$$\ln(y-\mu) + \ln C_1 = \frac{1}{2}\ln(1+y'^2)$$

or

$$C_1(y-\mu) = \sqrt{1+y'^2}.$$

From this, we get

$$\int \frac{dy}{\sqrt{C_1^2(y-\mu)^2 - 1}} = \int dx.$$

To integrate the left side, we substitute $\cosh v = C_1(y-v)$, since $\cosh^2 v - 1 = \sinh^2 v$. Then

$$dy = \frac{1}{C_1}\sinh v\, dv,$$

and therefore,

$$\frac{1}{C_1}\int dv = \int dx.$$

Integration yields

$$v = C_1(x + C_2)$$

or

$$y = \frac{1}{C_1}\cosh(C_1(x + C_2)) + \mu.$$

Thus, the solution is the catenary. The constants give the coordinates of the lowest point $(x_0, y_0) = (-C_2, (1/C_1) + \mu)$. They are determined by the given length l of the chain and by the suspension points P_1 and P_2.

Example 18.6: Exercise: Brachistochrone: Construction of an emergency chute

On board an aircraft, a fire breaks out after landing. The passengers must leave by an emergency chute on which they glide down without friction. Determine by variational calculus the form of the chute with the aim to evacuate the plane as fast as possible (height of the hatch y_0; distance to the bottom x_0). Find the time of gliding as compared to the harsh free fall, assuming $x_0 = (\pi/2)y_0$.

Hint: Use the substitution

$$y' = \frac{dy}{dx} = -\cot\frac{\Theta}{2}!$$

Remark: This problem is known as the "brachistochrone."

Solution The problem goes back to the Bernoulli brothers (brachistochrone, 1696).

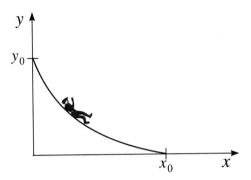

Figure 18.4. A passenger glides down a chute.

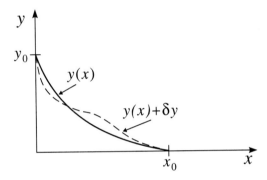

Figure 18.5. Illustration of various chutes.

Energy conservation yields

$$mgy_0 = \frac{1}{2}mv^2 + mgy,$$

$$g(y_0 - y) = \frac{1}{2}\left[\left(\frac{dx}{dt}\right)^2 + \left(\frac{dy}{dt}\right)^2\right],$$

$$(dt)^2 = \frac{(dx)^2 + (dy)^2}{2g(y_0 - y)}.$$

The total time T is then

$$T = \int_0^T dt = \int_0^{x_0} \sqrt{\frac{1 + (dy/dx)^2}{2g(y_0 - y)}}\, dx. \tag{18.33}$$

To get the minimum time, one has to solve a variational problem of the form

$$\delta \int_{x_1}^{x_2} F(x, y, y')\, dx = 0, \qquad \begin{aligned} y(x_1) &= y_0, \\ y(x_2) &= 0. \end{aligned}$$

GENERAL DISCUSSION OF VARIATIONAL PRINCIPLES

Because

$$0 = \int_{x_1}^{x_2} \left(\frac{\partial F}{\partial y} \delta y + \frac{\partial F}{\partial y'} \delta y' \right) dx = \int_{x_1}^{x_2} \left(\frac{\partial F}{\partial y} - \frac{d}{dx} \frac{\partial F}{\partial y'} \right) \delta y \, dx,$$

the Euler–Lagrange equation reads

$$\frac{d}{dx} \frac{\partial F}{\partial y'} - \frac{\partial F}{\partial y} = 0 \tag{18.34}$$

or

$$y'' \frac{\partial^2}{\partial y'^2} F + y' \frac{\partial^2}{\partial y \partial y'} F + \frac{\partial^2}{\partial x \partial y'} F - \frac{\partial F}{\partial y} = 0. \tag{18.35}$$

If the functional F is independent of x, (18.35) can be directly integrated. One finds

$$\frac{d}{dx}\left(y' \frac{\partial F}{\partial y'} - F \right) = y' \frac{d}{dx} \frac{\partial F}{\partial y'} + y'' \frac{\partial F}{\partial y'} - \frac{dF}{dx}$$

$$= y' \frac{d}{dx} \frac{\partial F}{\partial y'} + \left(y'' \frac{\partial F}{\partial y'} - y'' \frac{\partial F}{\partial y'} \right) - y' \frac{\partial F}{\partial y} - \underbrace{\frac{\partial F}{\partial x}}_{=0}$$

$$= y' \left(\frac{d}{dx} \frac{\partial F}{\partial y'} - \frac{\partial F}{\partial y} \right) = 0;$$

hence,

$$y' \frac{\partial F}{\partial y'} - F = \text{constant} \equiv \frac{1}{c}. \tag{18.36}$$

In our case, (18.33) is

$$F = \sqrt{\frac{1 + y'^2}{2g(y_0 - y)}}.$$

Then (18.36) reads

$$y' = \frac{1}{\sqrt{2g(y_0 - y)}} \cdot \frac{y'}{\sqrt{1 + y'^2}} - \frac{\sqrt{1 + y'^2}}{\sqrt{2g(y_0 - y)}} = \frac{1}{c},$$

$$\frac{1}{2g(y_0 - y)(1 + y'^2)} = \frac{1}{c^2}. \tag{18.37}$$

The transformation $y' = -\cot^2(\Theta/2)$ yields

$$\frac{c^2}{2g(y_0 - y)} = 1 + y'^2 = 1 + \cot^2\left(\frac{\Theta}{2}\right) = \frac{1}{\sin^2(\Theta/2)},$$

and thus,

$$y = y_0 - \frac{c^2}{4g}(1 - \cos \Theta).$$

By integration, one finds an equation for $x(\Theta)$, namely,

$$-\cot \frac{\Theta}{2} = \frac{dy}{dx} = -\frac{c^2}{2g} \sin\left(\frac{\Theta}{2}\right) \cos\left(\frac{\Theta}{2}\right) \frac{d\Theta}{dx}$$

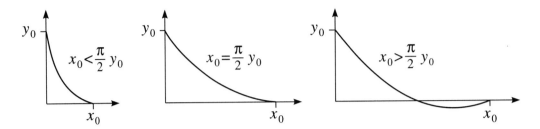

Figure 18.6. Possible types of solution.

$$\Rightarrow \quad x = \int_0^x dx' = \frac{c^2}{2g} \int_0^\Theta \sin^2\left(\frac{\Theta'}{2}\right) d\Theta' = \frac{c^2}{2g} \left|\left(\frac{1}{2}\Theta' - \frac{1}{2}\sin\Theta'\right)\right|_0^\Theta$$

$$x = \frac{c^2}{4g}(\Theta - \sin\Theta), \qquad y = y_0 - \frac{c^2}{4g}(1 - \cos\Theta). \tag{18.38}$$

This is just the parametric representation of a *cycloid*.

The maximum value of Θ is determined by x_0 and y_0, namely,

$$\frac{x_0}{y_0} = \frac{\Theta_0 - \sin\Theta_0}{1 - \cos\Theta_0}. \tag{18.39}$$

The transcendental equation (18.39) can be solved in general only numerically. Special cases:

$\Theta_0 = 0$	π	2π
$x_0/y_0 = 0$	$\pi/2$	∞

Calculation of the gliding time according to (18.33) and (18.37):

$$T = \int_0^{x_0} \sqrt{\frac{1 + y'^2}{2g(y_0 - y)}}\, dx = \int_{y_0}^0 \sqrt{\frac{(dx/dy)^2 + 1}{2g(y_0 - y)}}\, dy$$

$$= \int_0^{y_0} \sqrt{\frac{c^2}{2g(y_0 - y)(c^2 - 2g(y_0 - y))}}\, dy$$

$$= \frac{c}{2g} 2 \arctan \sqrt{\frac{c^2 - 2g(y_0 - y)}{2g(y_0 - y)}}\bigg|_0^{y_0}$$

$$= \frac{c}{g}\left(\frac{\pi}{2} - \arctan\sqrt{\frac{c^2 - 2gy_0}{2gy_0}}\right),$$

$$T = \frac{c}{g}\operatorname{arccot}\sqrt{\frac{c^2 - 2gy_0}{2gy_0}}.$$

GENERAL DISCUSSION OF VARIATIONAL PRINCIPLES

The integral can be found in tables.

$$x_0 = \frac{\pi}{2} y_0 \implies \Theta_0 = \pi \implies c = \sqrt{2gy_0} \implies T = \sqrt{\frac{2y_0}{g}} \frac{\pi}{2}.$$

For comparison, the time of free fall is

$$T' = \sqrt{\frac{2y_0}{g}}.$$

As is seen already from equation (18.25), according to the Hamilton principle the time is not being varied. The system passes a trace point and the appropriate varied trace point at the same time. Hence,

$$\delta t = 0.$$

Starting from the integral

$$\delta I = \delta \int_{t_1}^{t_2} L(q_\alpha(t), \dot{q}_\alpha(t), t) \, dt = 0, \qquad \alpha = 1, 2, \ldots, f, \tag{18.40}$$

where f is the number of degrees of freedom, we perform the variation according to the procedure described above and show that the Lagrange equations can be derived from the Hamilton principle. We describe the variation of a path curve $q_\alpha(t)$ by

$$q_\alpha(t) \longrightarrow q_\alpha(t) + \delta q_\alpha(t),$$

where the δq_α vanish at the endpoints,

$$\delta q_\alpha(t_1) = \delta q_\alpha(t_2) = 0.$$

Since time is not being varied, we have

$$\delta \int_{t_1}^{t_2} L \, dt = \int_{t_1}^{t_2} \delta L \, dt = \int_{t_1}^{t_2} \left(\sum_\alpha \frac{\partial L}{\partial q_\alpha} \delta q_\alpha + \sum_\alpha \frac{\partial L}{\partial \dot{q}_\alpha} \delta \dot{q}_\alpha \right) dt. \tag{18.41}$$

Because

$$\frac{d}{dt} \delta q_\alpha = \frac{d}{dt}(q_\alpha(t, \varepsilon) - q_\alpha(t, 0))$$
$$= \frac{d}{dt}(q_\alpha(t, \varepsilon)) - \frac{d}{dt}(q_\alpha(t, 0))$$
$$= \delta \frac{d}{dt} q_\alpha(t) = \delta \dot{q}_\alpha(t), \tag{18.42}$$

integration by parts of the second summand yields

$$\int_{t_1}^{t_2} \frac{\partial L}{\partial \dot{q}_\alpha} \delta \dot{q}_\alpha \, dt = \int_{t_1}^{t_2} \frac{\partial L}{\partial \dot{q}_\alpha} \frac{d}{dt} \delta q_\alpha \, dt$$

$$= \left[\frac{\partial L}{\partial \dot{q}_\alpha} \delta q_\alpha \right]_{t_1}^{t_2} - \int_{t_1}^{t_2} \left(\frac{d}{dt} \frac{\partial L}{\partial \dot{q}_\alpha} \right) \delta q_\alpha \, dt. \qquad (18.43)$$

Since δq_α vanishes at the endpoints (integration limits), we get for the variation of the integral

$$\delta I = \int_{t_1}^{t_2} \left(\sum_\alpha \left(\frac{\partial L}{\partial q_\alpha} - \frac{d}{dt} \frac{\partial L}{\partial \dot{q}_\alpha} \right) \delta q_\alpha \right) dt = 0. \qquad (18.44)$$

For holonomic constraints, we imagine that the dependent degrees of freedom were eliminated. We take the q_α as the independent coordinates. Hence, the δq_α are independent of each other, and the integral vanishes only if the coefficient of any δq_α vanishes. This means that the Lagrange equations hold:

$$\frac{d}{dt} \frac{\partial L}{\partial \dot{q}_\alpha} - \frac{\partial L}{\partial q_\alpha} = 0. \qquad (18.45)$$

Likewise, one can obtain the Hamilton equations by replacing L by $\sum_\alpha p_\alpha \dot{q}_\alpha - H$ and considering the variations δp_α and δq_α as independent. This will be worked out in the Example 18.7.

In order to show the equivalence of the Hamilton principle with the formulations of mechanics studied so far, we shall demonstrate its derivation from Newton's equations. We consider a particle in Cartesian coordinates. It moves along a certain path $\mathbf{r} = \mathbf{r}(t)$ between the positions $\mathbf{r}(t_1)$ and $\mathbf{r}(t_2)$. Now the path is varied by a virtual displacement $\delta \mathbf{r}$ that is compatible with the constraint:

$$\mathbf{r}(t) \longrightarrow \mathbf{r}(t) + \delta \mathbf{r}(t), \qquad \delta \mathbf{r}(t_1) = \delta \mathbf{r}(t_2) = 0.$$

The time is not varied. The work needed for the virtual displacement is

$$\delta A = \mathbf{F} \cdot \delta \mathbf{r} = \mathbf{F}^a \cdot \delta \mathbf{r},$$

if $\mathbf{F}^e$ is the external force and the constraint reaction does not perform work. If $\mathbf{F}^e$ is conservative, then

$$\mathbf{F}^e \cdot \delta \mathbf{r} = -\delta V,$$

and according to Newton

$$-\delta V = m\ddot{\mathbf{r}} \cdot \delta \mathbf{r}.$$

GENERAL DISCUSSION OF VARIATIONAL PRINCIPLES

The right-hand side can be transformed (the operator $(d/dt)\delta \mathbf{r} = \delta \dot{\mathbf{r}}$ is treated according to (18.42)):

$$\frac{d}{dt}(\dot{\mathbf{r}} \cdot \delta \mathbf{r}) = \dot{\mathbf{r}} \cdot \frac{d}{dt}\delta \mathbf{r} + \ddot{\mathbf{r}} \cdot \delta \mathbf{r} = \dot{\mathbf{r}} \cdot \delta \dot{\mathbf{r}} + \ddot{\mathbf{r}} \cdot \delta \mathbf{r} = \delta\left(\frac{1}{2}\dot{\mathbf{r}}^2\right) + \ddot{\mathbf{r}} \cdot \delta \mathbf{r}.$$

Multiplication by the mass m yields

$$m\ddot{\mathbf{r}} \cdot \delta \mathbf{r} = m\frac{d}{dt}(\dot{\mathbf{r}} \cdot \delta \mathbf{r}) - \delta\left(\frac{1}{2}m\dot{\mathbf{r}}^2\right),$$

and therefore,

$$\delta(T - V) = \delta L = m\frac{d}{dt}(\dot{\mathbf{r}} \cdot \delta \mathbf{r}).$$

Integration with respect to time leads to

$$\delta \int_{t_1}^{t_2} L\, dt = m\left[\dot{\mathbf{r}} \cdot \delta \mathbf{r}\right]_{t_1}^{t_2} = 0.$$

Thus, the Hamilton principle for a single particle has been derived from Newton's equations. The result can be directly extended to particle systems. This can be understood quite generally in the following way: If a particle system obeys the Lagrange equations (18.45) (which are equivalent to Newtonian mechanics), then we have (18.44) and from that—because of (18.43)—again (18.41) or (18.40), provided that $\delta q_\alpha(t_1) = \delta q_\alpha(t_2) = 0$. Thus, the Lagrange equations are equivalent to the Hamilton principle.

Example 18.7: Exercise: Derivation of the Hamiltonian equations

Derive the Hamilton equations from the Hamilton principle.

Solution   The Hamilton principle reads

$$\delta \int_{t_1}^{t_2} L\, dt = 0, \tag{18.46}$$

where the Lagrangian L is now expressed by the Hamiltonian H; hence,

$$L = \sum_\alpha p_\alpha \dot{q}_\alpha - H(p_\alpha, q_\alpha, t). \tag{18.47}$$

Then (18.46) becomes

$$\int_{t_1}^{t_2} \delta L\, dt = \int_{t_1}^{t_2} \sum_\alpha \left[\delta p_\alpha \dot{q}_\alpha + p_\alpha \delta \dot{q}_\alpha - \frac{\partial H}{\partial p_\alpha}\delta p_\alpha - \frac{\partial H}{\partial q_\alpha}\delta q_\alpha\right] dt. \tag{18.48}$$

The second term on the right-hand side can be transformed by integration by parts,

$$\int_{t_1}^{t_2} p_\alpha \delta \dot{q}_\alpha \, dt = \int_{t_1}^{t_2} p_\alpha \frac{d}{dt} \delta q_\alpha \, dt = p_\alpha \delta q_\alpha H \bigg|_{t_1}^{t_2} - \int_{t_1}^{t_2} \dot{p}_\alpha \delta q_\alpha \, dt. \tag{18.49}$$

The first term vanishes since the variations at the endpoints vanish: $\delta q_\alpha(t_1) = \delta q_\alpha(t_2) = 0$. Hence, (18.48) becomes

$$0 = \int_{t_1}^{t_2} \delta L \, dt = \int_{t_1}^{t_2} \sum_\alpha \left\{ \left[\dot{q}_\alpha - \frac{\partial H}{\partial p_\alpha} \right] \delta p_\alpha + \left[-\dot{p}_\alpha - \frac{\partial H}{\partial q_\alpha} \right] \delta q_\alpha \right\} dt. \tag{18.50}$$

The variations δp_α and δq_α are independent of each other because along a path in phase space the neighboring paths can have different coordinates or (and) different momenta. Thus, (18.50) leads to

$$\dot{q}_\alpha = \frac{\partial H}{\partial p_\alpha},$$
$$\dot{p}_\alpha = -\frac{\partial H}{\partial q_\alpha}, \tag{18.51}$$

which was to be demonstrated.

Phase space and Liouville's theorem

In the Hamiltonian formalism, the state of motion of a mechanical system with f degrees of freedom at a definite time t is completely characterized by the specification of the f generalized coordinates and f momenta $q_1, \ldots, q_f; p_1, \ldots, p_f$. These q_i and p_i can be understood as coordinates of a $2f$-dimensional Cartesian space, the *phase space*. The f-dimensional subspace of the coordinates q_i is the *configuration space*; the f-dimensional subspace of the momenta p_i is called *momentum space*. In the course of motion of the system the representative point describes a curve, the *phase trajectory*. If the Hamiltonian is known, then the entire phase trajectory can be uniquely calculated in advance from the coordinates of one point. Therefore to each point belongs only one trajectory, and two different trajectories cannot intersect each other. A path in phase space is given in parametric representation by $q_k(t), p_k(t)$ ($k = 1, \ldots, f$). Because of the uniqueness of the solutions of the Hamilton equations, the system develops from various boundary conditions along various trajectories. For conservative systems the point is bound to a $(2f - 1)$-dimensional hypersurface of the phase space by the condition $H(q, p) = E =$ constant.

Example 18.8: Phase diagram of a plane pendulum

If the angle φ is taken as a generalized coordinate, then we have for the plane pendulum (mass m, length l)

$$p_\varphi = ml^2 \dot{\varphi}.$$

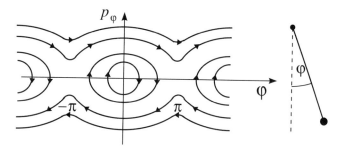

Figure 18.7. Phase space and phase diagram of the one-dimensional pendulum.

The Hamiltonian, which represents the total energy, reads

$$H = \frac{1}{2}m(l\dot{\varphi})^2 - mgl\cos\varphi = \frac{p_\varphi^2}{2ml^2} - mgl\cos\varphi = E.$$

The origin of the potential was put at the suspension point of the pendulum. One then gets the equation for the phase trajectory $p_\varphi = p_\varphi(\varphi)$:

$$p_\varphi = \pm\sqrt{2ml^2(E + mgl\cos\varphi)}.$$

Thus, we obtain a set of curves with the energy E as a parameter.

For energies $E < mgl$, the phase trajectories are closed (ellipse-like) curves; the pendulum oscillates forth and back (vibration). If the total energy E exceeds the value mgl, the pendulum still has kinetic energy at the highest point $\varphi = \pm\pi$ and continues its motion without reversal of direction (rotation).

We now consider a large number N of independent points that are mechanically identical, apart from the initial conditions, and are therefore described by the same Hamiltonian. As a specific example, we can imagine particles in the beam of an accelerator. If all points at time t_1 are distributed over a $2f$-dimensional phase space region G_1 with the volume

$$\Delta V = \Delta q_1 \cdots \Delta q_f \cdot \Delta p_1 \cdots \Delta p_f,$$

one can define the density

$$\varrho = \frac{\Delta N}{\Delta V}.$$

With the course of motion, G_1 transforms according to the Hamilton equations into the region G_2.

The statement of the *Liouville theorem*[3] is

---

[3] *Joseph Liouville*, b. March 24, 1809, St. Omer–d. Sept. 8, 1882, Paris. Liouville was professor of mathematics and mechanics in Paris, at the École Polytechnique, at the Collège de France, and at the Sorbonne. He was a member of the Bureau of Measures and of many scholarly societies. From 1840 to 1870, he was considered

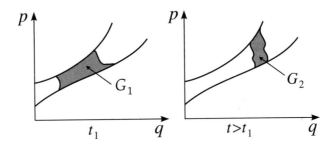

Figure 18.8. Evolution of a region in phase space.

The volume of an arbitrary region of phase space is conserved if the points of its boundary move according to the canonical equations.

Or, in other words, performing a limit transition:

The density of points in phase space in the vicinity of a point moving with the fluid is constant.

To prove that, we investigate the motion of system points through a volume element of the phase space. Let us first consider the components of the particle flux along the q_k- and p_k-direction.

The area $ABCD$ represents the projection of the $2f$-dimensional volume element dV onto the q_k, p_k-plane.

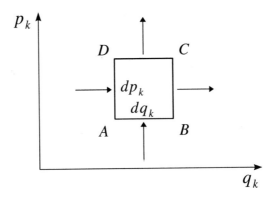

Figure 18.9. Projection of the volume element onto the q_k, p_k-plane.

---

the leading mathematician of France. He worked on statistical mechanics, boundary value problems, differential geometry, and special functions. His constructive proof of the existence of transcendental numbers and, in 1844, the proof that e and e^2 cannot be roots of a quadratic equation with rational coefficients, were of great significance.

The number of points entering the volume element per unit time through the "side face" (with the projection AD onto the q_k, p_k-plane) is

$$\varrho \dot{q}_k dp_k \cdot dV_k,$$

where

$$dV_k = \prod_{\substack{\alpha=1 \\ \alpha \neq k}}^{f} dq_\alpha dp_\alpha$$

is the $(2f - 2)$-dimensional remainder volume element; $dp_k \cdot dV_k$ is the magnitude of the lateral surface with the projection AD in the p_k, q_k-plane.

The Taylor expansion for the points leaving at BC in the first direction yields

$$\left(\varrho \dot{q}_k + \frac{\partial}{\partial q_k} (\varrho \dot{q}_k) dq_k \right) dp_k \cdot dV_k. \tag{18.52}$$

Analogously, for the flux in p_k-direction we have

entrance through AB: $\quad \varrho \dot{p}_k dq_k \cdot dV_k,$

exit through CD: $\quad \left(\varrho \dot{p}_k + \frac{\partial}{\partial p_k} (\varrho \dot{p}_k) dp_k \right) dq_k \cdot dV_k. \tag{18.53}$

From the flux components in p_k- and q_k-direction, the number of system points per unit time

$$-\left(\frac{\partial}{\partial q_k} (\varrho \dot{q}_k) + \frac{\partial}{\partial p_k} (\varrho \dot{p}_k) \right) dV \tag{18.54}$$

gets stuck in the volume element.

By summing over all $k = 1, \ldots, f$, one obtains the number of points that get stuck in total. This quantity just corresponds to the change with time (time derivative) of the density multiplied by dV. Hence, we can conclude

$$\frac{\partial \varrho}{\partial t} = -\sum_{k=1}^{f} \left(\frac{\partial}{\partial q_k} (\varrho \dot{q}_k) + \frac{\partial}{\partial p_k} (\varrho \dot{p}_k) \right). \tag{18.55}$$

We are dealing here with a *continuity equation* of the form

$$\text{div} (\varrho \dot{\mathbf{r}}) + \frac{\partial \varrho}{\partial t} = 0.$$

The divergence refers to the $2f$-dimensional phase space:

$$\nabla = \sum_{k=1}^{f} \frac{\partial}{\partial q_k} + \sum_{k=1}^{f} \frac{\partial}{\partial p_k}.$$

Continuity equations of this type often appear in flow physics (hydrodynamics, electrodynamics, quantum mechanics). They always express a conservation law.

Application of the product rule in (18.55) yields

$$\sum_{k=1}^{f} \left(\frac{\partial \varrho}{\partial q_k} \dot{q}_k + \varrho \frac{\partial \dot{q}_k}{\partial q_k} + \frac{\partial \varrho}{\partial p_k} \dot{p}_k + \varrho \frac{\partial \dot{p}_k}{\partial p_k} \right) + \frac{\partial \varrho}{\partial t} = 0. \qquad (18.56)$$

From the Hamilton equations, we have

$$\frac{\partial \dot{q}_k}{\partial q_k} = \frac{\partial^2 H}{\partial q_k \partial p_k} \quad \text{and} \quad \frac{\partial \dot{p}_k}{\partial p_k} = -\frac{\partial^2 H}{\partial q_k \partial p_k}.$$

If the second partial derivatives of H are continuous, then

$$\frac{\partial \dot{q}_k}{\partial q_k} + \frac{\partial \dot{p}_k}{\partial p_k} = 0,$$

and from this, it follows that

$$\sum_{k=1}^{f} \left(\frac{\partial \varrho}{\partial q_k} \dot{q}_k + \frac{\partial \varrho}{\partial p_k} \dot{p}_k \right) + \frac{\partial \varrho}{\partial t} = 0. \qquad (18.57)$$

This just equals the total derivative of the density with respect to time,

$$\frac{d}{dt} \varrho = 0, \qquad (18.58)$$

and hence, $\varrho = $ constant.

Example 18.9: Phase-space density for particles in the gravitational field

The system consists of particles of mass m in a constant gravitational field. For the energy, we have

$$H = E = \frac{p^2}{2m} - mgq.$$

The total energy of a particle remains constant.
The phase trajectories $p(q)$ are parabolas

$$p = \sqrt{2m(E + mgq)},$$

with the energy as a parameter. We consider a number of particles with momenta at time $t = 0$ between the limits $p_1 \leq p \leq p_2$, and with energies between $E_1 \leq E \leq E_2$. They cover the area F in phase space. At a later time t the points cover the area F'. They then have the momentum

$$p' = p + mgt,$$

so that F' is the area between the parabolas limited by $p_1 + mgt \leq p' \leq p_2 + mgt$. With

$$q = \frac{(p^2/2m) - E}{mg},$$

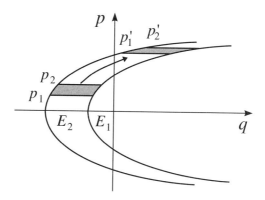

Figure 18.10. On Liouville's theorem: Phase space for particles in the gravitational field.

the size of the areas is calculated as

$$F = \int_{p_1}^{p_2} dp \int_{(1/mg)((p^2/2m)-E_2)}^{(1/mg)((p^2/2m)-E_1)} dq = \frac{E_2 - E_1}{mg} \int_{p_1}^{p_2} dp = \frac{E_2 - E_1}{mg}(p_2 - p_1),$$

and likewise,

$$F' = \frac{E_2 - E_1}{mg}(p'_2 - p'_1)$$
$$= \frac{E_2 - E_1}{mg}(p_2 - p_1).$$

This is just the statement of Liouville's theorem: $F = F'$ means that the density of the system points in phase space remains constant. The significance of Liouville's theorem lies in the field of statistical mechanics, where one considers ensembles because of lack of exact knowledge of the system.

A special application is the focusing of particle currents in accelerators where a large number of particles are subject to identical conditions. Here a reduction of the beam cross section must lead to an undesirable broadening of the momentum distribution.

The principle of stochastic cooling[4]

An essential implication of Liouville's theorem is that the phase space occupied by an ensemble of particles in the absence of friction behaves like an incompressible fluid.

We shall show in the following that the principle of stochastic cooling leads to a (seeming) contradiction to the theorem of Liouville. For this purpose it is necessary to expand on the

---

[4]This chapter was stimulated by a lecture given by Professor Herminghaus (Mainz), at the occasion of the sixtieth birthday of Professor P. Junior 1988 in Frankfurt. My thanks go to colleague Mr. Herminghaus for leaving his manuscript, which I found very useful when writing this section.

method of stochastic cooling of antiprotons developed by van der Meer.[5] The successful application of this method led to the proof of the existence of the *intermediate vector bosons* (IVB) W^+, W^-, and Z^0 predicted by the theory of weak interactions.

According to the predictions of the theory, these particles should be able to decay as follows:

$$\text{IVB} \longrightarrow \text{lepton} + \text{antilepton}, \tag{18.59}$$

$$\text{IVB} \longrightarrow \text{quark} + \text{antiquark}. \tag{18.60}$$

For the experimental proof of the IVB, one utilized the inverse reaction (18.60), by shooting high-energy beams of antiprotons onto protons in the proton synchrotron (PS) of the CERN. Since the protons consist only of three quarks (q) and the antiprotons of three antiquarks ($\overline{q}$), many quark-antiquark pairs are created by the violent collisions. The reactions between these quarks and antiquarks can generate the intermediate vector bosons (see Figure 18.11). In order to reach a high event rate, which is calculated according to

$$\text{event rate} = \text{cross section} \cdot \text{luminosity}, \tag{18.61}$$

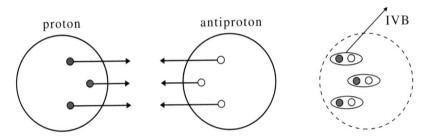

Figure 18.11. Schematic representation of a proton-antiproton collision. In the collision quark-antiquark pairs are created whose reactions can lead to the creation of intermediate vector bosons (• = quark, ○ = antiquark).

---

[5] *Simon van der Meer*, b. Nov. 24, 1925, Den Haag. He received the Nobel prize for physics in 1984. He studied mechanical and electrical engineering at the Technical University of Delft, took his diploma exams as engineer and worked at first in the Philips central laboratory in Eindhoven. In 1956 he got a position as a development engineer at CERN in Geneva. Here he soon earned a reputation for professional competence, imagination, and also for his talent for theory. He was appointed a "senior engineer." Meanwhile the Italian physicist Carlo Rubbia, a scientific coworker at CERN, had developed the idea to shoot 450 GeV protons from the just-finished super-high energy accelerator "SPS" onto their artificially produced "antiparticles"—antiprotons. The project was realized as a collider system. For the first time one could generate and demonstrate the so far only hypothetical intermediate W- and Z-bosons. Van der Meer, a "genuine puzzler," provided a genial invention: the stochastic cooling, which allowed researchers to collect antiprotons in sufficient quantity and to store them for the experiments. Only one year after their great success, which proved the predictions of theory in a brilliant way, van der Meer and Rubbia were awarded with the Nobel prize for physics "for decisive merits in the discovery of the field quanta of weak interaction."

THE PRINCIPLE OF STOCHASTIC COOLING

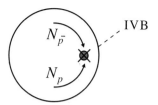

Figure 18.12. The existence of intermediate vector bosons could be proved for the first time at CERN by the collision of intense high-energy proton and antiproton beams ($N_{\bar{p}}$ = number of antiprotons in the beam, N_p = number of protons).

one needs both a large cross section and a high beam luminosity. Now one has

$$\text{luminosity} \sim \frac{N_p \cdot N_{\bar{p}}}{q}. \tag{18.62}$$

Here, N_p and $N_{\bar{p}}$ denote the number of protons (p) and antiprotons ($\bar{p}$) in the beam, and q represents the beam cross section. The higher the number of particles and the lower the beam cross section, the higher is the event rate for creating an intermediate vector boson. See also Figure 18.12.

An efficient *cooling mechanism* for the antiproton beams is therefore needed. Each particle of the beam moves by the action of magnetic fields in horizontal and vertical *vibrations about a closed pre-set trajectory*. In this context the term *cooling* means a reduction of the vibration amplitudes of the particles and thus of the beam cross section, or a reduction of the width of the momentum distribution of the particles about the mean value. This is illustrated by Figure 18.13. Already well-tried cooling methods are electron

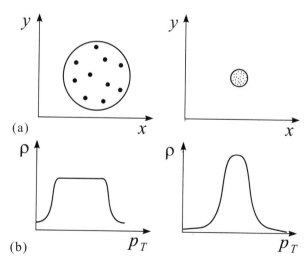

Figure 18.13. A beam before and after cooling, (a) in position space, (b) in momentum space. Part (b) essentially shows the particle density versus the transverse momentum.

cooling, cooling by synchrotron radiation, and the *stochastic cooling*, which will now be outlined in more detail.

The motion of each particle in the beam is described by a point in a 6-dimensional phase space spanned by the 3 spatial and the 3 momentum coordinates. This phase space point is surrounded by empty space. By an appropriate deformation of the phase space element the particle can be shifted toward the center of gravity of the distribution. This is the *principle of stochastic cooling*.

The experimental setup for cooling of antiproton beams is sketched in Figure 18.14.

In the ideal case, a probe (pick-up) measures the position or the momentum of a particle. This tiny signal is amplified and fed to the "kicker," which then corrects the transverse or the longitudinal momentum and thereby cools. Thus, the cooling can be interpreted as a one-particle effect, since each particle cools itself by emission of a self-generated signal (coherent effect). An essential prerequisite is that the particle and the signal reach the kicker simultaneously. Because of the finite resolving power of the probe in the real case, besides the desired signal the perturbing signals from other particles reach the kicker too. This noise causes a heating of the particles (incoherent effect) and thus counteracts the cooling effect. This interplay of cooling and heating mechanisms is illustrated by Figure 18.15 and will be discussed in Example 18.10. The cooling effect is directly proportional to the signal amplification, while the heating is proportional to the square of the amplification. The particle is cooled only in the hatched area (see Figure 18.15). Evidently there exists an optimum of amplification where the cooling effect reaches an extremum value. Thus, the greater the intensity of the beams, the greater is the noise and the heating effect, and the less is the factor of optimum amplification. Generation of an intense beam of antiprotons at CERN is therefore performed by stages and may last several hours. The principle is illustrated by Figure 18.15.

First, an antiproton pulse of low intensity is injected at the left border of the vacuum chamber (1). The corresponding momentum density distribution can be seen on the right. The beam and its momentum width are then compressed by cooling (2). A high-frequency voltage is used to shift the pulse to the right side of the chamber (3), thus giving space for a further antiproton pulse which is injected into the chamber (4). After cooling, the second pulse is shifted onto the already "deposited" pulse (5). This procedure is repeated

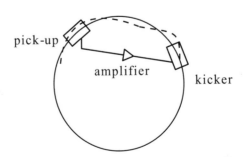

Figure 18.14. The cooling system consisting of "pick-up," amplifier, and "kicker."

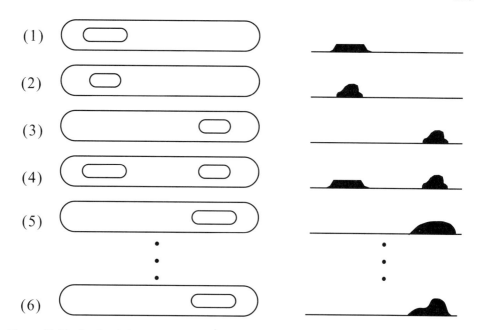

Figure 18.15. Sectional view of the vacuum chamber with beams in various stages of the accumulation process. The right-hand part shows the corresponding density distribution versus the momentum.

every 2 to 3 seconds for several hours. In this way, the longitudinal phase-space density is increased by accumulation of more and more particles into the same momentum interval (6). The final 6-dimensional phase-space density of the stack is higher than the density of a single pulse by a factor of $3 \cdot 10^8$. The intense antiproton beam generated this way can now be further accelerated and brought to collision with a proton beam. Only one year after the demonstration of the intermediate vector bosons, S. van der Meer and C. Rubbia[6] were awarded the Nobel Prize in physics for their achievements.

We now come back to the apparent contradiction between Liouville's theorem and the method of stochastic cooling. While according to the Liouville theorem only a single pulse can be accommodated in a ring, stochastic cooling allows one to accumulate about 36,000

---

[6]*Carlo Rubbia*, b. March 31, 1934, Goriza. He got his education as a physicist in Pisa at the Scuola Normale, a time-honored university. Here he got his doctorate in 1958, after which he worked for a year as a research scholar at the Columbia University in New York, and then as an assistant professor in Rome. In 1960, he came to CERN at Geneva as high-energy physicist. Since 1972, he has held a chair at Harvard University. In Geneva, Rubbia was inspired by the unified theory of weak and electromagnetic interactions developed by A. Salam, S. Glashow, and S. Weinberg (Nobel Prize for physics, 1979). In 1976, Rubbia proposed to CERN the construction of a new 450 GeV SPS accelerator for the purpose of proton-antiproton collision experiments. The accelerator achieved collision energies of 540 GeV, which were sufficient to create the (so far only predicted) W- and Z-bosons. Important for the success of the project was not only Rubbia, but also S. van der Meer, whose contributions made possible the generation of sharply bunched, pulsed antiproton currents. Both of them got the Nobel Prize for physics in 1984.

pulses in the course of a day. The final phase-space density is higher than that of a single pulse by a factor of $3 \cdot 10^8$.

However, stochastic cooling and the Liouville theorem are dealing with different situations. The former presupposes an ensemble of a finite number of *discrete* particles, while the Liouville theorem presupposes a phase-space continuum (see div **v** !). A discrete ensemble thus represents only a model *approximation* of this condition that works the better the more dense the occupation of the phase-space volume becomes.

This becomes clear by the example of the cooling rate (which will be calculated in the subsequent problem):

$$\frac{1}{\tau} = \frac{W}{N}(2g - g^2). \tag{18.63}$$

N denotes the number of particles in the beam, W the bandwidth of the system and g a gain factor that will be defined in problem 18.10. The essential point however is the dependence of the cooling rate on the inverse of the particle number of the beam, $1/N$. In the limit

$$\lim_{N \to \infty} \frac{1}{\tau} = 0,$$

cooling is no longer possible, as we would expect.

We note that the same restriction for applying Liouville's theorem basically also holds in thermodynamics, but there the approximation is better by 12 orders of magnitude ($10^{12} \longrightarrow 10^{24}$)!

Much more important, however, is the fact that Liouville's theorem holds on the condition that the particles obey the Hamilton equations, with a given Hamiltonian H. In this sense the particle system must be closed. But just this condition is violated by the reading off the particle position (coordinate, momentum pick-up) and by the corresponding correction (kicker; see Figure 18.14). This is a calculated interference from outside which cannot be described by a Hamiltonian. Hence, the Liouville theorem does not have to be fulfilled; moreover, it must not hold at all!

Example 18.10: Exercise: Cooling of a particle beam

(a) Calculate the cooling rate per second for a beam of N particles.

(b) When does maximum cooling occur?

(c) Calculate the cooling time for a beam of $N = 10^{12}$ particles. Let the bandwidth of the system be $W = 500$ MHz, and $g = 1$.

Solution (a) We first consider the case that the pick-up and the kicker are so fast that they seize each particle independently (see Figure 18.16). Let the displacement of this particle from the beam axis be x_k. After passing the distance $\lambda/4$ (λ is the wavelength of the x-vibration), the deviation is corrected electromagnetically in the kicker. Let the correction be

$$\Delta x_k = g x_k. \tag{18.64}$$

THE PRINCIPLE OF STOCHASTIC COOLING

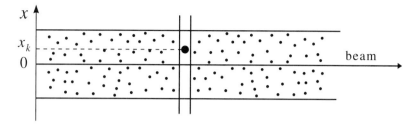

Figure 18.16. In the ideal case, the pick-up seizes *one* particle.

The corrected distance x'_k of the particle from the beam axis is thus given by

$$x'_k = x_k - \Delta x_k = (1-g)x_k \tag{18.65}$$

(Figure 18.17). For $g = 1$, the cooling would be ideal. However, in the real case there appears a noise in the pick-up which is due to further $N_s - 1$ particles passing the pick-up in the time interval T_s (see Figure 18.18). Thus, the pick-up measures not only the spatial displacement x_k of the kth particle from the beam axis ($x = 0$), but also that of the additional $N_s - 1$ particles located around the kth one. The recorded spatial displacement is therefore the *mean value* of all N_s seized particles (the kth and the $N_s - 1$ located around the kth particle):

$$\langle x_k \rangle = \frac{1}{N_s} \sum_{j=1}^{N_s} x_j. \tag{18.66}$$

For clarity, we will label the kth particle in the sum on the right-hand side, e.g., the numbering will be chosen so that $j = 1$ just denotes the kth particle. Moreover, it should be clear that the remaining

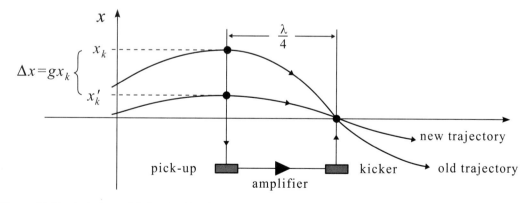

Figure 18.17. A single particle is seized by the pick-up and its momentum is corrected after a $\lambda/4$ wavelength at the kicker. After one revolution, the new trajectory leads to the corrected spatial displacement x'_k.

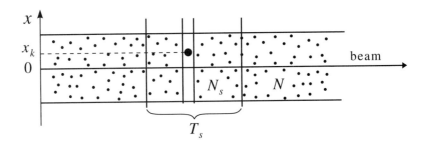

Figure 18.18. In the real case, the pick-up seizes several (N_s) particles that cause a noise.

$N_s - 1$ particles are closely located around the kth one when passing the pick-up. We therefore add the index k to the particular spatial displacement x_j:

$$\langle x_k \rangle = \frac{1}{N_s} \left\{ x_{1,k} + \sum_{j=2}^{N_s} x_{j,k} \right\}, \tag{18.67}$$

where $x_{1,k} \equiv x_k$ and $x_{j,k} \neq x_k$ for $2 \leq j \leq N_s$.

The correction for the kth particles is now

$$x'_k = x_k - g \langle x_k \rangle. \tag{18.68}$$

This means that there will be *no* kick if the sample of N_s particles on the *average* moves on the beam axis:

$$x'_k = x_k \quad \text{if } \langle x_k \rangle = 0. \tag{18.69}$$

In other words, the kicker will not be activated if the *center of gravity*

$$S_k = \frac{1}{mN_s} \sum_{j=1}^{N_s} m x_{j,k} = \frac{1}{N_s} \sum_{j=1}^{N_s} x_{j,k} \equiv \langle x_k \rangle \tag{18.70}$$

of the sample in the pick-up is already on the beam axis (all particles have the same mass m).

For real measurements in the pick-up, this will, of course, not be fulfilled in general. The probability that the center of gravity of the N_s particles that are statistically distributed over the beam just coincides with the beam axis is extremely low. In the realistic case the sample will always be "kicked."

We now want to know how the *mean value* of the spatial displacement of *all* N particles in the beam will change by the mechanism of stochastic cooling. This mean value is

$$E(x_k) = \frac{1}{N} \sum_{k=1}^{N} x_k \tag{18.71}$$

and will be denoted by $E(x_k)$ (expectation value) to distinguish it from the mean value for the sample of N_s particles, $\langle x_k \rangle$, which was defined in equation (18.66). It is however clear that the mean value of the positions of *all* the particles just defines the *beam axis*. Since we put the beam axis at the origin of the coordinate system, $x = 0$, the mean value of the *spatial displacement of all* the particles from the beam axis just vanishes:

$$E(x_k) \equiv 0. \tag{18.72}$$

THE PRINCIPLE OF STOCHASTIC COOLING

This always holds, independent of the mechanism of stochastic cooling. Thus, the mean value $E(x_k)$ is not an appropriate quantity for investigating the mechanism of stochastic cooling. It is evident that the *mean square* of the spatial displacement $E(x_k^2)$ is much better suited for this purpose. We therefore will investigate the *change* of $E(x_k^2)$ by the stochastic cooling mechanism. First we consider the mean square spatial displacement x_k^2 for the kth particle, and to this end, we square equation (18.68):

$$x_k'^2 = x_k^2 - 2g x_k \langle x_k \rangle + g^2 \langle x_k \rangle^2. \tag{18.73}$$

The change of x_k^2 for a single passage through the kicker is thus given by

$$\Delta(x_k^2) := x_k'^2 - x_k^2 = -2g x_k \langle x_k \rangle + g^2 \langle x_k \rangle^2. \tag{18.74}$$

Since there is one kick per revolution, this is also the change of x_k^2 *per revolution*. By averaging over all particles, one obtains

$$\begin{aligned} E(\Delta(x_k^2)) &\equiv E(x_k'^2 - x_k^2) = E(x_k'^2) - E(x_k^2) \\ &\equiv \Delta(E(x_k^2)) = -2g E(x_k \langle x_k \rangle) + g^2 E(\langle x_k \rangle^2). \end{aligned} \tag{18.75}$$

The second equals sign in the first line follows from the additivity of the expectation value $E(\ldots)$; compare equation (18.71).

To calculate the change of the expectation value of the mean square of the spatial displacement per revolution, $\Delta(E(x_k^2))$, the expectation values $E(x_k \langle x_k \rangle)$, $E(\langle x_k \rangle^2)$ must be expressed by $E(x_k^2)$. We then obtain $\Delta(E(x_k^2))$ as a function of $E(x_k^2)$, or a differential equation for $E(x_k^2)$, the solution of which allows us to calculate the desired quantities.

To evaluate $E(x_k \langle x_k \rangle)$, we write with equation (18.67)

$$\begin{aligned} E(x_k \langle x_k \rangle) &= \frac{1}{N} \sum_{k=1}^{N} x_k \frac{1}{N_s} \left\{ x_{1,k} + \sum_{j=2}^{N_s} x_{j,k} \right\} \\ &= \frac{1}{N_s} \left\{ \frac{1}{N} \sum_{k=1}^{N} x_k^2 + \frac{1}{N} \sum_{k=1}^{N} \sum_{j=2}^{N_s} x_{1,k} x_{j,k} \right\}. \end{aligned} \tag{18.76}$$

In the first term, we used $x_{1,k} \equiv x_k$, and in the second one $x_k \equiv x_{1,k}$.

We now realize that two *different* particles in the beam *cannot* be correlated (the particles are statistically distributed over the beam!). Even though they belong to the *same* sample of N_s particles around the kth particle, their spatial displacements $x_{i,k}$ and $x_{j,k}$, $i \neq j$, on the average must satisfy

$$E(x_{i,k} x_{j,k}) = \frac{1}{N} \sum_{k=1}^{N} x_{i,k} x_{j,k} = 0, \quad \text{for } i \neq j. \tag{18.77}$$

Thus, if $i = 1$ and $2 \leq j \leq N_s$, then

$$E(x_{1,k} x_{j,k}) = \frac{1}{N} \sum_{k=1}^{N} x_{1,k} x_{j,k} \equiv 0. \tag{18.78}$$

This is now utilized in the second term of equation (18.76), which then vanishes. The first term can immediately be rewritten using the definition of E, and we obtain

$$E(x_k \langle x_k \rangle) = \frac{1}{N_s} E(x_k^2). \tag{18.79}$$

Furthermore,

$$E(\langle x_k\rangle^2) = \frac{1}{N}\sum_{k=1}^{N}\frac{1}{N_s^2}\sum_{i,j=1}^{N_s} x_{i,k} x_{j,k}$$

$$= \frac{1}{N}\sum_{k=1}^{N}\frac{1}{N_s^2}\left\{\sum_{i=1}^{N_s} x_{i,k}^2 + \sum_{\substack{i,j=1 \\ i\neq j}}^{N_s} x_{i,k} x_{j,k}\right\}. \qquad (18.80)$$

The second term again vanishes by using equation (18.77). In the first term we first average by summing over all particles,

$$\frac{1}{N}\sum_{k=1}^{N}\frac{1}{N_s^2}\sum_{i=1}^{N_s} x_{i,k}^2 = \frac{1}{N_s^2}\sum_{i=1}^{N_s} E(x_{i,k}^2). \qquad (18.81)$$

The mean square spatial deviation $E(x_{i,k}^2)$ cannot depend on the label i of the particle from the sample of N_s particles, $E(x_{i,k}^2) \equiv E(x_k^2)$. The sum over i therefore yields only the additional factor N_s, and we obtain

$$E(\langle x_k\rangle^2) \equiv \frac{1}{N_s} E(x_k^2). \qquad (18.82)$$

Equation (18.75) with (18.79) and (18.82) thus turns into

$$\Delta(E(x_k^2)) = -\frac{2g - g^2}{N_s} E(x_k^2) \qquad (18.83)$$

for the change of the mean square spatial displacement per revolution. The "differential" change $dE(x_k^2)$ per "differential" revolution dn is

$$\frac{dE(x_k^2)}{dn} = -\frac{2g - g^2}{N_s} E(x_k^2). \qquad (18.84)$$

This differential equation is solved by the function

$$E(x_k^2) = C \exp\left(-n\frac{2g - g^2}{N_s}\right). \qquad (18.85)$$

For the *root of the mean square (rms) spatial displacement* $x_{\mathrm{rms}} := \sqrt{E(x_k^2)}$, we obtain

$$x_{\mathrm{rms}} = \sqrt{C} \exp\left(-n\frac{2g - g^2}{2N_s}\right). \qquad (18.86)$$

x_{rms} decreases to the eth fraction of its original value after

$$n_0 = \frac{2N_s}{2g - g^2} \qquad (18.87)$$

revolutions. Since each revolution takes the time T, it thus lasts for

$$\tau = n_0 T = \frac{2N_s T}{2g - g^2}, \qquad (18.88)$$

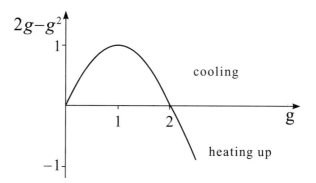

Figure 18.19.

to reduce x_{rms} to the fraction $1/e$ of its original value. Since among N particles orbiting in the time T, N_s particles are seized in the time T_s, for a homogeneous particle flux density we have

$$N_s = N \frac{T_s}{T}, \tag{18.89}$$

and therefore,

$$\tau = \frac{2N T_s}{2g - g^2}. \tag{18.90}$$

With the bandwidth $W = 1/2T_s$ (Nyquist's theorem) or

$$T_s = \frac{1}{2\omega}, \tag{18.91}$$

we finally obtain the *cooling rate*

$$\frac{1}{\tau} = \frac{W}{N}(2g - g^2). \tag{18.92}$$

(b) From the discussion of equation (18.92), it follows immediately that the cooling rate is maximized for $g = 1$. For $g > 2$, the particles are heated up.

(c) With the numerical values given in the problem, for the cooling rate we get

$$\frac{1}{\tau} = \frac{500\,\text{MHz}}{10^{12}} \cdot 1 = 5 \cdot 10^{-4} \frac{1}{\text{s}}, \tag{18.93}$$

and thus,

$$\tau = 2 \cdot 10^3\,\text{s} \approx \frac{1}{2}\,\text{h}. \tag{18.94}$$

19 Canonical Transformations

Given a Hamiltonian $H = H(q_i, p_i, t)$, the motion of the system is found by integration of the Hamilton equations:

$$\dot{p}_i = -\frac{\partial H}{\partial q_i} \quad \text{and} \quad \dot{q}_i = \frac{\partial H}{\partial p_i}.$$

For the case of a cyclic coordinate, we have, as we know,

$$\frac{\partial H}{\partial q_i} = 0, \quad \text{i.e.,} \quad \dot{p}_i = 0.$$

Hence, the corresponding momentum is constant: $p_i = \beta_i = $ constant.

Whether or not H contains cyclic coordinates depends in general on the coordinates adopted for describing a problem. This is immediately seen from the following example: If a circular motion in a central field is described in Cartesian coordinates, there is no cyclic coordinate. If we use polar coordinates (ϱ, φ), the angular coordinate is cyclic (angular momentum conservation).

A mechanical problem would therefore be greatly simplified if one could find a coordinate transformation from the set p_i, q_i to a new set of coordinates P_i, Q_i with

$$Q_i = Q_i(p_i, q_i, t), \qquad P_i = P_i(p_i, q_i, t), \tag{19.1}$$

where all coordinates Q_i for the problem were cyclic. Then all momenta are constant, $P_i = \beta_i$, and the new Hamiltonian $\mathcal{H}$ is then only a function of the constant momenta P_i; hence, $\mathcal{H} = \mathcal{H}(P_i)$. Then

$$\dot{Q}_i = \frac{\partial \mathcal{H}(P_i)}{\partial P_i} = \omega_i = \text{constant}, \qquad \dot{P}_i = -\frac{\partial \mathcal{H}(P_i)}{\partial Q_i} = 0.$$

Then integration with respect to time leads to

$$Q_i = \omega_i t + \omega_0, \qquad P_i = \beta_i = \text{constant}.$$

Here, we presupposed that the new coordinates (P_i, Q_i) again satisfy the (canonical) Hamilton equations, with a new Hamiltonian $\mathcal{H}(P_i, Q_i, t)$. This is an essential requirement for a coordinate transformation of the form (19.1) to make it *canonical*.

CANONICAL TRANSFORMATIONS

Just as p_i is the canonical momentum corresponding to q_i ($p_i = \partial L/\partial \dot{q}_i$), P_i shall be the canonical momentum to Q_i. A pair (q_i, p_i) is called *canonically conjugate* if the Hamilton equations hold for q_i and p_i. The transformation from one pair of canonically conjugate coordinates to another pair is called a *canonical transformation* (point transformation). Then

$$\dot{Q}_i = \frac{\partial \mathcal{H}}{\partial P_i}, \qquad \dot{P}_i = -\frac{\partial \mathcal{H}}{\partial Q_i}.$$

At the moment, we do not yet require that all Q_i be cyclic. This case will be considered later (Chapter 20).

In the new coordinates, the Hamilton principle must be fulfilled too, of course. Thus, we have both

$$\delta \int L(q_i, \dot{q}_i, t)\, dt = 0$$

and

$$\delta \int \mathcal{L}(Q_i, \dot{Q}_i, t)\, dt = 0.$$

Thus, the difference

$$\delta \int (L - \mathcal{L})\, dt = 0$$

also vanishes. Conversely, from this equation it follows that

$$\delta \int \mathcal{L}\, dt = \delta \int L\, dt = 0,$$

where the prerequisite $\delta \int L\, dt = 0$ holds.

Furthermore, we now conclude that the equation $\delta \int (L - \mathcal{L})\, dt = 0$ will then be fulfilled if the old and new Lagrangian differ only by a total differential:

$$L - \mathcal{L} = \frac{dF}{dt}, \quad \text{because} \quad \delta \int_1^2 \frac{dF}{dt}\, dt = \delta(F(2) - F(1)) = 0,$$

since the variation of a constant equals zero. As we shall see, the function F mediates the transformation (p_i, q_i) to (P_i, Q_i). F is therefore also called a *generating function*. In the general case, F will be a function of the old and the new coordinates; together with the time t it involves $4n + 1$ coordinates:

$$F = F(p_i, q_i, P_i, Q_i, t).$$

But since simultaneously there are $2n$ transformation equations

$$Q_i = Q_i(p_i, q_i, t), \qquad P_i = P_i(p_i, q_i, t), \tag{19.2}$$

F involves only $2n + 1$ independent variables. F must contain both a coordinate from the old coordinate set p_i (or q_i) and one of the new P_i (or Q_i) to enable us to establish a relation between the systems. Hence, there are *four possibilities for a generating function*:

$$F_1 = F(q_i, Q_i, t), \quad F_2 = F(q_i, P_i, t),$$
$$F_3 = F(p_i, Q_i, t), \quad F_4 = F(p_i, P_i, t).$$

Each of these functions has $2n + 1$ independent variables. The dependency must be selected in a suitable way, according to the actual problem. First, it is not yet evident how these generating functions F_i give transformations of the form (19.1). We will illustrate this here and consider F_1 as an example.

Because

$$L = \mathcal{L} + \frac{dF}{dt} \quad \text{and} \quad L = \sum p_i \dot{q}_i - H,$$

we have

$$\sum p_i \dot{q}_i - H = \sum P_i \dot{Q}_i - \mathcal{H} + \frac{dF}{dt}. \tag{19.3a}$$

If we use $F_1 = F(q_i, Q_i, t)$, for the total derivative of F_1 we then have

$$\frac{dF_1}{dt} = \sum \frac{\partial F_1}{\partial q_i} \dot{q}_i + \sum \frac{\partial F_1}{\partial Q_i} \dot{Q}_i + \frac{\partial F_1}{\partial t}. \tag{19.3b}$$

We insert the result into equation (19.3a). This yields

$$\sum p_i \dot{q}_i - \sum P_i \dot{Q}_i - H + \mathcal{H} = \sum \frac{\partial F_1}{\partial q_i} \dot{q}_i + \sum \frac{\partial F_1}{\partial Q_i} \dot{Q}_i + \frac{\partial F_1}{\partial t}.$$

By comparing the coefficients, we obtain

$$p_i = \frac{\partial F_1}{\partial q_i}(q_j, Q_j, t),$$
$$P_i = -\frac{\partial F_1}{\partial Q_i}(q_j, Q_j, t),$$
$$\mathcal{H} = H + \frac{\partial F_1}{\partial t}(q_j, Q_j, t). \tag{19.4}$$

Since it will be important for the next chapter, we derive the transformation equations for a generating function of the type F_2, which is denoted by S:

$$F_2 \equiv S = S(q_i, P_i, t).$$

For the derivation, we will use a comparison of coefficients as for F_1; therefore, we require that S be composed as follows:

$$S(q_i, P_i, t) = \sum_i P_i Q_i + F_1(q_i, Q_i, t),$$

CANONICAL TRANSFORMATIONS

since then we can consider the problem analogously to F_1. This is not evident at first, but will be made clear by the following. We imagine the Q_i as being expressed through the second of the equations (19.4), i.e., through

$$P_i = -\frac{\partial F_1(q_j, Q_j, t)}{\partial Q_j}$$

expressed by P_i and q_i. According to equation (19.3a), we have

$$\sum_i p_i \dot{q}_i - H = \sum_i P_i \dot{Q}_i - \mathcal{H} + \frac{d}{dt} F_1$$

$$= \sum_i P_i \dot{Q}_i - \mathcal{H} + \frac{d}{dt}\left(S(q_i, P_i, t) - \sum_i P_i Q_i\right).$$

This leads to

$$\sum_i p_i \dot{q}_i - \sum_i P_i \dot{Q}_i - H + \mathcal{H} = \frac{d}{dt}\left(S(q_i, P_i, t) - \sum_i P_i Q_i\right)$$

$$= \sum_i \frac{\partial S}{\partial q_i} \dot{q}_i + \sum_i \frac{\partial S}{\partial P_i} \dot{P}_i + \frac{\partial S}{\partial t}$$

$$- \sum_i \dot{P}_i Q_i - \sum_i P_i \dot{Q}_i$$

$$\sum_i p_i \dot{q}_i + \sum_i \dot{P}_i Q_i - H + \mathcal{H} = \sum_i \frac{\partial S}{\partial q_i} \dot{q}_i + \sum_i \frac{\partial S}{\partial P_i} \dot{P}_i + \frac{\partial S}{\partial t}.$$

The comparison of coefficients now yields the equations

$$p_i = \frac{\partial S(q_j, P_j, t)}{\partial q_i}, \qquad Q_i = \frac{\partial S(q_j, P_j, t)}{\partial P_i},$$

$$\mathcal{H}(P_i, Q_i, t) = H(p_i, q_i, t) + \frac{\partial S(q_i, P_i, t)}{\partial t}. \tag{19.5}$$

The first two relations allow us to determine the transformation equations $Q_i = Q_i(p_i, q_i, t)$, $P_i = P_i(p_i, q_i, t)$, which by insertion into the right side of equation (19.5) yield the new Hamiltonian $\mathcal{H} = \mathcal{H}(P_i, Q_i, t)$.

The transformation equations for the other types of generating functions are obtained analogously, by choosing an appropriate sum which enables us to use the methods of the first two problems.

From the equations (19.4) and (19.5), we now obtain the dependence of the new coordinates (P_i, Q_i) on the old (p_i, q_i) and vice versa. For the case F_1, from

$$p_i = \frac{\partial F_1(q_j, Q_j, t)}{\partial q_i}$$

follow the equations $p_i = p_i(q_j, Q_j)$, which can be solved for the Q_i:

$$Q_i = Q_i(p_j, q_j).$$

Insertion into the equations

$$P_i = -\frac{\partial F_1(q_j, Q_j, t)}{\partial Q_i}$$

then enables us to calculate

$$P_i = P_i(p_j, q_j).$$

We now understand the name *generating function* for F: The function F determines the canonical transformation

$$Q_i = Q_i(p_j, q_j, t), \qquad P_i = P_i(p_j, q_j, t)$$

through equations of the type (19.4) or (19.5).

Example 19.1: Example of a canonical transformation

Let the generating function be given by

$$F_1 = F_1(q_j, Q_j) = \sum q_j Q_j.$$

According to the equations (19.4) for F_1, it then follows that $p_i = Q_i$ and $P_i = -q_i$. The example shows that in the Hamiltonian formalism the momentum and position coordinates play equivalent parts.

Example 19.2: The harmonic oscillator

In the coordinates p, q, the kinetic and potential energy are given by

$$T = \frac{1}{2m}p^2, \qquad V = \frac{1}{2}kq^2 = \frac{1}{2}m\omega^2 q^2, \qquad \omega^2 = \frac{k}{m}. \tag{19.6}$$

We then have the Lagrangian and Hamiltonian

$$L = \frac{1}{2m}p^2 - \frac{1}{2}m\omega^2 q^2, \qquad H = \frac{1}{2m}p^2 + \frac{1}{2}m\omega^2 q^2. \tag{19.7}$$

The generating function for the transformation reads

$$F_1(q, Q) = \frac{m}{2}\omega q^2 \cot Q. \tag{19.8}$$

Thus, F is of the type F_1, and from this follows with the equations (19.4):

$$p = \frac{\partial F_1}{\partial q} = m\omega q \cot Q, \qquad P = -\frac{\partial F_1}{\partial Q} = \frac{m}{2}\frac{\omega q^2}{\sin^2 Q}. \tag{19.9}$$

Solving for the coordinates (p, q) yields

$$q = \sqrt{\frac{2}{m\omega}P} \sin Q, \qquad p = \sqrt{2m\omega P} \cos Q. \tag{19.10}$$

By inserting into H, we obtain

$$\mathcal{H} = P\omega(\cos^2 Q + \sin^2 Q), \quad \text{and thus} \quad \mathcal{H} = \omega P. \tag{19.11}$$

CANONICAL TRANSFORMATIONS

This means that Q is a cyclic coordinate.

Since the Hamiltonian does not explicitly depend on the time, H represents the total energy of the system, and therefore,

$$E = \omega P = \text{constant}. \tag{19.12}$$

Because $\dot{Q} = \partial \mathcal{H}/\partial P = \omega$, we also have

$$Q = \omega t + \varphi. \tag{19.13}$$

By insertion of (19.12) and (19.13) into (19.10), one gets the known dependence of the position q on the time:

$$q = \sqrt{\frac{2E}{m\omega^2}} \sin(\omega t + \varphi).$$

20 Hamilton–Jacobi Theory

In the preceding chapter, we tried to perform a transformation to coordinate pairs (q_i, $p_i = \beta_i$) for which the canonical momenta were constant. We now proceed one step further and look for a canonical transformation to coordinates $P_i = p_{i0}$ and $Q_i = q_{i0}$ which all are constant and are given by the initial conditions. When we have found such coordinates, the transformation equations are the solutions of the system in the normal position coordinates:

$$q_i = q_i(q_{i0}, p_{i0}, t), \qquad p_i = p_i(q_{i0}, p_{i0}, t).$$

The coordinates (P_i, Q_i) obey the Hamilton equations with the Hamiltonian $\mathcal{H}(Q_i, P_i, t)$. Since the time derivatives vanish by definition, we have

$$\dot{P}_i = 0 = -\frac{\partial \mathcal{H}}{\partial Q_i}, \qquad \dot{Q}_i = 0 = \frac{\partial \mathcal{H}}{\partial P_i}. \tag{20.1}$$

These conditions would certainly be fulfilled by the function $\mathcal{H} \equiv 0$. In order to perform the coordinate transformation, we need a generating function. For historical reasons—Jacobi made this choice—we adopt among the four possible types the type $F_2 = S(q_i, P_i, t)$, which already has been treated in the preceding chapter. It is generally known as the *Hamilton action function*. For this choice the equations (19.5) hold. We now require that the new Hamiltonian shall identically vanish. Then

$$\frac{\partial S}{\partial t} + H\left(q_1, \ldots, q_n; p_1 = \frac{\partial S}{\partial q_1}, \ldots, p_n = \frac{\partial S}{\partial q_n}; t\right) = 0. \tag{20.2}$$

Writing down this equation with the arguments, we obtain

$$\frac{\partial S(q_i, P_i = \beta_i, t)}{\partial t} + H\left(q_1, \ldots, q_n; \frac{\partial S}{\partial q_1}, \ldots, \frac{\partial S}{\partial q_n}; t\right) = 0. \tag{20.3}$$

This is the *Hamilton–Jacobi differential equation*.[1] The P_i denote constants that, as noted above, are fixed by the initial conditions p_{i0}. By means of this differential equation we can determine S. We note that this differential equation is a *nonlinear partial differential equation of first order* with $n+1$ variables q_i, t. It is nonlinear, since H depends quadratically on the momenta that enter as derivatives of the action function with respect to the position coordinates. There appear only first derivatives with respect to the q_i and the time.

To get the action function S, we have to integrate the differential equation $n+1$ times (each derivative $\partial S/\partial q_i$, $\partial S/\partial t$ requires one integration), and we thus obtain $n+1$ integration constants. But since S appears in the differential equation only as a derivative, S is determined only up to a constant a; i.e., $S = S' + a$. This means that one of the $n+1$ integration constants must be a constant additive to S. It is, however, not essential for the transformation. We thus obtain as a solution function

$$S = S(q_1, \ldots, q_n; \beta_1, \ldots, \beta_n; t),$$

where the β_i are integration constants. A comparison with equation (19.5) leads to the requirements

$$P_i = \beta_i; \qquad Q_i = \frac{\partial S}{\partial P_i} = \frac{\partial S(q_1, \ldots, q_n; \beta_1, \ldots, \beta_n; t)}{\partial \beta_i} = \alpha_i. \qquad (20.4)$$

The β_i, α_i can be determined from the initial conditions.

The original coordinates result from the transformation equations (19.5) as follows: From

$$\alpha_i = \frac{\partial S(q_j, \beta_j, t)}{\partial \beta_i}$$

follow the position coordinates

$$q_i = q_i(\alpha_j, \beta_j, t).$$

Insertion into

$$p_i = \frac{\partial S(q_j, P_j, t)}{\partial q_i} = p_i(q_i, \beta_i, t)$$

finally yields

$$p_i = p_i(\alpha_i, \beta_i, t).$$

Now the $q_i(\alpha_j, \beta_j, t)$ and $p_i(\alpha_j, \beta_j, t)$ are known as functions of the time and of the integration constants α_j, β_j. This simply means the complete solution of the many-body problem characterized by the Hamiltonian $H(q_i, p_i, t)$.

---

[1] *Carl Gustav Jacob Jacobi*, b. Dec. 18, 1804, Potsdam, son of a banker–d. Feb. 18, 1851, Berlin. After his studies (1824), Jacobi became a lecturer in Berlin and in 1827 to 1842 held a chair as a professor in Königsberg (Kaliningrad). After an extended travel through Italy to restore his weak health, Jacobi lived in Berlin. Jacobi became known for his work *Fundamenta Nova Theoria Functiorum Ellipticarum* (1829). In 1832, Jacobi discovered that hyperelliptic functions can be inverted by functions of several variables. Jacobi also made fundamental contributions to algebra, to elimination theory, and to the theory of partial differential equations, e.g., in his *Lectures on Dynamics* (1842 to 1843), published in 1866.

We can separate off the time dependence in S. If H is not an explicit function of the time, H represents the total energy of the system:

$$-\frac{\partial S}{\partial t} = H = E. \tag{20.5}$$

From this, it follows that S can be represented as

$$S(q_i, P_i, t) = S_0(q_i, P_i) - Et.$$

To explain the meaning of S, we form the total derivative of S with respect to time:

$$\frac{dS}{dt} = \sum \frac{\partial S}{\partial q_i} \dot{q}_i + \sum \frac{\partial S}{\partial P_i} \dot{P}_i + \frac{\partial S}{\partial t}.$$

But, since $\dot{P}_i = 0$, we have

$$\frac{dS(q_i, P_i = \beta_i, t)}{dt} = \sum \frac{\partial S}{\partial q_i} \dot{q}_i + \frac{\partial S}{\partial t}.$$

Because

$$\frac{dS(q_j, P_j = \beta_j, t)}{dq_i} = p_i \quad \text{and} \quad \frac{\partial S}{\partial t} = -H,$$

it further follows that

$$\frac{dS(q_i, P_i(p_\alpha, q_\alpha), t)}{dt} = \sum p_i \dot{q}_i - H(q_i, p_i, t) = L(q_i, p_i, t). \tag{20.6}$$

H and L are not bound by restrictions; in particular they can be time-dependent. This means that S is given by the time integral over the Lagrangian:

$$S = \int L \, dt + \text{constant}. \tag{20.7}$$

Since this integral physically represents an action (energy · time), the term *action function* for S is obvious. The action function differs from the time integral over the Lagrangian by at most an additive constant. However, this last relation cannot be used for a practical calculation, since as long as the problem is not yet solved, one does not know L as a function of time. Moreover, $L(q_i, p_i, t)$ in equation (20.6) depends on the original coordinates q_i, p_i, while the S-function is needed in the coordinates $q_i, P_i(q_\alpha, p_\alpha)$.

Equation (20.7) is not unknown to us: The action function S turned up before when formulating the Hamilton principle (18.25). Before further continuing this discussion, we will illustrate the Hamilton–Jacobi method by an example.

Example 20.1: The Hamilton–Jacobi differential equation

We start again with the harmonic oscillator. The Hamiltonian is

$$H = \frac{p^2}{2m} + \frac{k}{2} q^2.$$

HAMILTON–JACOBI THEORY

The Hamilton action function then has the form (compare equations (19.5) and (20.3))

$$S = S(q, P, t) \quad \text{and} \quad p = \frac{\partial S}{\partial q}.$$

From this, we obtain the Hamilton–Jacobi differential equation:

$$\frac{\partial S}{\partial t} + \frac{1}{2m}\left(\frac{\partial S}{\partial q}\right)^2 + \frac{k}{2}q^2 = 0.$$

For solving the problem, we make a separation *ansatz* into a space and a time variable. A product *ansatz* would not work here, since the differential equation is not linear. We therefore set a sum:

$$S = S_1(t) + S_2(q).$$

For the partial derivatives, we then get

$$\frac{\partial S}{\partial q} = \frac{dS_2(q)}{dq}, \quad \frac{\partial S}{\partial t} = \frac{dS_1(t)}{dt}.$$

This leads to

$$-\dot{S}_1(t) = \frac{1}{2m}\left(\frac{dS_2(q)}{dq}\right)^2 + \frac{k}{2}q^2 = \beta,$$

where β is the separation constant. (The left side depends only on the time t, the right side only on the coordinate q: Therefore, both sides can only be equal if they are equal to a common constant β.) For the time-dependent function, we then have

$$\dot{S}_1(t) = -\beta,$$

which leads to

$$S_1(t) = -\beta t.$$

For the space-dependent part, there remains the following equation:

$$\frac{1}{2m}\left(\frac{dS_2(q)}{dq}\right)^2 + \frac{k}{2}q^2 = \beta, \quad \frac{dS_2}{dq} = \sqrt{2m\beta - mkq^2}.$$

As sum of the two parts, we then obtain for S

$$S(q, \beta, t) = \sqrt{mk}\int \sqrt{\frac{2\beta}{k} - q^2}\, dq - \beta t.$$

For the constant $Q \equiv \alpha$, we then have

$$\alpha = Q = \frac{\partial S}{\partial \beta} = \frac{\sqrt{mk}}{k}\int \left(\frac{2\beta}{k} - q^2\right)^{-1/2} dq - t.$$

The integral can easily be evaluated, and we obtain

$$Q + t = \sqrt{\frac{m}{k}}\arcsin\left(\sqrt{k/(2\beta)}\, q\right).$$

With the usual abbreviation $\omega^2 = k/m$, we obtain the equation

$$q = \sqrt{\frac{2\beta}{k}}\sin\omega(t + Q).$$

A comparison with the known equation of motion of the harmonic oscillator shows that β corresponds to the total energy E, and Q to an initial time t_0. Energy and time are therefore *canonically conjugate variables*. Both the energy and the time t_0 (which corresponds to an initial phase) are given by the initial conditions.

The separation of the Hamilton–Jacobi equation represents a general (often the only feasible) way of solving it. If the Hamiltonian does not explicitly depend on the time, then

$$\frac{dS}{dt} + H\left(q_1, \ldots, q_n; \frac{\partial S}{\partial q_1}, \ldots, \frac{\partial S}{\partial q_n}\right) = 0, \tag{20.8}$$

and the time can be separated off immediately. We set for S a solution of the form

$$S = S_0(q_i, P_i) - \beta t.$$

The constant β then equals H and normally represents the energy. After this separation, there remains the equation

$$H\left(q_1, \ldots, q_n; \frac{\partial S_0}{\partial q_1}, \ldots, \frac{\partial S_0}{\partial q_n}\right) = E. \tag{20.9}$$

To achieve a separation of the position variables, we make the *ansatz*

$$S_0(q_1, \ldots, q_n; P_1, \ldots, P_n) = \sum_i S_i(q_i, P_i) = S_1(q_1, P_1) + \cdots + S_n(q_n, P_n). \tag{20.10}$$

This means that the Hamilton action function splits into a sum of partial functions S_i, each depending only on *one* pair of variables. The Hamiltonian then becomes

$$H\left(q_1, \ldots, q_n; \frac{dS_1}{dq_1}, \ldots, \frac{dS_n}{dq_n}\right) = E. \tag{20.11}$$

To ensure that this differential equation also separates into n differential equations for the $S_i(q_i, P_i)$, H must obey certain conditions. For example, if H has the form

$$H(q_1, \ldots, q_n, p_1, \ldots, p_n) = H_1(q_1, p_1) + \cdots + H_n(q_n, p_n), \tag{20.12}$$

the separation is certainly possible. A Hamiltonian of this form describes a system of independent degrees of freedom; i.e., in (20.12) there are *no* interaction terms, e.g., of the form $H(q_i, p_i, q_j, p_j)$, which describe an interaction between the ith and the jth degree of freedom.

With equation (20.12), equation (20.11) reads

$$H_1\left(q_1, \frac{\partial S_1}{\partial q_1}\right) + \cdots + H_n\left(q_n, \frac{\partial S_n}{\partial q_n}\right) = E. \tag{20.13}$$

This equation can be satisfied by setting each term H_i separately equal to a constant β_i; hence,

$$H_1\left(q_1, \frac{\partial S_1}{\partial q_1}\right) = \beta_1, \ldots, \quad H_n\left(q_n, \frac{\partial S_n}{\partial q_n}\right) = \beta_n, \tag{20.14}$$

HAMILTON–JACOBI THEORY

where

$$\beta_1 + \beta_2 + \cdots + \beta_n = E. \tag{20.15}$$

Thus, there are n integration constants β_i in total.

Since the kinetic energy term of the Hamiltonian involves the momentum $p_i = dS_i/dq_i$ quadratically, these differential equations are of first order and second degree. As solutions, we then obtain the n action functions

$$S_i = S_i(q_i, \beta_i), \tag{20.16}$$

which, apart from the separation constants β_i, depend only on the coordinate q_i. According to (19.5), S_i immediately leads to the conjugate momentum $p_i = dS_i/dq_i$ to the coordinate q_i. The essential point is (see (20.12)) that the coordinate pair (q_i, p_i) is not coupled to other coordinates $(q_k, p_k, i \neq k)$, so that the motion in these coordinates can be considered fully independent of the other ones.

We now restrict ourselves to periodic motions and define the *phase integral*

$$J_i = \oint p_i \, dq_i, \tag{20.17}$$

which is to be taken over a full cycle of a rotation or vibration. The phase integral has the dimension of an action (or of an angular momentum). It is therefore also referred to as an *action variable*. If we replace the momentum by the action function

$$J_i = \oint \frac{dS_i}{dq_i} \, dq_i, \tag{20.18}$$

we see from equation (20.16) that J_i depends only on the constants β_i, since q_i is only an integration variable. We therefore can move from the constants β_i to the likewise constant J_i and use them as new canonical momenta. Hence, one performs the transformation

$$J_i = J_i(\beta_i) \longrightarrow \beta_i = \beta_i(J_i).$$

The total energy E which corresponds to the Hamiltonian can also be recalculated by (20.15) to the J_k:

$$H = E = \sum_{i=1}^{n} \beta_i(J_i). \tag{20.19}$$

The Hamiltonian is therefore only a function of the action variables, which take the role of the momenta. All corresponding conjugate coordinates are cyclic. The conjugate coordinates belonging to the J_i are called *angle variables* and are denoted by φ_i. The generating function $S_i(q_i, \beta_k)$ turns with $\beta_k(J_k)$ into $S(q_i, J_i)$. The J_i are the new momenta. We therefore can apply (19.5), and for the related new coordinates, we have

$$\varphi_j = \frac{\partial S(q_i, J_i)}{\partial J_j}.$$

By transforming to the action variables and angle variables, we thus have performed a canonical transformation, mediated by the generating function

$$S_i(q_i, \beta_i) \longrightarrow S_i(\varphi_i, J_i). \tag{20.20}$$

This transformation from one set of constant momenta to another set actually does not give new insights. The meaning for periodic processes lies in the angle variable φ_i. Since we performed only canonical transformations, we have

$$\dot{\varphi}_i = \frac{\partial H}{\partial J_i} = \nu_i(J_i) = \text{constant}. \tag{20.21}$$

One can show that ν_i is the frequency of the periodic motion in the coordinate i. This relation thus offers the advantage that the frequencies, which are often of primary interest, can be determined without solving the full problem. We briefly demonstrate this point by the following example:

Example 20.2: Angle variable

We again consider the harmonic oscillator. The expression for the total energy

$$E = \frac{p^2}{2m} + \frac{kq^2}{2}$$

is transformed so that we get the representation of an ellipse in phase space:

$$\frac{p^2}{2mE} + \frac{q^2}{2E/k} = 1.$$

The phase integral is the area enclosed by the ellipse in phase space:

$$J = \oint p\, dq = \pi ab.$$

The two half-axes of the ellipse are

$$a = \sqrt{2mE} \quad \text{and} \quad b = \sqrt{\frac{2E}{k}}.$$

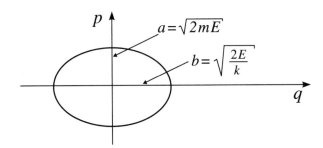

Figure 20.1. Ellipses in phase space.

HAMILTON–JACOBI THEORY

We therefore obtain

$$J = 2\pi E \sqrt{\frac{m}{k}}, \quad \text{or} \quad E = H = \frac{J}{2\pi}\sqrt{\frac{k}{m}}.$$

This leads to the frequency

$$\nu = \frac{dH}{dJ} = \frac{1}{2\pi}\sqrt{\frac{k}{m}}.$$

Example 20.3: Exercise: Solution of the Kepler problem by the Hamilton–Jacobi method

Use the Hamilton–Jacobi method for solving the Kepler problem in a central force field of the form

$$V(r) = -\frac{K}{r}.$$

Solution We adopt plane polar coordinates (r, Θ) as generalized coordinates. The Hamiltonian reads

$$H = \frac{1}{2m}\left(p_r^2 + \frac{p_\Theta^2}{r^2}\right) - \frac{K}{r}. \tag{20.22}$$

H is cyclic in Θ, and hence, $p_\Theta = \text{constant} = l$. The p_i can be expressed by the Hamilton action function S:

$$p_i = \frac{\partial S}{\partial q_i} \implies p_r = \frac{\partial S}{\partial r}, \quad p_\Theta = \frac{\partial S}{\partial \Theta} = \text{constant} = \beta_2.$$

Thus, we obtain the Hamilton–Jacobi differential equation

$$\frac{\partial S}{\partial t} + \frac{1}{2m}\left\{\left(\frac{\partial S}{\partial r}\right)^2 + \frac{1}{r^2}\left(\frac{\partial S}{\partial \Theta}\right)^2\right\} - \frac{K}{r} = 0. \tag{20.23}$$

For the action function, we adopt a separation *ansatz*

$$S = S_1(r) + S_2(\Theta) + S_3(t), \tag{20.24}$$

which is inserted into (20.23):

$$\frac{1}{2m}\left\{\left(\frac{\partial S_1(r)}{\partial r}\right)^2 + \frac{1}{r^2}\left(\frac{\partial S_2(\Theta)}{\partial \Theta}\right)^2\right\} - \frac{K}{r} = -\frac{\partial S_3(t)}{\partial t}. \tag{20.25}$$

Equation (20.25) can be satisfied only if both sides are constant. The constant is the total energy of the system, because

$$-\frac{\partial S}{\partial t} = H = E \implies -\frac{\partial S_3}{\partial t} = \text{constant} = \beta_3 = E. \tag{20.26}$$

We remember that

$$P_i = \beta_i, \quad Q_i = \frac{\partial S}{\partial P_i} = \frac{\partial S}{\partial \beta_i} = \alpha_i,$$

where α_i, β_i are constants that follow from the initial conditions.

We insert (20.26) into equation (20.25), and solve for $\partial S_2/\partial \Theta$:

$$\left(\frac{\partial S_2}{\partial \Theta}\right)^2 = r^2 \left\{ 2m\beta_3 + \frac{2mK}{r} - \left(\frac{\partial S_1}{\partial r}\right)^2 \right\}. \tag{20.27}$$

The same argument that led us to equation (20.26) now yields

$$\frac{\partial S_2}{\partial \Theta} = \frac{dS_2(\Theta)}{d\Theta} = \text{constant} = \beta_2, \tag{20.28}$$

and therefore,

$$\frac{\partial S_1}{\partial r} = \frac{dS_1(r)}{dr} = \sqrt{2m\beta_3 + \frac{2mK}{r} - \frac{\beta_2^2}{r^2}}. \tag{20.29}$$

The Hamilton action function can now be written down as follows:

$$S = \int \sqrt{2m\beta_3 + \frac{2mK}{r} - \frac{\beta_2^2}{r^2}}\, dr + \beta_2 \Theta - \beta_3 t. \tag{20.30}$$

We now define β_2 and β_3 as new momenta P_Θ and P_r. The quantities Q_i conjugate to the P_i are also constant.

$$Q_r = \frac{\partial S}{\partial \beta_3} = \frac{\partial}{\partial \beta_3} \int \sqrt{2m\beta_3 + \frac{2mK}{r} - \frac{\beta_2^2}{r^2}}\, dr - t = \alpha_3, \tag{20.31}$$

$$Q_\Theta = \frac{\partial S}{\partial \beta_2} = \frac{\partial}{\partial \beta_2} \int \sqrt{2m\beta_3 + \frac{2mK}{r} - \frac{\beta_2^2}{r^2}}\, dr + \Theta = \alpha_2. \tag{20.32}$$

If we identify α_2 with Θ', which follows from the initial conditions, we obtain

$$\int \frac{\beta_2 dr}{r^2 \sqrt{2m\beta_3 + 2mK/r - \beta_2^2/r^2}} = \Theta - \Theta'. \tag{20.33}$$

Insertion of the constants and substitution $u \equiv 1/r$ leads to

$$-\int \frac{du}{\sqrt{(2mE)/l^2 + (2mKu/l^2) - u^2}} = \Theta - \Theta'. \tag{20.34}$$

This integral of the form

$$\int \frac{dx}{\sqrt{ax^2 + bx + c}} = \cdots$$

can according to integral tables be written as a closed expression, with

$$\Delta = 4ac - b^2 = -4\frac{2mE}{l^2} - \frac{4m^2K^2}{l^4} < 0,$$

$$\Theta = \Theta' + \arcsin\left(\frac{-2u + (2mK/l^2)}{\sqrt{(4m/l^2)(2E + (mK^2/l^2))}}\right)$$

$$= \Theta' - \arcsin\left(\frac{(l^2 u/mK) - 1}{\sqrt{1 + (2El^2/mK^2)}}\right) \tag{20.35}$$

HAMILTON–JACOBI THEORY

$$\Leftrightarrow \quad r = \frac{l^2}{mK} \frac{1}{(1+\sqrt{1+(2El^2/mK^2)}\cos(\Theta-\Theta'+\pi/2))}. \tag{20.36}$$

This is the solution of the Kepler problem, known from *Classical Mechanics: Point Particles and Relativity*. The types of trajectories follow from the discussion of conic sections in the representation $r = p/(1+\varepsilon \cos\varphi)$:

$\varepsilon = 1 \cong E = 0$: parabolas;

$\varepsilon < 1 \cong E < 0$: ellipses;

$\varepsilon > 1 \cong E > 0$: hyperbolas.

Compare also *Classical Mechanics: Point Particles and Relativity*, Chapter 26.

Equation (20.31) could be rewritten further, by pulling the differentiation into the integral and transforming the resulting equation in such a way that the position r becomes a function of the time. We skip that here.

Example 20.4: **Exercise: Formulation of the Hamilton–Jacobi differential equation for particle motion in a potential with azimuthal symmetry**

Let a particle of mass m move in a force field that in spherical coordinates has the form $V = -K\cos\Theta/r^2$. Write down the Hamilton–Jacobi differential equation for the particle motion.

Solution We first need the Hamiltonian operator as a function of the conjugate momenta in spherical coordinates. For this purpose we first write the kinetic energy T in spherical coordinates:

$$\dot{\mathbf{r}} = \dot{r}\mathbf{e}_r + r\dot{\Theta}\mathbf{e}_\Theta + r\sin\Theta\dot{\varphi}\mathbf{e}_\varphi \tag{20.37}$$

$$\Longrightarrow \quad T = \frac{1}{2}m\dot{\mathbf{r}}\cdot\dot{\mathbf{r}} = \frac{1}{2}m(\dot{r}^2 + r^2\dot{\Theta}^2 + r^2\sin^2\Theta\dot{\varphi}^2). \tag{20.38}$$

The Lagrangian then reads

$$L = T - V = \frac{1}{2}m(\dot{r}^2 + r^2\dot{\Theta}^2 + r^2\sin^2\Theta\,\dot{\varphi}^2) - V(r,\Theta,\varphi). \tag{20.39}$$

We now assume that $V(r,\Theta,\varphi)$ is velocity-independent (which is indeed fulfilled) and form the canonical conjugate momenta:

$$p_r = \frac{\partial L}{\partial \dot{r}} = m\dot{r}, \quad p_\Theta = \frac{\partial L}{\partial \dot{\Theta}} = mr^2\dot{\Theta}, \quad p_\varphi = \frac{\partial L}{\partial \dot{\varphi}} = mr^2\sin^2\Theta\dot{\varphi}. \tag{20.40}$$

From this, we obtain

$$\dot{r} = \frac{p_r}{m}, \quad \dot{\Theta} = \frac{p_\Theta}{mr^2}, \quad \dot{\varphi} = \frac{p_\varphi}{mr^2\sin^2\Theta}.$$

Hence, H can be given in the desired form

$$H = \sum_\alpha p_\alpha \dot{q}_\alpha - L$$

$$= p_r\dot{r} + p_\Theta\dot{\Theta} + p_\varphi\dot{\varphi} - \frac{1}{2}m\left(\frac{p_r^2}{m^2} + \frac{p_\Theta^2}{r^2 m^2} + \frac{p_\varphi^2}{m^2 r^2 \sin^2\Theta}\right) + V(r,\Theta,\varphi)$$

$$= \frac{p_r^2}{2m} + \frac{p_\Theta^2}{2mr^2} + \frac{p_\varphi^2}{2mr^2\sin^2\Theta} + V(r,\Theta,\varphi), \tag{20.41}$$

and for the actual potential (see the formulation of the problem),

$$H = \frac{p_r^2}{2m} + \frac{p_\Theta^2}{2mr^2} + \frac{p_\varphi^2}{2mr^2 \sin^2 \Theta} - \frac{K \cos \Theta}{r^2}. \tag{20.42}$$

The p_i as functions of the Hamilton action variables read

$$p_r = \frac{\partial S}{\partial r}, \quad p_\Theta = \frac{\partial S}{\partial \Theta}, \quad p_\varphi = \frac{\partial S}{\partial \varphi}. \tag{20.43}$$

Therefore, the Hamilton–Jacobi equation has the form

$$\frac{\partial S}{\partial t} + \frac{1}{2m}\left\{\left(\frac{\partial S}{\partial r}\right)^2 + \frac{1}{r^2}\left(\frac{\partial S}{\partial \Theta}\right)^2 + \frac{1}{r^2 \sin^2 \Theta}\left(\frac{\partial S}{\partial \varphi}\right)^2\right\} - \frac{K \cos \Theta}{r^2} = 0. \tag{20.44}$$

Example 20.5: Exercise: Solution of the Hamilton–Jacobi differential equation of Example 20.4

(a) Find the complete solution of the Hamilton–Jacobi differential equation from the preceding Example 20.4, and

(b) sketch how to determine the motion of the particle.

Solution (a) The approach is analogous to Example 20.3. We adopt the separation *ansatz* for S,

$$S = S_1(r) + S_2(\Theta) + S_3(\varphi) - Et, \tag{20.45}$$

and insert this into equation (20.44):

$$\frac{1}{2m}\left(\frac{\partial S_1}{\partial r}\right)^2 + \frac{1}{2mr^2}\left(\frac{\partial S_2}{\partial \Theta}\right)^2 + \frac{1}{2mr^2 \sin^2 \Theta}\left(\frac{\partial S_3}{\partial \varphi}\right)^2 - \frac{K \cos \Theta}{r^2} = E \tag{20.46}$$

$$\Leftrightarrow$$

$$r^2\left(\frac{\partial S_1(r)}{\partial r}\right)^2 - 2mEr^2 = -\left(\frac{\partial S_2(\Theta)}{\partial \Theta}\right)^2 - \frac{1}{\sin^2 \Theta}\left(\frac{\partial S_3(\varphi)}{\partial \varphi}\right)^2 + 2mK \cos \Theta. \tag{20.47}$$

Equations (20.46) and (20.47) can only be satisfied if both sides are constant:

$$r^2\left(\frac{\partial S_1(r)}{\partial r}\right)^2 - 2mEr^2 = \text{constant} = \beta_1, \tag{20.48}$$

$$-\left(\frac{\partial S_2(\Theta)}{\partial \Theta}\right)^2 - \frac{1}{\sin^2 \Theta}\left(\frac{\partial S_3(\varphi)}{\partial \varphi}\right)^2 + 2mK \cos \Theta = \beta_1. \tag{20.49}$$

To separate Θ from φ, we multiply (20.49) by $\sin^2 \Theta$:

$$\left(\frac{\partial S_3(\varphi)}{\partial \varphi}\right)^2 = 2mK \cos \Theta \sin^2 \Theta - \beta_1 \sin^2 \Theta - \left(\frac{\partial S_2(\Theta)}{\partial \Theta}\right)^2 \sin^2 \Theta. \tag{20.50}$$

The separation constant is denoted by β_3, since

$$\frac{\partial S}{\partial \varphi} = p_\varphi \quad \text{and thus} \quad \left(\frac{\partial S_3(\varphi)}{\partial \varphi}\right)^2 = \beta_3^2 = p_\varphi^2. \tag{20.51}$$

Therefore,

$$2mK \cos\Theta \sin^2\Theta - \beta_1 \sin^2\Theta - \sin^2\Theta \left(\frac{\partial S_2}{\partial \Theta}\right)^2 = p_\varphi. \quad (20.52)$$

Integration of equations (20.48), (20.51), and (20.52) yields

$$S_1 = \int \sqrt{2mE + \frac{\beta_1}{r^2}}\, dr + c_1,$$

$$S_2 = \int \sqrt{2mK \cos\Theta - \beta_1 - \frac{p_\varphi^2}{\sin^2\Theta}}\, d\Theta + c_2, \quad (20.53)$$

$$S_3 = \varphi p_\varphi + c_3.$$

The complete solution of the Hamilton–Jacobi differential equation is obtained from (20.53) and the ansatz (20.45) for S:

$$S = \int \sqrt{2mE + \frac{\beta_1}{r^2}}\, dr + \int \sqrt{2mK \cos\Theta - \beta_1 - \frac{p_\varphi^2}{\sin^2\Theta}}\, d\Theta + \varphi p_\varphi - Et + C. \quad (20.54)$$

(b) The explicit equations for the motion of the particle follow from the requirement

$$Q_i = \frac{\partial S}{\partial P_i} \iff \alpha_i = \frac{\partial S}{\partial \beta_i},$$

since Q_i, P_i are constants that are denoted by α_i, β_i, and thus,

$$\frac{\partial S}{\partial \beta_1} = \alpha_1, \qquad \frac{\partial S}{\partial E} = \alpha_2, \qquad \frac{\partial S}{\partial p_\varphi} = \alpha_3. \quad (20.55)$$

The α_i follow from the initial conditions; for example,

$$\frac{\partial S}{\partial E} = \int \frac{m}{\sqrt{2mE + \beta_1/r^2}}\, dr - t = \alpha_2, \quad (20.56)$$

$$\int \frac{m}{\sqrt{2mE + \beta_1/r^2}}\, dr = \sqrt{\frac{m}{2E}} \int \frac{r}{\sqrt{r^2 + \beta_1/2mE}}\, dr$$

$$= \sqrt{\frac{mr^2}{2E} + \frac{\beta_1}{4E^2}} + c$$

$$\implies \alpha_2 + t = \sqrt{\frac{mr^2}{2E} + \frac{\beta_1}{4E^2}} + c, \quad (20.57)$$

$$r(t=0) = r_0 \implies \alpha_2 - c = \sqrt{\frac{mr_0^2}{2E} + \frac{\beta_1}{4E^2}},$$

and as the solution for t,

$$t = \sqrt{\frac{mr^2}{2E} + \frac{\beta_1}{4E^2}} - \sqrt{\frac{mr_0^2}{2E} + \frac{\beta_1}{4E^2}}. \quad (20.58)$$

The $\partial S/\partial \beta_1$ and $\partial S/\partial p_\varphi$ are treated likewise. The evaluation of the elliptic integrals requires numerical methods.

We come back once again to our discussion in the context of equations (20.8) and (20.9). If the Hamiltonian does not explicitly depend on the time, as is the case for conservative scleronomic systems, the Hamilton–Jacobi differential equation can be brought to a simpler form, since S can only linearly depend on t. We therefore transform to

$$S = S_0 - Et,$$

where $S_0 = S_0(q-1, \ldots, q_n, \beta_1, \ldots, \beta_n)$. One then obtains the so-called *reduced* Hamilton–Jacobi equation:

$$H\left(q_1, \ldots, q_n, \frac{\partial S_0}{\partial q_1}, \ldots, \frac{\partial S_0}{\partial q_n}\right) = E.$$

The solution of this differential equation yields arbitrary constants, one of them, e.g., β_1, is additive ($S_0 + c$ solves the above Hamilton–Jacobi equation too) and can be omitted. But the reduced Hamilton–Jacobi equation now involves the total energy, so that S_0 will also depend on E, and therefore, in

$$S_0 = S_0(q_1, \ldots, q_n, E, \beta_2, \ldots, \beta_n)$$

β_1 is replaced by E. We can express this in the following way: Just as in the original Hamilton–Jacobi equation, the reduced form also has n integration constants, one of them the total energy E.

Example 20.6: Exercise: Formulation of the Hamilton–Jacobi differential equation for the slant throw

Use the reduced Hamilton–Jacobi differential equation to formulate the equation of motion for the slant throw.

Solution Let the coordinates of the throw plane be x (abcissa) and y (ordinate), which will also be used as generalized coordinates.

$$H = T + V = \frac{m}{2}(\dot{x}^2 + \dot{y}^2) + mgy, \tag{20.59}$$

$$p_x = \frac{\partial H}{\partial \dot{x}} = m\dot{x}, \qquad p_y = \frac{\partial H}{\partial \dot{y}} = m\dot{y}. \tag{20.60}$$

The conjugate momentum $p_x = m\dot{x}$ to the cyclic coordinate x ($\partial H/\partial x = 0$) is a conserved quantity. We recalculate (20.59) as

$$H(x, y, p_x, p_y) = \frac{1}{2m}(p_x^2 + p_y^2) + mgy. \tag{20.61}$$

Since H does not depend explicitly on the time and the system is conservative, the reduced Hamilton–Jacobi differential equation can be applied.

$$\frac{1}{2m}\left[\left(\frac{\partial S_0}{\partial x}\right)^2 + \left(\frac{\partial S_0}{\partial y}\right)^2\right] + mgy = E. \tag{20.62}$$

By inserting the separation *ansatz* $S_0 = S_1(x) + S_2(y)$ into equation (20.62), we obtain

$$\frac{1}{2m}\left[\left(\frac{\partial S_1(x)}{\partial x}\right)^2 + \left(\frac{\partial S_2(y)}{\partial y}\right)^2\right] + mgy = E \tag{20.63}$$

HAMILTON–JACOBI THEORY

or

$$\left(\frac{\partial S_1(x)}{\partial x}\right)^2 = 2mE - 2m^2gy - \left(\frac{\partial S_2(y)}{\partial y}\right)^2. \tag{20.64}$$

This is satisfied only if both sides of the equation are constant, since x and y are independent coordinates.

$$\left(\frac{\partial S_1(x)}{\partial x}\right)^2 \beta_2, \qquad \left(\frac{\partial S_2(y)}{\partial y}\right)^2 = (2mE - \beta_2) - 2m^2gy. \tag{20.65}$$

Integration yields the solutions

$$S_1(x) = \sqrt{\beta_2}x + c_1, \tag{20.66}$$

$$S_2(y) = -\frac{1}{3m^2g}[(2mE - \beta_2) - 2m^2gy]^{3/2} + c_2. \tag{20.67}$$

The complete solution of the reduced Hamilton–Jacobi differential equation is of the form

$$S_0(x, y, E, \beta_2) = \sqrt{\beta_2}x - \frac{1}{3m^2g}[(2mE - \beta_2) - 2m^2gy]^{3/2} + c, \tag{20.68}$$

$$\frac{\partial S_0}{\partial E} = t + \alpha_1, \qquad \frac{\partial S_0}{\partial \beta_2} = \alpha_2,$$

where the first relation holds because

$$\alpha_1 = \frac{\partial S_1}{\partial E} = \frac{\partial S_0}{\partial E} - t.$$

From this, we obtain $y = y(t)$ as

$$-\frac{1}{mg}[(2mE - \beta_2) - 2m^2gy]^{1/2} = t + \alpha_1 \tag{20.69}$$

$$\Leftrightarrow \quad 2mE - \beta_2 - 2m^2gy = m^2g^2(t + \alpha_1)^2$$
$$\Leftrightarrow \quad y = -\frac{1}{2}g(t + \alpha_1)^2 + \frac{2mE - \beta_2}{2m^2g}$$
$$\Leftrightarrow \quad y = -\frac{1}{2}gt^2 + c_1t + c_2. \tag{20.70}$$

In the last step, we renamed the constants.
The analogous procedure with $\partial S/\partial \beta_2 = \alpha_2$ yields

$$y = -c_1x^2 + c_2x + c_3, \tag{20.71}$$

i.e., the familiar throw parabola. For the case of the slant throw the Hamilton–Jacobi equation may appear clumsy for establishing the equation of motion. A certain advantage of the method shows up in complicated problems, e.g., in the Kepler problem in Example 20.3.

Visual interpretation of the action function S

In the preceding problems, the Hamilton–Jacobi differential equation proved successful for establishing the equations of motion, in particular for complex mechanical problems. There remains the question about the visual meaning of the action function S. We consider the motion of a single mass point in a time-independent potential and write

$$S = S_0(q_i, P_i) - Et,$$

where, as already indicated, $S_0(q_i, P_i)$ describes a spatial field which is time-independent. With the labeling $q_1 = x$, $q_2 = y$, $q_3 = z$ and $p_1 = p_x$, $p_2 = p_y$, $p_3 = p_z$, for the momentum components we have according to (19.5)

$$p_x = \frac{\partial S}{\partial x} = \frac{\partial S_0}{\partial x}, \quad p_y = \frac{\partial S}{\partial y} = \frac{\partial S_0}{\partial y}, \quad p_z = \frac{\partial S}{\partial z} = \frac{\partial S_0}{\partial z}.$$

Written as a vector equation, this is

$$\mathbf{p} = \text{grad}\, S = \nabla S_0.$$

Since grad S is always perpendicular to the equipotential surfaces of S, we realize that in a representation of the S-field by $S =$ constant, the orbits are represented by trajectories orthogonal to this set of surfaces. Accordingly, to a given field S belong all motions with trajectories perpendicular to the equipotential surfaces of S ($S =$ constant), and moreover along each trajectory all motions starting at an arbitrary moment (see Figure 20.2).

The time behavior of the S-field can be seen from the representation $S = S_0 - Et$. For $t = 0$, the surfaces $S(q_i, P_i) = 0$ and $S_0(q_i, P_i) = 0$ are identical. For $t = 1$, the surface $S = 0$ coincides with the surface $S_0 = E$, $S = E$ with $S_0 = 2E$, etc. This means graphically that surfaces of constant S-values move across surfaces of constant S_0-values, i.e., that surfaces of constant S move through space. The formal meaning of S follows from the action integral. One has

$$\int L\, dt = \int (p_x\, dx + p_y\, dy + p_z\, dz - H\, dt),$$

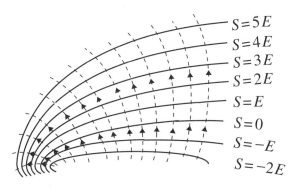

Figure 20.2. Surfaces $S =$ constant. Trajectories are dotted.

VISUAL INTERPRETATION OF THE ACTION FUNCTION S

$$\int_{t_1}^{t_2} L\, dt = \int_{t_1}^{t_2} \left(\frac{\partial S}{\partial x} dx + \frac{\partial S}{\partial y} dy + \frac{\partial S}{\partial z} dz + \frac{\partial S}{\partial t} dt\right) = S_2 - S_1.$$

S therefore represents an action (energy · time). It is the time integral over the Lagrangian. The Hamilton principle $\delta \int L\, dt = 0$ therefore states that a motion proceeds with the boundary condition of minimum action.

Example 20.7: Illustration of the action waves

To illustrate the action waves, we consider the throw or fall motion in the gravitational field of the earth, where the equation of motion is well known. In analogy to Example 20.6, we obtain the following Hamilton–Jacobi differential equation:

$$\frac{1}{2m}\left\{\left(\frac{\partial S}{\partial x}\right)^2 + \left(\frac{\partial S}{\partial y}\right)^2 + \left(\frac{\partial S}{\partial z}\right)^2\right\} + mgz + \frac{\partial S}{\partial t} = 0. \tag{20.72}$$

With the separation *ansatz* $S = S_x(x) + S_y(y) + S_z(z) - Et$, we obtain

$$S_x = x\beta_x, \qquad S_y = y\beta_y$$

up to additive constants, and

$$\frac{1}{2m}\left(\frac{\partial S_z}{\partial z}\right)^2 + mgz = E - \frac{\beta_x^2 + \beta_y^2}{2m} = \beta_z. \tag{20.73}$$

The quantities β_x and β_y are separation constants, just like β_z. Integration over z yields, up to a constant,

$$S_z = -\frac{2}{3g}\sqrt{\frac{2}{m}}(\beta_z - mgz)^{3/2}. \tag{20.74}$$

We write the constant β_z as $\beta_z = mgz_0$ and thereby can express the total energy as

$$E = \frac{p_x^2 + p_y^2}{2m} + mgz_0. \tag{20.75}$$

By insertion, one gets the action function

$$S = x\beta_x + y\beta_y - \frac{2m\sqrt{2g}}{3}(z_0 - z)^{3/2} - \left(\frac{\beta_x^2 + \beta_y^2}{2m} + mgz_0\right)t, \tag{20.76}$$

and by the familiar scheme the equations of motion,

$$\begin{aligned}
Q_x = \alpha_x &= \frac{\partial S}{\partial \beta_x} = x - \frac{\beta_x}{m}t, \\
Q_y = \alpha_y &= \frac{\partial S}{\partial \beta_y} = y - \frac{\beta_y}{m}t, \\
Q_z = \alpha_z &= \frac{\partial S}{\partial z_0} = -m\sqrt{2g}(z_0 - z)^{1/2} - mgt.
\end{aligned} \tag{20.77}$$

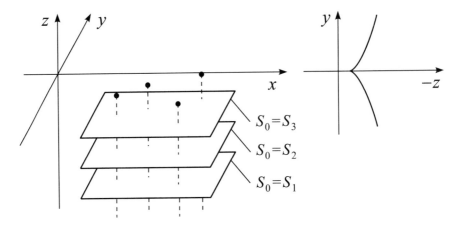

Figure 20.3.

Among the possible motions of a body in the gravitational field we pick the ensemble with $\beta_x = 0$, $\beta_y = 0$, $z_0 = 0$. From equation (20.76), we get for the surfaces with $S_0 =$ constant:

$$\text{constant} = -\frac{2m\sqrt{2g}}{3}(-z)^{3/2}.$$

These are planes parallel to the x,y-plane.

The possible trajectories are shown in the Figure 20.3 by dashed vertical straight lines. Since the action function is real only for $z \leq 0$, there are only such throw conditions that ascend up to the plane $z = 0$ and then return again. In the present example, the action waves are planes parallel to the x,y-plane that propagate in z-direction. This is easily seen from equation (20.76) for $z_0 =$ constant $\neq 0$. Any vertical throw up to the height z_0 thus belongs to the same S-field or to the same action wave, respectively. Here it is not essential at which space point the throw motion begins.

As a further ensemble of motions, we consider

$$\beta_x = 0, \quad z_0 = 0, \quad \beta_y = \frac{2m\sqrt{2g}}{3}, \tag{20.78}$$

so that from equation (20.76) we obtain

$$S_0 = \frac{2m\sqrt{2g}}{3}\{y - (-z)^{3/2}\} \tag{20.79}$$

$$\Leftrightarrow \quad y = \frac{3}{2m\sqrt{2g}} S_0 + (-z)^{3/2}. \tag{20.80}$$

Equation (20.80) represents a Neil or semicubic parabola ($y = ax^{3/2}$) in the y,z-plane. The surfaces with $S_0 =$ constant are therefore surfaces parallel to the x-axis: $F(y, z) = 0$, i.e., cylindrical surfaces intersecting the y,z-plane in a set of Neil parabolas with the top on the y-axis (Figure 20.4).

With increasing S_0 the tops of the Neil parabolas move in the y-direction. The related trajectories are throw parabolas in the y, z-plane which have no velocity component along the x-direction and reach their highest point at $z = 0$ (dashed curves in the Figure 20.5).

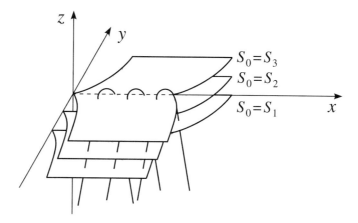

Figure 20.4.

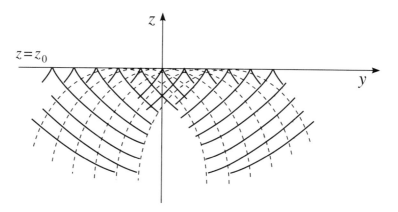

Figure 20.5. Projection onto the y, z-plane of the preceding Figure 20.4.

The velocity component in y-direction is the same for all throws:

$$v_y = \frac{2}{3}\sqrt{2g}.$$

In this case, the action waves consist of cylindrical surfaces parallel to the x-axis that propagate with increasing time in z-direction.

The starting point of the motion in the x,y-plane is again arbitrary; the turning points of the trajectories are at $z = 0$. All throw parabolas parallel to the x-axis belong to the same action wave; i.e., any throw described by equation (20.79) can be represented by a set of action waves propagating in the z-direction and parallel to the x-axis.

We see from this example that the simple throw in the gravitational field can be correctly represented by the Hamilton–Jacobi formalism but becomes hopelessly complicated. This confirms our thesis:

Although the Hamilton–Jacobi method contains beautiful formal ideas, it is hardly practicable, too clumsy, and too abstract for physicists.

Example 20.8: Periodic and multiply periodic motions

In this example, the peculiarities of periodic motions shall be compiled and extended to multiply periodic motions.[2]

1. Periodic motions: Here, one distinguishes two kinds, namely, the *properly periodic motion*, for which

$$q_i(t + \tau) = q_i(t),$$
$$p_i(t + \tau) = p_i(t),$$
(20.81)

i.e., both the coordinates and the momenta have *the same period* τ. This motion is also called *libration*. Two-dimensional examples are the (nondamped) harmonic oscillator or the (nondamped) vibrating pendulum. The phase-space diagram (the *phase trajectory*) is a closed curve (see Figure 20.6).

The other type of periodic motion is the *rotation*. Here one has (e.g., in the two-dimensional case)

$$p(q + q_0) = p(q),$$
(20.82)

i.e., the momentum takes for $q + q_0$ the same value as for q. The coordinate q is mostly an angle variable and $q_0 = 2\pi$. One might imagine for example a circulating pendulum; in this case q is the pendulum angle. The phase-space trajectory is then not closed but periodic with the period q_0 (see Figure 20.7).

The limiting case between rotation and libration is called *limitation motion*. The pendulum, which is almost circulating, is an example for this type of motion. The coordinate period q_0 is then $q_0 = 2\pi$ as before, but the time period is $\tau = \infty$. (The pendulum then comes to rest in the upper vertical position (unstable point), i.e., the function graph terminates at the point q_0). If the system is conservative and is described by the Hamiltonian $H(p, q)$, we have the equations

$$H(p, q) = E,$$
$$H\left(q, \frac{\partial S}{\partial q}\right) = E.$$
(20.83)

The first equation yields $p = p(q, E)$, i.e., for a given energy E the phase trajectory. The second equation is the (reduced) Hamilton–Jacobi equation from which the action function (generating

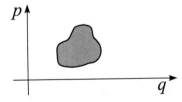

Figure 20.6. Two-dimensional phase diagram of a properly periodic motion. A closed phase trajectory occurs, e.g., for a nondamped vibrating pendulum.

---

[2] Here, we follow A. Budo, *Theoretische Mechanik*, Deutscher Verlag der Wissenschaften, Berlin (1956).

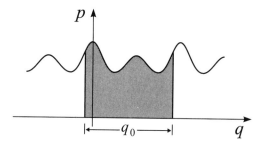

Figure 20.7. Phase-space diagram of the rotation as a periodic motion. The trajectory is open but has the period q_0. In other words, the momentum p is a periodic function of the coordinate q with the period q_0.

function) $F_2(q, P) = S(q, E)$ can be calculated. If that is done, one can calculate the *phase-space integral*

$$J = \oint p\, dq = \oint \frac{\partial S}{\partial q}\, dq. \tag{20.84}$$

Here, $\oint$ means the integration over a closed trajectory in the case of libration, or over a full period $q_1 \leq q \leq q_1 + q_0$ in the case of rotation. Hence, the phase integral J exactly corresponds to the shaded areas in Figures 20.6 and 20.7.

The phase integral $J = J(E)$ depends only on E and is constant in time, since the total energy is constant in time. Hence, equation (20.84) leads to the relations

$$J = J(E) \quad \text{or} \quad E = E(J). \tag{20.85}$$

As a consequence, the function $S(q, E)$ changes to

$$S(q, E) \quad \Rightarrow \quad S(q, E(J)) \equiv S'(q, J). \tag{20.86}$$

The function $S'(q, J)$ can serve as the generator of a canonical transformation. The new momentum is now identified with J, i.e.,

$$P = J. \tag{20.87}$$

The canonically conjugate variable belonging to $P = J$ is denoted by $Q = \varphi$. It is also called an angle variable and is calculated according to equation (20.4) as

$$\varphi = \frac{\partial S'(q, J)}{\partial J}. \tag{20.88}$$

According to equation (20.83), the Hamiltonian is

$$\mathcal{H}(\varphi, J) = E(J). \tag{20.89}$$

The Hamilton equations in the new coordinates then read

$$\begin{aligned}\dot{Q} &= \frac{\partial \mathcal{H}}{\partial P} \quad \text{or} \quad \dot{\varphi} = \frac{\partial E(J)}{\partial J} = \text{constant}, \\ \dot{P} &= -\frac{\partial \mathcal{H}}{\partial Q} \quad \text{or} \quad \dot{J} = 0.\end{aligned} \tag{20.90}$$

$\partial E(J)/\partial J$ depends only on J, which is constant in time. Hence, $\dot{\varphi}$ is also constant in time, and then

$$\varphi = \left(\frac{\partial E}{\partial J}\right) t + \delta. \tag{20.91}$$

Here, the phase constant δ appears. If we had not selected $S'(q, J)$ as the generating function but rather the complete time-dependent action function

$$W(q, J, t) = S'(q, J) - E(J)t, \tag{20.92}$$

the coordinate conjugate to J would be according to (20.88)

$$\frac{\partial W(q, J)}{\partial J} = \frac{\partial S'(q, J)}{\partial J} - \frac{\partial E(J)}{\partial J} t$$

$$= \varphi - \frac{\partial E(J)}{\partial J} t = \delta, \tag{20.93}$$

i.e., just the phase constant from (20.91). Equation (20.91) states that the angle variable φ linearly increases with the time. It is a cyclic coordinate as is evident from (20.89), since the Hamiltonian $\mathcal{H}(\varphi, J) = E(J)$ does not depend on φ. The change of φ during a period τ is found from (20.91) to be

$$\Delta\varphi = \left(\frac{\partial E}{\partial J}\right) \tau, \tag{20.94}$$

which can be specified more precisely by means of (20.88). We have

$$\Delta\varphi = \oint \frac{\partial \varphi}{\partial q} dq = \oint \frac{\partial^2 S'(q, J)}{\partial q \partial J} dq$$

$$= \frac{\partial}{\partial J} \oint \frac{\partial S'(q, J)}{\partial q} dq = \frac{\partial J}{\partial J} = 1. \tag{20.95}$$

Hence, the angular coordinate increases during a period after which the system returns to its initial configuration, exactly by 1. We therefore can state that the motion of the system is periodic in φ with the period 1. Combining (20.94) and (20.95) yields

$$\tau \frac{\partial E}{\partial J} = 1 \quad \Leftrightarrow \quad \frac{\partial E}{\partial J} = \frac{1}{\tau} = \nu. \tag{20.96}$$

ν is the *frequency* of the periodic motion. Obviously the complete solution of the equations of motion is not needed for calculating ν. It is sufficient to express E as a function of J and to differentiate with respect to J. This is the advantage of introducing the action (J) and angle variables (φ). The approach is illustrated in Example 20.2 for the case of the harmonic oscillator.

2. Separable multiply periodic systems: We imagine a conservative system with f degrees of freedom, which is described by the f coordinates $q_1, \ldots, q_f$ and is separable. This means that the solution of the reduced Hamilton–Jacobi equation

$$H\left(q_1, \ldots, q_f; \frac{\partial S}{\partial q_1}, \ldots, \frac{\partial S}{\partial q_f}\right) = E \tag{20.97}$$

can be written in the form

$$S(q_1, \ldots, q_f; E, \beta_2, \ldots, \beta_f) = S_1(q_1; E, \beta_2, \ldots, \beta_f) + \cdots + S_f(q_f; E, \beta_2, \ldots, \beta_f). \tag{20.98}$$

The f integration constants

$$E, \beta_2, \beta_3, \ldots, \beta_f \tag{20.99}$$

characterize the constant momenta $P_1, \ldots, P_f$. If the Hamiltonian decomposes into a sum of terms $H(q_i, p_i)$, only one constant appears in the functions $S_k(q_k; E, \beta_2, \ldots, \beta_f)$ in (20.98); the functions then have the form $S_k(q_k, \beta_k)$ – see (20.10). When can we classify a motion as periodic? The answer is simple: If *any* pair of conjugate variables (q_i, p_i) always behaves as discussed in the first part of this example, the motion is periodic. More precisely: The projection of the phase trajectory onto *each* q_i, p_i-plane of the phase space must be either a libration or a rotation, to guarantee the periodicity of the entire motion of the system.

The procedure is analogous to that outlined in the first section. First one defines the action variables

$$J_i = \oint p_i\, dq_i = \oint \frac{\partial S_i(q_i; E, \beta_2, \ldots, \beta_f)}{\partial q_i}\, dq_i$$
$$= J_i(E, \beta_2, \ldots, \beta_f), \qquad i = 1, \ldots, f. \tag{20.100}$$

They are constant in time, since $E, \beta_2, \ldots, \beta_f$ are constant. The f equations (20.100) can be solved for $E, \beta_2, \ldots, \beta_f$ and yield

$$E = E(J_1, \ldots, J_f),$$
$$\beta_2 = \beta_2(J_1, \ldots, J_f),$$
$$\vdots \tag{20.101}$$
$$\beta_f = \beta_f(J_1, \ldots, J_f).$$

By inserting (20.101) into (20.98), we obtain

$$S_i(q_i, E(J_k), \beta_2(J_k), \ldots, \beta_f(J_k)) = S'(q_i, J_1, \ldots, J_f). \tag{20.102}$$

This is a generating function with the constant momenta

$$P_i = J_i. \tag{20.103}$$

The relation (20.102) is fully analogous to the relation (20.86), and (20.103) corresponds to (20.87). The canonically conjugate angle variables result—like (20.88)—from

$$\varphi_i = \frac{\partial S'}{\partial J_i} = \sum_{k=1}^{f} \frac{\partial S'_k(q_k, J_1, \ldots, J_f)}{\partial J_i}, \qquad i = 1, \ldots, f. \tag{20.104}$$

Among the canonical variables $(Q_i, P_i) = (\varphi_i, J_i)$ is the Hamiltonian

$$\mathcal{H}(\varphi_i, J_i) = E(J_i), \tag{20.105}$$

since the Hamiltonian is independent of time (see equation (20.83)). From this follow the Hamilton equations

$$\dot{\varphi}_i = \frac{\partial \mathcal{H}}{\partial P_i} = \frac{\partial E(J_k)}{\partial J_i} = \text{constant} \equiv \nu_i,$$
$$\dot{J}_i = -\frac{\partial \mathcal{H}}{\partial \varphi_i} = 0; \tag{20.106}$$

hence,

$$\varphi_i = \nu_i t + \delta_i,$$
$$J_i = \text{constant}. \tag{20.107}$$

We are now interested in the change of the angle variables φ_i over a period (full revolution or back-and-forth motion of a coordinate q_i with the remaining coordinates kept fixed). It is given by

$$\Delta_k \varphi_i = \oint \frac{\partial \varphi_i}{\partial q_k} dq_k = \oint \frac{\partial^2 S'}{\partial J_i \partial q_k} dq_k$$

$$= \frac{\partial}{\partial J_i} \oint \frac{\partial S'}{\partial q_k} dq_k = \frac{\partial J_k}{\partial J_i} = \delta_{ki}. \qquad (20.108)$$

According to (20.107),

$$\Delta_k \varphi_k = \nu_k \tau_\kappa, \qquad (20.109)$$

if τ_k is the "vibration time" (time interval of the period) of q_k. A comparison of (20.109) and (20.108) yields

$$\nu_k \tau_k = 1. \qquad (20.110)$$

Thus,

$$\nu_k = \frac{1}{\tau_k} \qquad (20.111)$$

obviously are the frequencies of the q_k-motion. In other words, according to (20.106) the (fundamental) frequency ν_k for the coordinate q_k is $\nu_k = \partial E(J_1, \ldots, J_f)/\partial J_k$.

The equations (20.104) can also be inverted, which yields the original coordinates q_n with

$$q_k = q_k(\varphi_1, \ldots, \varphi_f), \qquad k = 1, \ldots, f \qquad (20.112)$$

as functions of the new angle variable φ_i. When increasing φ_i by $\Delta \varphi_i = 1$ (keeping the values of all other φ_k with $k \neq i$ fixed), q_i (and only this!) must run through a period. This follows from equation (20.108): If q_k (with $k \neq i$) ran through a period when φ_i changes to $\varphi_i + \Delta \varphi_i = \varphi_i + 1$, then according to (20.108) the variable φ_k also should increase by $\Delta \varphi_k = 1$. But this shall not occur by assumption. Therefore if φ_i increases to $\varphi_i + 1$, q_i changes as follows:

$$\begin{aligned} \varphi_i &\to \varphi_i + 1, \\ q_i &\to q_i \qquad \text{for libration,} \\ q_i &\to q_i + q_{i0} \qquad \text{for rotation.} \end{aligned} \qquad (20.113)$$

For a libration, q_i is periodic; for a rotation,

$$q_i - \varphi_i q_{i0} \qquad (20.114)$$

is a periodic function of φ_i. Actually,

$$\begin{aligned} \varphi_i &\to \varphi_i + 1, \\ q_i - \varphi_i q_{i0} &\to q_i + q_{i0} - (\varphi_i + 1)q_{i0} = q_i - \varphi_i q_{i0}. \end{aligned} \qquad (20.115)$$

We therefore can expand the separation coordinates q_i (for libration) or $q_i - \varphi_i q_{i0}$ (for rotations) in a Fourier series and write

$$\left. \begin{aligned} q_i(\varphi_1(t), \ldots, \varphi_f(t)) \\ q_i - \varphi_i q_{i0}(\varphi_1(t), \ldots, \varphi_f(t)) \end{aligned} \right\} = \sum_{n=-\infty}^{+\infty} a_n^{(i)} e^{i 2\pi \varphi_i n}$$

VISUAL INTERPRETATION OF THE ACTION FUNCTION S

$$= \sum_{n=-\infty}^{+\infty} a_n^{(i)} e^{i2\pi n(\nu_i t + \delta_i)}, \qquad (20.116)$$

where

$$a_n^{(i)}(\varphi_1, \ldots, \varphi_{i-1}, \varphi_{i+1}, \ldots, \varphi_f) = \int_0^1 q_i(\varphi_1, \ldots, \varphi_f) e^{-i2\pi n \varphi_i} \, d\varphi_i. \qquad (20.117)$$

The Fourier coefficients $a_n^{(i)}(\varphi_1, \ldots, \varphi_{i-1}, \varphi_{i+1}, \ldots, \varphi_f)$ in general still depend on all angle variables, except for φ_i.

We now imagine other variables x_l which describe the system and are useful for certain problems. They shall unambiguously depend on the $q_i(t)$ and therefore are also functions of the time. Then we can write

$$x_l(q_1(t), \ldots, q_f(t))$$

$$= \sum_{n_1=-\infty}^{+\infty} \cdots \sum_{n_f=-\infty}^{+\infty} A_{n_1,\ldots,n_f}^{(l)} e^{i2\pi(n_1\varphi_1 + \cdots + n_f\varphi_f)}$$

$$= \sum_{n_1,\ldots,n_f=-\infty}^{\infty} A_{n_1,\ldots,n_f}^{(l)} e^{i2\pi[(n_1\nu_1 + \cdots + n_f\nu_f)t + (\delta_{n_1} + \cdots + \delta_{n_f})]}, \quad l = 1, \ldots, f. \qquad (20.118)$$

In the second step, we used $\varphi_i = \nu_i t + \delta_i$. The coordinates x_l can be represented only by a multiple Fourier series. Equation (20.118) now suggests that the motion $x_l(t)$ is *in general not periodic* in time. For example, if t increases by $\Delta t = 1/\nu_1$, the first exponential factor in (20.118) does not change because $e^{i2\pi n_1 \nu_1 \Delta t} = e^{i2\pi n_1 \nu_1 (1/\nu_1)} = e^{i2\pi n_1} = 1$, but the other exponential factors in (20.118) vary. The system is therefore called *multiply periodic* in the coordinates x_l. If the system is *simply periodic*, the frequencies $\nu_1, \ldots, \nu_f$ must be correlated by $f - 1$ equations of the type

$$C_{i1}\nu_1 + C_{i2}\nu_2 + \cdots + C_{if}\nu_f = 0, \qquad i = 1, \ldots, f - 1. \qquad (20.119)$$

These are $f - 1$ equations for the f unknown quantities $\nu_1, \ldots, \nu_f$. It is evident that besides ν_i, also $\nu\nu_i$ (ν an arbitrary factor) is also a solution of (20.119). Let $\nu_i = n_i/m_i$; thus, the ν_i can be represented by the fraction n_i/m_i. Then

$$\nu_i' = (m_1 m_2 \cdots m_f)\nu_i = (m_1 \cdots m_f)\frac{n_i}{m_i} \qquad (20.120)$$

are also solutions of (20.119). Hence, the ν_i' are integers. Since the solutions of (20.119) can be determined only up to a common factor ν, the general solution reads

$$\nu_i = a_i \nu, \qquad (20.121)$$

where the $a_i = (m_1 m_2 \cdots m_f) n_i / m_i$ are integers, and ν is a common factor. Thus, the system is periodic if and only if all frequencies are commensurable. The fundamental frequency ν_0 is then the largest common divisor of all frequencies $\nu_1, \ldots, \nu_f$. If there exist only s (with $s \leq f - 1$) relations of the form (20.119), s frequencies can be rationally expressed by the remaining ones. The system (the motion) is then called *s-fold degenerate* or *$(f - s)$-fold periodic*. Special cases are

$s = 0$: the motion is f-fold periodic or nondegenerate;

$s = f - 1$: the motion is single-periodic or fully degenerate.

Transition to quantum mechanics

In the last chapters, we have emphasized the formal aspects of mechanics. Although for solving practical problems sometimes no advantages could be achieved, the insights in the structure of mechanics provided by the Hamiltonian formalism contributed essentially to the development of quantum mechanics. For example, the concept of the phase integral was of fundamental importance for the transition to quantum mechanics. The first clear formulation of the *quantum hypothesis* consisted of the requirement that the *phase integral take only discrete values*; hence,

$$J = \oint p\, dq = nh, \qquad n = 1, 2, 3, \ldots, \tag{20.122}$$

where h is Planck's action quantum, which has the value $h = 6.6 \cdot 10^{-34}\, J \cdot s$. We again consider the case of the harmonic oscillator. In Example 20.2, we evaluated the phase integral

$$J = 2\pi E \sqrt{\frac{m}{k}}. \tag{20.123}$$

$\nu = (1/2\pi)\sqrt{k/m}$ was the frequency. With the quantum hypothesis, we then obtain

$$E_n = nh\nu. \tag{20.124}$$

Thus, the quantum hypothesis leads to the conclusion that the vibrating mass point can take only discrete energy values E_n. For the motion, this means that only certain trajectories in the phase space are allowed. We therefore get ellipses for the phase-space trajectories (compare Example 20.2), whose areas (the phase integral) always differ by the amount h. In this way, the phase space acquires a grid structure that is defined by the allowed trajectories.

Each trajectory corresponds to an energy E_n. In a transition between two trajectories the mass point receives (or releases) the energy $E_n - E_m = (n-m)h\nu$. The smallest transferred amount of energy is given by $h\nu$.

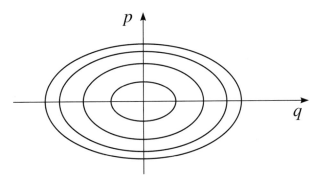

Figure 20.8. In quantum mechanics, the phase-space trajectories of the harmonic oscillator are ellipses that differ by an area of h.

TRANSITION TO QUANTUM MECHANICS

Since the action quantum h is so small, the discrete structure of the phase space is significant only for atomic processes. For macroscopic processes the trajectories in the phase space are so dense that one can consider the phase space as a continuum. The energy quanta $h\nu$ are so small that they have no meaning for macroscopic processes. For example, the energy emitted in a transition in the hydrogen atom is $h\nu = 13.6\,\text{eV}$ (electron volt). Expressed in the (macroscopic) unit of Watt seconds $h\nu = 2 \cdot 10^{-18}\,\text{Ws}$. The quantum hypothesis was confirmed by the explanation of the spectra of radiating atoms.

Example 20.9: Exercise: The Bohr–Sommerfeld hydrogen atom

At the beginning of the development of modern quantum mechanics, N. Bohr and A. Sommerfeld formulated a "quantization prescription" for periodic motions. Accordingly, only such trajectories in phase space are admitted for which the phase integral

$$\oint p_\alpha \, dq_\alpha = n_\alpha h, \qquad n_\alpha = 0, 1, 2, \ldots \tag{20.125}$$

is a multiple of Planck's action quantum $h = 6.626 \cdot 10^{-34}\,\text{J s}$. The integral extends over a period of motion. q_α and p_α are the generalized coordinates and the canonically conjugate momenta, respectively.

(a) Write the Lagrangian, the Hamiltonian, the Hamilton equations, and the constants of motion for a particle in the potential $v(r) = -e^2/r$.

(b) Calculate the bound energy states of the hydrogen atom from the condition (20.125).

Solution (a) The Lagrangian is $L = T - V = (1/2)mv^2 + e^2/r$. The Hamiltonian then follows as

$$H = \sum_\alpha \dot{x}_\alpha \frac{\partial L}{\partial \dot{x}_\alpha} - L = \frac{1}{2}mv^2 - \frac{e^2}{r} = \frac{1}{2}m\dot{r}^2 + \frac{1}{2}mr^2\dot{\varphi}^2 - \frac{e^2}{r} \tag{20.126}$$

in polar coordinates. The canonical momenta are

$$p_\varphi = \frac{\partial L}{\partial \dot{\varphi}} = mr^2\dot{\varphi} = \mathcal{L} \quad \text{and} \quad p_r = \frac{\partial L}{\partial \dot{r}} = m\dot{r}. \tag{20.127}$$

$\mathcal{L}$ is the angular momentum of the particle.
Then

$$H(p, q) = \frac{p_r^2}{2m} + \frac{p_\varphi^2}{2mr^2} - \frac{e^2}{r}. \tag{20.128}$$

Constants of motion are

(i) $H = E$, since $H(q, p)$ does not explicitly depend on the time; and

(ii) $p_\varphi = \mathcal{L}$, since φ is a cyclic variable.

$\mathcal{L}$ represents the constant angular momentum. The Hamilton equations read

$$\dot{p}_\varphi = -\frac{\partial H}{\partial \varphi} = 0, \qquad \dot{\varphi} = \frac{\partial H}{\partial p_\varphi} = \frac{p_\varphi}{mr^2},$$
$$\dot{p}_r = -\frac{\partial H}{\partial r} = -\frac{e^2}{r} + \frac{p_\varphi^2}{mr^3}, \qquad \dot{r} = \frac{\partial H}{\partial p_r} = \frac{p_r}{m}. \qquad (20.129)$$

(b) The quantization conditions for the *angular motion* are

$$lh = \oint p_\varphi \, d\varphi = \int_0^{2\pi} \mathcal{L} \, d\varphi = 2\pi \mathcal{L}$$

$$\Rightarrow \quad \mathcal{L} = l\hbar, \qquad \hbar = \frac{h}{2\pi}, \qquad l = 0, 1, 2, \ldots, \qquad (20.130)$$

i.e., the orbital angular momentum can take only integer multiples of $\hbar$. For the *radial motion*, the phase integral equals

$$kh = \oint p_r \, dr = 2 \int_{r_{\min}}^{r_{\max}} \sqrt{2m\left(E + \frac{e^2}{r}\right) - \frac{\mathcal{L}^2}{r^2}} \, dr, \qquad k = 0, 1, 2, \ldots. \qquad (20.131)$$

The limits of integration are determined from the condition

$$p_r = 0; \qquad (20.132)$$

thus,

$$r_m^2 + \frac{e^2}{E} r_m - \frac{\mathcal{L}^2}{2mE} = 0,$$

$$r_m = -\frac{e^2}{2E} \mp \frac{\sqrt{-\Delta}}{4mE} \quad \text{with} \quad \Delta = -4m(2E\mathcal{L}^2 + me^4). \qquad (20.133)$$

The integral in (20.131) is of the type

$$\int \frac{\sqrt{ar^2 + br + c}}{r} \, dr \equiv \int \frac{\sqrt{X(r)}}{r} \, dr, \qquad (20.134)$$

and one easily verifies by differentiation that

$$\int \frac{\sqrt{X(r)}}{r} \, dr = \sqrt{X(r)} + \frac{b}{2} \int \frac{dr}{\sqrt{X(r)}} + c \int \frac{dr}{r\sqrt{X(r)}}. \qquad (20.135)$$

Here, $X(r) = ar^2 + br + c$. Furthermore,

$$\int \frac{dr}{\sqrt{X(r)}} = -\frac{1}{\sqrt{-a}} \arcsin\left(\frac{2ar + b}{\sqrt{-\Delta}}\right) \qquad \text{for } a < 0 \text{ (since } E < 0\text{)}$$

and

$$\int \frac{dr}{r\sqrt{X(r)}} = \frac{1}{\sqrt{-c}} \arcsin\left(\frac{br + 2c}{r\sqrt{-\Delta}}\right) \qquad \text{for } c < 0 \text{ and } \Delta < 0. \qquad (20.136)$$

Here,
$$\Delta = 4ac - b^2. \tag{20.137}$$

This leads to
$$\int \frac{\sqrt{X(r)}}{r} dr = \sqrt{ar^2 + br + c} - \frac{b}{2\sqrt{-a}} \arcsin\left(\frac{2ar + 2b}{\sqrt{-\Delta}}\right)$$
$$+ \frac{c}{\sqrt{-c}} \arcsin\left(\frac{br + 2c}{r\sqrt{-\Delta}}\right) \tag{20.138}$$

for $a < 0$, $c < 0$, $\Delta < 0$. In our case (see (20.131)),
$$a = 2mE, \quad b = mc^2, \quad c = -\mathcal{L}^2,$$
$$\Delta = -4m(2E\mathcal{L}^2 + me^4). \tag{20.139}$$

For the integral, one gets (with $E < 0$)
$$\int \frac{\sqrt{2mEr^2 + 2me^2r - \mathcal{L}^2}}{r} dr$$
$$= \sqrt{2mEr^2 + 2me^2r - \mathcal{L}^2}$$
$$+ \frac{me^2}{\sqrt{-2mE}} \arcsin\left(\frac{4mEr + 2me^2}{\sqrt{-\Delta}}\right) - \mathcal{L} \arcsin\left(\frac{2me^2r - 2\mathcal{L}^2}{r\sqrt{-\Delta}}\right). \tag{20.140}$$

Insertion of the integration limits yields
$$\sqrt{\frac{2\pi^2 me^4}{-E}} - 2\pi \mathcal{L} = kh. \tag{20.141}$$

If one defines the "principal quantum number" $n = l + k = 0, 1, 2, \ldots$, the formula for the binding energy reads
$$E_n = -\frac{me^4}{2\hbar^2 n^2}. \tag{20.142}$$

This formula for the discrete energy levels in the hydrogen atom agrees exactly with the quantum mechanical result. Only the value $n = 0$, which was allowed in this consideration, is excluded in the quantum mechanical approach. The underlying classical picture (electron moves in an elliptic orbit with the eccentricity $\varepsilon = \sqrt{1 - (l/n)^2}$) leads however to contradictions and must be modified in quantum mechanics. Because $n = l + k$, the energy levels with $n = 1, 2, \ldots$, are twofold, threefold, $\ldots$, degenerate.

Example 20.10: Exercise: On Poisson brackets

If the functions F and G depend on the coordinates q_α, the momenta p_α, and the time t, the *Poisson bracket* of F and G is defined as follows:
$$[F, G] = \sum_\alpha \left(\frac{\partial F}{\partial q_\alpha} \frac{\partial G}{\partial p_\alpha} - \frac{\partial F}{\partial p_\alpha} \frac{\partial G}{\partial q_\alpha} \right).$$

Show the subsequent properties of this Poisson bracket:

(a) $[F, G] = -[G, F]$,

(b) $[F_1 + F_2, G] = [F_1, G] + [F_2, G]$,

(c) $[F, q_r] = -\dfrac{\partial F}{\partial p_r}$, and

(d) $[F, p_r] = \dfrac{\partial F}{\partial q_r}$.

Solution (a)

$$[F, G] = \sum_\alpha \left(\frac{\partial F}{\partial q_\alpha} \frac{\partial G}{\partial p_\alpha} - \frac{\partial F}{\partial p_\alpha} \frac{\partial G}{\partial q_\alpha} \right) = -\sum_\alpha \left(\frac{\partial G}{\partial q_\alpha} \frac{\partial F}{\partial p_\alpha} - \frac{\partial G}{\partial p_\alpha} \frac{\partial F}{\partial q_\alpha} \right)$$
$$= -[G, F].$$

We see that the *Poisson bracket is not commutative*.

(b)

$$[F_1 + F_2, G] = \sum_\alpha \left(\frac{\partial (F_1 + F_2)}{\partial q_\alpha} \frac{\partial G}{\partial p_\alpha} - \frac{\partial (F_1 + F_2)}{\partial p_\alpha} \frac{\partial G}{\partial q_\alpha} \right)$$
$$= \sum_\alpha \left(\frac{\partial F_1}{\partial q_\alpha} \frac{\partial G}{\partial p_\alpha} - \frac{\partial F_1}{\partial p_\alpha} \frac{\partial G}{\partial q_\alpha} \right) + \sum_\alpha \left(\frac{\partial F_2}{\partial q_\alpha} \frac{\partial G}{\partial p_\alpha} - \frac{\partial F_2}{\partial p_\alpha} \frac{\partial G}{\partial q_\alpha} \right)$$
$$= [F_1, G] + [F_2, G].$$

Therefore, *the Poisson bracket is distributive*.

(c)

$$[F, q_r] = \sum_\alpha \left(\frac{\partial F}{\partial q_\alpha} \frac{\partial q_r}{\partial p_\alpha} - \frac{\partial F}{\partial p_\alpha} \frac{\partial q_r}{\partial q_\alpha} \right)$$

$$\frac{\partial q_r}{\partial p_\alpha} = 0 \quad \Rightarrow \quad [F, q_r] = \sum_\alpha \left(-\frac{\partial F}{\partial p_\alpha} \delta_{r\alpha} \right) = -\frac{\partial F}{\partial p_r}.$$

Here, $\delta_{r\alpha}$ is the Kronecker symbol:

$\delta_{r\alpha} = 1$ for $r = \alpha$,

$\delta_{r\alpha} = 0$ for $r \neq \alpha$.

(d) Analogously, we find

$$[F, p_r] = \sum_\alpha \left(\frac{\partial F}{\partial q_\alpha} \frac{\partial p_r}{\partial p_\alpha} - \frac{\partial F}{\partial p_\alpha} \frac{\partial p_r}{\partial q_\alpha} \right) = \sum_\alpha \left(\frac{\partial F}{\partial q_\alpha} \delta_{r\alpha} \right) = \frac{\partial F}{\partial q_r},$$

since $\partial p_r / \partial q_\alpha = 0$.

We shall meet the rules on Poisson brackets in quantum mechanics again, since the transition to quantum mechanics (the so-called canonical quantization) is performed by the transition to operators and by replacing the Poisson bracket $[\ ,\ ]$ by the commutator $(1/i\hbar)\{\ ,\ \}$, where

$$\{A, B\}_- = AB - BA.$$

TRANSITION TO QUANTUM MECHANICS

If we form, e.g., $[q_i, p_j]$, we obtain

$$[q_i, p_j] = \sum_\alpha \left(\frac{\partial q_i}{\partial q_\alpha} \frac{\partial p_j}{\partial p_\alpha} - \frac{\partial q_i}{\partial p_\alpha} \frac{\partial p_j}{\partial q_\alpha} \right) = \delta_{ij}. \tag{20.143}$$

In the canonical quantization, one passes from the classical momenta p_j to operator momenta $\widehat{p}_j$, and from the classical Poisson bracket [,] to the quantum mechanical Poisson bracket $(1/i\hbar)\{ , \}_-$. Thus, in the canonical quantization one substitutes the relation (20.143) by

$$\{q_i, \widehat{p}_j\}_- = i\hbar \delta_{ij}. \tag{20.144}$$

Equation (20.144) is satisfied if $\widehat{p}_j = -i\hbar \partial/\partial q_j$:

$$\{q_i, \widehat{p}_j\}_- = -i\hbar \left\{ q_i, \frac{\partial}{\partial q_j} \right\}_-,$$

where the commutator operates on a function $f(q_1, \ldots, q_\alpha)$. For example,

$$-i\hbar \left\{ \frac{\partial}{\partial q_j}, q_i \right\}_- f(q_1, \ldots, q_\alpha) = -i\hbar \left[\frac{\partial}{\partial q_j}(q_i f(q_1, \ldots, q_\alpha)) - q_i \frac{\partial}{\partial q_j} f(q_1, \ldots, q_\alpha) \right]$$
$$= -i\hbar \delta_{ij} \cdot f(q_1, \ldots, q_\alpha),$$

where the product rule was used and thus equation (20.144) is verified. The rules for the quantum mechanical commutators are identical with those for the Poisson brackets. One might say that quantum mechanics is another algebraic realization of the Poisson brackets. As will be seen in quantum mechanics, this conclusion is premature and in this form not correct.

Example 20.11: **Exercise: Total time derivative of an arbitrary function depending on q, p, and t**

Let H denote the Hamiltonian. Show that for an arbitrary function depending on q_i, p_i, and t we have

$$\frac{df}{dt} = \frac{\partial f}{\partial t} + [f, H].$$

Solution The total differential of the function $f(p_i, q_i, t)$ reads

$$df = \frac{\partial f}{\partial t} dt + \sum_\alpha \left(\frac{\partial f}{\partial q_\alpha} dq_\alpha + \frac{\partial f}{\partial p_\alpha} dp_\alpha \right) \tag{20.145}$$

$$\Rightarrow \quad \frac{df}{dt} = \frac{\partial f}{\partial t} + \sum_\alpha \left(\frac{\partial f}{\partial q_\alpha} \dot{q}_\alpha + \frac{\partial f}{\partial p_\alpha} \dot{p}_\alpha \right). \tag{20.146}$$

By means of the Hamilton equations

$$\frac{\partial H}{\partial p_\alpha} = \dot{q}_\alpha, \qquad \frac{\partial H}{\partial q_\alpha} = -\dot{p}_\alpha,$$

we can rewrite equation (20.146) as

$$\frac{df}{dt} = \frac{\partial f}{\partial t} + \sum_\alpha \left(\frac{\partial f}{\partial q_\alpha} \frac{\partial H}{\partial p_\alpha} - \frac{\partial f}{\partial p_\alpha} \frac{\partial H}{\partial q_\alpha} \right) = \frac{\partial f}{\partial t} + [f, H]. \tag{20.147}$$

Thus, the Poisson brackets enter automatically. Equation (20.147) reminds us even more of the results of quantum mechanics than the analogies of the last problem. In quantum mechanics we shall find the following expression for the time derivative of an operator $\widehat{F}$:

$$\frac{d\widehat{F}}{dt} = \frac{\partial \widehat{F}}{\partial t} + \frac{1}{i\hbar}\{\widehat{F}, \widehat{H}\}_-, \tag{20.148}$$

where $\widehat{H}$ represents the Hamiltonian operator of the quantum mechanical problem. It is, e.g., of the form

$$\widehat{H} = \widehat{H}(x, \widehat{p}) \quad \text{with } \widehat{p} = -i\hbar\frac{\partial}{\partial x}$$

and depends in general on the coordinates, momentum operators, and possibly even further quantities, e.g., spin.

PART VII

NONLINEAR DYNAMICS

The treatment of mechanics in these lectures would not be complete if we did not deal at least in brief with a topic which recently has attracted much attention: nonlinear dynamics, and thereof the "theory of chaos" as a special topic.

The starting point is the observation that ordered and regular motions like those occurring in the harmonic oscillator, the pendulum, or the Kepler problem of planetary motion are more an exception in nature than the standard case. One frequently encounters erratic phenomena and phenomena that are unpredictable in the details. A particularly striking example is the occurrence of turbulence in the flow of liquids.

Toward the end of the nineteenth century, the "father of nonlinear dynamics," Henri Poincaré,[1] for the first time pointed out that an irregular behavior in mechanics is not at all an unusual feature if the system being studied involves a *nonlinear interaction*. Closely related is the—at first sight astonishing—insight that also very simple systems may exhibit a highly complex dynamics. A simple deterministic differential equation involving nonlinearities may have solutions the behavior of which over longer time periods evolves quite irregularly and practically cannot be predicted. This is one of the characteristic features of *chaotic* systems. The meaning of this concept, which may be precisely defined in the frame of nonlinear dynamics, extends far beyond mechanics, since the phenomenon of chaos arises in many fields not only of physics but also of chemistry, biology, etc.

---

[1] *Jules-Henri Poincaré*, French mathematician and physicist, b. April 29, 1854, Nancy–d. July 17, 1912, Paris. Poincaré studied at the École Polytechnique and the École des Mines and was a scholar of Ch. Hermite. Soon after he received his doctorate, he obtained a chair at the Sorbonne in 1881, which he held until his death. In pure mathematics, he became famous as the founder of algebraic topology and of the theory of analytic functions of several complex variables. Further essential fields of work were algebraic geometry and number theory. But Poincaré also dealt with applications of mathematics to numerous physical problems, e.g., in optics, electrodynamics, telegraphy, and thermodynamics. Together with Einstein and Lorentz, he founded the special theory of relativity. Poincaré's work on celestial mechanics, in particular, on the three-body problem, culminated in a monograph in three volumes (1892–1899). In this context, he was the first who discovered the appearance of chaotic orbits in planetary motion. Poincaré has been called "the last universalist in mathematics" because of the unusually broad scope of his interests.

In the following sections, we shall learn quite a lot about general properties of nonlinear dynamic systems. The time dependence and stability of their solutions will be discussed, and concepts like attractors, bifurcations, and chaos will be introduced. However, a detailed treatment of nonlinear dynamics, its manifold problems, and interdisciplinary applications exceeds the scope of this book.[2] In particular, we cannot deal in more detail with the important topic of chaos in Hamiltonian systems.

---

[2]Some textbooks from the very extensive literature on nonlinear dynamics:
H. Schuster, *Deterministic Chaos*, VCH Verlag (1989).
G. Faust, M. Haase, J. Argyris, *Die Erforschung des Chaos*, Vieweg (1995). (This was also translated into English and published as G. Faust, M. Haase, J. Argyris, *An Exploration of Chaos*, North Holland (1994).)
H.-O. Peitgen, H. Jürgens, D. Saupe, *Chaos and Fractals: New Frontiers of Science*, Springer (1992).
R.C. Hilborn, *Chaos and Nonlinear Dynamics*, Oxford University Press (1994).
G. Jetschke, *Mathematik der Selbstorganisation*, Deutscher Verlag der Wiss., Berlin (1989).

21 Dynamical Systems

A unified theoretical description may be given for many of the systems of interest. A system is described by a finite set of dynamic variables that will be combined to a column vector $\mathbf{x} = (x_1, \ldots, x_N)^T \in I\!R^N$. The state of the system at a given time t is uniquely described by such a point $\mathbf{x}$ in phase space. The x_i are generalized coordinates that may represent a variety of quantities. Note that the vector $\mathbf{x}$ shall also comprise the velocities (or momenta, respectively). We now assume that the system behaves *deterministically*. Thus, the entire time evolution $\mathbf{x}(t)$ is determined if an initial value $\mathbf{x}(t_0)$ is given. The time evolution shall be described by a differential equation of first order with respect to time:

$$\frac{d}{dt}\mathbf{x}(t) = \mathbf{F}(\mathbf{x}(t), t; \lambda). \tag{21.1}$$

Here, $\mathbf{F}$ is in general a *nonlinear* function of the coordinates $\mathbf{x}$ (also called the velocity field or *vector field*). Moreover, $\mathbf{F}$ may also still explicitly depend on the time t, for example if varying external forces are acting on the system. If there is no such dependence, the system is called *autonomous*. Finally, the third argument in (21.1) shall indicate that possibly there exist one or several *control parameters* λ. These are fixed given constants whose values affect the dynamics of the system and may possibly change the character of the dynamics. Typical control parameters are, e.g., the coupling strength of an interaction, or the amplitude or frequency of an external perturbation imposed onto the system.

Note: A possible explicit time dependence in (21.1) may be eliminated by a simple trick. For this purpose we consider a system with one additional degree of freedom,

$$\tilde{\mathbf{x}} = (x_1, \ldots, x_N, x_{N+1})^T \in I\!R^{N+1},$$

and postulate for the additional vector component the differential equation

$$\frac{d}{dt} x_{N+1} = 1.$$

With the initial condition $\mathbf{x}_{N+1}(0) = 0$, this simply implies $x_{N+1}(t) = t$. Hence, the time on the right side of t may be replaced by x_{N+1}, and we are dealing with an *autonomous* system with one additional dimension.

The equation of motion (21.1) is very far-reaching, in spite of its simple shape. In particular, it incorporates the *Hamiltonian mechanics* as a special case: For a system with N degrees of freedom described by the generalized coordinates $q_1, \ldots, q_N$ and the associated canonical momenta $p_1, \ldots, p_N$, the Hamiltonian equations of motion (see Chapter 18) read as follows:

$$\dot{q}_i = \frac{\partial H}{\partial p_i}, \qquad \dot{p}_i = -\frac{\partial H}{\partial q_i}. \tag{21.2}$$

If the coordinates and momenta are combined according to $\mathbf{x} = (q_1, \ldots q_N; p_1, \ldots, p_N)^\mathrm{T}$ to a $2N$-dimensional vector, then (21.2) may be written as a combined matrix equation of the form (21.1):

$$\frac{d}{dt}\mathbf{x} = \mathbf{J}\nabla_\mathbf{x} H. \tag{21.3}$$

$\nabla_\mathbf{x} H$ stands for the gradient vector of the Hamiltonian function,

$$\nabla_\mathbf{x} H = \left(\frac{\partial H}{\partial q_1}, \ldots, \frac{\partial H}{\partial q_N}; \frac{\partial H}{\partial p_1}, \ldots, \frac{\partial H}{\partial p_N}\right)^\mathrm{T}, \tag{21.4}$$

and the $2N \times 2N$-matrix $\mathbf{J}$ provides both the permutation of the components as well as the correct signs:

$$\mathbf{J} = \begin{pmatrix} 0 & +\mathbf{I} \\ -\mathbf{I} & 0 \end{pmatrix}, \tag{21.5}$$

where $\mathbf{I}$ denotes the $N \times N$ unit matrix. By the way, $\mathbf{J}$ has the following useful properties:

$$\mathbf{J}^{-1} = \mathbf{J}^\mathrm{T} = -\mathbf{J}, \qquad \mathbf{J}^2 = -\mathbf{I}, \qquad \det \mathbf{J} = 1. \tag{21.6}$$

Moreover, *dissipative systems* may also be described by the equation (21.1) by introducing velocity-dependent friction terms; see, e.g., Example 21.2.

Obviously, the solutions of (21.1) may be highly manifold. For a given starting vector $\mathbf{x}(t = 0) = \mathbf{x}_0$, a *trajectory* $\mathbf{x}(t)$, also called an *orbit*, may be calculated (which in nonlinear systems, as a rule, is of course not feasible in an analytic manner) the mathematical existence and uniqueness of which is guaranteed under very general conditions by the theory of differential equations. Of particular interest is the asymptotic behavior of the trajectory at large times: Does it reach a stationary state (a *fixed point*) or a periodic vibration (a *limit cycle*), or does it behave irregularly?

The connection between $\mathbf{x}(t)$ and $\mathbf{x}_0$ is mathematically a mapping $\Phi_t : \mathbb{R}^N \to \mathbb{R}^N$, namely,

$$\Phi_t(\mathbf{x}_0) = \mathbf{x}(t). \tag{21.7}$$

This mapping that depends on the time t as a parameter is called the *phase flow* or simply the *flow* of the vector field $\mathbf{F}(\mathbf{x})$. For $t = 0$, the flow obviously reduces to the identical mapping

$$\Phi_{t=0} = \mathit{I\!I}. \tag{21.8}$$

Furthermore, for autonomous (not explicitly time-dependent) systems, we have for a subsequent performance of two time shifts

$$\Phi_{t_1} \Phi_{t_2} = \Phi_{t_1+t_2}. \tag{21.9}$$

For a thorough understanding of the dynamical system, it is not sufficient to inspect individual trajectories. Of much more interest is the behavior of the ensemble of all trajectories in phase space. This may be interpreted as a query for the global properties of the mapping Φ_t. Important questions are as follows: Can the flow be characterized universally "on the large scale"? Are there regions with qualitatively distinct behavior (ordered vs. disordered motion)? How does the flow vary with the value of a possibly existing control parameter λ (are there critical threshold values at which a new type of behavior arises)?

The answer to these questions depends of course on the system being considered. Nevertheless, one may find general criteria in the frame of nonlinear dynamics, and it turns out that seemingly very distinct systems display amazing similarities in their dynamics.

Dissipative systems: Contraction of the phase-space volume

Conservative systems are characterized by a *volume-conserving* dynamical flow. *Liouville's theorem*, proved in Chapter 18, states that the volume of a cell in the $2N$-dimensional phase space $(q_1, \ldots, q_N; p_1, \ldots, p_N)$ does not vary with time if the points contained therein are moving according to the Hamiltonian equations. In *dissipative systems*, on the contrary, the cells in phase space are shrinking with time. We shall now derive a quantitative measure of this phenomenon for a general autonomous dynamical system, the trajectories of which obey the equation of motion

$$\frac{d}{dt}\mathbf{x}(t) = \mathbf{F}(\mathbf{x}(t)) \tag{21.10}$$

in an N-dimensional phase space. To this end, we consider a small volume element $\Delta V(\mathbf{x})$ that at time $t = t_0$ shall be at the position $\mathbf{x} = \mathbf{x}_0$ and shall move with the flow. In Cartesian coordinates, the volume is given by the product of the edge lengths,

$$\Delta V(\mathbf{x}) = \prod_{i=1}^{N} \Delta x_i(\mathbf{x}). \tag{21.11}$$

The time derivative of this quantity is, according to the chain rule, given by

$$\begin{aligned}\frac{d}{dt}\Delta V(\mathbf{x}) &= \sum_{i=1}^{N} \frac{d\Delta x_i(\mathbf{x})}{dt} \prod_{j \neq i}^{N} \Delta x_j(\mathbf{x}) \\ &= \underbrace{\prod_{j=1}^{N} \Delta x_j(\mathbf{x})}_{=\Delta V(\mathbf{x})} \sum_{i=1}^{N} \frac{1}{\Delta x_i(\mathbf{x})} \frac{d\Delta x_i(\mathbf{x})}{dt}, \end{aligned} \tag{21.12}$$

where the extension $\Delta x_i/\Delta x_i$ has been added. Hence, the relative change (= logarithmic time derivative) of the volume is

$$\frac{1}{\Delta V(\mathbf{x})}\frac{d}{dt}\Delta V(\mathbf{x}) = \sum_{i=1}^{N}\frac{1}{\Delta x_i(\mathbf{x})}\frac{d\Delta x_i(\mathbf{x})}{dt}. \quad (21.13)$$

The change of the edge lengths of the volume[1] may be calculated from the equation of motion (21.10). Let us consider the distance between two edges of the cube along the i-direction which are determined by the trajectories $\mathbf{x}_0(t)$ with $\mathbf{x}_0(t_0) = \mathbf{x}_0$ and $\mathbf{x}(t)$ with $\mathbf{x}(t_0) = \mathbf{x}_0 + \mathbf{e}_i \Delta x_i$:

$$\begin{aligned}\left.\frac{d\Delta x_i}{dt}\right|_{t_0} &= \left.\frac{d}{dt}\Big(x_i(t) - x_{0i}(t)\Big)\right|_{t_0} \\ &= F_i(\mathbf{x}(t_0)) - F_i(\mathbf{x}_0(t_0)) \\ &= F_i(\mathbf{x}_0 + \mathbf{e}_i \Delta x_i) - F_i(\mathbf{x}_0).\end{aligned} \quad (21.14)$$

For small deviations Δx_i, the Taylor expansion of $\mathbf{F}(\mathbf{x})$ yields to first order

$$\left.\frac{d\Delta x_i}{dt}\right|_{t_0} = \left.\frac{\partial F_i}{\partial x_i}\right|_{\mathbf{x}_0} \Delta x_i. \quad (21.15)$$

At the point $t_0 = t$, $\mathbf{x}_0 = \mathbf{x}$ (21.13) thus yields

$$\Lambda(\mathbf{x}) := \frac{1}{\Delta V(\mathbf{x})}\frac{d}{dt}\Delta V(\mathbf{x}) = \sum_{i=1}^{N}\frac{\partial F_i}{\partial x_i} = \nabla \cdot \mathbf{F}. \quad (21.16)$$

The rate of change Λ of the phase-space volume is therefore determined by the *divergence of the velocity field* $\mathbf{F}$.

Liouville's theorem is included in (21.16) as a special case. According to (21.3) to (21.5), the velocity field of a Hamiltonian system with the coordinates $\mathbf{x} = (q_1, \ldots, q_N; p_1, \ldots, p_N)^T$ reads

$$\mathbf{F}(\mathbf{x}) = \left(\frac{\partial H}{\partial p_1}, \ldots, \frac{\partial H}{\partial p_N}; -\frac{\partial H}{\partial q_1}, \ldots, -\frac{\partial H}{\partial q_N}\right)^T. \quad (21.17)$$

This leads to the volume change

$$\begin{aligned}\Lambda = \nabla \cdot \mathbf{F} &= \sum_{i=1}^{N}\frac{\partial}{\partial q_i}F_i + \sum_{i=1}^{N}\frac{\partial}{\partial p_i}F_{N+i} \\ &= \sum_{i=1}^{N}\frac{\partial}{\partial q_i}\frac{\partial H}{\partial p_i} - \sum_{i=1}^{N}\frac{\partial}{\partial p_i}\frac{\partial H}{\partial q_i} = 0,\end{aligned} \quad (21.18)$$

which confirms that conservative systems are volume conserving.

---

[1] Strictly speaking, the shape of ΔV is distorted, and the edges do not remain orthogonal to each other. But this is of no meaning when calculating the volume to lowest order.

If the flow in phase space is *contracting*, i.e., if $\Lambda = \nabla \cdot \mathbf{F} < 0$, the system is called *dissipative*. This is so far a local statement holding at a point $\mathbf{x}$ in phase space. In order to get a global estimate of the dynamics, $\Lambda(\mathbf{x})$ has to be averaged over a trajectory $\mathbf{x}(t)$. If Λ thereby changes sign, then there is no simple method of finding out whether the system is dissipative; one actually has to evaluate the mean value.

In dissipative systems, the volume filled by neighboring trajectories shrinks with increasing time; asymptotically it even tends to zero. This *may* happen in a trivial manner if the trajectories are converging. In the simplest case they are moving towards an equilibrium point and the motion comes to rest (see the section on limit cycles). There is, however, also the possibility that the volume is shrinking and the distance between the trajectories is being reduced only along certain directions, while they are diverging in other directions. In this case the resulting distance even increases with time. An originally localized region in phase space is so to speak "rolled out" and widely distributed by the dynamic flow. The shrinking of the volume towards zero then means that an originally N-dimensional hypercube in phase space changes over to a geometric object with *lower dimension $D < N$*. D may even take a nonintegral value, as will be explained in Chapter 24.

Attractors

The dynamics of a nonlinear system may be highly complicated. It is convenient to distinguish between *transient* and *asymptotic* behavior. A transient process denotes the initial behavior of a system after starting from a given point $\mathbf{x}_0$ in phase space. Naturally, it is particularly difficult to make general statements, since the transients depend on the particular initial condition. Theorists therefore tend to ignore this part of the trajectory, even if it may play an important role in practice, depending on the dominant time scales. Only recently has the study of transients gotten more attention.

The systematic treatment of the asymptotic or stationary behavior of a system is somewhat simpler. "Stationary" shall not mean here that the system is at rest but only that possible transient phenomena have faded away. In dissipative systems, which will be treated here, the trajectories will asymptotically approach a subset of the phase space of lower dimension, a so-called *attractor*.

The definition and correct mathematical classification of attractors is not quite simple. Actually there are several concepts in literature that differ from each other in detail. Here we first give a mathematical definition[2] but shall also illustrate the concept of the attractor by various examples in the subsequent chapters.

Let us consider a vector field $\mathbf{F}(\mathbf{x})$ on a space M (e.g., $M = \mathbb{R}^N$) with an associated phase flow Φ_t. A subset $A \subset M$ is denoted as an *attractor* if it fulfills the following criteria:

(1) A is *compact*.

(2) A is *invariant* under the phase flow Φ_t.

---

[2] F. Scheck, *Mechanik*, Springer (1992). This book is also available in English: F. Scheck, *Mechanics: From Newton's Laws to Deterministic Chaos*, 3rd edition, Springer (1999).

(3) *A* has an open environment *U* that *contracts to A* under the flow.

This statement needs several explanations:

(1) A set is called compact if it is *closed* and *restricted*. This means that any limit value of an infinite sequence belongs itself to the set, and the set cannot extend up to infinity. "Exploding" solutions where for example particles escape to infinity therefore cannot be attractors.

(2) Invariance under the phase flow means that

$$\Phi_t(A) = A \quad \text{for all } t. \tag{21.19}$$

Hence, a point on the attractor never may leave this attractor.

(3) This may be formulated in two steps. First, the environment $U \supset A$ is larger than the attractor itself, since we are dealing with an open range that includes the compact *A*. *U* shall be *positively invariant*, i.e.,

$$\Phi_t(U) \subseteq U \quad \text{for all } t \geq 0. \tag{21.20}$$

If a point once lies within *U*, then it cannot leave it. It will, on the contrary, even be pulled toward *A*, which may be formulated as follows: For any open environment *V* of *A* that lies completely within *U*, i.e., $A \subset V \subset U$, one can find a time t_V after passing that the image of *U* lies entirely within *V*:

$$\Phi_t(U) \subset V \quad \text{for all } t > t_V. \tag{21.21}$$

Since *V* may be chosen arbitrarily "close" about *A*, this means that for large time values *U* is shrinking toward the attractor *A*.

Frequently, the definition of an attractor is still extended by the requirement that it shall consist of one piece only.

(4) *A* cannot be separated into several closed nonoverlapping invariant subsets.

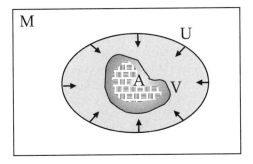

Figure 21.1. Visualization of the definition of an attractor *A* (hatched). In the course of time evolution, the environment *U* is shrinking such that for $t > t_V$ it is enclosed in any arbitrary smaller environment *V*.

An important property of an attractor is its domain of attraction. The maximum environment U that contracts to A is called the *basin of attraction B*. In correct mathematical formulation, B is the union of all open environments of A that fulfill the conditions (21.20) and (21.21).

The introduction of the concept of attractor given here is rather complex. This is justified, however, by the fact that attractors may have very complex properties. Of central importance for nonlinear dynamics are the concepts of *strange* and *chaotic* attractors, which sometimes—not quite correctly—are used as synonyms. These concepts will become fully transparent only in the subsequent chapters and by examples. But we shall present the definitions now:

Chaotic attractor: The motion is extremely sensitive with respect to the initial conditions. The distance between two initially closely neighboring trajectories increases exponentially with time. For more details see Chapter 24.

Strange attractor: The attractor has a strongly rugged geometrical shape that is described by a fractal. For more details see Chapter 24.

Both of these properties arise, as a rule, in common. There are, however, also examples[3] where an attractor is chaotic but not strange, or is strange but not chaotic.

Equilibrium solutions

A particularly simple case arises if the system is in stationary equilibrium, i.e.,

$$\mathbf{F}(\mathbf{x}_0) = 0 \quad \text{such that} \quad \mathbf{x}(t) = \mathbf{x}_0 = \text{constant.} \tag{21.22}$$

Such an $\mathbf{x}_0$ is also called a *critical point* or *fixed point*. Of particular interest is the question of whether or not the system is moving toward such a fixed point and—if several ones exist—to which of them. A fixed point that attracts the trajectories is the simplest example of an attractor. In this case the set A defined in the previous section is trivial and consists of a single point.

We are therefore interested in the *stability* of equilibrium solutions. To this end we consider the trajectories $\mathbf{x}(t)$ in the vicinity of a critical point $\mathbf{x}_0$. We thus require that the distance

$$\boldsymbol{\xi}(t) = \mathbf{x}(t) - \mathbf{x}_0 \tag{21.23}$$

be a small quantity. Under this condition, the problem may be greatly simplified, since it usually suffices to take only the lowest term of the Taylor expansion of $\mathbf{F}(\mathbf{x})$ into account. The *linearized equation of motion* then reads

$$\frac{d}{dt}\boldsymbol{\xi}(t) = \mathbf{M}\boldsymbol{\xi}(t), \tag{21.24}$$

---

[3]C. Grebogi, E. Ott, S. Pelikan, J.A. Yorke, *Physica 13D*, 261 (1984).

where terms of quadratic or higher order in $\boldsymbol{\xi}$ were neglected. $\mathbf{M}$ denotes the *Jacobi matrix* (functional matrix) of the function $\mathbf{F}(\mathbf{x})$ evaluated at the position $\mathbf{x}_0$. This matrix has the elements

$$M_{ik} = \left.\frac{\partial F_i}{\partial x_k}\right|_{\mathbf{x}_0}. \tag{21.25}$$

Contrary to the original nonlinear equation of motion (21.1), the solution of the linearized problem (21.24) is in principle simple; it may be given analytically. Let us first consider the trivial special case of a one-dimensional system ($N = 1$). The Jacobi matrix then has only a single element, say, μ, and (21.24) is solved by

$$\xi(t) = e^{\mu t}\,\xi(0). \tag{21.26}$$

The character of the solution is determined by the *sign* of μ: For $\mu < 0$, x_0 is a *stable* equilibrium point, since small perturbations decay exponentially. For $\mu > 0$, the equilibrium is unstable, since even the smallest displacements from the equilibrium position "explode" exponentially. For $\mu = 0$, the limit case of *indifferent* or *neutral* equilibrium arises. The behavior of the system under perturbations is then determined by the higher derivatives of the function $F(x)$ at the point x_0.

The general case ($N > 1$) may be treated as in Chapter 8 by the method of normal vibrations. One then constructs solution vectors $\mathbf{u}(t)$ normalized to 1, all components of which show the same (exponential) time dependence:

$$\mathbf{u}(t) = e^{\mu t}\,\mathbf{u}. \tag{21.27}$$

Using (21.24), one arrives at the eigenvalue problem

$$\mathbf{M}\mathbf{u} = \mu\,\mathbf{u}. \tag{21.28}$$

This N-dimensional linear system of equations only has nontrivial solutions if the determinant

$$\det\!\left(M_{ij} - \mu\,\delta_{ij}\right) = 0 \tag{21.29}$$

vanishes. This *characteristic equation* (secular equation) has as a polynomial of Nth order in general N eigenvalues μ_n with the associated *eigenvectors* $\mathbf{u}_n$. The general solution of (21.24) may then be written as a superposition

$$\boldsymbol{\xi}(t) = \sum_{n=1}^{N} c_n\, e^{\mu_n t}\,\mathbf{u}_n, \tag{21.30}$$

where the expansion coefficients c_n may be determined from the initial condition at $t = 0$. The eigenvalues μ_n may be real or complex. Complex eigenvalues thereby arise always *pairwise*: If μ_n solves the equation (21.29), then the complex-conjugate μ_n^* obviously also solves the equation, since the Jacobi matrix M_{ij} is real.

The real parts of the eigenvalues of the characteristic equation are decisive for characterizing an equilibrium point $\mathbf{x}_0$. We now define a tightened form of the condition of stability:

EQUILIBRIUM SOLUTIONS

An equilibrium point $\mathbf{x}_0$ with $\mathbf{F}(\mathbf{x}_0) = 0$ is called *asymptotically stable* if there exists an environment $U \ni \mathbf{x}_0$ within which all trajectories are running toward $\mathbf{x}_0$ for large times:

$$\lim_{t \to \infty} \mathbf{x}(t) = \mathbf{x}_0 \quad \text{for } \mathbf{x}(0) \in U. \tag{21.31}$$

If the function (the vector field) $\mathbf{F}$ is sufficiently smooth so that it can be described by the linear approximation, one may immediately give a sufficient *condition for asymptotic stability*: The point $\mathbf{x}_0$ is asymptotically stable if all eigenvalues of the Jacobi matrix have a negative real part, i.e., if

$$\Re \mu_n \leq c < 0 \quad \text{for all } n = 1, \ldots N \tag{21.32}$$

with a positive constant c.

A glance at (21.30) shows that under this condition all contributions to the displacement $\xi(t)$ exponentially tend to zero, such that asymptotically

$$||\mathbf{x}(t) - \mathbf{x}_0|| < \text{constant} \cdot e^{-(\min_n |\Re \mu_n|)t}. \tag{21.33}$$

Conversely, if at least one of the eigenvalues has a positive real part, $\Re \mu_n > 0$, then x_0 is an *unstable* fixed point, since displacements along $\mathbf{u}_n$ are increasing exponentially.

With the knowledge of the eigenvectors $\mathbf{u}_n$, the total phase space may be spanned in *partial spaces*. The stable (or unstable) partial space is spanned by all vectors $\mathbf{u}_n$ satisfying $\Re \mu_n < 0$ (or > 0). In addition, a partial space may occur with the special value $\Re \mu_n = 0$. If this happens, one speaks of a *degenerate* fixed point. (The associated partial space is also called the *center*; but we shall not deal here in more detail with the related problems.) If one considers a general perturbation of a trajectory, it will have components in all partial spaces. After a sufficiently long time the contribution with the maximum $\Re \mu_n$ will dominate.

Finally, we note that the linear stability analysis holds only in the vicinity of a critical point $\mathbf{x}_0$. It may be shown mathematically that the topological behavior of the flow does not change there under the influence of the nonlinearity. But this vicinity may be very small, such that one cannot make a statement about the global behavior of the flow by this way.

Example 21.1: Linear stability in two dimensions

The stability analysis becomes particularly transparent for the case $N = 2$ that corresponds to a dynamic system with one degree of freedom $x_1 = q$ and the associated momentum $x_2 = p$. In the vicinity of a fixed point $\dot{\mathbf{x}} = \mathbf{F}(\mathbf{x}_0) = 0$, the motion is determined in a linear approximation by the four elements of the Jacobi matrix M_{ij}. The characteristic equation (21.29)

$$\begin{vmatrix} M_{11} - \mu & M_{12} \\ M_{21} & M_{22} - \mu \end{vmatrix} = 0 \tag{21.34}$$

is a quadratic polynomial

$$\mu^2 - (M_{11} + M_{22})\mu + M_{11}M_{22} - M_{12}M_{21} = 0 \tag{21.35}$$

or

$$\mu^2 - 2s\mu + d = 0 \tag{21.36}$$

with

$$s = \frac{1}{2}(M_{11} + M_{22}) = \frac{1}{2}\operatorname{Tr} M, \qquad d = M_{11}M_{22} - M_{12}M_{21} = \det M. \tag{21.37}$$

The two solutions of (21.36) may be given explicitly:

$$\mu_{1/2} = s \pm \sqrt{s^2 - d}. \tag{21.38}$$

Depending on the magnitude and sign of the two constants s and d, there are many distinct possibilities for the eigenvalues μ_1, μ_2:

(a) μ_1, μ_2 real and both negative (if $s < 0$ and $0 < d < s^2$)   *stable node*
(b) μ_1, μ_2 real and both positive (if $s > 0$ and $0 < d < s^2$)   *unstable node*
(c) μ_1, μ_2 real with distinct signs (if $d < 0$)   *saddle*
(d) $\mu_1 = \mu_2^*$, negative real part (if $s < 0$ and $d > s^2$)   *stable spiral*
(e) $\mu_1 = \mu_2^*$, positive real part (if $s > 0$ and $d > s^2$)   *unstable spiral*
(f) $\mu_1 = \mu_2^*$, purely imaginary (if $s = 0$ and $d > 0$)   *rotor*

The ranges are represented in Figure 21.2 in the s,d-plane. To these alternatives, there correspond distinct types of trajectories $\boldsymbol{\xi}(t) = \mathbf{x}(t) - \mathbf{x}_0$ according to equation (21.30).

Figure 21.3 illustrates how the trajectories in the vicinity of a *stable node* are running into the fixed point:

$$\boldsymbol{\xi}(t) = c_1 e^{-|\mu_1|t} \mathbf{u}_1 + c_2 e^{-|\mu_2|t} \mathbf{u}_2, \tag{21.39}$$

where $\mathbf{u}_1$ and $\mathbf{u}_2$ are the (not necessarily orthogonal) eigenvectors. The curvature of the trajectories arises if $\mu_1 \neq \mu_2$. These curves are parabola-like, with a common tangent at the origin (in $\mathbf{u}_1$- or $\mathbf{u}_2$-direction depending on whether μ_2 or μ_1 is larger). The trajectories for the *unstable node*, Figure 21.3(b), have the same shape but are passed in the opposite direction (exponential "explosion"). For the case of a *saddle* the trajectories are running in the $\mathbf{u}_1$-direction (let $\mu_1 < \mu_2$ without restriction

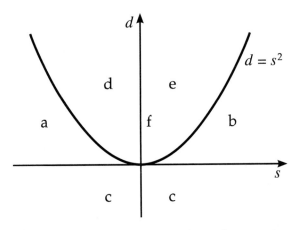

Figure 21.2. Ranges of distinct stability depending on the parameters s and d.

EQUILIBRIUM SOLUTIONS

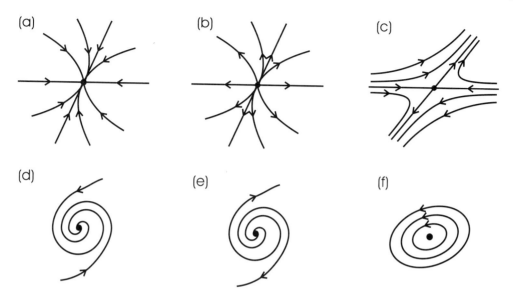

Figure 21.3. Various types of stability of a fixed point in two dimensions. Upper row: Stable and unstable node, saddle. Lower row: Stable and unstable spiral, rotor.

of generality) toward the fixed point but are pushed off in the $\mathbf{u}_2$-direction, which results in the hyperbola-like trajectories of Figure 21.3(c).

If the eigenvalues are *complex*,

$$\mu_1 = \mu_r + i\mu_i, \qquad \mu_2 = \mu_r - i\mu_i, \tag{21.40}$$

this will hold, because of (21.28), for the eigenvectors too:

$$\mathbf{u}_1 = \mathbf{u}_r + i\mathbf{u}_i, \qquad \mathbf{u}_2 = \mathbf{u}_r - i\mathbf{u}_i. \tag{21.41}$$

The general solution (21.30) then has the form

$$\boldsymbol{\xi}(t) = c_1 e^{\mu_1 t}\mathbf{u}_1 + c_2 e^{\mu_2 t}\mathbf{u}_2 \tag{21.42}$$
$$= c_1 e^{\mu_1 t}\mathbf{u}_1 + c_1^* e^{\mu_1^* t}\mathbf{u}_1^*$$
$$= 2 e^{\mu_r t} \Re\left(c_1 e^{i\mu_i t}\mathbf{u}_1\right), \tag{21.43}$$

where $c_2 = c_1^*$ to get $\boldsymbol{\xi}$ real. If the constant c_1, the value of which is fixed by the initial condition $\boldsymbol{\xi}(0)$, is split into magnitude and phase, and the same is done for the Cartesian components of the complex eigenvector $\mathbf{u}_1$,

$$c_1 = \rho e^{i\phi}, \qquad \mathbf{u}_1 = a e^{i\alpha}\mathbf{e}_x + b e^{i\beta}\mathbf{e}_y, \quad \text{with } a^2 + b^2 = 1, \tag{21.44}$$

then (21.42) may be rewritten as follows:

$$\boldsymbol{\xi}(t) = 2\rho e^{\mu_r t}\left[a\cos(\mu_i t + \phi + \alpha)\mathbf{e}_x + b\cos(\mu_i t + \phi + \beta)\mathbf{e}_y\right]. \tag{21.45}$$

The factor in brackets describes harmonic vibrations shifted in phase relative to each other (if $\alpha \neq \beta$). One thus has the parametric representation of an *ellipse*. Due to the prefactor, the size of the ellipse varies exponentially with time. Thus, the trajectories are *logarithmic spirals* moving toward the fixed point or away from it, depending on the sign of the real part of μ; see Figure 21.3(d,e)—hence, the name *spiral*. The case of the rotor with $\Re\mu = 0$ plays a particular role, since the trajectories in the vicinity of $\mathbf{x}_0$ are periodic functions (concentric ellipses). This means that the equilibrium point is *stable* (small displacements are not amplified) but *not asymptotically stable* (the trajectory does not run into the fixed point), and hence this point is not an attractor.

Example 21.2: Exercise: The nonlinear oscillator with friction

Let a one-dimensional system be described by the following equation of motion:

$$\ddot{x} + \alpha\dot{x} + \beta x + \gamma x^3 = 0. \tag{21.46}$$

Show that the system is dissipative. Interpret the individual terms and discuss the possible fixed points and their stability.

Solution

We are dealing with a harmonic oscillator involving friction and nonlinearity. Besides the linear backdriving force of the harmonic oscillator (third term), there acts a friction force proportional to the velocity (second term). Moreover, a cubic nonlinearity (fourth term) becomes important. This force law corresponds to a potential

$$V(x) = \frac{m}{2}\beta x^2 + \frac{m}{4}\gamma x^4, \tag{21.47}$$

where m denotes the mass. We obtain various types of motion, depending on the magnitude and sign of the constants in (21.46). We first rewrite the equation of motion (21.46) in the standard form. For this purpose, we introduce the velocity as a second coordinate, $\mathbf{x} = (x, y) = (x, \dot{x})$, which leads to the coupled differential equations of first order:

$$\dot{\mathbf{x}} = \frac{d}{dt}\begin{pmatrix} x \\ y \end{pmatrix} = \begin{pmatrix} y \\ -\alpha y - \beta x - \gamma x^3 \end{pmatrix} \equiv \mathbf{F}(\mathbf{x}). \tag{21.48}$$

For $\alpha > 0$, the system is dissipative, since the divergence of the velocity field is

$$\Lambda = \nabla \cdot \mathbf{F} = \frac{\partial}{\partial x}y + \frac{\partial}{\partial y}(-\alpha y - \beta x - \gamma x^3) = -\alpha < 0. \tag{21.49}$$

The equilibrium condition $\mathbf{F}(\mathbf{x}_0)$ reads

$$y = 0, \qquad x(\beta + \gamma x^2) = 0. \tag{21.50}$$

Hence, besides the equilibrium position $\mathbf{x}_0 = (0, 0)$ without displacement, there still occur two further symmetrically positioned fixed points $\mathbf{x}_0 = (\pm\sqrt{-\beta/\gamma}, 0)$, provided that the constants β and γ have distinct signs. Figure 21.4 shows the associated potential functions $V(x)$ for all combinations of signs.

To discuss the linear stability, we need the Jacobi matrix

$$\mathbf{M} = \begin{pmatrix} \partial F_1/\partial x & \partial F_1/\partial y \\ \partial F_2/\partial x & \partial F_2/\partial y \end{pmatrix} = \begin{pmatrix} 0 & 1 \\ -\beta - 3\gamma x^2 & -\alpha \end{pmatrix}. \tag{21.51}$$

EQUILIBRIUM SOLUTIONS

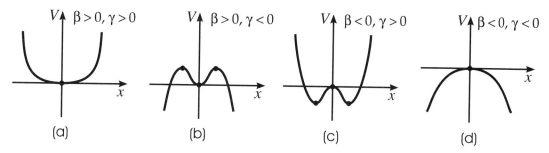

Figure 21.4. Potentials of the quadratic oscillator for various signs of the parameters β and γ.

The characteristic equation (21.36) in example 21.1 for the eigenvalues involves the following coefficients:

$$\text{For } \mathbf{x}_0 = (0,0): \quad s = -\frac{1}{2}\alpha, \quad d = \beta,$$

$$\text{For } \mathbf{x}_0 = (0, \pm\sqrt{-\beta/\gamma}): \quad s = -\frac{1}{2}\alpha, \quad d = -2\beta.$$

Obviously, asymptotic stability may occur only for a positive sign of the constant $\alpha > 0$. Only then is one dealing physically with a damping friction term. For the fixed point in the rest position $x_0 = (0, 0)$, we get the alternatives

(1) $\beta > \frac{1}{4}\alpha^2$ stable spiral
(2) $0 < \beta < \frac{1}{4}\alpha^2$ stable node
(3) $\beta < 0$ saddle

In the first case, there arise weakly damped vibrations, in the second case the oscillator is overdamped, and the displacement monotonically tends to zero. For $\beta < 0$, the equilibrium position is unstable, as may be seen from the potential plots in Figure 21.4c,d. The analogous considerations for the fixed points $\mathbf{x}_0 = \left(\pm\sqrt{-\beta/\gamma}, 0\right)$ lead to (assuming $\alpha > 0$):

(1) $-2\beta > \frac{1}{4}\alpha^2$ stable spiral
(2) $0 < -2\beta < \frac{1}{4}\alpha^2$ stable node
(3) $\beta > 0$ saddle

The factor 2 arises because the curvature of the potential (21.47) in the equilibrium positions with finite displacement is twice as large as in the rest position. Only the double-oscillator potential (Figure 21.4c) allows stable displaced fixed points ($\beta < 0$ and $\gamma > 0$).

It is instructive to plot the position of the fixed points as a function of the parameter β. As is seen from Figure 21.5, for $\beta = 0$ there occurs a square-root branching. For $\gamma > 0$, a stable equilibrium position bifurcates into two new stable solutions. Such *bifurcations* (Lat. *furca* = fork) frequently occur in nonlinear systems; see also Chapter 23.

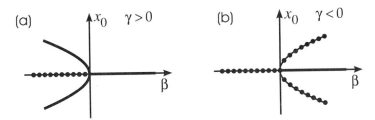

Figure 21.5. Position of the stable (continuous) and unstable (dotted) fixed points depending on the parameters β and γ.

Limit cycles

Besides the simple stationary equilibrium points studied in detail in the section on attractors, a dynamic system may exhibit still other types of stable solutions. These are the so-called *limit cycles* that are characterized by periodically oscillating closed trajectories. Similar to the fixed points discussed already, limit cycles may also act as *attractors* of motion; compare the section on attractors. Then there exists a more or less extended range in phase space (the "basin of attraction" of the attractor): trajectories starting from there move toward the limit cycle, which is approached for $t \to \infty$. For limit cycles, one may also perform a mathematical stability analysis as for fixed points which by its very nature is somewhat more difficult.

We shall concentrate ourselves here to a special but typical example, namely a *harmonic oscillator with a nonlinear friction term*. The associated differential equation has the general form

$$\frac{d^2x}{dt^2} + f(x)\frac{dx}{dt} + \omega^2 x = 0. \tag{21.52}$$

If the middle term is absent, we obtain a harmonic oscillator with angular frequency ω. The case of a constant coefficient, $f(x) = \alpha =$ constant, leads to a linear differential equation that may be solved easily. The character of the solution is determined by an exponential factor $\exp(-\alpha t/2)$. The solution thus decreases exponentially toward the fixed point at $x = \dot{x} = 0$ if α is positive. A negative value of α means that a force is acting along the same direction as that of the instantaneous velocity which leads to an unlimited amplification of the solution (negative damping). Physically one of course no longer deals with a friction force; rather an external source must exist that pumps energy into the system.

If allowance is made for more general functions $f(x)$, it may happen that the damping coefficient takes partly positive, partly negative values, depending on the displacement. Of particular interest is the case when $f(x)$ is negative for small magnitudes of x, and positive for large displacements. The simplest *ansatz* providing such a behavior is a quadratic polynomial

$$f(x) = \alpha(x^2 - x_0^2), \tag{21.53}$$

where α determines the strength of the damping/excitation and two zeros are at $x = \pm x_0$. The zeros may be set to the value 1 without loss of generality by rescaling the variables to $x' = x/x_0$ with $\alpha' = \alpha x_0^2$. For convenience one may also choose the value 1 for the frequency by rescaling the time: $t' = \omega_0 t$ with $\alpha'' = \alpha'/\omega$. The standard form of the equation of motion then reads (dropping the primes again):

$$\frac{d^2 x}{dt^2} + \alpha(x^2 - 1)\frac{dx}{dt} + x = 0. \tag{21.54}$$

This differential equation has been set up and discussed in 1926 by the Dutch engineer B. van der Pol. It served first for describing an electronic oscillator circuit with feedback (at that time still with valves), but it was already clear to the author that his equation could be applied to a variety of vibrational processes. Actually the origin of this equation may be traced back even further, since around 1880 Lord Rayleigh investigated the following differential equation in the context of nonlinear vibrations:

$$\frac{d^2 v}{dt^2} + \alpha\left[\frac{1}{3}\left(\frac{dv}{dt}\right)^3 - \frac{dv}{dt}\right] + v = 0. \tag{21.55}$$

One easily sees the relation between (21.54) and (21.55). We have only to differentiate the Rayleigh equation (21.55) with respect to time and then substitute

$$\frac{dv}{dt} = x \tag{21.56}$$

to get the van der Pol equation (21.54). Thus, both equations are essentially equivalent to each other.

We now discuss the solutions of the van der Pol equation (21.54). It may be transformed as usual to the standard form (21.1) of two coupled differential equations of first order for the vector $\mathbf{x}(t) = (x, y)^T$:

$$\frac{dx}{dt} = y, \tag{21.57}$$

$$\frac{dy}{dt} = -x - \alpha(x^2 - 1)y. \tag{21.58}$$

It is now advantageous to transform to polar coordinates in the x, y-phase space:

$$x = r\cos\theta, \qquad y = r\sin\theta. \tag{21.59}$$

The time derivatives of r and θ may be expressed by those of x and y. For the radius coordinate the relation follows immediately from the differentiation of $r^2 = x^2 + y^2$:

$$r\frac{dr}{dt} = x\frac{dx}{dt} + y\frac{dy}{dt}. \tag{21.60}$$

An analogous relation for the angle coordinate may be obtained from the time derivatives of (21.59):

$$\frac{dx}{dt} = \frac{dr}{dt}\cos\theta - r\frac{d\theta}{dt}\sin\theta,$$

$$\frac{dy}{dt} = \frac{dr}{dt}\sin\theta + r\frac{d\theta}{dt}\cos\theta. \tag{21.61}$$

By multiplying the first of these equations by y and the second one by x and subtracting both equations, one obtains

$$r^2\frac{d\theta}{dt} = x\frac{dy}{dt} - y\frac{dx}{dt}. \tag{21.62}$$

Using (21.60) and (21.62), the van der Pol system of equations in polar coordinates reads as follows:

$$\frac{dr}{dt} = -\alpha(r^2\cos^2\theta - 1)r\sin^2\theta, \tag{21.63}$$

$$\frac{d\theta}{dt} = -1 - \alpha(r^2\cos^2\theta - 1)\sin\theta\cos\theta. \tag{21.64}$$

The nonlinear terms on the right side have a rather complex shape, but one may give some qualitative statements on the solutions to be expected. In the limit $\alpha = 0$, one has of course a normal harmonic oscillator. The trajectories in phase space are circles which are travelled through uniformly with the frequency 1, such that

$$x(t) = \rho\sin(t - t_0) \tag{21.65}$$

with arbitrary ρ and t_0. Due to the nonlinearity in (21.63) and (21.64), the behavior of the solution is modified. As long as $\alpha \ll 1$, the influence on the revolution frequency remains small: Since the function $\sin\theta\cos\theta$ changes its sign twice in each period, the influence of the nonlinear term in (21.64) cancels out on the average. It is quite different, however, for the radial motion: Here $\sin^2\theta$ is positive definite, and small changes of the radius may accumulate from period to period. The evolution direction of the effect is determined by the sign of $-\alpha(r^2\cos^2\theta - 1)$. In the following we shall discuss the (more interesting) case $\alpha > 0$ (the set of solutions for $\alpha < 0$ may be obtained by inversion of the time coordinate $t \to -t$).

For *small displacements* $r < 1$, the factor $-\alpha(r^2\cos^2\theta - 1)$ is then always *positive*, and the radius increases slowly but monotonically. For *large displacements* $r \gg 1$, on the contrary, the factor is predominantly *negative* (except for the vicinity of the zeros of $\cos\theta$), and the mean radius decreases from cycle to cycle. A more detailed investigation as is performed in Example 21.3 shows that the trajectory in the course of time approaches a periodic one, independent of the initial conditions, which for given α is uniquely determined. This is the *limit cycle* of the system.

As long as α is very small, the limit cycle resembles a harmonic vibration as in (21.65). The crucial difference is, however, that the amplitude ρ now has a sharply determined value, namely, $\rho = 2$. If one starts from a smaller or larger value, the trajectory is a spiral approaching the limit cycle. The result of a numeric calculation for the value $\alpha = 0.1$ is represented in Figure 21.6. One can follow the spiraling motion towards the limit cycle. Moreover, deviations from the purely harmonic vibration become visible.

Even more interesting is the solution in the opposite limit $\alpha \gg 1$, in which the nonlinearity plays a dominant role. Here also a limit cycle evolves for the same reasons, the shape of which, however, strongly differs from a harmonic vibration. Figure 21.7 shows the

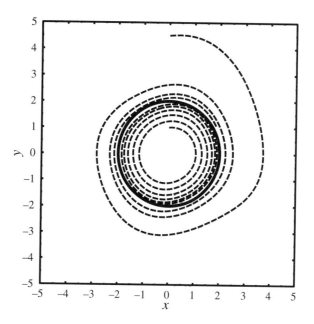

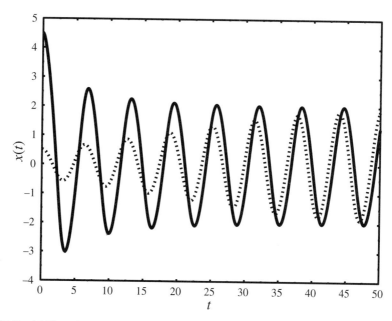

Figure 21.6. (a) The solutions of the system of differential equations (21.57) in the x,y-plane approach the limit cycle (bold curve), a slightly deformed circle, independent of the initial condition. The nonlinearity parameter is $\alpha = 0.1$. (b) The solutions $x(t)$ of the van der Pol oscillator after a transient motion coincide with an approximately harmonic vibration.

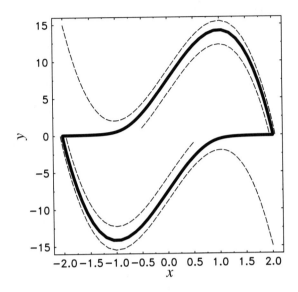

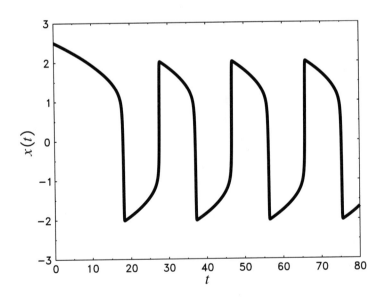

Figure 21.7. Solutions of the van der Pol oscillator with strong nonlinearity, $\alpha = 10$. The meaning of the curves is the same as in Figure 21.6.

phase-space plot and the trend of the amplitude of the limit cycle for the case $\alpha = 10$. One notices that the displacement remains in the range of the maximum amplitude $x = 2$ and slowly decreases toward $x = 1$. Subsequently, a sudden "flip-over" sets in, and the displacement drops to the value $x = -2$. Then the game repeats with opposite sign. The period length of this kind of vibration is no longer determined by the oscillator frequency (here $\omega = 1$) but takes a much larger value. An analytic investigation (see Example 21.4) shows that it increases proportional to the "friction" parameter α:

$$T \simeq (3 - 2\ln 2)\alpha. \tag{21.66}$$

A motion performed by a van der Pol- or Rayleigh oscillator for large α is also called *relaxation vibration*. The name indicates that a tension builds up slowly which then equilibrates via a sudden relaxation process. Such relaxation vibrations frequently occur in nature. For example, the vibration of a string excited by a bow, the squeak of a brake, and even the rhythm of a heartbeat or the time variation of animal populations may be classified in this way.

An important and also practically useful property of nonlinear oscillators with a limit cycle lies in the fact that *self-exciting* vibrations occur that are well defined and independent of the initial conditions. As a somewhat nostalgic example, we quote the balance of a mechanical clock, the vibrations of which are largely independent of the strength of the driving force. Finally, we quote without proof a mathematical theorem[4] stating that the possible types of motion of a two-dimensional system (corresponding to a mechanical system with one degree of freedom: one coordinate plus one velocity) are completely governed by fixed points and limit cycles.

The theorem of Poincaré and Bendixson: Let a two-dimensional dynamic system $\mathbf{x}(t) = \big(x(t), y(t)\big)^\mathrm{T}$ be described by the differential equation

$$\frac{d\mathbf{x}}{dt} = \mathbf{F}(\mathbf{x})$$

with a continuous function $\mathbf{F}$. Let B be a closed and restricted range of the x,y-plane. If a trajectory lies in B for any time $t > 0$, $\mathbf{x}(t) \in B$, there are three possibilities:

(i) $\mathbf{x}(t)$ is a periodic function,

(ii) $\mathbf{x}(t)$ asymptotically approaches a stationary equilibrium point, and

(iii) $\mathbf{x}(t)$ asymptotically approaches a periodic function (limit cycle).

The general theorem says of course nothing about the number and shape of the fixed points and limit cycles. However, it excludes the existence of more complicated nonperiodic types of solutions! It is important that the statement holds *only for two-dimensional systems*. Two trajectories are not allowed to intersect each other in phase space, which in the

---

[4]See, e.g., J. Guckenheimer and P. Holmes, *Nonlinear Oscillations, Dynamical Systems and Bifurcations of Vector Fields*, Springer (1983).

two-dimensional plane leads to considerable restrictions. But in more than two dimensions the trajectories may "evade" each other, and more complex patterns of motion are possible. In this case, the already-mentioned *strange attractors* with a complicated shape may also occur. This will be treated in the next chapters.

Example 21.3: **Exercise: The van der Pol oscillator with weak nonlinearity**

Show that the solutions of the van der Pol oscillator for small values of α are spirals that approach a circle (the limit cycle) with the radius 2. *Hint:* It is a good idea to introduce new variables that are averaged over one oscillation period.

Solution

We start from the plausible assumption that for $\alpha \ll 1$ the solution of the system of differential equations (21.63) and (21.64) differs only slightly from that of the harmonic oscillator if it is considered for short time intervals. In order to calculate a long-term drift of the variables, it is efficient to average over one vibrational period in each case. We define the *averaged amplitude* $\bar{r}(t)$ as

$$\bar{r}(t) := \frac{\oint d\tau \, r(t+\tau)}{\oint dt}. \tag{21.67}$$

The integration thereby extends over a full revolution of the angle, i.e., from θ to $\theta - 2\pi$ (the minus sign arises because of $d\theta/dt \simeq -1$). The corresponding time interval runs from t to approximately (for $\alpha = 0$ exactly) the value $t + 2\pi$.

We are interested in the time variation of the averaged amplitude for which according to (21.63) we have

$$\frac{d\bar{r}}{dt} = -\alpha \frac{1}{2\pi} \oint d\theta \, r \sin^2\theta (r^2 \cos^2\theta - 1)$$

$$= -\alpha \int_0^{2\pi} \frac{d\theta}{2\pi} r \left(r^2 \frac{1}{4} \sin^2 2\theta - \sin^2\theta \right). \tag{21.68}$$

For small α, the quantity $r(t)$ considered over a period varies only slowly, and hence may be pulled out of the integral and replaced by $\bar{r}(t)$. The remaining angle integration is trivial, since the mean value of both $\sin^2\theta$ and $\sin^2 2\theta$ just equals $1/2$. Hence, the averaged amplitude satisfies the differential equation

$$\frac{d\bar{r}}{dt} = \frac{1}{2} \alpha \bar{r} \left(1 - \frac{1}{4} \bar{r}^2 \right), \tag{21.69}$$

which is correct up to the order $O(\alpha^2)$. The circulation frequency, on the contrary, does not change to first order:

$$\frac{d\bar{\theta}}{dt} = \int_0^{2\pi} \frac{d\theta}{2\pi} \left[-1 - \alpha(r^2 \cos^2\theta - 1) \sin\theta \cos\theta \right] = -1. \tag{21.70}$$

The angular integral vanishes here, since the integrand is an odd function with respect to $\theta = \pi$. The differential equation (21.69) for the averaged amplitude may be solved in closed form. We write

$$\frac{d\bar{r}}{dt} = a\bar{r} - b\bar{r}^3 \tag{21.71}$$

and transform to the new variable

$$u = \frac{1}{\bar{r}^2} \quad \text{such that} \quad du = -2\frac{d\bar{r}}{\bar{r}^3}. \tag{21.72}$$

Obviously, (21.71) reduces to the simple linear differential equation

$$-\frac{1}{2}\frac{du}{dt} = au - b, \tag{21.73}$$

the solution of which is a shifted exponential function:

$$u(t) = \frac{b}{a} + c\,e^{-2at}, \tag{21.74}$$

where the free constant c is to be determined from the initial condition: $c = u(0) - b/a$. Insertion of $a = \alpha/2$ and $b = \alpha/8$ finally yields

$$\bar{r}(t) = \frac{2r(0)}{\sqrt{r^2(0) + (4 - r^2(0))e^{-at}}}. \tag{21.75}$$

Thus, it is proved that the trajectories are spirals which approach a circle with radius 2 from inside ($r(0) < 2$) or outside ($r(0) > 2$). This is the limit cycle of the van der Pol oscillator for small values of α.

Example 21.4: Exercise: Relaxation vibrations

Discuss the solutions of the Rayleigh oscillator (21.55) qualitatively for large values of the parameter $\alpha \gg 1$. Find an approximate solution for the period length of the resulting relaxation vibration.

Solution The differential equation (21.55) of the Rayleigh oscillator written in standard form reads

$$\frac{dv}{dt} = x,$$
$$\frac{dx}{dt} = -v - \alpha\left(\frac{1}{3}x^3 - x\right). \tag{21.76}$$

In order to discuss the behavior of the solution for large values of α, it is convenient to rescale the amplitude to a new variable $z = v/\alpha$:

$$\frac{dz}{dt} = \frac{1}{\alpha}x,$$
$$\frac{dx}{dt} = -\alpha\bigl[z + f(x)\bigr], \tag{21.77}$$

with the abbreviation $f(x) = (x^3/3) - x$. From this quantity one may read off the direction of the trajectory for any point of the z,x-plane:

$$\frac{dx}{dz} = \frac{dx/dt}{dz/dt} = -\alpha^2\frac{z + f(x)}{x}. \tag{21.78}$$

This means that for $\alpha \gg 1$, the trajectories are almost vertical. Other directions may occur only near the curve $z(x) = -f(x)$. This cubic limit curve subdivides the z,x-plane into two halves (see Figure 21.8). In the right half, the derivative dx/dt is negative, according to (21.77), and the trajectories are running (almost) vertically downward. In the left half, they are running upward.

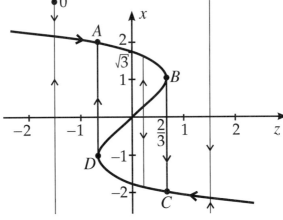

Figure 21.8. The cubic limit curve z(x) determines the asymptotic behavior of the relaxation vibrations of the van der Pol oscillator.

From this knowledge, the motion for large α may be constructed graphically. Beginning from an arbitrary initial point, e.g., the point O in the figure, the trajectory at first falls almost vertically down to the curve $z(x) = -f(x)$. The further motion proceeds with significantly lower velocity near this curve (directly on the curve the velocity would vanish, $dx/dt = 0$). Finally, the point of inversion B is reached at $(z, x) = (2/3, 1)$.

Because $dz/dt > 0$, the trajectory cannot follow the backward-running branch of the curve but "falls down" to the point C at $(2/3, -2)$. Now the game is repeating with inverse sign. The curve $ABCD$ forms the *limit cycle* of the Rayleigh oscillator. It consists of two slowly passed parts ($x = 2 \ldots 1$ and $-2 \ldots -1$) and two fast jumps ($x = 1 \ldots -2$ and $-1 \ldots 2$). This discussion immediately applies, of course, to the van der Pol oscillator, since according to (21.56) its displacement just corresponds to the velocity x of the Rayleigh oscillator introduced in (21.76).

The *period length* T of the relaxation vibration may be evaluated easily:

$$T = \oint dt = \alpha \oint \frac{dz}{x}, \tag{21.79}$$

where the integral extends over a full period. Since the motion along the partial branches BC and DA proceeds very quickly, it is sufficient to calculate the contribution of AB:

$$T \simeq \alpha \int_{AB} \frac{dz}{x} + \alpha \int_{CD} \frac{dz}{x}$$

$$= 2\alpha \int_{AB} \frac{dz}{x} = 2\alpha \int_{2}^{1} dx \, \frac{dz/dx}{x}. \tag{21.80}$$

The derivative dz/dx is to be formed on the curve AB, i.e., $dz/dx = -df/dx$:

$$T \simeq 2\alpha \int_{2}^{1} dx \, \frac{-df/dx}{x} = 2\alpha \int_{1}^{2} dx \, \frac{x^2 - 1}{x} = 2\alpha \left(\frac{1}{2} x^2 - \ln x \right) \Big|_{1}^{2}$$

$$= \left(3 - 2\ln 2\right)\alpha \simeq 1.614\,\alpha. \tag{21.81}$$

The period length calculated numerically in the Figure 21.7b for $\alpha = 10$ amounts to about 19, hence the asymptotic range is not yet fully reached.

22 Stability of Time-Dependent Paths

In Chapter 21, we have considered the stability of an equilibrium point x_0 by investigating the behavior of the trajectories in the vicinity of this point. Now we are interested in the environment of a *time-dependent* reference trajectory $\mathbf{x}_r(t)$. The former case of a stationary fixed point $\mathbf{x}_r(t) = \mathbf{x}_0 =$ constant is of course included too. One may again distinguish between various kinds of stability. *Stability in the sense of Lyapunov* exists if a point on a neighboring trajectory $\mathbf{x}(t)$ remains close to $\mathbf{x}_r(t)$ for all times. The formal expression of this concept reads as follows: A path $\mathbf{x}_r(t)$ is called *Lyapunov-stable* if for any $\epsilon > 0$ a value $\delta(\epsilon) > 0$ can be found such that any solution with $|\mathbf{x}(t_0) - \mathbf{x}_r(t_0)| < \delta$ satisfies the condition $|\mathbf{x}(t) - \mathbf{x}_r(t)| < \epsilon$ for all times $t > t_0$.

Figure 22.1(a) shows that the "perturbed" paths are confined within an $(N + 1)$-dimensional tube of radius ϵ about the reference trajectory. This does not yet mean that the trajectories are approaching each other with time. If the latter happens, one speaks as before (see equation (21.31)) of asymptotic stability; see Figure 22.1(b).

A path $\mathbf{x}_r(t)$ is called *asymptotically stable* if it is Lyapunov-stable and if for the neighbor trajectories $\lim_{t \to \infty} |\mathbf{x}(t) - \mathbf{x}_r(t)| = 0$.

It is of interest that for time-dependent trajectories $\mathbf{x}_r(t)$ there exists still another concept of stability that was not needed for stationary fixed points $\mathbf{x}_0$. It may happen that although the shape of two paths in phase space is very similar, these paths are nevertheless passed

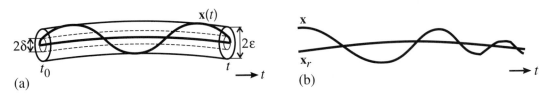

Figure 22.1. (a) The neighboring paths $\mathbf{x}(t)$ of a Lyapunov-stable path $\mathbf{x}_r(t)$ remain in its vicinity. (b) In the case of asymptotic stability, neighboring paths are attracted such that the distance decreases to zero with increasing time.

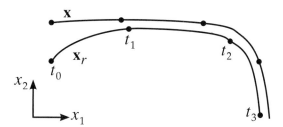

Figure 22.2. Example of a path that is asymptotically stable but not orbitally stable. Neighboring paths are passed with distinct speeds.

with distinct speeds; see Figure 22.2. Thereby a time shift may evolve, $|\mathbf{x}(t) - \mathbf{x}_r(t)|$ increases, and the definitions of stability given so far do not apply. One therefore introduces a weakened version: A path $\mathbf{x}_r(t)$ is called *orbitally stable* if for any $\epsilon > 0$ a value $\delta(\epsilon) > 0$ can be found such that any solution with $|\mathbf{x}(t_0) - \mathbf{x}_r(t_0)| < \delta$ is confined for all times $t > t_0$ within a tube of radius ϵ about the path $\mathbf{x}_r(t)$.

Using this definition, we consider the geometric position $\mathbf{x}(t)$. The time as a parameter of this curve, on the contrary, does not play a role.

Periodic solutions

The investigation of the stability of time-dependent solutions is by its very nature more difficult than in the case of stationary fixed points. Here we shall treat the important special case of *periodic solutions*. Then we may apply a formalism that goes back to the French mathematician Floquet (1883). Thus, we assume that the reference trajectory $\mathbf{x}_r(t)$ repeats itself after a period length T,

$$\mathbf{x}_r(t+T) = \mathbf{x}_r(t). \tag{22.1}$$

This may originate in two distinct manners. First, in an *autonomous system* vibrations may arise by themselves, e.g., in the harmonic oscillator. Here the right-hand side of the equation of motion does not depend explicitly on the time, $\dot{\mathbf{x}} = \mathbf{F}(\mathbf{x})$. On the other hand, there are also *periodically externally excited systems* which are under the action of a time-dependent external drive with the periodicity $\mathbf{F}(\mathbf{x}, t+T) = \mathbf{F}(\mathbf{x}, t)$ that reflects itself in the trajectory. An advantage is here that the period length T is imposed from outside, while the vibrational frequency of an autonomous system is not known from the beginning and must be determined—except for simple special cases—by numerical solution of the equation of motion.

To discuss the stability of $\mathbf{x}_r(t)$, one investigates, as in (21.23), the neighboring trajectories

$$\mathbf{x}(t) = \mathbf{x}_r(t) + \boldsymbol{\xi}(t), \tag{22.2}$$

where the deviation $\boldsymbol{\xi}(t)$ is assumed to be small. From the equations of motion

$$\dot{\mathbf{x}}(t) = \mathbf{F}(\mathbf{x}, t) \quad \text{and} \quad \dot{\mathbf{x}}_r(t) = \mathbf{F}(\mathbf{x}_r, t), \tag{22.3}$$

we find

$$\dot{\mathbf{x}}_r + \dot{\boldsymbol{\xi}} = \mathbf{F}(\mathbf{x}_r + \boldsymbol{\xi}, t) = \mathbf{F}(\mathbf{x}_r, t) + \dot{\boldsymbol{\xi}}, \tag{22.4}$$

or

$$\dot{\boldsymbol{\xi}} = \mathbf{F}(\mathbf{x}_r + \boldsymbol{\xi}, t) - \mathbf{F}(\mathbf{x}_r, t) \equiv \mathbf{G}(\boldsymbol{\xi}, t). \tag{22.5}$$

This equation can be *linearized* by expanding the right side in a Taylor series and neglecting higher terms:

$$\dot{\boldsymbol{\xi}}(t) = \mathbf{M}(t)\,\boldsymbol{\xi}(t) \tag{22.6}$$

with the Jacobi matrix (here written in abstract without giving the indices)

$$\mathbf{M}(t) = \left.\frac{\partial \mathbf{G}}{\partial \boldsymbol{\xi}}\right|_{\boldsymbol{\xi}=0} = \left.\frac{\partial \mathbf{F}}{\partial \mathbf{x}}\right|_{\mathbf{x}_r(t)}. \tag{22.7}$$

Equation (22.6) is, like (21.24), a linear system of differential equations, but now the matrix of coefficients is *periodically time-dependent*, $\mathbf{M}(t+T) = \mathbf{M}(t)$, while formerly it was constant. This periodicity also holds for autonomous systems: Although the function $\mathbf{F}(\mathbf{x})$ does not involve the time explicitly, the reference trajectory $\mathbf{x}_r(t)$ by itself nevertheless induces a periodic time dependence.

Discretization and Poincaré cuts

There exists a mathematical tool that is useful for the stability analysis of time-dependent paths but also in general for the qualitative understanding of dynamic systems. The basic idea is to perform a *discretization of the time dependence* of a trajectory. This may be done in two somewhat different ways.

An obvious possibility is the *stroboscopic mapping*. Instead of the continuous function $\mathbf{x}(t)$, one considers a discrete sequence of "snapshots" $\mathbf{x}_n = \mathbf{x}(t_n)$, $n = 0, 1, 2, \ldots$. The time points of support of the spectroscopic method are chosen as equidistant, hence $t_n = t_0 + nT$ with a scanning interval T. Of course one should choose a value of T that is appropriate for the problem. If an oscillating driving force is acting, one will use its period length for T. The stroboscopic method becomes particularly simple if the trajectory $\mathbf{x}(t)$ itself is *periodic* and T coincides with the period length; then all $\mathbf{x}_n$ are of course identical. The stroboscopic mapping of the path consists of a *single point* $\mathbf{x}_n = \mathbf{x}_0$ in phase space. One should note that the position of the point $\mathbf{x}_0$ depends of course on the selected reference time t_0 and thereby may be shifted arbitrarily along the orbit.

As was described in the preceding section, for a stability analysis one investigates neighboring trajectories $\mathbf{x}(t)$ that are in general no longer strictly periodic. The thin line in Figure 22.3 shows such an example. The first three stroboscopic snapshots $\mathbf{x}_0, \mathbf{x}_1, \mathbf{x}_2$ are marked by dots and the distance vectors $\boldsymbol{\xi}_n = \mathbf{x}_n - \mathbf{x}_{r0}$ are plotted.

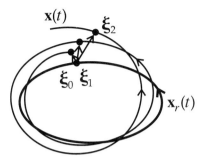

Figure 22.3. The stroboscopic scanning of the distance $\xi(t) = \mathbf{x}(t) - \mathbf{x}_r(t)$ yields information on the stability of a path.

An alternative method of discretization of trajectories, which is not oriented to the periodicity and is of advantage particularly for autonomous systems having no fixed eigenfrequency, is the *Poincaré cut*. When changing over to the discretized sequence $\mathbf{x}_n$, one again chooses momentary snapshots of the continuous orbit $\mathbf{x}(t)$. As a criterion one now adopts not any fixed equidistant time distances, but rather a geometric property of the orbit itself, namely the *piercing of a given hypersurface* Σ. One thereby selects an $(N-1)$-dimensional hypersurface in phase space and marks all points x_n at which the trajectory intersects the hypersurface. One further requires that Σ is not only touched but properly pierced. Mathematically this means that the surface shall be transverse to the dynamic flow, $\mathbf{n}(\mathbf{x}) \cdot \mathbf{F}(\mathbf{x}) \neq 0$ everywhere on Σ, where $\mathbf{n}$ is the surface normal. One therefore speaks of a *transverse cut*. In a transverse cut one usually marks only points with a definite sign of $\mathbf{F} \cdot \mathbf{n}$, i.e., only piercings of Σ that proceed in the same direction.

This method of discretization of trajectories was invented by Henri Poincaré and is called the *Poincaré cut*. Figure 22.4 shows as an example a trajectory in an $(N = 3)$-dimensional

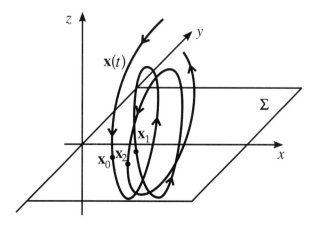

Figure 22.4. The piercing points of a trajectory through a given surface constitute the Poincaré cut.

space, with the x,y-plane as the cut surface Σ. Three piercings in the negative z-direction are marked as $\mathbf{x}_0, \mathbf{x}_1, \mathbf{x}_2$. The use of Poincaré cuts makes sense only then if the trajectory is moving largely or completely in a restricted range of the phase space, such that the cut surface is pierced again and again. Many systems with approximately periodic or also chaotic motion satisfy this condition. Examples of nontrivial Poincaré cuts will be given in Chapter 25.

An advantage of the Poincaré cut is first of all the *reduction of dimension* of the phase space from N to $N-1$, which may be very helpful for the qualitative discussion. For detailed studies, not only does one want to know the ensemble of points $\mathbf{x}_n$, but also their detailed sequence. Mathematically, this connection is mediated by a mapping

$$P : \mathbf{x}_n \to \mathbf{x}_{n+1} \quad \text{with } \mathbf{x}_n, \mathbf{x}_{n+1} \in \Sigma. \tag{22.8}$$

This *Poincaré mapping* thus connects every point of the sequence $\mathbf{x}_0, \mathbf{x}_1, \mathbf{x}_2, \ldots$ with its successor. Note that P has no index. It is a single mapping of the plane Σ onto itself, which according to (22.8) is "scanned" at individual points. The individual points of the Poincaré cut arise by successive *iteration of the Poincaré mapping*

$$\mathbf{x}_1 = P(\mathbf{x}_0), \quad \mathbf{x}_2 = P(\mathbf{x}_1) = P^2(\mathbf{x}_0), \quad \ldots, \quad \mathbf{x}_n = P^n(\mathbf{x}_0). \tag{22.9}$$

Hence, the long-term behavior of a trajectory may be derived from the properties of the iterated Poincaré mapping P^n, $n \to \infty$. If the time evolution of the dynamic system is determined by a differential equation $\dot{\mathbf{x}} = \mathbf{F}(\mathbf{x}, t)$, the Poincaré mapping is *unique* and also *reversible* (possibly except for singular points), since trajectories are not allowed to intersect each other.

The problem of describing a dynamic system is of course not yet solved by defining the Poincaré mapping but is only postponed, since P must also be constructed explicitly. In most cases this cannot be achieved analytically, and one is finally left with a numerical integration of the differential equation of the system. It turns out, however, that the exact Poincaré mapping P frequently has amazing common features with very simply constructed analytic discrete mappings. As an example we shall discuss the "logistic mapping" in Chapter 25.

Let us return to the problem of *stability of periodic paths*. As was outlined in the preceding section, it is sufficient to investigate the small deviations $\boldsymbol{\xi}(t) = \mathbf{x}(t) - \mathbf{x}_r(t)$ from the reference trajectory in the linear approximation. In this approximation the Poincaré mapping simplifies to a *linear mapping*, i.e., the multiplication by a matrix $\mathbf{C}$:

$$\boldsymbol{\xi}_{n+1} = \mathbf{C}\boldsymbol{\xi}_n; \quad \text{hence}, \quad \mathbf{x}_n = \mathbf{C}^n \boldsymbol{\xi}_0. \tag{22.10}$$

Decisive for the long-term behavior of the deviation $\boldsymbol{\xi}(t)$ are the eigenvalues $\lambda_1, \ldots, \lambda_N$ of the matrix $\mathbf{C}$. If all eigenvalues satisfy the condition $|\lambda_i| < 1$, the mapping is contracting, and the sequence converges toward zero. In this case the periodic solution $\mathbf{x}_r(t)$ is thus asymptotically stable. If at least one of the eigenvalues $|\lambda| > 1$, the perturbations are increasing along the direction of the associated eigenvector, and the path is unstable.

In the subsequent example, the mathematical theory of the stability of periodic solutions developed by Floquet will be presented in more detail.

Example 22.1: Floquet's theory of stability

As described at the beginning of this chapter, we are interested in the long-term behavior of the path deviations $\boldsymbol{\xi}(t) = \mathbf{x}(t) - \mathbf{x}_r(t)$ which approximately obey a linear differential equation

$$\frac{d}{dt}\boldsymbol{\xi}(t) = \mathbf{M}\,\boldsymbol{\xi}(t) \tag{22.11}$$

with a periodic matrix of coefficients $\mathbf{M}(t+T) = \mathbf{M}(t)$. Since we are dealing with a linear problem, any solution may be expanded in terms of a fundamental system of linearly independent basic solutions $\boldsymbol{\phi}_1(t), \ldots, \boldsymbol{\phi}_N(t)$. The basic solutions are not uniquely determined, and for sake of clarity we choose them in such a way that at the time $t = 0$ (we might choose also $t = t_0$) they just coincide with the unit vectors in the N-dimensional space:

$$\boldsymbol{\phi}_1(0) = (1, 0, \cdots, 0)^{\mathrm{T}} \quad \text{to} \quad \boldsymbol{\phi}_N(0) = (0, 0, \cdots, 1)^{\mathrm{T}}, \tag{22.12}$$

where the transposition symbol T indicates that these vectors shall be column vectors. Geometrically all of these vectors are lying on a (hyper-) spherical surface of radius unity. The superposition of a general solution $\boldsymbol{\xi}(t)$ reads

$$\boldsymbol{\xi}(t) = \sum_{i=1}^{N} c_i\, \boldsymbol{\phi}_i(t), \tag{22.13}$$

which may also be written in matrix form:

$$\boldsymbol{\xi}(t) = \boldsymbol{\Phi}(t)\,\mathbf{c}. \tag{22.14}$$

$\mathbf{c}$ is a column vector formed out of the expansion coefficients, and $\boldsymbol{\Phi}$ is a $N \times N$-matrix containing one of the basic vectors in each column,

$$\boldsymbol{\Phi}(t) = \left[\boldsymbol{\phi}_1(t), \cdots, \boldsymbol{\phi}_N(t)\right] \quad \text{and} \quad \mathbf{c} = (c_1, \cdots, c_N)^{\mathrm{T}}. \tag{22.15}$$

Due to (22.12), the matrix $\boldsymbol{\Phi}$ satisfies the initial condition

$$\boldsymbol{\Phi}(0) = \mathbf{I}. \tag{22.16}$$

How does the periodicity of the differential equation (22.11) manifest itself in the matrix $\boldsymbol{\Phi}$? To see this, one should realize that any solution $\boldsymbol{\xi}$ of (22.11) at the time $t + T$ satisfies the same differential equation as at the time t. This does of course not mean that the solution will be periodic; in general $\boldsymbol{\xi}(t+T) \neq \boldsymbol{\xi}(t)$. But it may be expanded both in terms of the basic solutions $\boldsymbol{\phi}_i(t+T)$ as well as in terms of the $\boldsymbol{\phi}_i(t)$,

$$\boldsymbol{\xi}(t+T) = \boldsymbol{\Phi}(t+T)\,\mathbf{c} \quad \text{and also} \quad \boldsymbol{\xi}(t+T) = \boldsymbol{\Phi}(t)\,\mathbf{c}'. \tag{22.17}$$

This implies a linear relation

$$\boldsymbol{\Phi}(t+T) = \boldsymbol{\Phi}(t)\,\mathbf{C}. \tag{22.18}$$

The constant $N \times N$-matrix $\mathbf{C}$ is called the *monodromy matrix*. This quantity governs how the solutions develop from one period to the next. Using the initial condition (22.16), one may immediately read off the value of the monodromy matrix from (22.18):

$$\mathbf{C} = \boldsymbol{\Phi}(T). \tag{22.19}$$

Hence, the monodromy matrix may be calculated by integrating the differential equation (22.11) N times with distinct initial conditions over a period from 0 to T and writing the resulting solution vectors $\phi_i(T)$ into the columns. The evolution of the matrix $\boldsymbol{\Phi}(t)$ for arbitrarily large times is obtained by iteration of (22.18). For full periods in particular, we have

$$\boldsymbol{\Phi}(2T) = \boldsymbol{\Phi}(T)\mathbf{C} = \boldsymbol{\Phi}(0)\mathbf{CC} = \mathbf{C}^2, \tag{22.20}$$

and generally,

$$\boldsymbol{\Phi}(nT) = \boldsymbol{\Phi}^n(T) = \mathbf{C}^n. \tag{22.21}$$

According to (22.14) and (22.21), the evolution of the solutions $\boldsymbol{\xi}(t)$ for large times is thus determined by the powers of the monodromy matrix $\mathbf{C}$. What happens thereby may be read off from the N eigenvalues λ_i of this matrix, which are called characteristic multipliers or *Floquet multipliers*,

$$\boldsymbol{\Phi}(T)\mathbf{u}_i = \lambda_i(T)\mathbf{u}_i. \tag{22.22}$$

If $\mathbf{u}_i$ is an eigenvector of the matrix $\boldsymbol{\Phi}(T)$, then it keeps this property for the iterated mapping as well, e.g.,

$$\begin{aligned}\boldsymbol{\Phi}(2T)\mathbf{u}_i &= \boldsymbol{\Phi}(T)\boldsymbol{\Phi}(T)\mathbf{u}_i = \boldsymbol{\Phi}\lambda_i(T)\mathbf{u}_i = \lambda_i^2(T)\mathbf{u}_i \\ &= \lambda_i(2T)\mathbf{u}_i, \quad \text{etc.}\end{aligned} \tag{22.23}$$

This leads to the following functional equation for the eigenvalues of the iterated mapping:

$$\lambda_i(nT) = \lambda_i^n(T). \tag{22.24}$$

This behavior is characteristic for the exponential function; i.e., equation (22.24) is solved by

$$\lambda_i(T) = e^{\sigma_i T} \tag{22.25}$$

with an (in general complex) constant σ_i that is called the *Floquet exponent*. We still note that (22.21) may be considered a functional equation like (22.24), with the same kind of solution

$$\boldsymbol{\Phi}(T) = e^{\mathbf{S}T}. \tag{22.26}$$

Here, a matrix $\mathbf{S}$ stands in the argument of the exponential function, and the resulting function value is again a matrix. Such a matrix function is mathematically defined simply through its power series expansion. One can show that the eigenvalues of the matrix $\mathbf{S}$ introduced by (22.26) are just the Floquet exponents σ_i of (22.25). If one is interested in the evolution matrix at any times (not only multiples of the period T), then (22.26) still has to be generalized:

$$\boldsymbol{\Phi}(t) = \mathbf{U}(t)\,e^{\mathbf{S}t} \quad \text{with} \quad \mathbf{U}(0) = \mathbf{I}. \tag{22.27}$$

The matrix $\mathbf{U}(t)$ may exhibit a complicated time dependence but must be periodic. Because of (22.18), we have

$$\mathbf{U}(t+T)\,e^{\mathbf{S}(t+T)} = \boldsymbol{\Phi}(t)\boldsymbol{\Phi}(T) = \mathbf{U}(t)\,e^{\mathbf{S}t}e^{\mathbf{S}T}. \tag{22.28}$$

The product of the exponential functions on the right side may be combined to $\exp(\mathbf{S}t)\exp(\mathbf{S}T) = \exp\mathbf{S}(t+T)$ (for noncommuting matrices in the exponent this would in general not be correct), and we find

$$\mathbf{U}(t+T) = \mathbf{U}(t). \tag{22.29}$$

Hence, for the long-term behavior of the solutions $\boldsymbol{\xi}(t)$, $\mathbf{U}$ does not play a role. This behavior is only determined by the magnitude of the Floquet multipliers λ_i. Beginning with an eigenvector $\boldsymbol{\xi}(0) = \mathbf{u}_i$,

DISCRETIZATION AND POINCARÉ CUTS

this solution according to (22.24) will increase as $\boldsymbol{\xi}(nT) = \mathbf{u}_i \exp(\sigma_i nT)$. From that, we conclude the following: The trajectory $\mathbf{x}_r(t)$ is *asymptotically stable* if for all Floquet multipliers we have $|\lambda_i| < 1$; i.e., Re $\sigma_i < 0$. It is *unstable* if for at least one eigenvalue we have $|\lambda_i| > 1$; i.e., Re $\sigma_i > 0$.

These statements, which were obtained by linearizing the equation of motion, transfer also to the stability behavior of the nonlinear system. The limit of marginal stability $|\lambda_i| = 1$ may be cleared up only by additional investigations.

For an *autonomous* periodically vibrating system, a peculiarity arises: In this case, one of the eigenvalues always has the value $\lambda = 1$ and must not be considered in the stability analysis. To prove this assertion, we consider the function $\dot{\mathbf{x}}_r(t)$. The mode under consideration is namely the motion *tangential* to the reference orbit. For an autonomous system, one obtains by differentiating the nonlinear equation of motion

$$\dot{\mathbf{x}}_r(t) = \mathbf{F}(\mathbf{x}_r) \quad \longrightarrow \quad \ddot{\mathbf{x}}_r(t) = \left.\frac{\partial \mathbf{F}}{\partial \mathbf{x}}\right|_{\mathbf{x}_r} \dot{\mathbf{x}}_r = \mathbf{M}(t)\dot{\mathbf{x}}_r, \tag{22.30}$$

which agrees with the linearized equation of motion (22.16). The time evolution of the solutions of this differential equation is determined by the matrix $\mathbf{\Phi}(t)$; hence,

$$\dot{\mathbf{x}}_r(t) = \mathbf{\Phi}(t)\dot{\mathbf{x}}_r(0). \tag{22.31}$$

The reference orbit $\mathbf{x}_r(t)$ and therefore also its derivative $\dot{\mathbf{x}}_r(t)$ are however (contrary to the case of general perturbations $\boldsymbol{\xi}(t)$) periodic; hence,

$$\dot{\mathbf{x}}_r(T) = \mathbf{\Phi}(T)\dot{\mathbf{x}}_r(0) = \dot{\mathbf{x}}_r(0), \tag{22.32}$$

which proves that the monodromy matrix has an eigenvector, namely, $\dot{\mathbf{x}}_r(0)$, with the eigenvalue $\lambda = 1$. This is vividly clear: A reference orbit that is shifted in the tangential direction simply corresponds to a shift of the time coordinate $t \to t + \delta t$. Since the absolute value of the time does not play a role in autonomous systems, $\mathbf{x}_r(t)$ and $\mathbf{x}(t) = \mathbf{x}_r(t + \delta t)$ are always running with unchanged distance one behind the other. Hence, the associated Floquet multiplier must have the value unity.

Example 22.2:   Exercise: Stability of a limit cycle

Let a nonlinear system be described by the following equation of motion:

$$\dot{x} = -y + x\left(\rho - (x^2 + y^2)\right), \tag{22.33}$$

$$\dot{y} = x + y\left(\rho - (x^2 + y^2)\right). \tag{22.34}$$

Investigate the stable solutions and find the Floquet multipliers of the limit cycle.

Solution   A stationary *fixed point* exists at $\mathbf{x}_0 = (0, 0)^\mathrm{T}$. Its stability is governed by the Jacobi matrix (22.7)

$$\mathbf{M}(\mathbf{x}) = \frac{\partial \mathbf{F}}{\partial \mathbf{x}} = \begin{pmatrix} \rho - 3x^2 - y^2 & -1 - 2xy \\ 1 - 2xy & \rho - 3y^2 - x^2 \end{pmatrix}, \tag{22.35}$$

which at the fixed point $\mathbf{x}_0$ takes the form

$$\mathbf{M}(\mathbf{x}_0) = \begin{pmatrix} \rho & -1 \\ 1 & \rho \end{pmatrix}. \tag{22.36}$$

The two eigenvalues are according to Example 21.1:

$$\mu_{1/2} = s \pm \sqrt{s^2 - d} = \rho \pm \sqrt{\rho^2 - (\rho^2 + 1)} = \rho \pm i. \tag{22.37}$$

Hence, for $\rho < 0$ one has a stable spiral and for $\rho > 0$ an unstable one; $\rho = 0$ represents the special case of a rotor.

By inspecting (22.33), one immediately finds a periodic solution for $\rho > 0$, since for constant $x^2 + y^2 = \rho$ the system reduces to a harmonic oscillator:

$$\mathbf{x}_r(t) = (\sqrt{\rho} \cos t, \ \sqrt{\rho} \sin t)^\mathrm{T}. \tag{22.38}$$

The Jacobi matrix (22.12) evaluated at the limit cycle (22.38) reads

$$\mathbf{M}(t) = \mathbf{M}(\mathbf{x}_r) = \begin{pmatrix} -2\rho \cos^2 t & -1 - 2\rho \sin t \cos t \\ 1 - 2\rho \sin t \cos t & -2\rho \sin^2 t \end{pmatrix}. \tag{22.39}$$

For the linearized system of equations (22.33) with this matrix $\mathbf{M}(t)$, the normalized fundamental solutions (22.12) from Example 22.1 may be given explicitly. One finds

$$\boldsymbol{\phi}_1(t) = e^{-2\rho t} \begin{pmatrix} \cos t \\ \sin t \end{pmatrix} \quad \text{and} \quad \boldsymbol{\phi}_2(t) = \begin{pmatrix} -\sin t \\ \cos t \end{pmatrix}. \tag{22.40}$$

Combining these vectors to the matrix $\boldsymbol{\Phi}(t)$ and evaluating at $T = 2\pi$ leads to the monodromy matrix

$$\mathbf{C} = \boldsymbol{\Phi}(T) = \begin{pmatrix} e^{-4\pi\rho} & 0 \\ 0 & 1 \end{pmatrix}. \tag{22.41}$$

Hence, the basic solutions (22.40) are also already eigenvectors of the monodromy matrix, with the eigenvectors

$$\lambda_1 = e^{-4\pi\rho} \quad \text{and} \quad \lambda_2 = 1. \tag{22.42}$$

As expected, one of the Floquet multipliers has the value unity (the corresponding eigensolution $\boldsymbol{\phi}_2(t)$ is tangential to $\mathbf{x}_r(t)$). The value λ_1 determines the stability of the limit cycle: It is asymptotically stable, since $\lambda_1 < 1$ for $\rho > 0$. For $\rho < 0$, no limit cycle exists.

The nonlinear system (22.33) is so simple that it allows also a closed analytic solution. We change to polar coordinates $x = r \cos \varphi$, $y = r \sin \varphi$. The differential equations (22.33) lead to the decoupled system

$$\dot{r} = r(\rho - r^2), \qquad \dot{\varphi} = 0. \tag{22.43}$$

Hence, the angle simply increases linearly with time, $\varphi(t) = t + \varphi_0$. The radial equation may be integrated as follows:

$$\int_{r_0}^{r} \frac{dr}{r(\rho - r^2)} = \int_{0}^{t} dt \quad \longrightarrow \quad \frac{1}{2\rho} \ln \frac{r^2}{\rho - r^2} \bigg|_{r_0}^{r} = t, \tag{22.44}$$

or solved for r

$$r(t) = \frac{\sqrt{\rho}}{\sqrt{\left(\rho/r_0^2 - 1\right) e^{-2\rho t} + 1}}. \tag{22.45}$$

For $\rho > 0$, the solution asymptotically approaches the limit cycle $r(t) \to \sqrt{\rho}$. For $\rho < 0$,

$$r(t) = \frac{\sqrt{|\rho|}}{\sqrt{\left(|\rho|/r_0^2 + 1\right)e^{2|\rho|t} - 1}} ; \qquad (22.46)$$

any trajectory asymptotically spirals to the stable fixed point $r_0 = 0$.

23 Bifurcations

As a rule, the behavior of dynamical systems is influenced by the value of one or several control parameters μ. These may be the strength of an interaction, the magnitude of friction, the amplitude and frequency of a periodic perturbation, or many other quantities. One then frequently observes that the long-term behavior of the trajectories changes qualitatively when passing a critical value $\mu > \mu_c$ of the parameter. For example, instead of a stable equilibrium position, suddenly two such positions may occur, or a system initially at rest may begin to oscillate. The phenomenon of additionally arising solutions or of solutions that suddenly change their character is called *branching* or *bifurcation*; the value μ_c is called the *branching value*.

The general theory of bifurcations is difficult and not yet completely worked out mathematically. Here we shall confine ourselves to easily tractable but important cases. This means for the present that we consider only *local* bifurcations, where the behavior of the system in the neighborhood of an equilibrium solution is changing. There exist also *global* bifurcations; here the topological structure of the solutions is modified "on a large scale," e.g., the shape of the ranges of attraction of attractors. Furthermore, we consider only the *one-dimensional* case, which means the bifurcation shall arise when varying a single control parameter. (The phase space of the dynamical system, however, may be multidimensional.) Even under these restricting conditions a series of various types of bifurcations are possible. In the following, these types shall be classified and illustrated by simple examples.

Static bifurcations

For the present, we concentrate on the stability of stationary fixed points $\mathbf{x}_0$ characterized by

$$\dot{\mathbf{x}} = \mathbf{F}(\mathbf{x}_0, \mu) = 0. \tag{23.1}$$

This may be interpreted as an implicit equation for the position of the fixed point depending on the parameter μ, $\mathbf{x}_0 = \mathbf{x}_0(\mu)$. A premise for the existence and continuity of this function is, according to the theorem from analysis on implicit functions, a nonsingular Jacobi matrix $\mathbf{M} = \partial \mathbf{F}/\partial \mathbf{x}|_{\mathbf{x}_0}$. A discontinuous behavior, thus bifurcations, may therefore be expected if the determinant of $\mathbf{M}$ vanishes; this means if one of the eigenvalues of this matrix depending

STATIC BIFURCATIONS

on μ takes the value zero. The meaning of the eigenvalues of the Jacobi matrix for stability has been discussed in Chapter 22.

We now consider the typical cases in the simplest possible form, namely for a one-dimensional system. Without restriction of generality the fixed point is set to $x_0 = 0$, and let the branching value be $\mu_c = 0$. This may always be achieved by appropriate coordinate transformations. Moreover, the one-dimensional bifurcation may be embedded into a higher-dimensional space.

(a) The saddle-node branching: Saddle-node branching occurs in the dynamical system

$$\dot{x} = F(x, \mu) = \mu - x^2. \tag{23.2}$$

This system has fixed points at

$$x_{01} = +\sqrt{\mu}, \qquad x_{02} = -\sqrt{\mu}. \tag{23.3}$$

But, since x is real, there is no fixed point for negative μ. If μ passes the critical value $\mu_c = 0$, the number of fixed points jumps from 0 to 2. The lower fixed point is however *unstable*, as is shown by the linear stability analysis according to Chapter 21. The eigenvalue γ of the Jacobi "matrix"

$$M = \left.\frac{\partial F}{\partial x}\right|_{x_0} = -2x_0, \quad \text{thus,} \quad \gamma = -2x_0, \tag{23.4}$$

is, namely, positive for the solution $x_{02} = -\sqrt{\mu}$. Figure 23.1 shows the stable (solid curve) and unstable (dashed curve) fixed points as a function of the control parameter μ. The arrows indicate the direction of motion, which may be immediately read off from (23.2). One may state that a stable and an unstable solution meet each other at the critical point and annihilate each other.

We still have to clarify the origin of the name saddle-node branching. For this purpose, the branching is embedded in a two-dimensional space, which may be achieved, e.g., by

$$\dot{x} = \mu - x^2,$$

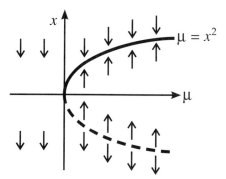

Figure 23.1. Stable (solid) and unstable (dashed) fixed points for the saddle-node branching.

$$\dot{y} = -y. \tag{23.5}$$

The variables x and y are decoupled, and the solutions $y(t)$ tend asymptotically to zero. The fixed points are $\mathbf{x}_{01} = (+\sqrt{\mu}, 0)$ and $\mathbf{x}_{02} = (-\sqrt{\mu}, 0)$. The Jacobi matrix has the form

$$\mathbf{M} = \begin{pmatrix} \partial F_1/\partial x & \partial F_1/\partial y \\ \partial F_2/\partial x & \partial F_2/\partial y \end{pmatrix}\bigg|_{\mathbf{x}_0} = \begin{pmatrix} -2x & 0 \\ 0 & -1 \end{pmatrix}\bigg|_{\mathbf{x}_0} \tag{23.6}$$

and is, of course, diagonal, with the eigenvalues $\gamma_1 = -2x_0$, $\gamma_2 = -1$. The two eigenvalues for the solution $\mathbf{x}_{02}$ are $2\sqrt{\mu}$ and -1; i.e., they have distinct signs, which—according to the nomenclature from Example 21.1—corresponds to a *saddle point*. For $x_0 = +\sqrt{\mu}$, both eigenvalues are negative, and there arises a *stable node*. Saddle and node coalesce with each other at the critical point. The distinct dynamical flows at $\mu < 0$, $\mu = 0$, $\mu > 0$, are represented in Figure 23.2. One clearly notes that after coalescing of saddle and node, there is no longer a fixed point and thus the flow continues to infinity (left diagram).

(b) The pitchfork branching: The simplest example of this kind of branching arises if one uses a cubic polynomial for $F(x, \mu)$:

$$\dot{x} = F(x, \mu) = \mu x - x^3. \tag{23.7}$$

Since this polynomial originates from (23.2) by multiplying by the factor x, a zero as solution simply adds to the former fixed points,

$$x_{01} = +\sqrt{\mu}, \qquad x_{02} = -\sqrt{\mu}, \qquad x_{03} = 0. \tag{23.8}$$

At the critical point $\mu_c = 0$, the number of fixed points therefore jumps from 1 to 3, whereby one of the latter solutions turns out to be unstable. The Jacobi matrix

$$M = \frac{\partial F}{\partial x}\bigg|_{x_0} = \mu - 3x_0^2 \tag{23.9}$$

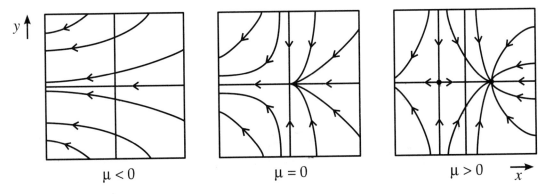

Figure 23.2. The dynamical flow at a saddle-node branching which is embedded in a two-dimensional space.

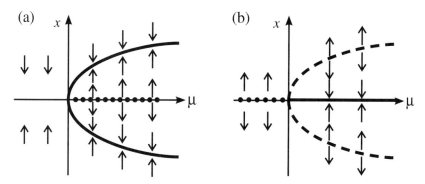

Figure 23.3. Stable (solid) and unstable (dashed) fixed points of the pitchfork branching: (a) A supercritical branching. (b) A subcritical branching.

has for the three fixed points the eigenvalues

$$\gamma_1 = \gamma_2 = -2\mu^2, \qquad \gamma_3 = \mu. \tag{23.10}$$

Thus, the solution x_{03} is stable below μ_c, but looses this property at the critical point to the two branches x_{01} and x_{02} of the "fork," as is represented in Figure 23.3. As compared with Figure 23.1, the arrows representing the direction of flow have changed their orientation in the lower half-plane $x < 0$. This is obvious because $F(x, \mu)$ contains the additional factor x.

The pitchfork bifurcation may still arise in a second version, with the stability properties exactly inverted. An example for that is

$$\dot{x} = F(x, \mu) = \mu x + x^3, \tag{23.11}$$

which differs from (23.7) by the sign. Since the flow direction and the signs of the eigenvalues are inverted, the branching diagram changes as represented in Figure 23.3(b). In this case, there remains only a single stable branch. The pitchfork bifurcation of Figure 23.3(a) is *supercritical*, and that of Figure 23.3(b) is *subcritical*. The two remaining combinations of signs of the linear and cubic term in $F(x, \mu)$ do not yield any qualitatively new features, they just correspond to the reflection $\mu \to -\mu$.

(c) The transcritical branching: In this case, we consider a polynomial with linear and quadratic term

$$\dot{x} = F(x, \mu) = \mu x - x^2 \tag{23.12}$$

with the fixed points

$$x_{01} = \mu, \qquad x_{02} = 0. \tag{23.13}$$

Thus, there always exist two fixed points in the entire parameter space. The eigenvalues of the stability matrix are

$$\gamma_1 = -\mu, \qquad \gamma_2 = \mu. \tag{23.14}$$

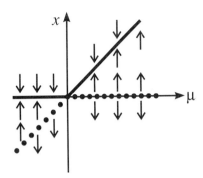

Figure 23.4. Stable (solid) and unstable (dashed) fixed points of the transcritical branching.

In each case, one of the solutions is stable, the other one is unstable. At the branching point the two solutions change their roles; see Figure 23.4.

(d) The Hopf branching: The branchings considered so far were characterized by the fact that a *real eigenvalue* of the Jacobi matrix takes the value zero when varying the control parameter. However, branchings may also occur for *complex eigenvalues*. Since the eigenvalues always occur in the form of complex-conjugate pairs, the system must have at least the dimension 2. As an example, we consider the system of equations

$$\dot{x} = -y + x(\mu - (x^2 + y^2)), \tag{23.15}$$
$$\dot{y} = x + y(\mu - (x^2 + y^2)). \tag{23.16}$$

This system has been investigated already in Example 22.2 in the context of the stability of limit cycles. For all values of μ the origin is a fixed point, $\mathbf{x}_0 = (0, 0)$. At this position the Jacobi matrix has the value

$$\mathbf{M}(\mathbf{x}_0) = \begin{pmatrix} \mu & -1 \\ 1 & \mu \end{pmatrix} \tag{23.17}$$

with the eigenvalues $\gamma_1 = \mu + i$, $\gamma_2 = \mu - i$. This means that the fixed point for $\mu < 0$ is a stable spiral, and for $\mu > 0$ an unstable spiral. The fate of the stable solution at the critical point differs from that in the cases considered so far. From the stationary attractor $\mathbf{x}_0$ there evolves for $\mu > 0$ a periodically oscillating solution $\mathbf{x}_r(t) = (\sqrt{\mu} \cos t, \sqrt{\mu} \sin t)$, which turns out to be a stable limit cycle.

As is seen from Figure 23.5, the bifurcation diagram resembles that of the pitchfork branching, Figure 23.3. Actually, the system of equations (23.15) may be decoupled by changing to polar coordinates $x = r \cos \phi$, $y = r \sin \phi$. The equation of motion for the radius $r(t)$ then exactly coincides with (23.7); see equation (22.43) in Example 22.2. The new state is, however, independent of time only in a "convected rotating" coordinate system, since the phase angle increases linearly, $\phi(t) = t$. In the original phase space the behavior of the attractor therefore changes qualitatively: A *static* solution becomes a

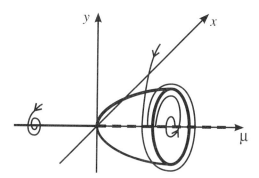

Figure 23.5. For a Hopf branching, a fixed point converts into a limit cycle.

dynamic solution. Branchings of this kind were first investigated in a systematic way[1] in 1942 by the mathematician Eberhard Hopf.[2]

As for the pitchfork branching, there are also two versions of the Hopf branching. In addition to the *supercritical* case represented here, there exists also a *subcritical Hopf bifurcation* with the opposite stability properties. In practice this version is of less interest, due to the fact that no stable limit cycle arises as an attractor.

One might consider the branchings presented in (a) to (d) as particularly simple special cases. But it has been shown under rather general premises that the change of stability (characterized by the zero passage of eigenvalues of the Jacobi matrix) for variation of a single control parameter always proceeds according to one of these four scenarios. We cannot deal here with the more detailed mathematical foundation, but refer for an introduction, e.g., to G. Faust, M. Haase, J. Argyris, *Die Erforschung des Chaos*, Vieweg, 1995, Chapter 6.[3]

Bifurcations of time-dependent solutions

It may also happen for periodic trajectories that their character changes stepwise under variation of a parameter μ. We shall only briefly touch on the bifurcation theory of periodic solutions and vividly illustrate some interesting aspects. The mathematical tool is the Poincaré mapping introduced in Chapter 22. A periodic orbit $\mathbf{x}_r(t)$ is characterized by a fixed point $\mathbf{x}_{r0}$ in the Poincaré cut. The discretization of the neighboring path $\mathbf{x}(t)$ consists

---

[1] E. Hopf, *Abh. der Sächs. Akad. der Wiss., Math. Naturwiss. Klasse* **94**, 1 (1942).

[2] *Eberhard Friedrich Ferdinand Hopf*, b. April 17, 1902, Salzburg–d. July 24, 1983, Bloomington, Indiana. Hopf studied mathematics in Berlin and taught at MIT (1932 to 1936), and at the universities of Leipzig (1936 to 1944) and Munich (1944 to 1948), as well as at Indiana University, Bloomington (from 1948). His main research fields were differential and integral equations, variational calculus, ergodic theory, and celestial mechanics.

[3] This was also translated into English and published as G. Faust, M. Haase, J. Argyris, *An Exploration of Chaos*, North Holland (1994).

of a sequence of points $x_0, x_1, x_2, \ldots$. For the distance between the two orbits, we have according to (22.10)

$$x_n - x_{r0} = C^n(x_0 - x_{r0}), \qquad (23.18)$$

where the matrix C represents the linearized approximation of the Poincaré mapping P. As long as all eigenvalues λ_i of C fall into the complex unit circle, $|\lambda_i| < 1$, the neighboring orbits are attracted and the solution $x_r(t)$ is a *stable limit cycle*.

The approach $x_n \to x_{r0}$ may proceed in different ways, as will be illustrated for a selected eigenvalue, say, λ_1. If the eigenvalue is real and positive, $0 < \lambda_1 < 1$, the x_n approach the fixed point x_{r0} monotonically, as is shown in Figure 23.6(a). The axis in this diagram corresponds to the direction of the eigenvector of C belonging to λ_1. If the eigenvalue is real and negative, $-1 < \lambda_1 < 0$, the x_n form an alternating sequence; see Figure 23.6(b). Finally, for a pair of complex eigenvalues, $\lambda_1 = \lambda_2^*$, $|\lambda_1| < 1$, the x_n lie on a spiral converging toward x_{r0}; see Figure 23.6(c).

Branchings occur if an eigenvalue λ leaves the unit circle at a critical value of the control parameter μ_c. One distinguishes three possible cases:

(a) $\lambda_1 = +1$.

According to (23.18), in this case the distance of a neighboring trajectory from the reference trajectory does not change. (A deviation along the direction of the eigenvector ξ_1 of the matrix C is multiplied by $\lambda_1 = 1$ for each cycle.) This indicates that for $\mu > \mu_c$, new limit cycles may arise that have the same period length T as the reference orbit x_r. Similar to the bifurcation of a stable fixed point, a periodic solution may also undergo a pitchfork bifurcation and split into two separated solutions. This is sketched in Figure 23.7(a). The other bifurcations from the last section are also possible. Which of these cases is actually realized cannot be read off from the criterion $\lambda_1 = +1$ alone.

(b) $|\lambda_1| = |\lambda_2^*| = 1$.

In this case as well, the sequence of points x_n of the Poincaré mapping remains at a constant distance from x_{r0} but now rotates on a circle. In the Poincaré cut itself, a limit cycle evolves. This means geometrically that the topology of the periodic orbit changes: It now lies on a *torus* which envelops the originally closed orbit, as is shown in Figure 23.7(b). If the two circulation frequencies on the torus mantle are incommensurable, i.e.,

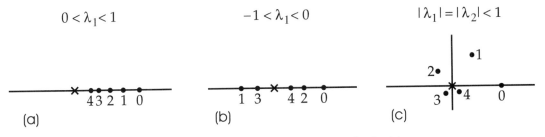

Figure 23.6. Various kinds of approaches to a fixed point.

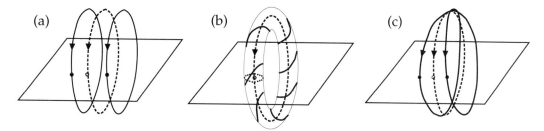

Figure 23.7. Typical branchings of a periodic trajectory (dashed).

do not form a rational ratio p/q, $p, q \in N$, then one speaks of a *quasiperiodic motion*, since the orbit for infinitely large times never closes. It thereby approaches any point on the torus surface to arbitrarily close distance.

(c) $\lambda_1 = -1$.

This case is of particular interest, since the distance of the points $\mathbf{x}_n$ remains constant, but the direction is alternating. This means that the neighboring orbit $\mathbf{x}$ *after each second passage* returns to its old position. There arises a periodic orbit with twice the period length $2T$, as is indicated in Figure 23.7(c). The phenomenon is therefore called a *period-doubling* or *subharmonic bifurcation*. Bifurcations of this kind play an important role in the transition from periodic to chaotic motion. An explicit example will be discussed in Example 25.1 in the context of logistic mapping.

24 Lyapunov Exponents and Chaos

In Chapter 22, the concept of stability of time-dependent orbits was discussed, and in Example 22.1, Floquet's theory of stability, which may be applied to periodic paths, was explained in more detail. Building on the works of Floquet and Poincaré, the Russian mathematician Lyapunov[1] published in 1892 an even more general study of the stability problem in which arbitrary and also nonperiodic motions were admitted. The characteristic exponents introduced by Lyapunov have played a central role in the theory of nonlinear systems.

One-dimensional systems

We first consider the case of a *one-dimensional discrete mapping*

$$x_{n+1} = f(x_n). \tag{24.1}$$

Physically, this may be for example the Poincaré mapping of a dynamical system. We now ask, how does the point sequence $x_0, x_1, x_2, \ldots$ differ from the point sequence $\tilde{x}_0, \tilde{x}_1, \tilde{x}_2, \ldots$ that evolves from a slightly modified initial condition $\tilde{x}_0 = x_0 + \delta x_0$? We have, in general,

$$\begin{aligned}\tilde{x}_n &= x_n + \delta x_n = f(x_{n-1} + \delta x_{n-1}) \\ &= f(x_{n-1}) + f'(x_{n-1})\delta x_{n-1} + \ldots;\end{aligned} \tag{24.2}$$

---

[1] *Alexander Mikhailovich Lyapunov*, Russian mathematician, b. June 6, 1857, Yaroslavl–d. November 3, 1918, Odessa. Lyapunov was a scholar of Chebyshev. He investigated the stability of equilibrium and the motion of mechanical systems and the stability of rotating liquids. Lyapunov also worked in the fields of potential theory and probability theory.

ONE-DIMENSIONAL SYSTEMS

thus, the deviation in the nth step in linear approximation is

$$\delta x_n = f'(x_{n-1})\delta x_{n-1}. \tag{24.3}$$

This equation can be applied n times recursively, with the result

$$\delta x_n = f'(x_{n-1})f'(x_{n-2})\cdots f'(x_0)\,\delta x_0 \tag{24.4}$$

$$= \prod_{l=0}^{n-1} f'(x_l)\,\delta x_0. \tag{24.5}$$

Obviously, the values $f'(x_l)$ are a measure of how fast the neighboring solutions x_n and $\tilde{x}_n$ go away from each other (or move towards each other). In the special case of a periodic motion which was studied by Floquet the point sequence x_l (interpreted as a Poincaré mapping) is constant and all factors in (24.4) have the same value. This yields the exponential relation

$$|\delta x_n| = |f'(x_0)|^n\,|\delta x_0| = e^{n\sigma}|\delta x_0| \quad \text{with } \sigma = \ln|f'(x_0)|, \tag{24.6}$$

compare (22.25) in Example 22.1. (A difference in the definition of σ is that in (24.6) one considers magnitudes. This implies that σ here is always real.) The characteristic exponent σ determines the "growth rate" of a perturbation. Even if the points x_l differ from each other (nonperiodic orbit), the exponent σ may be further used by forming a *mean value*. The mathematical definition of the *Lyapunov exponent* of a point sequence x_n is

$$\sigma = \lim_{n\to\infty}\lim_{\delta x_0 \to 0} \frac{1}{n}\ln\left|\frac{\delta x_n}{\delta x_0}\right|. \tag{24.7}$$

Using (24.4) and the multiplication rule for the logarithm, we can also write this as

$$\sigma = \lim_{n\to\infty} \frac{1}{n}\sum_{l=0}^{n-1}\ln|f'(x_l)|. \tag{24.8}$$

If all x_l are equal, the quantity σ of (24.7) or (24.8) obviously reduces to the special case (24.6).

The Lyapunov exponent is a logarithmic measure for the mean expansion rate per iteration (i.e., per unit time) of the distance between two infinitesimally close trajectories.

The case $\sigma > 0$ is of particular interest. A dynamical system with a *positive Lyapunov exponent* is called *chaotic*. The paths of such a system are extremely *sensitive to changes of the initial conditions*. Because of the exponential dependence there is no need for long waiting, and an initially small deviation δx_0 explodes to arbitrary magnitude. More strictly speaking, it is sufficient that the product $n\sigma$ be a number not very much larger than unity; compare (24.6).

This property of chaotic systems is very significant, both practically as well as conceptually. The behavior of a chaotic system is *not predictable*, at least not over a long time period. Since physical quantities can always be determined only with a limited precision, δx_0 inevitably has a value that differs from zero. Therefore, it is hopeless to aim at predicting the state of a chaotic system for times that are significantly larger than $1/\sigma$. The attempt to reach that by more and more precise fixing of the initial conditions is doomed to failure. Ultimately, the exponential increase of the deviation will always win.

The fascinating point is that this effect arises in a completely *deterministic* system. The dynamics of such a system is "in principle" mathematically uniquely fixed by the basic equation of motion—be it a differential equation or a discrete mapping as in (24.1). Nevertheless, an exact knowledge of the equation of motion cannot help in the attempt to find the solutions. In this way a new kind of uncertainty is brought into physics, in addition to the more familiar sources of the statistical fluctuations (noise) and the quantum fluctuations (uncertainty relation). The first to clearly understand and state this phenomenon was Henri Poincaré toward the end of the nineteenth century. Considering the rapid development of other branches of physics, nonlinear dynamics has lived in the shadows for a long time. The current interest in the chaos theory has been inspired significantly by the proliferation of electronic computing facilities, which allow for exploring the dynamics of systems resisting the analytic treatment.

An important impetus for dealing with chaotic systems was given by the meteorologist Edward N. Lorenz, whose work[2] received little attention for a long time. In 1963, he derived from the basic equations of hydrodynamics, under strongly simplifying assumptions, a system of three coupled nonlinear differential equations for describing a meteorological system. When working on their numerical solution, he discovered the signature of deterministic chaos: nonpredictability due to sensitive dependence on the initial conditions. Nobody will be surprised that meteorology offers an appropriate example for that phenomenon. Finally, weather may be predicted reliably even with massive computer effort for at most a few days in advance. The attempt to extend this space of time requires an exponentially increasing effort.

Lorenz has also coined the concept of the "butterfly effect" in this context, which entered public awareness: The flapping of wings of a butterfly in Brazil decides whether Texas is hit by a tornado some days later. Of course one should not just flog this picture to death. Even all butterflies together cannot change the structure of the "climatic attractor" by their flapping. Tropical whirlwinds will occur over and over again, at least as long as no global change in climate happens. But at what time and at which place a thunderstorm occurs depends, however, so sensitively on subtle details that even a butterfly may affect it.

Multidimensional systems

The definition of the Lyapunov exponent given here for one-dimensional systems with discrete dynamics may be generalized to several dimensions and to a continuous time evolution. The discussion of the stability of periodic orbits in Chapter 22 may serve as a starting point. There we have discussed trajectories $\mathbf{x}(t) = \mathbf{x}_t(t) + \boldsymbol{\xi}(t)$ that differ from a reference orbit $\mathbf{x}_r(t)$ by a small perturbation $\boldsymbol{\xi}(t)$. To first approximation the perturbation obeys a linear differential equation; compare equation (22.6),

$$\dot{\boldsymbol{\xi}}(t) = \mathbf{M}(t)\boldsymbol{\xi}(t), \tag{24.9}$$

---

[2] E.N. Lorenz, *J. Atmos. Sci.* **20**, 130 (1963).

with the Jacobi determinant $\mathbf{M} = \partial \mathbf{F}/\partial \mathbf{x}|_{\mathbf{x}_r(t)}$. In agreement with (24.7), as a measure for the time evolution of the perturbation one may define the quantity

$$\sigma_{\mathbf{x}_r, \boldsymbol{\xi}_0} = \lim_{t \to \infty} \ln \frac{|\boldsymbol{\xi}(t)|}{|\boldsymbol{\xi}(t_0)|}. \tag{24.10}$$

This definition of the Lyapunov exponent raises several mathematical problems that can only be touched on here. First of all, $\sigma_{\mathbf{x}_r, \boldsymbol{\xi}_0}$ depends on the reference trajectory $\mathbf{x}_r(t)$ and therefore on the position of the starting point. But if the system has an attractor, then the value of σ in the range of attraction of the attractor is independent of the reference orbit. Moreover, there exists the important class of *ergodic systems* for which the mean values with respect to the time (taken along an orbit) can be replaced by mean values in phase space. One can show also for ergodic systems that the Lyapunov exponents defined according to (24.10) exist and are independent of the special reference trajectory. (More strictly speaking, there may occur "pathological" orbits with differing σ, but these form a set of measure zero).[3]

Furthermore, the value of $\sigma_{\mathbf{x}_r, \boldsymbol{\xi}_0}$ depends on the direction of the perturbation $\boldsymbol{\xi}_0 = \boldsymbol{\xi}(t_0)$. In an N-dimensional space, one may construct N linearly independent vectors $\mathbf{e}_i$ that lead to a set of N Lyapunov exponents

$$\sigma_i = \sigma_{\mathbf{x}_r, \mathbf{e}_i}. \tag{24.11}$$

The indices are chosen such that the σ_i are in descending order,

$$\sigma_1 \geq \sigma_2 \geq \ldots \geq \sigma_N, \tag{24.12}$$

where degeneracies (equality of several values) are possible.

The value of the *largest Lyapunov exponent* σ_1 is the most easily determined one. For this purpose the linearized equation of motion (24.9) is integrated with an initial perturbation $\boldsymbol{\xi}(t_0)$ chosen at random. Such a vector may be decomposed according to

$$\boldsymbol{\xi}(t_0) = \sum_{i=1}^{N} c_i \, \mathbf{e}_i \tag{24.13}$$

and will always have also a component along the vector $\mathbf{e}_1$. When tracing the solution over a sufficiently long time interval, the most rapidly increasing component of the perturbation (or, if all σ_i are negative, the most slowly decreasing component) will dominate. Performing the limit in (24.10) then guarantees that the calculation yields the *maximum Lyapunov exponent*. In practice one has to take care in such a calculation that the trajectory $\mathbf{x}(t)$ for chaotic systems moves away from the reference trajectory $\mathbf{x}_r(t)$ very rapidly. It is therefore recommended that one performs a *rescaling* of the perturbation $\boldsymbol{\xi}'(t_n) = c\,\boldsymbol{\xi}(t_n)$ in regular time intervals, with a constant $c \ll 1$; see Figure 24.1. The value of σ is then obtained by averaging over many time intervals.

---

[3] More information on these questions may be found, e.g., in D. Ruelle, *Chaotic Evolution and Strange Attractors*, Cambridge University Press, Cambridge (1989).

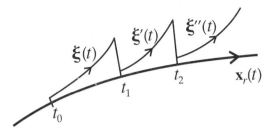

Figure 24.1. In the numerical calculation of Lyapunov exponents one performs a rescaling in regular time steps, because of the divergence of the trajectories.

The full set of the N Lyapunov exponents may be calculated by following the time evolution of all N linearly independent perturbations $\boldsymbol{\xi}_i$ with $\boldsymbol{\xi}_i(t_0) = \mathbf{e}_i$. From the N volumes $V^{(p)}$ of the parallelepipeds spanned by the $\boldsymbol{\xi}_1, \boldsymbol{\xi}_2, \ldots, \boldsymbol{\xi}_p$ (that must be calculated for all values $p = 1, 2, \ldots, N$), one then may obtain successively all σ_i.[4]

We still note that for *periodic* orbits $\mathbf{x}_r(t+T) = \mathbf{x}_r(t)$, the Lyapunov coefficients coincide with the real part of the Floquet exponents introduced in Example 22.1. Thus, one has a generalization of this concept.

The Lyapunov exponents are decisive for the long-term evolution of a dynamical system. As discussed already, positive σ imply a rapid divergence of neighboring trajectories and nonpredictability. Of particular interest are trajectories which attract neighboring solutions and have (at least) one positive Lyapunov exponent. They are called *chaotic attractors*.

Let us consider for illustration an autonomous system with three degrees of freedom. Depending on the combination of signs of $(\sigma_1, \sigma_2, \sigma_3)$, various kinds of attractors may occur:

$$(\sigma_1, \sigma_2, \sigma_3) = \begin{cases} (-,-,-), & \text{fixed point,} \\ (0,-,-), & \text{limit cycle,} \\ (0,0,-), & \text{torus,} \\ (+,0,-), & \text{chaotic attractor.} \end{cases}$$

These cases are illustrated in Figure 24.2.

(a) If all Lyapunov exponents are negative, there arises a stable fixed point to which the neighboring trajectories from all directions are converging.

(b) The vanishing of a Lyapunov exponent, $\sigma_1 = 0$, indicates the existence of a *periodic motion*. This has been demonstrated explicitly in Chapter 22, based on the equation of motion. The vector $\mathbf{e}_1$ associated with σ_1 points along the direction of the tangent of the orbit. The attractor is a *limit cycle*, i.e., a one-dimensional object with the topology (but not necessarily the geometric shape) of a circle.

---

[4]G. Bennettin, C. Froeschle, J.P. Scheidecker, *Phys. Rev.* **A19**, 2454 (1979); T.S. Parker and L.O. Chua, *Practical Numerical Algorithms for Chaotic Systems*, Springer, New York (1989).

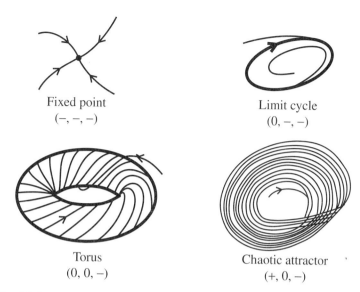

Figure 24.2. Distinct types of attractors in a dynamical system with three degrees of freedom. The pictures differ by the signs of the Lyapunov exponents.

(c) If two of the Lyapunov exponents vanish, $\sigma_1 = \sigma_2 = 0$, there exists a periodic motion in two directions. Therefore the attractor is two-dimensional and has the topology of a *torus* about which the trajectory is winding up. Whether or not the trajectory is periodic in total depends on the circulation frequencies ω_1 and ω_2 for the two degrees of freedom of the torus. If the values ω_1 and ω_2 are incommensurable, i.e., the ratio ω_1/ω_2 is not a fraction of integer numbers but an irrational number, the orbit will never close. Such an orbit that with increasing time covers the torus more and more densely is called *quasiperiodic*.

(d) If the largest Lyapunov exponent is positive, $\sigma_1 > 0$, there arises a *chaotic attractor* with the already discussed properties of irregular motion that depends strongly on the initial conditions. The typical combination is $\sigma_1 > 0$, $\sigma_2 = 0$, $\sigma_3 < 0$, but chaotic attractors with several positive Lyapunov exponents may also occur. Geometrically the chaotic attractors usually are also *strange attractors*, as mentioned already in Chapter 21. They have the strange property that they are objects with a *broken dimension*. These are neither lines nor surfaces (or higher-dimensional hypersurfaces) but "something in between." That such objects are not pure mathematical inventions but also occur in nature has been noted only recently, in particular by B. Mandelbrot,[5] who named them

---

[5]*Benoit B. Mandelbrot*, b. 1924, Warsaw. After the emigration of his family to France (1936), Mandelbrot studied in Lyon, at the California Institute of Technology, and in Paris, where he did his doctorate in 1952. He worked at CRNS, in Geneva, and at the École Polytechnique before he went in 1958 to the IBM Watson Research Center, where he was appointed as Research Fellow. He served as visiting professor among others at Harvard and Yale. Mandelbrot's interests are extraordinarily broad and oriented interdisciplinarily. Building on the work of

fractals.[6] A beautiful example of a fractal attractor will be given in Example 25.4 in the context of a periodically driven pendulum; compare Figure 25.19.

Stretching and folding in phase space

How can a strange attractor with infinitely branched internal structure evolve in a dynamic system? The answer lies in the mechanism of the *stretching* and *folding* in phase space, which may be visualized as follows.

We consider a dynamical system with a positive Lyapunov exponent σ_1. Moreover, let the system be organized in such a way that the accessible part of phase space is restricted (in Chapter 21 this property was included in the definition of attractors). We now consider a small connected domain ΔV in phase space, e.g. a cube, and follow its deformation under the phase flow Φ_t. Neighboring points rapidly drift apart in the direction belonging to the Lyapunov exponent σ_1. Since the phase-space volume is conserved in conservative systems, or even shrinks in dissipative systems (see Chapter 21), a contraction must take place in the other directions. Since on the other hand the available domain of phase space is restricted, the trajectories must bend back again. The test volume ΔV which is stretched at first will thus be folded back, as is schematically shown in Figure 24.3.

If one follows the history of ΔV still further, the game of stretching and folding will continue again and again. One may imagine that in this way a infinitely fine subdivided puff-pastry-like structure develops, with the geometrical properties of a fractal. Two initially closely neighboring points end up after a while on two distinct layers of the "pastry," after which any recognizable correlation between their positions is lost. A simple mathematical model for the mechanism described here is the *baker transformation* (called so in analogy to the kneading of pastry); see Example 24.1.

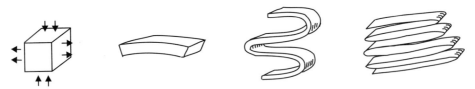

Figure 24.3. A strange attractor may develop if the dynamical flow continuously stretches a domain in phase space and then folds it back again.

G.M. Julia (1893–1978) on iterated rational functions, he demonstrated the properties of fractals using computer graphics and pointed out their manifold occurrence in nature. Besides many other awards, Mandelbrot received the Wolf Prize in physics in 1993.

[6]For more on this point, see, e.g., B. Mandelbrot, *The Fractal Geometry of Nature*, Freeman (1982); H.-O. Peitgen, H. Jürgens, D. Saupe, *Chaos and Fractals: New Frontiers of Science*, Springer (1992).

Fractal geometry

The long-term behavior of the trajectories of dissipative systems is characterized by the approach toward attractors that are embedded in the phase space as geometric forms of lower dimension. The simplest cases include fixed points (dimension $D = 0$), limit cycles ($D = 1$), and tori ($D = 2$) for quasiperiodic motion with two incommensurable frequencies. As new types of attractors in nonlinear systems, *strange attractors* with more complicated geometric structure may arise which are characterized by a *broken dimension*. The first *fractals* of this type, still without this name, were introduced by mathematicians as abstract constructions. For a long time, they seemed to belong into the mathematical cabinet of oddities, until it was discovered that fractal structures actually frequently occur in nature.

The geometric construction of a fractal is usually based on a (mostly simple) *iteration rule* that is applied repeatedly. In the limit of infinitely many iteration steps, the fractal arises which thus has an infinitely fine resolved internal structure. We shall briefly consider several familiar examples.

(1) The Cantor set: The construction begins with a line, namely, the set of all real numbers in the unit interval $I_0 = [0, 1]$. The iteration rule reads: Remove the mean third in each interval. In the first iteration step there arise two disjunct partial intervals $I_1 = [0, 1/3] \cup [2/3, 1]$ which then split further into $I_2 = [0, 1/9] \cup [2/9, 1/3] \cup [2/3, 7/9] \cup [7/9, 1]$ etc. The first iteration steps are represented in Figure 24.4(a). In the limit $n \to \infty$ from the I_n, there arises the Cantor set[7] as a kind of finely distributed dust of points in the unit interval.

To get a measure of extension of the Cantor set, we consider the magnitude of its complementary set, i.e., the total length of all parts cut out. This leads to a geometric series:

$$L = 1\frac{1}{3} + 2\frac{1}{9} + 4\frac{1}{27} + \ldots = \frac{1}{3} \sum_{n=0}^{\infty} \left(\frac{2}{3}\right)^n = \frac{1}{3} \left(\frac{1}{1 - 2/3}\right) = 1. \tag{24.14}$$

The parts cut out thus add up to the total length of the unit interval, and the length of the Cantor set is therefore zero! But it consists of infinitely many points, and one can show that its cardinal number (power) is the same as that of the real numbers.[8]

(2) The Koch curve: This construction again begins with a straight line of length unity. Instead of cutting out parts, something is added, and the straight line is built up to a toothed

---

[7] *Georg Cantor*, German mathematician, b. March 3, 1845, St. Petersburg, Russia–d. January 6, 1918, Halle. Cantor studied at the universities of Zurich and Berlin under Weierstraß, Kummer, and Kronecker. From 1869 to 1913, he was a professor at Halle. Cantor is the founder of set theory. He invented the notion of cardinal numbers and the concept of infinite (transfinite) numbers and dealt with the definition of the continuum. He proved the nondenumerability of real numbers. Furthermore, he contributed to the theory of geometric series.

[8] This may be understood in a rather simple manner: Every point of the Cantor set may be characterized by an infinite sequence of "left-right decisions"; i.e., for every iteration step in Figure 24.4 one must say in which of the two partial intervals the point lies. But this sequence may also be interpreted as the binary representation of a real number in the unit interval [0, 1], whereby the assertion becomes clear.

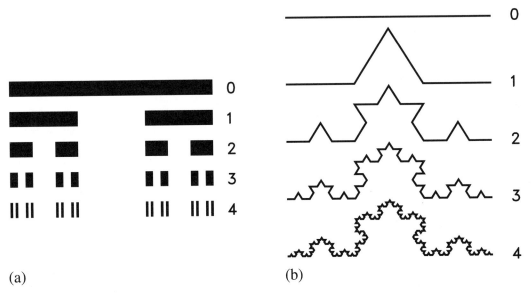

Figure 24.4. (a) Iterative construction of the Cantor set: In each step, the mean third of an interval is cut out. (b) Iterative construction of the Koch curve. It originates by continued adding of triangles.

curve in the two-dimensional plane. The iteration rule reads: Remove the mean third of every straight partial piece and replace it by the sides of an equilateral triangle. The first steps of the iteration are shown in Figure 24.4(b) (they remind one of the bulwarks of old fortresses).

The peculiar feature of the Koch[9] curve and its related ones is that it is everywhere continuous but *nowhere differentiable*. One cannot give a tangent to the curve, because of the infinitely many sharp corners. The calculation of the length of the Koch curve also leads to a remarkable result: In every partial step, always 3 partial pieces are replaced by 4 of the same length. The total length L is therefore

$$L = \lim_{n \to \infty} L_n = \lim_{n \to \infty} \left(\frac{4}{3}\right)^n = \infty, \qquad (24.15)$$

and thus diverges. This cannot be seen directly from the graph of the curve, due to the conversion to smaller and smaller length scales. Here the *self-similarity* becomes apparent

---

[9]*Niels Fabian Helge von Koch*, Swedish mathematician, b. January 25, 1870, Stockholm–d. March 11, 1924, Stockholm. Koch was a scholar and successor of Mittag-Leffler at the University of Stockholm. His main research fields were systems of linear equations of infinite dimension and with infinitely many unknowns. His name became familiar mainly from the curve named after him.

which is a typical feature of fractals: On any length scale, a linear magnification of a detail of the object again resembles the entire object.[10]

(3) The Sierpinski gasket: The basic element of the Sierpinski[11] gasket is a two-dimensional area, namely, an equilateral triangle. Iteration rule: Subdivide each triangle into 4 congruent parts and remove the central triangle. Figure 24.5 shows the first steps of this iteration. The resulting object is something between an area and a curve. It has again the property of self-similarity.

Nature offers a variety of fractal objects. Example from the organic world are the branchings of plants (trees, cauliflower, particularly beautiful ferns) or vessels. Inorganic fractal shapes are observed in clouds, mountains, snowflakes, lightning discharges, etc.

A classical example studied by Mandelbrot that resembles the Koch curve are coastlines. For the length of one and the same coast the geographers may give quite different values, depending on the length scale adopted in the measurement. The smaller the scale, the better is the scanning of the bays and windings of the coast, corresponding to the higher iterations L_n of the Koch curve. Ultimately the coastline should wind about each individual grain of sand on the beach, which would blow up the length enormously. But here also shows up that the application of fractal geometry to natural objects is meaningful only in a certain range. Ultimately on the atomic scale when the granularity of matter becomes apparent, the mathematical limit $n \to \infty$ loses its meaning. Nevertheless, the fractal scale behavior (see below) frequently can be traced over many orders of magnitude.

The fractal dimension: An important tool for characterizing geometric objects is their dimension. For ordinary objects (smooth curves, areas, spheres, ...), the dimension is clear and coincides with the visual conception. Fractals, however, behave differently. To

Figure 24.5. Iterative construction of the Sierpinski gasket. In each step, the middle of 4 partial triangles is removed.

---

[10] Self-similarity alone is, however, not a sufficient criterion for fractals. For example, a straight line is self-similar in a trivial manner.

[11] *Waclaw Sierpinski*, Polish mathematician, b. August 20, 1882, Warsaw–d. May 14, 1969, Warsaw. Sierpinski was a professor of mathematics in Lwow (now Ukraine, 1908–1914), Moscow (1915–1918), and later in Warsaw. His main fields of work were the theory of sets (here, in particular, the selection axiom and the continuum hypothesis), the topology of point sets, and number theory.

characterize them, the concept of *fractal dimension* was invented. We will restrict ourselves mainly to the so-called *capacity dimension* (or box counting dimension).[12]

The dimension shall be a measure of how dense a point set fills the space into which it is embedded. We start from the assumption that the space is a metric (normally Euclidean) space in which one can measure distances. This space is then, as illustrated in Figure 24.6, covered by a grid of lateral length ϵ (i.e., distances for $n = 1$, boxes for $n = 2$, cubes for $n = 3$, hypercubes for $n > 3$) and one counts how many of the boxes contain one or several points of the point set under consideration. This number $N(\epsilon)$ will in general depend on the box size ϵ. For an improved resolution the object for sure will spread over more boxes, but the question is how fast that happens. If the scaling behavior in the limit $\epsilon \to \infty$ follows a power law

$$N(\epsilon) = V(\epsilon)\epsilon^{-D_f} \quad \text{with } V(\epsilon) \to \text{constant} \quad \text{for } \epsilon \to 0, \quad (24.16)$$

then the value of the exponent defines the *fractal dimension* or *capacity dimension* D_f. Solving yields

$$D_f = \lim_{\epsilon \to 0} \frac{\ln N(\epsilon) - \ln V(\epsilon)}{\ln(1/\epsilon)} = \lim_{\epsilon \to 0} \frac{\ln N(\epsilon)}{\ln(1/\epsilon)}, \quad (24.17)$$

since $V(\epsilon)$ remains finite in the limit and therefore does not contribute.

For nonfractal geometric objects, the dimension determined in this way coincides with the normal Euclidean dimension, and $V(0)$ corresponds to the Euclidean volume. For example, for sufficiently high resolution a circular disk of radius R overlaps $N(\epsilon) \simeq \pi R^2/\epsilon^2$ boxes,

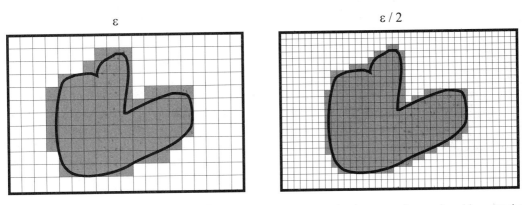

Figure 24.6. The capacity dimension of an object is determined by counting how many boxes of a grid are touched by the object. The dependence of the number $N(\epsilon)$ on the box size ϵ determines the dimension D_f. This is illustrated here for a "normal" two-dimensional object.

---

[12]The situation is complicated by the fact that a multitude of distinct mathematical concepts of dimension are available. The calculated dimension values partly agree with each other but partly do not, depending on the considered object. For details, see K. Falconer, *Fractal Geometry, Mathematical Foundations and Applications*, Wiley, New York (1990).

and each further doubling of the resolution increases the number of boxes by the factor 4. Hence, one finds for the circle $N(\epsilon) \simeq 2\pi R/\epsilon$, and so on.

For fractals, however, the power of the scaling law (24.16) differs from the naively expected value and is in general not an integer. For the Cantor set the determination of D_f is particularly simple. Here it suffices to take the unit interval ($n = 1$) as the embedding space. It is most convenient to consider a sequence of subdivisions of the length ϵ_i that always differ by the factor 3; thus, $\epsilon_0 = 1$, $\epsilon_1 = 1/3$, $\epsilon_2 = 1/9$, etc. The Cantor set is constructed in such a way that then the number of "occupied" boxes (partial intervals) always doubles; thus, $N(\epsilon_0) = 1$, $N(\epsilon_1) = 2$, $N(\epsilon_2) = 4$, etc. Hence, the fractal dimension of the Cantor set is, according to equation (24.17), given by

$$D_\text{f} = \lim_{\epsilon \to 0} \frac{\ln N(\epsilon)}{\ln(1/\epsilon)} = \lim_{n \to \infty} \frac{\ln N(\epsilon_n)}{\ln(1/\epsilon_n)}$$
$$= \lim_{n \to \infty} \frac{\ln 2^n}{\ln 3^n} = \frac{\ln 2}{\ln 3} \simeq 0.6309. \tag{24.18}$$

The result is independent of the manner of performing the passage to the limit $\epsilon \to 0$.

Similarly, one finds the dimension of the Koch curve. To cover it, in the first step one needs $N(\epsilon_1) = 4$ intervals of length $\epsilon_1 = 1/3$, in the next one $N(\epsilon_2) = 16$ intervals of length $\epsilon_1 = 1/9$, etc. As in (24.18), this implies

$$D_\text{f} = \lim_{n \to \infty} \frac{\ln 4^n}{\ln 3^n} = \frac{\ln 4}{\ln 3} \simeq 1.2618. \tag{24.19}$$

Finally, for the Sierpinski gasket we have

$$D_\text{f} = \frac{\ln 3}{\ln 2} \simeq 1.5850, \tag{24.20}$$

since here for each bisection of the box size the number of partial objects increases by the factor 3.

The results (24.18) to (24.20) are not implausible. They quantify how the considered fractals by their properties stand between the familiar objects point, line, area, ... The comparison of (24.19) and (24.20) shows that the Sierpinski gasket is "more space-filling" than the Koch curve, but does not come up to a normal area. An extreme example in this respect is the area-covering curve that was discovered in 1890 by G. Peano[13] and investigated in modified form by D. Hilbert. The Peano curve may also be obtained iteratively:

---

[13] *Giuseppe Peano*, Italian mathematician, b. August 27, 1858, Cuneo (province Piemont)–d. April 20, 1932, Torino. Peano studied mathematics at the University of Torino, where he taught beginning in 1880 as a lecturer and beginning in 1890 as a professor. His early works concerned analysis, the initial-value problem of differential equations, and recursive functions. Peano emerged mainly as a founder of the mathematical logic (with G. Frege). The Peano axioms (1889) define the natural numbers via the properties of sets. His aim was the axiomatization of all of mathematics. Later, Peano moved beyond mathematics and developed an universal world language (Interlingua) that, however, did not gain acceptance.

In a subdivision of the scale length into three sections there arise nine partial distances of equal length; see Figure 24.7. Accordingly, the dimension is

$$D_\mathrm{f} = \frac{\ln 9}{\ln 3} = 2. \tag{24.21}$$

This is the dimension $n = 2$ of the embedding space and hence the largest value that may be taken by the capacity dimension D_f. $N(\epsilon)$ takes the maximum value, since all boxes include parts of the object. The definition of the capacity dimension is very clear and has the advantage that it provides immediately an operative calculation rule. One only has to span grids and to count the boxes, which may easily be done on a computer. If the function $N(\epsilon)$ in a doubly logarithmic representation yields a straight line (at least over a larger range of scale), then one can immediately read off D_f from its slope. A related but more subtle definition of the dimension which refrains from equidistant grids and works with overlaps of variable magnitude was developed in 1918 at Bonn University by the mathematician Felix Hausdorff.[14] We shall not enter into the details here and mention only that the Hausdorff dimension D_H in the most cases coincides with D_f,[15] but there also exist exceptional cases. Generally, $D_\mathrm{H} \leq D_\mathrm{f}$.

Just for the classification of strange attractors, other dimension measures may also be meaningful. The Poincaré mapping yields a possibly very *inhomogeneously distributed*

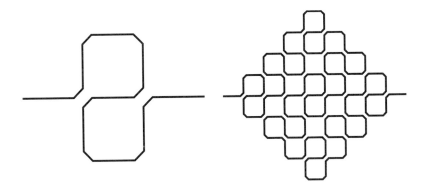

Figure 24.7. The first two iteration steps in the construction of the area-filling curve of Peano. In the limit there results a continuous mapping of the unit square onto the interval [0, 1]. To make the construction transparent, the rectangular edges were cut off in the plot.

---

[14] *Felix Hausdorff*, German mathematician, b. November 8, 1869, Breslau–d. January 26, 1942, Bonn. Hausdorff studied and taught mathematics in Leipzig and beginning in 1910 in Bonn, until his forced retirement in 1935. He was persecuted because of his Jewish origin. In 1942, he and his wife committed suicide, shortly before deportation to the concentration camp. Hausdorff's main research fields were topology and group theory. He introduced the concept of partly ordered sets and dealt with Cantor's continuum hypothesis. He founded a theory of topological and metric spaces and about 1919 defined the concepts of dimension and measure named after him.

[15] J.D. Farmer, E. Ott, J.A. Yorke, *Physica* **7D**, 153 (1983).

cloud of points in phase space. In the calculation of the capacity dimension the information on the frequency distribution of the points is ignored. There it has no meaning whether a box is occupied by a single point or by thousands. To take this quantity into account, one defines an *information dimension*, which includes the density distribution of the points.

The construction is similar to that for the capacity dimension D_f. The embedding space is again subdivided into boxes, but now the contribution of every cell is weighted by the probability p_i of meeting points there. Practically, this value is determined by generating a very large number N of points and counts how many of them fall in the cell number i; thus, $p_i = N_i/N$ with $\sum_i N_i = N$. The weighting is performed with a logarithmic measure. The information dimension D_I is defined as

$$D_I = \lim_{\epsilon \to 0} \frac{\sum_i p_i \ln(1/p_i)}{\ln(1/\epsilon)}. \tag{24.22}$$

The factor $f(p) = p \ln(1/p)$ has the following meaning: Take an arbitrary point from the distribution and ask "Does the point lie in cell i?" The answer to this question yields the information set $f(p_i)$. This function vanishes for $p_i \to 0$ and $p_i \to 1$, since in these cases the answer is trivial (always "no" or always "yes"). The information gain is maximum for $p_i = 1/2$; then the answer is in any case a surprise.

The exact foundation for the function $f(p)$ is provided by the statistical mechanics, or by information theory (Shannon's information measure).[16] But at least we may easily see that (24.22) turns into the formula (24.17) if the point distribution is homogeneous, i.e., if all probabilities in the in total $N(\epsilon)$ covered cells have the same value $p_i = p = 1/N(\epsilon)$. For the remaining cells $p_i = 0$. Thereby, (24.22) reduces to

$$D_I = \lim_{\epsilon \to 0} \frac{N(\epsilon) p \ln(1/p)}{\ln(1/\epsilon)} = \lim_{\epsilon \to 0} \frac{\ln N(\epsilon)}{\ln(1/\epsilon)} = D_f. \tag{24.23}$$

For a less homogeneous point distribution, the informational content is lower and one can show that then $D_I < D_f$.[17]

Example 24.1: Exercise: The baker transformation

The process of stretching and folding, which is characteristic for the phase flow in chaotic systems, can be illustrated by a simple two-dimensional discrete mapping. Let the phase space be the unit square $[0, 1] \times [0, 1]$. The motion of a point proceeds according to the transformation

$$x_{n+1} = 2x_n \bmod 1, \tag{24.24}$$

$$y_{n+1} = \begin{cases} a y_n, \\ \frac{1}{2} + a y_n, \end{cases} \quad \text{for} \quad \begin{cases} 0 \le x_n < \frac{1}{2}, \\ \frac{1}{2} \le x_n \le 1, \end{cases} \tag{24.25}$$

with a parameter $0 < a \le 1/2$. Interpret this transformation, and calculate the fractal dimension of the set that is generated by application of (24.24) on the unit square.

---

[16] C.E. Shannon, W. Weaver, *The Mathematical Theory of Communication*, Univ. of Ill. Press, Urbana (1949).

[17] See, e.g., H.-O. Peitgen, H. Jürgens, P. Richter, op. cit. p. 735.

Figure 24.8. The baker transformation.

Solution

The effect of this transformation may be simply visualized geometrically. All lengths in x-direction are stretched by the factor 2. If a point thereby passes over the right boundary of the unit interval, then it is set back to the left by one unit length. This mapping is also called the Bernoulli shift; compare Example 25.2. Simultaneously, all lengths in y-direction are compressed by the factor a. Points from the right half of the unit interval are additionally shifted by $1/2$ in y-direction upward. This reminds us of the work of a baker: The pastry is rolled out to twice its length. Then the part sticking out is cut off and put on as a second layer onto the pastry. (If $a < 1/2$, there still is an "air cushion" of thickness $1/2 - a$ between the layers.)

The volume change of a phase space domain ΔV under (24.24) is composed of the product of the stretching- and compression factor in x- and y-direction; thus,

$$\Delta V_{n+1} = 2a\,\Delta V_n\,. \tag{24.26}$$

For $a = 1/2$, the volume is conserved. However, a connected domain in phase space is rapidly torn up by the repeated "back-folding" (here it is more a back-shifting) and distorted beyond recognition. The figure represents the first steps of the transformation of a circle of radius $1/2$.

For $a < 1/2$, the volume element shrinks in each iteration step and goes asymptotically to zero. In the sense of Chapter 21, one also speaks of the *dissipative* baker transformation. The resulting geometrical object is a fractal which in y-direction displays the structure of a Cantor set. Just as for the latter one may calculate the fractal dimension according to equation (24.17). In the y-direction in the first transformation step the unit interval is converted into 2 parts of length a, in the next step 4 parts of length a^2, and generally 2^n parts of length a^n. If one selects a sequence of overlaps of the lateral length $\epsilon_n = a^n$, then $N(\epsilon_n) = 2^n(1/\epsilon^n)$, where the second factor originates from the overlapping of the x-axis. The fractal dimension is therefore

$$\begin{aligned}D_f &= \lim_{n\to\infty}\frac{\ln N(\epsilon_n)}{\ln(1/\epsilon_n)} = \lim_{n\to\infty}\frac{\ln(2^n/\epsilon_n)}{\ln(1/\epsilon_n)} = \lim_{n\to\infty}\frac{n\ln(2/a)}{\ln(1/a)} = \frac{\ln 2 - \ln a}{-\ln a}\\ &= 1 + \frac{\ln 2}{|\ln a|}.\end{aligned} \tag{24.27}$$

25 Systems with Chaotic Dynamics

In this chapter, we shall get to know various dynamical systems that may display highly complex forms of motion despite their very simple, even trivial structure. We shall meet previously discussed concepts such as bifurcations and periodic and strange attractors (limit cycles and chaotic trajectories) in a series of specific examples. For sake of simplicity, we shall begin with the investigation of systems with discrete dynamics. The examples considered will gradually become physically more and more realistic.

Dynamics of discrete systems

So far, we have been interested in dynamical systems that are described by continuous differential equations with respect to time. In the Poincaré mapping we have seen the possibility of reducing a continuous system to a time-discrete system. But since the Poincaré mapping of a realistic physical system as a rule cannot be given in a closed form, it is instructive instead of these to consider simple mathematical model mappings. As it turns out, such models share many properties with systems which from a physical point of view are more interesting.

Let us consider a sequence of vectors $\mathbf{x}_0, \mathbf{x}_1, \mathbf{x}_2, \ldots$ which is generated by a simple *iterative mapping*, i.e., by repeated application of a continuous function $\mathbf{f}(\mathbf{x})$,

$$\mathbf{x}_{n+1} = \mathbf{f}(\mathbf{x}_n). \tag{25.1}$$

The asymptotic behavior of the sequence of the $\mathbf{x}_n$ may be characterized as in Chapter 22 by the appearance of *fixed points* or *periodic limit cycles*. The stability of a fixed point $\mathbf{x}^p$ is again governed by the eigenvalues of the Jacobi matrix $\mathbf{D} = \partial \mathbf{f}/\partial \mathbf{x}|_{\mathbf{x}^p}$. We shall not go again into these details but still note an interesting relation between stationary and periodic solutions. Consider e.g. a periodic solution alternating between two values: $\mathbf{x}_1, \mathbf{x}_2, \mathbf{x}_1, \mathbf{x}_2, \ldots$. Then obviously $\mathbf{x}_2 = \mathbf{f}(\mathbf{x}_1)$ and $\mathbf{x}_1 = \mathbf{f}(\mathbf{x}_2)$, from which it follows that $\mathbf{x}_1 = \mathbf{f}(\mathbf{f}(\mathbf{x}_1)) \equiv \mathbf{f}^2(\mathbf{x}_1)$ and also $\mathbf{x}_2 = \mathbf{f}^2(\mathbf{x}_2)$. Here $\mathbf{f}^2(\mathbf{x})$ must not be confused with the power $(\mathbf{f}(\mathbf{x}))^2$. This procedure may also of course be transferred to periodic solutions of length m, $\mathbf{x}_1, \mathbf{x}_2, \ldots, \mathbf{x}_m, \mathbf{x}_1, \mathbf{x}_2, \ldots$. Correspondingly, a periodic solution of the iteration function **f** reduces to a set of m constant *fixed points of the m-times iterated function* $\mathbf{f}^m$. For continuous (not time-discretized) solutions $\mathbf{x}(t)$, such a relation does not exist.

One-dimensional mappings

For further simplification, we now concentrate on one-dimensional mappings $x_{n+1} = f(x_n)$. The sequence of the x_n can then be simply constructed graphically. The function $f(x)$ is plotted in a two-dimensional coordinate system with x_n as the abscissa and x_{n+1} as the ordinate, together with the bisector of the angle $x_{n+1} = x_n$. Starting with the first point $x_n = x_1$, the associated function value $x_{n+1} = f(x_1) = x_2$ is marked. Since this value shall serve as the initial value for the next iteration step, it must be transferred to the x_n-axis. This is done by drawing a horizontal line to the bisector. The intersection point $x_n = x_2$ is the base for marking again the function value $x_{n+1} = f(x_2) = x_3$, etc.

This geometrical construction is represented in Figure 25.1, where the various possibilities of stability of a fixed point $x_s = f(x_s)$ of the mapping are illustrated simultaneously. As is known, the derivative of the function at the fixed point, $\lambda = f'(x_s)$, governs the stability

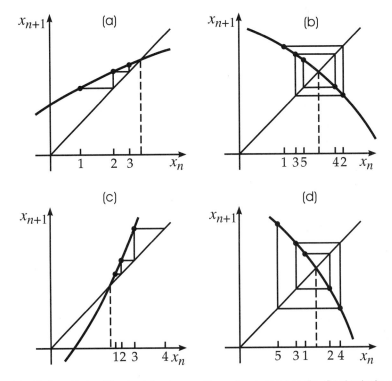

Figure 25.1. Various kinds of fixed points of a one-dimensional mapping. The fixed point is stable ((a) and (b)) or unstable ((c) and (d)), depending on the slope of the function f(x).

ONE-DIMENSIONAL MAPPINGS   477

(eigenvalue of the Jacobi matrix). For $|\lambda| < 1$, the fixed point is *stable* (point attractor). The sequence of points approaches the fixed point either *monotonically* (for $0 < \lambda < 1$, (a)) or *alternating* (for $-1 < \lambda < 0$, (b)). A value $|\lambda| > 1$, on the contrary, implies an *unstable* fixed point where the point sequence escapes, again either monotonically (for $\lambda > 1$, (c)) or alternating (for $\lambda < -1$, (d)). In the limit $\lambda = -1$, there results a periodic sequence of length 2, while for $\lambda = +1$ also each point in the vicinity of x_s is a fixed point. In order to determine the stability for $|\lambda| = 1$, the first derivative (linear approximation) is not sufficient.

In Examples 25.1 to 25.3, various discrete one-dimensional mappings are presented and their dynamics are studied in detail.

Example 25.1: The logistic mapping

One of the simplest but not least impressive examples of an iterative mapping is generated by the *logistic function*.[1] This function is defined as an inverted parabola

$$f(x) = \alpha x(1-x), \qquad x \in [0, 1], \tag{25.2}$$

with zeros at the border of the unit interval and a maximum at $f(1/2) = \alpha/4$. The logistic function depends on a real parameter α which shall lie in the range $1 < \alpha \leq 4$, since otherwise the mapping either would lead beyond the unit interval $[0, 1]$ ($\alpha < 0, \alpha > 4$) or would become trivial ($0 < \alpha < 1$, all solutions converge toward $x = 0$).

The mapping (25.2) has the two fixed points

$$x_{s1} = 0 \quad \text{and} \quad x_{s2} = 1 - \frac{1}{\alpha}, \tag{25.3}$$

with the derivatives

$$f'(x_{s1}) = \alpha \quad \text{and} \quad f'(x_{s2}) = 2 - \alpha. \tag{25.4}$$

Since α shall be > 1, the first fixed point is always *unstable*. The second one is *stable* (point attractor) if $1 < \alpha < 3$. In this parameter range all solutions look like that represented in Figure 25.2 for $\alpha = 2.8$; i.e., x_n converges toward the fixed point x_{s2}, for small starting value initially monotonically increasing, later on alternating. If α exceeds the value $\alpha_1 = 3$, then according to (25.4), the fixed point x_{s2} becomes unstable, too.

The geometric or numerical construction of the solution shows that a stable limit cycle of period 2 evolves, as is illustrated in Figure 25.2(b) for the case $\alpha = 3.3$. For $\alpha = \alpha_1$, a *period doubling* of the solution arises. One faces a pitchfork bifurcation as is illustrated in Figure 23.3. The old solution becomes unstable, and there appears a pair of new stable solution branches. (Note that here both branches simultaneously belong to the solution, since the latter one periodically jumps back and forth

---

[1] The logistic mapping may be interpreted with some effort as the model equation of a particular physical system; see the remark at the end of Example 25.3. It is also related to other fields of science. The logistic mapping was introduced first in 1845 in biological population dynamics by the Belgian biomathematician P.F. Verhulst. There it describes the evolution of a population of animals or plants in a restricted environment whereby each iteration corresponds to a new generation (the old one thereby dies off). For a low population density, the reproduction rate is positive (if $\alpha > 1$). But, if one approaches the saturation density $x = 1$, the reproduction rate decreases by overpopulation such that $x_{n+1} < x_n$. The simple parabola *ansatz* is sufficient to generate complex dynamics.

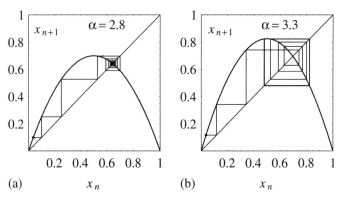

Figure 25.2. Typical trajectories of the logistic mapping. For $\alpha = 2.8$, the attractor is a fixed point, and for $\alpha = 3.3$, a limit cycle of period 2.

between the branches. This behavior differs from the case studied in Chapter 23, where both solution branches were independent of each other.)

The period doubling can also be understood by inspecting the iterated mapping $f^2(x) = f(f(x))$. As already mentioned, for this mapping a periodic solution reduces to two separated stationary fixed points. The iterated function

$$f^2(x) = \alpha^2 x(1-x)(1-\alpha x + \alpha x^2) \tag{25.5}$$

is a polynomial of fourth order with zeros at $x = 0$ and $x = 1$. Figure 25.3 shows this function for various values of the parameter α. For $\alpha < \alpha_1$, the function f^2 intersects the bisector x just as f does only at the two points x_{s1} and x_{s2} given in (25.3). As a polynomial of fourth order, (25.5) allows however four intersection points with a straight line, and just this happens for $\alpha > \alpha_1$. As shown in Figure 25.3, f^2 has a minimum at $x = 1/2$ for $\alpha > 2$ and two maxima at $x = 1/2 \pm \sqrt{(1/4) - 1/(2\alpha)}$. This can be seen without much calculation from

$$f^{2\prime}(x) = f'(f(x)) f'(x). \tag{25.6}$$

A zero of the derivative (extreme value of f^2) follows from $f'(x) = 0$; thus $x = 1/2$. For the two other zeros one finds

$$x = f^{-1}\left(\frac{1}{2}\right), \quad \text{since } f'\big(f(x)\big) = f'\left(f\left(f^{-1}(\tfrac{1}{2})\right)\right) = f'\left(\frac{1}{2}\right) = 0, \tag{25.7}$$

which as a quadratic equation yields the two roots mentioned above. For the slope of $f^2(x)$ at the position of the fixed point x_{s2}, one obtains

$$f^{2\prime}(x_{s2}) = (\alpha - 2)^2. \tag{25.8}$$

For $\alpha < \alpha_1 = 3$, this slope is smaller than 1, and therefore x_{s2} is a stable fixed point of f^2 too. Of course this must be so because f^2 simply picks every second value of the series of the x_n, and thus it takes over the stability from the solution of the mapping f. For $\alpha = \alpha_1$, $f^2(x)$ touches the bisector, and for $\alpha > \alpha_1$ two new intersection points x_1, x_2 arise. These are two stable fixed points of the iterated mapping f^2, i.e., $f^2(x_1) = x_1$, $f^2(x_2) = x_2$, which are related by $x_2 = f(x_1)$ and $x_1 = f(x_2)$.

ONE-DIMENSIONAL MAPPINGS

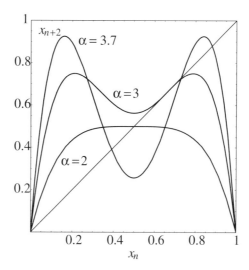

Figure 25.3. The iterated logistic function $f(f(x))$ displays two maxima for $\alpha > 0$.

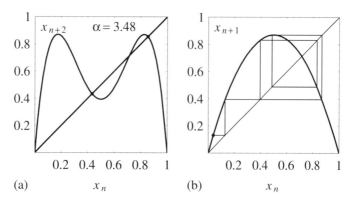

Figure 25.4. (a) For $\alpha = 3.48$, the iterated logistic function $f(f(x))$ has one unstable and two stable fixed points at $x \simeq 0.43298$ and $x \simeq 0.85437$. (b) The trajectory shows a period of length 4.

If the parameter α increases further, these fixed points also become unstable at a critical value $\alpha = \alpha_2$, and the game repeats for the mapping $f^2(x)$. A further bifurcation arises there with period doubling, and the new stable solution has a *period length of* 4. Figure 25.4 shows this solution for the example $\alpha = 3.48$.

The critical value of α_2 still can be given analytically, but with some effort. To this end, one first has to find the position of the fixed points x_1 and x_2. The condition $f^2(x) = x$ together with (25.5) leads to a quartic equation. But we already know two of the nodes, namely, the fixed points x_{s1} and x_{s2} from (25.3), which may be factored out by polynomial division:

$$f^2(x) - x = x\left(x - 1 + 1/\alpha\right)\left(\alpha^3 x^2 - \alpha(1+\alpha)x + \alpha + 1\right) = 0. \tag{25.9}$$

Setting the last bracket to zero yields a quadratic equation that determines the two new fixed points, namely,

$$x_{1,2} = \frac{1}{2}\left(1 + \frac{1}{\alpha}\right) \pm \frac{1}{2}\sqrt{\left(1 + \frac{1}{\alpha}\right)\left(1 - \frac{3}{\alpha}\right)}. \tag{25.10}$$

The second bifurcation occurs when these fixed points become unstable. This is decided by the magnitude of the slope of the function $f^2(x)$ at the points x_1, x_2. The slope begins with the value $+1$ at $\alpha = \alpha_1$, and then decreases continuously to -1. At this point, the next bifurcation arises. One therefore has to solve the equation

$$f^{2\prime}(x_{1,2}, \alpha_2) = -1. \tag{25.11}$$

Taking into account (25.6) and (25.10), equation (25.11) seems to depend on α in a very complicated manner. After some elementary transformations, (25.11) reduces however to a simple quadratic equation for both fixed points in common,

$$\alpha_2^2 - 2\alpha_2 - 5 = 0. \tag{25.12}$$

Hence, the critical bifurcation parameter for which the 2-cycle turns over to the 4-cycle is

$$\alpha_2 = 1 + \sqrt{6} = 3.4495\ldots. \tag{25.13}$$

That both fixed points simultaneously become unstable is plausible and immediately follows from (25.6), since the slope of $f^2(x)$ has the same value at both points:

$$f^{2\prime}(x_1) = f'\left(f(x_1)\right)f'(x_1) = f'(x_2)f'\left(f(x_2)\right) = f^{2\prime}(x_2). \tag{25.14}$$

It is not surprising that with increasing α the cycle of the period 4 becomes unstable too. There arises a full *cascade of period doublings* at $\alpha_1, \alpha_2, \alpha_3, \ldots$. In the interval $\alpha_k < \alpha < \alpha_{k+1}$, there exists a stable limit cycle of period 2^k. Mathematically one may consider instead of the limit cycle the iterated function $f^{2^k}(x)$ which displays a set of 2^k distinct stationary fixed points.

One should note that the critical points α_k are more and more closely spaced. As was found at first empirically and then analytically, the α_k obey the law of a geometric sequence that converges toward a cluster point α_∞:

$$\alpha_k \simeq \alpha_\infty - \frac{1}{\delta^k}. \tag{25.15}$$

The number δ is a constant that may be determined from the ratio

$$\delta = \lim_{k \to \infty} \frac{\alpha_k - \alpha_{k-1}}{\alpha_{k+1} - \alpha_k}. \tag{25.16}$$

For its value, one gets numerically

$$\delta = 4.669201\ldots, \tag{25.17}$$

and the accumulation point is at

$$\alpha_\infty = 3.569944\ldots. \tag{25.18}$$

ONE-DIMENSIONAL MAPPINGS

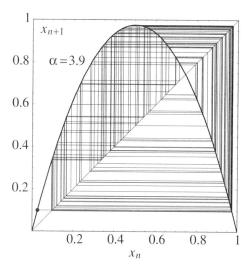

Figure 25.5. A chaotic trajectory of the logistic mapping for $\alpha = 3.9$.

The cascade of period doublings was considered first by Großmann and Thomae.[2] Feigenbaum[3] showed that the behavior (25.15) is not restricted to the logistic mapping but is *universally valid* for a large class of iterative mappings.[4]

Surprisingly, the numerical value (25.17) is also universally valid, and δ is therefore called the *Feigenbaum constant*. Essentially it suffices that the iterated function be smooth and display a quadratic maximum. The mathematical properties of the bifurcation cascade have been thoroughly studied; for a survey see, e.g., H. Schuster, *Deterministic Chaos*, VCH Publishing Company, 1989.

Of particular interest is what happens beyond α_∞. For $\alpha = \alpha_\infty$ obviously a cycle of "infinite period" arises, i.e., an aperiodic, never-repeating solution. $\alpha > \alpha_\infty$ is the domain of *chaos* that is characterized by irregular and seemingly random trajectories. As an example, Figure 25.5 shows a fraction from such a chaotic solution calculated for $\alpha = 3.9$.

The attractor diagram of the logistic mapping: Figure 25.6 presents an attractor diagram of the logistic mapping that is highly instructive and of amazing complexity. For each value of α on the abscissa the values passed by a trajectory are plotted as a point cloud along the ordinate direction.

---

[2]S. Großmann and S. Thomae, *Z. Naturforsch.* **32a**, 1353 (1977).

[3]*Mitchell Feigenbaum*, American physicist and mathematician, b. 1945, Philadelphia. Feigenbaum studied electrical engineering at the City College of New York and physics at MIT, where he did his doctorate in 1970 in high-energy physics. In 1974, he went to Los Alamos National Laboratory, where he was involved with problems of turbulence and nonlinear dynamics. In 1982, Feigenbaum was appointed as a professor at Cornell University, and presently, he heads the laboratory of mathematical physics at Rockefeller University in New York. Around 1976, he discovered, at first in numerical experiments on the logistic mapping, the universality of the period-doubling cascade and the constant that is named after him. In 1986, Feigenbaum (in common with Albert Libchaber, who experimentally demonstrated the period doubling in the flow of liquids) was awarded the Wolf Prize in physics.

[4]M.J. Feigenbaum, *J. Stat. Phys.* **19**, 25 (1978).

The first iteration steps (here 500) were omitted in order to filter out "transient processes" (transients) and to represent the asymptotic attractor itself. The subsequent (200 each) iterations are then plotted in the diagram.

The left part of Figure 25.6 shows the first segments of the bifurcation cascade: At $\alpha_1, \alpha_2, \alpha_3$, the attractor splits in 2, 4, and 8 branches, respectively. According to (25.15), the further critical points α_k follow each other so closely that they are no longer resolved in the figure. Beyond α_∞ one sees continuous bands more or less uniformly grey. The grey domains are passed through by some kind of "scars." Obviously these places are more frequently passed by the trajectory, and thus, the probability of finding $P(x)$ of the attractor is increased here.

One notes that in the chaotic domain $\alpha > \alpha_\infty$, the orbits are confined to a partial interval of $[0, 1]$. The borders of this interval are obtained as $x_{\max}(\alpha) = f(1/2) = \alpha/4$ and $x_{\min}(\alpha) = f^2(1/2) = (1/16)\alpha^2(4-\alpha)$. In the limit $\alpha = 4$, the full unit interval is passed. Thereby, the attractor covers the interval completely. More precisely, for any point $x \in [0, 1]$ and any $\epsilon > 0$, one can find an $n(\epsilon)$ such that $|x_n - x| < \epsilon$. Thus, the trajectory approaches each point arbitrarily closely if one waits for a sufficiently long time.

If the parameter is reduced from $\alpha = 4$, there arise various interesting phenomena. For example, when going below $\alpha'_1 \simeq 3.6785$ a *band splitting* is observed. The chaotic attractor which formerly covered a connected interval splits at $\alpha = \alpha'_1$ into two disjunct domains, as is clearly seen in Figure 25.6. The trajectory thereby alternates back and forth between the two "partial bands." For a further reduction of α, the partial bands in turn split again, into 4 parts at $\alpha'_2 \simeq 3.5926$, and so on. Similar to the bifurcation cascade $\alpha_1, \alpha_2, \ldots, \alpha_\infty$ of the regular attractor, there also exists a kind of reversed

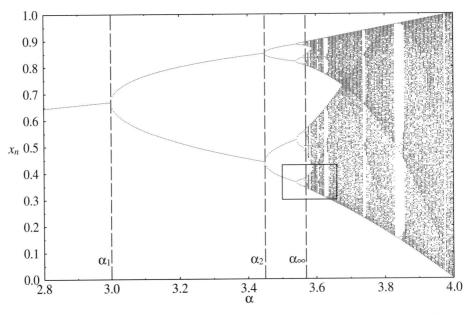

Figure 25.6. The attractor diagram of the logistic mapping (Feigenbaum diagram) shows the iterated values x_n versus the parameter α. One notes a cascade of period doublings, and above $\alpha_\infty \simeq 3.569944$ a chaotic domain that however is infiltrated by windows of regular solutions. The marked rectangular domain is shown in linear magnification in Figure 25.7.

bifurcation cascade $\alpha'_1, \alpha'_2, \ldots, \alpha'_\infty$ of the chaotic attractor. Both cascades meet at a common limit $\alpha_\infty = \alpha'_\infty$.

A closer inspection of the attractor diagram Figure 25.6 reveals that there are also *windows of periodic solutions* embedded in the chaotic domain. Particularly striking is the domain with solutions of period 3 which occur above $\alpha \simeq 3.8283$. This is connected with a fixed point of the triply iterated mapping that arises at the position of the maximum of $f(x)$, i.e., at $x = 1/2$:

$$f^3\left(\frac{1}{2}\right) = \frac{1}{2}. \tag{25.19}$$

One easily confirms that this happens at $\alpha_{s3} = 1 + \sqrt{8} \simeq 3.8284$. The point $x = 1/2$ therefore has a particular meaning, since here the derivative vanishes; $f'(1/2) = 0$. Because

$$\frac{d}{dx} f^3(x) = f'(f^2(x)) f'(f(x)) f'(x), \tag{25.20}$$

this property transfers also to all iterated mappings. This guarantees that the fixed point (25.19) is stable. Here, one even speaks of a *superstable cycle*, since the magnitude of the derivative of $f^n(x)$ which determines the stability takes the smallest possible value, namely zero. Due to the continuity of the mapping as a function of the parameter, the period-3 cycle is still stable in a finite environment of α_{s3}. In the downward direction the window of stability borders on the chaotic region.

If α increases beyond α_{s3}, one again finds a cascade of bifurcations with period doubling, i.e., the 3-cycle turns into a 6-cycle, etc. The bifurcations follow each other more and more closely until chaotic solutions emerge again.

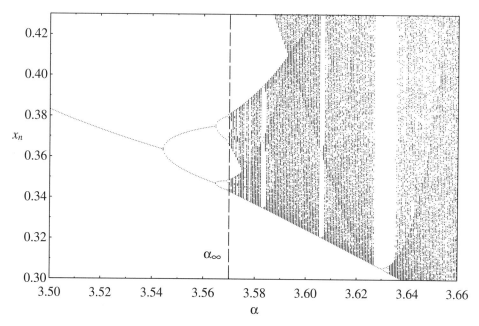

Figure 25.7. A small section of the attractor diagram of the logistic mapping shows quite a similar structure to the full diagram in Figure 25.6.

This consideration is not restricted to the cycle of period 3 but holds also for arbitrarily larger period lengths. There exist superstable cycles for any natural number m (their number even increases exponentially with m) with period length m which are defined by

$$f^m\left(\frac{1}{2}\right) = \frac{1}{2}. \tag{25.21}$$

They are always enclosed by a small window of regular cyclic orbits in the otherwise chaotic domain.

It is highly instructive to study the attractor diagram in detail. Figure 25.7 shows a small section $\alpha = 3.52$ to $\alpha = 3.65$ at the border of the 3-cycle window. The agreement with the full attractor diagram Figure 25.6 is amazing. Although there are tiny differences in the shape of both geometrical objects, their structure is nevertheless very similar. This property—a partial section looks just as the entire diagram—is called *self-similarity* (in the parameter space). For the attractor diagram of the logistic mapping, the process of successive magnification of a section may be continued ad infinitum: In Figure 25.7, one again finds partial domains that resemble the entire figure and so on, without an end.

The Lyapunov exponent of the logistic mapping: In the nonchaotic range, the Lyapunov exponent may be calculated rather simply analytically. We first consider the parameter range $\alpha < \alpha_1 = 3$. Here, a stable fixed point exists at $x_{s2} = 1 - 1/\alpha$; see (25.3). The quantity to be calculated is

$$\sigma = \lim_{n \to \infty} \frac{1}{n} \sum_{l=0}^{n-1} \ln |f'(x_l)|. \tag{25.22}$$

The influence of transients is excluded by the limit process. Because $x_l \to x_{s2}$, only the fixed point contributes, and thus the sum reduces to a single term:

$$\sigma = \ln |f'(x_{s2})| = \ln |\alpha(1 - 2x_{s2})| = \ln |\alpha(1 - 2 + 2/\alpha)|$$
$$= \ln |2 - \alpha|. \tag{25.23}$$

The result shows that the Lyapunov exponent diverges logarithmically at $\alpha \to 2$, $\sigma \to -\infty$. This of course happens just there where the derivative of the function f at the fixed point vanishes, which corresponds to a particularly fast approach to the attractor. This situation is called superstable. When approaching the parameter value $\alpha \to \alpha_1 = 3$, *the Lyapunov exponent vanishes*. This is a general feature of bifurcation points, since here the old attractor loses its stability, and the new attractor is not yet born. Beyond α_1, the period-2 cycle becomes an attractor, and therefore, σ again decreases to negative values.

It is not difficult to calculate the Lyapunov exponent in the interval $\alpha_1 < \alpha < \alpha_2$. Since now a periodic attractor exists that alternates between the points x_1 and x_2 from equation (25.10), (25.22) reduces to a sum of two terms:

$$\sigma = \frac{1}{2} \ln |f'(x_1)| + \frac{1}{2} \ln |f'(x_2)| = \frac{1}{2} \ln |f'(x_1) f'(x_2)|. \tag{25.24}$$

For the derivatives, we get

$$f'(x_{1,2}) = \alpha(1 - 2x_{1,2}) = -1 \mp \sqrt{(\alpha + 1)(\alpha - 3)}, \tag{25.25}$$

and the Lyapunov exponent is

$$\sigma = \frac{1}{2} \ln |(1 - (\alpha + 1)(\alpha - 3)|. \tag{25.26}$$

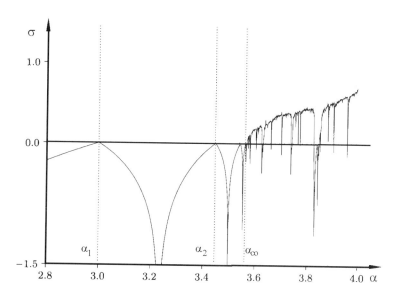

Figure 25.8. Numerically calculated values of the Lyapunov exponent σ of the logistic mapping. For $\sigma > 0$, chaotic trajectories arise.

This function begins with the value $\sigma = 0$ at $\alpha = \alpha_1$, then decreases monotonically and diverges at $\alpha = 1 + \sqrt{5}$. This is again the superstable point of the 2-cycle, since one easily confirms that

$$f^2\left(\frac{1}{2}\right) = \frac{1}{2} \quad \text{at} \quad \alpha = 1 + \sqrt{5}. \tag{25.27}$$

σ then increases again and finally reaches the value $\sigma = 0$. This happens when the argument of the logarithm in (25.27) takes the value -1, and thus, $1 - (\alpha + 1)(\alpha - 3) = -1$, which leads to the quadratic equation (25.12). The solution $\alpha = \alpha_2 = 1 + \sqrt{6}$ is the bifurcation point at which the 2-cycle becomes unstable. In the further course of the bifurcation cascade $\alpha_2 < \alpha < \alpha_\infty$, the game is repeating, and the Lyapunov exponent oscillates in the interval $0 \geq \sigma > -\infty$.

A qualitatively new feature arises in the chaotic range $\alpha > \alpha_\infty$, since here σ takes *positive* values. This is represented in Figure 25.8, for which the Lyapunov exponent was determined numerically. To the left of $\alpha = \alpha_\infty$, one faces the range with negative σ that has been discussed already. For $\alpha > \alpha_\infty$, σ becomes positive and on the average increases with increasing α. In the limit $\alpha = 4$, there results the maximum value $\sigma = \ln 2 \simeq 0.6931$, as will be shown in Example 25.2. The chaotic domain is repeatedly interspersed with windows corresponding to regular solutions where σ becomes negative. The figure reflects the complexity of the function $\sigma(\alpha)$ only imperfectly. The windows may be perceived only approximately, due to the limited resolution. The indicated peaks pointing down in the figure are actually poles extending to $\sigma \to -\infty$. A realistic representation of $\sigma(\alpha)$, when plotted with finite line width, would display only a largely homogeneous black block that extends to $-\infty$. There are infinitely many windows with stable cycles, and one may even show that they *lie densely* over the entire real interval $0 < \alpha < 4$: Any arbitrarily small vicinity of each point α still includes stable cycles!

The function $\sigma(\alpha)$ also has the property of *self-similarity*, just like the attractor in Figure 25.6. For each step of magnification, the partial sections look like the full figure.

Example 25.2: Exercise: Logistic mapping and the Bernoulli shift

The mapping

$$y_{n+1} = 2 y_n \pmod 1 \quad \text{with} \quad y_n \in [0, 1] \tag{25.28}$$

is called the saw-tooth mapping, or the *Bernoulli shift*.

(a) Discuss the trajectories y_n generated by the Bernoulli shift as a function of the start value y_0.

(b) Show that for the parameter $\alpha = 4$, i.e., in the region of the fully developed chaos, the logistic mapping and the Bernoulli shift are equivalent.
   Hint: Use the transformation of variables

$$x_n = \frac{1}{2}\left(1 - \cos(2\pi y_n)\right) = \sin^2(\pi y_n). \tag{25.29}$$

(c) Find the frequency distribution $P(x)$ of a "typical" trajectory x_n of the logistic mapping for passing the various points in the interval $0 < x < 1$, and calculate the Lyapunov exponent σ from this distribution.

Solution

(a) The graph of the function $f(y) = 2y \pmod 1$ consists of two pieces of a straight line with the slope 2 that are shifted relative to each other, as is represented in Figure 25.9. Since everywhere $f'(y) = 2 > 1$, stable fixed points cannot exist. The iterated solution of (25.28) is simple,

$$y_n = 2^n y_0 \pmod 1. \tag{25.30}$$

Hence, all iterated solutions of an initial value y_0 are explicitly known, nevertheless the mapping is chaotic! This is due to the factor 2^n, which implies an exponential enhancement of smallest deviations in the initial value y_0. As is represented in the figure, an interval Δy is stretched by the Bernoulli shift by a factor 2 to the length $2\Delta y$. Values falling into the range $y > 1$ are folded back to the unit interval by the modulo operation. This repeated *sequence of stretching and folding* is characteristic of chaotic mappings. In this way there results a thorough mixing of the trajectories and a sensitive dependence on the initial conditions.

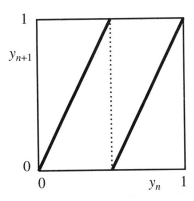

Figure 25.9. The mapping function of the Bernoulli shift.

ONE-DIMENSIONAL MAPPINGS

The Bernoulli shift mapping may be visualized by writing y as a number in *binary representation*:

$$y = \sum_{k=1}^{\infty} b_k \, 2^{-k} \quad \text{i.e.,} \quad y = 0.b_1 b_2 b_3 b_4 \ldots \quad \text{with} \quad b_k = 0 \text{ or } 1. \tag{25.31}$$

A doubling of y corresponds to a *left shift* (therefore the name) of the binary digits b_k, and the modulo operation ensures that the digit before the decimal point is cut off:

$$\begin{aligned} f(y) &= b_1.b_2 b_3 b_4 \ldots \pmod{1} \\ &= 0.b_2 b_3 b_4 b_5 \ldots . \end{aligned} \tag{25.32}$$

Now the effect of the Bernoulli shift becomes clear: Each iteration enforces the "back digits" in the binary expansion. If the initial value y_0 is known to an accuracy of 2^{-m}, this information is exhausted after m iteration steps. Later on there remains only "numerical noise," the trajectory wanders around through the unit interval in a nonpredictable way.

Mathematically, there still remains a subtle difference between rational and irrational values of the initial condition. A *rational* number (a fraction p/q) y_0 has a binary representation which (after a finite number of steps) becomes *periodic*. Consequently the trajectory y_n formed according to (25.28) with (25.32) also becomes periodic. A simple example is provided by the cycle $2/3, 1/3, 2/3, \ldots$, since

$$y_0 = 0.101010\ldots = \frac{1}{2}\left(1 + \frac{1}{2^2} + \frac{1}{2^4} + \ldots\right) = \frac{1}{2}\left(\frac{1}{1-1/4}\right) = \frac{2}{3}$$

$$y_1 = 0.010101\ldots = \frac{1}{4}\left(1 + \frac{1}{2^2} + \frac{1}{2^4} + \ldots\right) = \frac{1}{4}\left(\frac{1}{1-1/4}\right) = \frac{1}{3}$$

$$y_2 = 0.101010\ldots .$$

$$\vdots$$

The rational numbers lie *densely* on the real axis. Thus, there are infinitely many start solutions leading to periodic trajectories, and in any arbitrarily small vicinity ϵ of a point y one finds such solutions. On the other hand, the set of rational numbers has the *measure zero*; these are therefore "atypical" numbers. If one adopts a "typical" initial value y_0, e.g., a random number in the interval $[0, 1]$, then the probability for y_0 being rational and thus leading to a periodic trajectory is arbitrarily small. For a physicist, rational initial conditions do not play a role because of the finite precision of measurement. In the numerical simulation, the situation is different: If the computer used stores the numbers with a precision to m bits, then after at most m steps the simulation of the Bernoulli shift becomes meaningless. Fortunately for many purposes it makes no difference whether one is dealing with a cyclic solution with a very long period or with a genuine nonperiodic solution.

(b) The logistic equation

$$x_{n+1} = 4 x_n (1 - x_n) \tag{25.33}$$

is transformed to the new variable y_n. The right-hand side then becomes

$$4 \frac{1}{2}(1 - \cos(2\pi y_n)) \left(1 - \frac{1}{2} + \frac{1}{2}\cos(2\pi y_n)\right)$$

$$= 1 - \cos^2(2\pi y_n) = 1 - \frac{1}{2}\big(1 + \cos(4\pi y_n)\big)$$

$$= \frac{1}{2}\big(1 - \cos(4\pi y_n)\big), \tag{25.34}$$

and hence, (25.33) reads

$$\frac{1}{2}\left(1 - \cos(2\pi y_{n+1})\right) = \frac{1}{2}\left(1 - \cos(4\pi y_n)\right) \tag{25.35}$$

or

$$\cos(2\pi y_{n+1}) = \cos(4\pi y_n). \tag{25.36}$$

This solution is solved by

$$y_{n+1} = 2y_n \ (\text{mod } 1), \tag{25.37}$$

i.e., the Bernoulli shift mapping. When transformed back to the variable x, the solution (25.37) reads

$$x_n = \frac{1}{2}\left(1 - \cos(2\pi 2^n y_0)\right) = \sin^2(\pi 2^n x_0)$$
$$= \sin^2\left(2^n \arcsin \sqrt{x}\right). \tag{25.38}$$

As was discussed in (a), this leads for almost all initial values x_0 to chaotic trajectories which cannot be calculated even numerically if n becomes large.

(c) In the range of definition of the Bernoulli shift mapping (25.28), no point is particularly distinguished. For typical, i.e., irrationally chosen initial conditions, the solution (25.30) will meet all numbers in the interval (0, 1) with the same probability, hence $P(y) = 1$. This implies a corresponding probability of the logistic mapping of

$$P(x) = 2P(y)\frac{dy}{dx} = 2\frac{1}{2\pi}\frac{1}{\sin(\pi y)\cos(\pi y)}$$
$$= \frac{1}{\pi\sqrt{x(1-x)}}. \tag{25.39}$$

The factor 2 arises because the transformation equation has two solutions y symmetrical about $y = 1/2$ for any value of x. The probability of finding $P(x)$ is minimum at $x = 1/2$ and increases toward the borders of the unit interval. At $x \to 0$ and $x \to 1$, the function diverges but remains integrable. It is normalized to unity.

For the calculation of the Lyapunov exponent (24.8), we replace the mean value over the time sequence by a mean value over the probability distribution $P(x)$:

$$\sigma = \lim_{n \to \infty} \frac{1}{n} \sum_{l=0}^{n-1} \ln|f'(x_l)|$$
$$= \int_0^1 dx \, P(x) \ln|f'(x)|. \tag{25.40}$$

Systems for which this substitution *time average* $\leftrightarrow$ *phase-space average* is permissible are called *ergodic*. The proof of ergodicity of a system is in general not simple. Since $f'(x) = 2$ for all x and since the probability distribution $P(x)$ is normalized to unity, the Lyapunov exponent of the logistic mapping at $\alpha = 4$ is obtained as

$$\sigma = \ln 2 = 0.6931\ldots, \tag{25.41}$$

which agrees with the numerical result from Figure 25.8.

ONE-DIMENSIONAL MAPPINGS

Example 25.3: The periodically kicked rotator

In this section, we shall meet another example of a discrete mapping that, despite of its simple shape, leads to complex solutions and to chaotic behavior. Thereby, several new concepts arise, and a way is described that leads from a quasiperiodic to a chaotic motion.

In contrast to the logistic mapping, which was introduced as a purely mathematical example, we now consider the motion of a specific mechanical system, namely, a damped rotator that is under the influence of an external force. The corresponding equation of motion for the rotational angle θ reads

$$\ddot{\theta} + \beta\dot{\theta} = M(\theta, t). \tag{25.42}$$

Here, β is a friction parameter, and $M(\theta, t)$ describes the imposed time-dependent torque divided by the moment of inertia of the rotator. The external force shall depend *periodically* on the time. The problem simplifies if the force is acting in short pulse-like impacts spaced in time by the period T:

$$M(\theta, t) = M(\theta) \sum_{n=0}^{\infty} \delta(t - nT). \tag{25.43}$$

The nonautonomous differential equation of second order

$$\ddot{\theta} + \beta\dot{\theta} - M(\theta) \sum_{n=0}^{\infty} \delta(t - nT) = 0 \tag{25.44}$$

can be rewritten as usual in an autonomous system of *three* coupled differential equations of first order. With $x = \theta$, $y = \dot{\theta}$, and $z = t$, we have

$$\dot{x} = y,$$
$$\dot{y} = -\beta\dot{\theta} + M(\theta) \sum_{n=0}^{\infty} \delta(z - nT), \tag{25.45}$$
$$\dot{z} = 1.$$

Because of the specific form of the force (25.43), the rotator is again and again accelerated by the pulse, but otherwise moves freely influenced only by friction. Therefore, the equations of motion can be integrated exactly. For this purpose, we introduce the discretized variables

$$x_n = \lim_{\epsilon \to 0} x(nT - \epsilon), \qquad y_n = \lim_{\epsilon \to 0} y(nT - \epsilon). \tag{25.46}$$

Thus, position and velocity are scanned shortly before the individual kicks. We now consider the momentum number n and integrate the equation of motion over the nth time interval $nT - \epsilon < t < (n+1)T - \epsilon$. Since only one kick contributes in this interval, the equation of motion for y reads

$$\dot{y} = -\beta y + M(x)\delta(t - nT). \tag{25.47}$$

The solution of the homogeneous differential equation holding between the kicks may be given immediately:

$$y(t) = a\, e^{-\beta t} \quad \text{for } t \neq nT. \tag{25.48}$$

The impact of force causes a sudden increase of velocity according to

$$y(nT + \epsilon) - y(nT - \epsilon) = \int_{nT-\epsilon}^{nT+\epsilon} dt\, \left(-\beta y + M(x)\,\delta(t - nT)\right) \tag{25.49}$$
$$= -2\epsilon\beta\, y(nT) + M(x(nT))$$

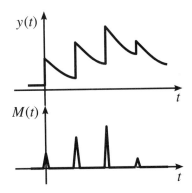

Figure 25.10. Periodic impacts of force (lower part) cause sudden changes of velocity (upper part). The influence of friction between the impacts is perceptible.

$$\simeq M(x(nT)), \tag{25.50}$$

as is schematically represented in Figure 25.10. With (25.48) and (25.49), one obtains

$$y(t) = \Big(y_n + M(x_n)\Big)e^{-\beta(t-nT)} \quad \text{for } nT < t < (n+1)T. \tag{25.51}$$

The angle coordinate $x(t)$ results from this by integration:

$$x(t) = x_n + \int_{nT}^{t} dt'\, y(t') = x_n - \frac{1}{\beta}\Big(e^{-\beta(t-nT)} - 1\Big)\Big(y_n + M(x_n)\Big). \tag{25.52}$$

The equations (25.51) and (25.52) taken at $t = (n+1)T - \epsilon$ thus yield a *two-dimensional discrete mapping*

$$x_{n+1} = x_n + \frac{1}{\beta}\Big(1 - e^{-\beta T}\Big)\Big(y_n + M(x_n)\Big) \pmod{2\pi}, \tag{25.53}$$

$$y_{n+1} = e^{-\beta T}\Big(y_n + M(x_n)\Big). \tag{25.54}$$

By performing the modulo operation, one takes into account that x is a periodic angular coordinate. The system (25.53) describes a Poincaré mapping of the periodically kicked rotator.

The angular dependence of the torque $M(x)$ is determined by the specific physical system. If a body is moving on a circular orbit and is under a force pointing always in the same direction (see Figure 25.11), then the torque is proportional to the sine of the angle, $\mathbf{M} = \mathbf{r} \times \mathbf{F} = -rF\sin\theta\, \mathbf{e}_z$.

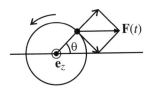

Figure 25.11. The force $\mathbf{F}(t)$ always points in the same direction, independent of the displacement angle θ of the rotator.

In addition, we take into account an angle-independent torque M_0, i.e., with $K_0 = r\,F$:

$$M(x) = M_0 + K_0 \sin x. \tag{25.55}$$

The system (25.53) and (25.54) with the nonlinear force law (25.55) is called the *dissipative circular mapping* ("dissipative" since we are dealing with the limit of strong damping) and displays very interesting dynamic properties.

The equations may be written still more clearly by introducing the following abbreviations:

$$b = e^{-\beta T}, \qquad \Omega = \frac{1}{\beta} M_0,$$
$$K = \frac{1}{\beta}\left(1 - e^{-\beta T}\right) K_0, \tag{25.56}$$
$$r_n = \frac{1}{b\beta}\left(1 - e^{-\beta T}\right) y_n - \Omega.$$

r_n represents a rescaled velocity coordinate. Insertion into (25.53) leads to the following form of the dissipative circular mapping:

$$x_{n+1} = x_n + b r_n + \Omega - K \sin x_n \ (\text{mod } 2\pi), \tag{25.57}$$
$$y_{n+1} = e^{-\beta T}\left(y_n + M(x_n)\right). \tag{25.58}$$

One can iterate this equation numerically and study the solutions for various values of the parameters b, K, and Ω. A further simplification results in the *limit of strong damping*, i.e., $\beta T \gg 1$ or $b \ll 1$. (To overcome the friction, K_0 must, according to (25.56), increase linearly with c.) In this case, the velocity y_n is decelerated to zero immediately after each "kick." Consequently, the equation for the angle decouples to

$$\begin{aligned}x_{n+1} &= f(x_n) \ (\text{mod } 2\pi) \\ &= x_n + \Omega - K \sin x_n \ (\text{mod } 2\pi).\end{aligned} \tag{25.59}$$

This equation is called the *one-dimensional circular mapping* or the *standard mapping*. Its mathematical properties were intensely studied, in particular by the Russian mathematician V.I. Arnold.[5] Its interesting properties are due to the nonlinearity of the sine function.

Let us consider for a moment the trivial limit $K = 0$, i.e., the *linear circular mapping*

$$x_{n+1} = x_n + \Omega \ (\text{mod } 2\pi). \tag{25.60}$$

The rotator is moving forward in equidistant steps Ω. If it reaches the old position again after a finite number of steps, the motion is *periodic*. This obviously happens if $\Omega/2\pi$ is a *rational number*,

$$\Omega = 2\pi \frac{p}{q}, \qquad p \text{ and } q \quad \text{coprime numbers.} \tag{25.61}$$

This means that after q time steps the rotator has performed p full turns, i.e., takes again (modulo 2π) its original position. One deals with a solution with the period q. If however $\Omega/2\pi$ is an *irrational number*, the initial point x_0 is not reached again even after an arbitrarily long time. But there occur values x_n in any arbitrarily small vicinity of x_0. In such a case the motion is called *quasiperiodic*.

---

[5] V.I. Arnold, *Trans. of the Am. Math. Soc.* **42**, 213 (1965).

In order to characterize the motion, one may define a *winding number*:

$$W = \frac{1}{2\pi} \lim_{n \to \infty} \frac{f^n(x_0) - x_0}{n}. \tag{25.62}$$

Here, $f^n(x_0)$ is the n-fold iterated mapping function (without performing the modulo operation) from equation (25.59). The winding number thus represents the mean shift per stroke interval. $W = W(\Omega, K)$ depends on both parameters of the circular mapping. In the linear case, $K = 0$, W just coincides with the fraction p/q defined in (25.61). Since the rational numbers, although densely located on the real axis, form only a set of measure zero, the "typical" trajectories are quasiperiodic.

What happens now if the nonlinearity becomes efficient (i.e., for $K \neq 0$) in the circular mapping (25.60)? Let us consider a *periodic solution* with a rational winding number $W = p/q$. The angular coordinate passes a cycle $x_1, x_2, \ldots, x_q$ of length q that causes a final shift of $x_q = x_1 + 2\pi p \bmod 2\pi = x_1$; hence,

$$f^q(x_1) = x_1 + 2\pi p. \tag{25.63}$$

At the beginning of the present chapter, we met the criterion for the *stability* of a discrete mapping, in the case at hand, of the q-fold iterated function $f^q(x)$: The magnitude of the derivative of the function must be smaller than unity, and thus,

$$|f^{q\prime}(x_1)| = \left| \frac{d}{dx_1} f\left(f\left(\cdots f(x_1)\right)\right) \right| = \left| \prod_{i=1}^{q} f'(x_i) \right|$$
$$= \left| \prod_{i=1}^{q} (1 - K \cos x_i) \right| < 1. \tag{25.64}$$

If this condition is fulfilled, then x_1 (and since all points in the cycle are on equal base, all other x_i too) belongs to a stable periodic attractor. In the linear case, $K = 0$, the derivative has always the value $f^{q\prime}(x) = 1$. There exists marginal stability where neighboring orbits are neither attracted nor repelled. If however $0 < K < 1$, then each of the solutions $\Omega_{p,q} = 2\pi \, p/q$ becomes a *periodic attractor* with a domain of attraction of finite width $\Delta\Omega_{p,q}$. This is an example for a very interesting phenomenon that arises in many branches of physics. Vibrating systems which are characterized by two distinct frequencies adjust themselves—provided that there is a corresponding interaction—in such a manner that the frequencies are *synchronized*, i.e., are in an integral ratio to each other. The phenomenon is also called *mode locking*. The two frequencies of the system considered here are determined on one hand by the stroke length T, and on the other hand by the magnitude of the torque M_0.

Possibly the earliest experimental evidence for the phenomenon of mode locking is ascribed to the Dutch physicist Christian Huygens. He observed that a series of pendulum watches (Huygens played a decisive role in their discovery) which were suspended in a row began to vibrate in the same rhythm, although their limited accuracy of movement would rather have suggested a drifting apart. Huygens recognized that the weak coupling of the watches via their common back wall was responsible for the synchronization.

The extension of the mode-locking ranges can be calculated from (25.63). For larger period lengths q, this may be performed only numerically. We therefore restrict ourselves here to the simplest case $q = 1$. If there is a complete synchronization with winding number $W = 1$, then (25.63) becomes

$$f(x_1) = x_1 + 2\pi \quad \text{or} \quad \Omega = K \sin x_1 + 2\pi. \tag{25.65}$$

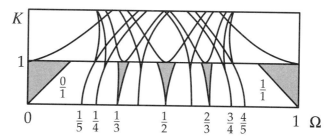

Figure 25.12. Several domains of stability for the one-dimensional circular mapping depending on the parameters Ω and K. The hatched regions are called "Arnold tongues." On the lines in the upper half the condition of superstability is fulfilled.

The stability condition (25.64) reads

$$\left|1 - K \cos x_1\right| < 1, \tag{25.66}$$

which is fulfilled for angles $0 < x_1 < \pi/2$ or $3\pi/2 < x_1 < 2\pi$. The associated values of Ω may be read off from (25.65):

$$2\pi - K < \Omega_{1,1} < 2\pi + K. \tag{25.67}$$

The range of mode locking with just one turn of the rotator per stroke interval thus has the shape of a triangle that opens with increasing K. Just the same consideration leads to the attraction domain of the attractor with winding number $W = 0$, and thus, $p = 0, q = 1$:

$$-K < \Omega_{0,1} < K. \tag{25.68}$$

Because of the periodicity of the angle coordinate, it suffices to investigate the interval $0 \leq \Omega \leq 2\pi$. Values beyond this range mean only that the rotator for each kick performs additional full turns, which does not change the dynamics significantly. In Figure 25.12, the borders of the ranges (25.67) and (25.68) are drawn as straight lines. The parameter values for which a periodic synchronized motion arises are shaded in the diagram.

Analogous considerations may be made for any rational winding number $W = p/q$. Some of the stability ranges (there are, of course, infinitely many ones) are represented in Figure 25.12. The width of these ranges, which are also called *Arnold tongues*, increases monotonically with K. In the linear limit ($K = 0$), the summed-up total width of all Arnold tongues is equal to zero (measure of the rational numbers), as was already mentioned. It can be shown[6] that for $K = 1$ the Arnold tongues cover the entire Ω-range:

$$\sum_{p,q} \Delta\Omega_{p,q} = 2\pi \quad \text{for } K = 1. \tag{25.69}$$

Hence, the situation is exactly complementary to the case $K = 0$; the "typical" solutions are now periodic (mode locking), while the quasiperiodic solutions have the measure zero.

The function $W(\Omega)$, i.e., the winding number as a function of the frequency parameter Ω at $K = 1$, is called the *devil's staircase* (see Figure 25.13). It is a function that is everywhere continuous but

---

[6]M.H. Jensen, P. Bak, T. Bohr, *Phys. Rev.* **A30**, 1960 (1984).

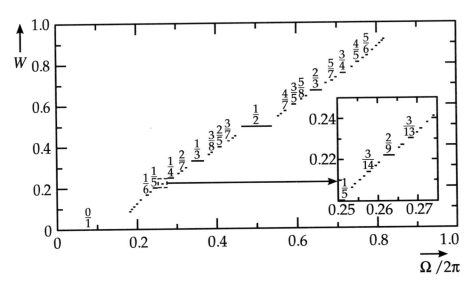

Figure 25.13. The steps of the "devil's staircase" are ranges where the winding number "clicks into place," i.e., is independent of the frequency parameter Ω. The sectional magnification on the right indicates the self-similar structure of the devil's staircase.

nowhere differentiable. To any rational number p/q belongs a step; the width of the steps decreases with increasing period length q; its total width according to (25.69) covers the entire interval. The devil's staircase has the property of *self-similarity*, i.e., any sectional magnification again resembles the entire object.

If the parameter of the nonlinear coupling exceeds the value $K = 1$, then the Arnold tongues coalesce. This is connected with a conversion of the quasiperiodic solutions into *chaotic* ones, as may be seen from the occurrence of positive Lyapunov exponents. At the same time, for $K > 1$ there still exist domains of periodic solutions with a negative Lyapunov exponent. Both types of solutions are interwoven in a complex manner. For the logistic mapping, we observed a mixing of regular and chaotic motion. But now the relations are even more complicated, since now *two* parameters, Ω and K, may be varied independently. An interesting feature of the chaotic solutions is that they don't have a well-defined winding number. The solution moves so irregularly that the limit in (25.62) does not exist.

In the upper half of Figure 25.12, for $K > 1$, the centers of several periodic ranges are plotted as lines. Here, the condition of *superstability* introduced on page 483 is fulfilled, i.e., the derivative of the iterated mapping function of a q-cycle vanishes, $f^{q\prime}(x) = 0$. For the period $q = 1$, this condition may be evaluated easily. The solution with winding number $W = 0$ has as its fixed-point condition

$$f(x_1) = x_1; \quad \text{thus}, \quad \Omega - K \sin x_1 = 0. \tag{25.70}$$

This 1-cycle is superstable if (see also (25.59))

$$f'(x_1) = 0; \quad \text{thus}, \quad 1 - K \cos x_1 = 0. \tag{25.71}$$

With the value x_1 following from (25.70), equation (25.71) yields the condition for superstability

$$K = \sqrt{1 + \Omega^2} \tag{25.72}$$

as is plotted in Figure 25.12. The condition for a superstable fixed point with the winding number $W = 1$ follows analogously as

$$K = \sqrt{1 + (2\pi - \Omega)^2}. \tag{25.73}$$

The crossing of the curves indicates that for equal values of Ω and K distinct stable solutions may coexist. In the range $K > 1$, there is no longer a straightforward relation between the parameters K, Ω, and the winding number W. Which of the solutions is realized then depends on the initial condition for x.

The occurrence of chaotic solutions is associated with a qualitative change of the mapping function $f(x)$. As Figure 25.14 shows, for $K < 1$, $f(x)$ is a monotonically increasing function. For $K > 1$, the nonlinear coupling is so strong that $f(x)$ reflects the shape of the sine function; i.e., there arise (quadratic) maxima and minima. Similar to the previously treated logistic mapping, $f(x)$ is *not invertible* for $K > 1$. This is a necessary (but not sufficient) condition for the occurrence of chaos in one-dimensional mappings.

The complex behavior in the mode locking expressed by the devil's staircase is well confirmed by experiment. Even simpler than mechanical oscillators are nonlinear electric circuits. For example, mode locking was investigated in an externally periodically driven circuit involving a superconducting Josephson junction and an induction which may be described mathematically by the circular mapping.[7]

Supplement: It should be noted that the logistic mapping from Example 25.1 may be interpreted as the motion of a periodically kicked rotator. For this purpose, the angular dependence of the torque is not represented by (25.55) but rather by the (somewhat artificially constructed) function

$$M(x) = K_0\left((\alpha - 1)x - \frac{\alpha}{2\pi}x^2\right). \tag{25.74}$$

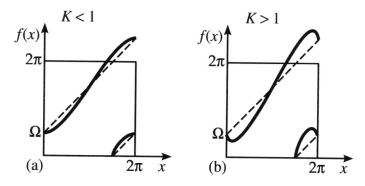

Figure 25.14. (a) The mapping function of the circular function for $K < 1$ increases monotonically. (b) For $K > 1$, a maximum develops, and $f(x)$ is no longer invertible. The function $f(x) = x + \Omega$ (mod 2π) is drawn as a dashed line.

---

[7]M. Bauer, U. Krüger, W. Martienssen, *Europhys. Lett.* **9**, 191 (1989).

If we consider the dissipative Poincaré mapping (25.53) in the limit of strong damping, $\beta T \gg 1$, the equation for x_n again decouples,

$$x_{n+1} = x_n + \frac{K_0}{\beta}\left((\alpha - 1)x_n - \frac{\alpha}{2\pi}x_n^2\right) \pmod{2\pi}. \tag{25.75}$$

The choice $K_0 = \beta$ leads to

$$x_{n+1} = \alpha x_n\left(1 - \frac{1}{2\pi}x_n\right) \pmod{2\pi}, \tag{25.76}$$

which after rescaling of the angular variable to the unit interval by $x'_n = x_n/2\pi$ yields the logistic mapping

$$x'_{n+1} = \alpha x'_n(1 - x'_n) \pmod{2\pi}. \tag{25.77}$$

For parameter values $\alpha \leq 4$, the values of x_n are automatically bound to the unit interval, and hence the modulo operation may be omitted.

Example 25.4: The periodically driven pendulum

In the preceding examples, we have studied systems the dynamics of which could be described by the iteration of simple, analytically known discrete mappings. The logistic mapping from Example 25.1 with its extremely simple structure served as a "test laboratory" for investigating many aspects of nonlinear dynamics but has no plausible physical analog. For the "periodically kicked" damped rotator of Example 25.3, the dynamics could also be reduced to the iteration of discrete equations of motion (in the limit of strong damping to the one-dimensional circular mapping (25.57)), because of the pulse-like nature of the acting force. Another model example, possibly even more realistic and appropriate for a clear illustration of the characteristic phenomena of nonlinear dynamics, is the *periodically driven pendulum*.[8]

Let the pendulum, a nonlinear oscillator with a backdriving force proportional to the sine of the displacement angle θ, be under the influence of an additional external force with a harmonic time dependence. Moreover, let the system be damped by friction which is proportional to the velocity. Mathematically, these system properties are described by the following equation of motion:

$$\frac{d^2\theta}{dt^2} + \beta\frac{d\theta}{dt} + \sin\theta = f\cos(\Omega t). \tag{25.78}$$

Here, β is the friction parameter, and f and Ω denote the strength and frequency of the driving force, respectively. The eigenfrequency of the pendulum has been set to the value $\omega_0 = 1$, which may always be achieved by rescaling the time and the parameters β and f. As was described in Chapter 21, this explicitly time-dependent differential equation of second order may be rewritten in a system of three coupled autonomous (i.e., not time-dependent) differential equations of first order:

$$\frac{d\omega}{dt} = -\beta\omega - \sin\theta + f\cos\phi,$$
$$\frac{d\theta}{dt} = \omega, \tag{25.79}$$

---

[8] See, e.g., G.L. Baker and J.P. Gollub, *Chaotic Dynamics*, Cambridge University Press, Cambridge (1996). In this book, extensive use is made of the example of the driven pendulum. We also refer to H. Heng, R. Doerner, B. Huebinger, W. Martienssen, *Int. Journ. of Bif. and Chaos* **4**, 751, 761, 773 (1994).

$$\frac{d\phi}{dt} = \Omega.$$

For $\beta > 0$, this is obviously a dissipative system, since the divergence of the velocity field $\mathbf{F}$ (compare (21.16)) is then negative:

$$\Lambda = \nabla \cdot \mathbf{F} = -\beta. \tag{25.80}$$

The equations of motion of the driven pendulum are too complicated to allow analytic solutions. Their numerical integration by the computer does not cause any trouble, however.[9]

Depending on the parameters Ω, β, f, the driven pendulum displays many distinct types of motion. Here we shall investigate only a small section of the parameter space, namely the dependence on the driving strength f for fixed values of the frequency Ω and of the friction constant β. As an example *we choose a frequency $\Omega = 2/3$ for all of the subsequent investigations*, i.e., a value somewhat below the natural vibration frequency of the pendulum, *and a friction parameter $\beta = 0.5$*.

The system of differential equations (25.79) is integrated numerically for various values of the parameter f, beginning with selected initial conditions $\theta(0)$ and $\omega(0)$. To avoid needless effort, transient processes, i.e., the initial solutions of, e.g., the first 20 vibrational periods, will be ignored.

The effect of dissipation in the system ensures that the solution after some finite time approaches an attractor. The shape of this attractor may be analyzed in different manners. One may directly consider the time dependence of the displacement angle $\theta(t)$ or plot the trajectory in the three-dimensional phase space θ, ω, ϕ, where the third coordinate because of $\phi = \Omega t$ just corresponds to the time. More transparent than this three-dimensional representation are reduced two-dimensional phase-space diagrams where the time is considered as a parameter and the trajectory is plotted in the θ, ω-plane. Contrary to the full three-dimensional phase space, the projected orbits may intersect here. Since θ is a periodic angular variable, we restrict it to the interval $-\pi < \theta \leq \pi$ by the modulo operation. A trajectory that leaves the diagram at the right or left edge, corresponding to a loop of the pendulum, therefore enters again at the opposite edge.

Figure 25.15 shows a gallery of selected phase-space diagrams arranged by increasing value of the driving force f. The value of f is always given at the top left in the partial figures.

For weak perturbations, e.g., $f = 0.9$, the pendulum performs approximately *harmonic librations* about the zero position. The limit cycle $\theta(t)$ is a slightly distorted sinusoidal vibration with the frequency Ω, and correspondingly the path in phase space is approximately an ellipse.

With increasing perturbation strength a bifurcation arises at about $f = 1.07$ with a *period doubling*, as is represented in Figure 25.15 for $f = 1.075$: Two slightly different vibrational tracks are alternating. After further period doublings, one then finds a *libration with twice the amplitude* (represented for $f = 1.12$) in which the pendulum performs a loop but then moves back. The frequency of this oscillation is $\Omega/3$.

In the range $f \simeq 1.15 \ldots 1.3$, there arise *chaotic solutions*. The trajectory for $f = 1.2$ in Figure 25.15 fluctuates in an erratic manner between librations and rotations in both directions, and correspondingly the path densely covers a domain in phase space. For comparison, Figure 25.16 shows a regular ($f = 1.12$) and a chaotic ($f = 1.2$) trajectory, which correspond to the third and fourth partial figure in Figure 25.15.

---

[9]Readers are advised to explore the dynamics of the driven pendulum by their own computer experiments. For the integration of the differential equation a Runge-Kutta approach is recommended; see, e.g., W.H. Press et al., *Numerical Recipes*, Cambridge University Press (1989).

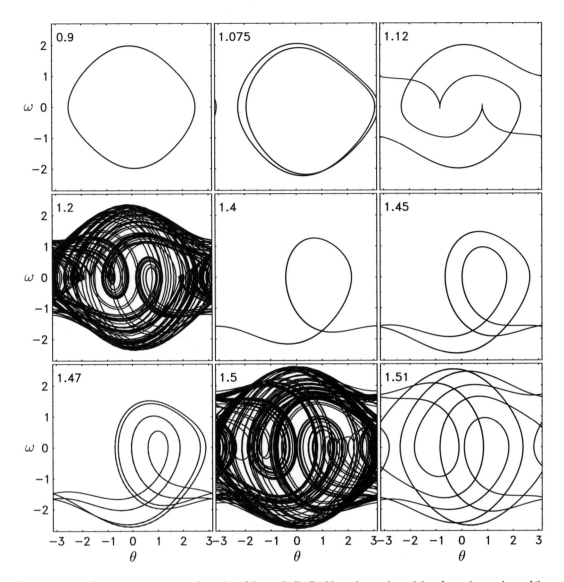

Figure 25.15. Typical phase-space trajectories of the periodically driven damped pendulum for various values of the parameter f. The parameters $\Omega = 2/3$ and $\beta = 0.5$ are kept fixed.

For even stronger coupling f, the chaotic range is left again, and there occur *rotating periodic solutions*, as is represented for $f = 1.4$. The angle increases linearly with time, according to $\theta(t) \propto \pm \Omega t$, superimposed by local fluctuations. For $f = 1.45$, a bifurcation with *period doubling of the rotating solution* occurs. On the average, the angle is unchanged, but now the local deviations alternate from period to period. At $f = 1.47$, one obtains a *second period doubling*.

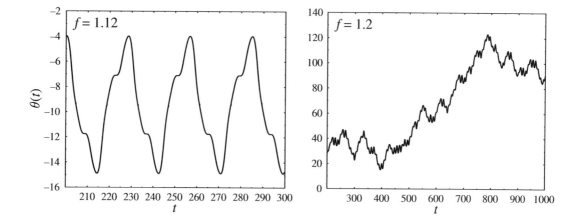

Figure 25.16. Two trajectories $\theta(t)$ for distinct values of the parameter f. For $f = 1.12$ (left) a periodic motion arises, for $f = 1.2$ (right) a chaotic one. Note the distinct scales!

After a bifurcation cascade, there again occurs a range with *chaotic solutions*, as is represented for the example $f = 1.5$. But soon the chaos is followed by regular motion, as is demonstrated by the beautiful phase-space trajectory for $f = 1.51$ in the last shown example. Here, one faces a periodic libration motion with two loops (angular range $3 \cdot 2\pi$) and the period 5 (i.e., the frequency $\Omega/5$).

A global survey of the behavior of the system is obtained from the *attractor diagram*. One of the coordinates is scanned in regular time intervals, and the result is plotted along the ordinate as a function of a system parameter. Figure 25.17 shows the angular velocity $\omega(t_n)$ scanned at the time points $t_n = t_0 + n \, 2\pi/\Omega$ versus the strength f of the driving force. For any value on the abscissa, 150 values of ω are plotted. The upper margin marks the nine values of f for which the associated phase-space trajectories are shown in Figure 25.15.

The attractor diagram clearly exhibits the previously discussed alternating ranges of regular and chaotic motion. The chaotic window at $f = 1.15 \ldots 1.28$ is followed by a broad domain of periodic (rotating) solutions at $f = 1.28 \ldots 1.48$, showing several pronounced period-doubling bifurcations (a "subharmonic cascade"). In the ranges $f = 1.11 \ldots 1.15$ and $f > 1.54$, one finds solutions of period 3. The structural similarity of the attractor diagram for the driven pendulum with its counterpart for the logistic mapping, Figure 25.6, is obvious.

One should note that the attractor diagram represented in Figure 25.17 is not complete. This is due to the fact that our system of differential equations (25.79) is *invariant under reflection* because the pendulum has no preferred direction of oscillation. Each solution is accompanied by a reflected trajectory

$$\theta \to -\theta \;(\text{mod } 2\pi), \qquad \omega \to -\omega, \qquad \phi \to \phi + \pi \;(\text{mod } 2\pi), \tag{25.81}$$

which also satisfies the equation of motion. Angle and velocity are inverted, and the phase is shifted by a half-period. In general, the solutions always occur pairwise, which is obvious for rotational solutions because the pendulum may run "clockwise" or "anti-clockwise." Which attractor is actually reached depends in a complicated manner on the initial conditions $\theta(0)$, $\omega(0)$. When plotting Figure 25.17, only one path has been calculated, hence the "reflected" branches are missing. (The points are

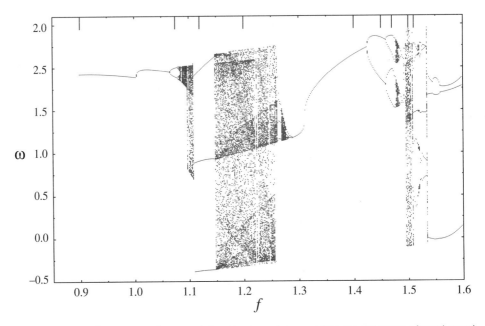

Figure 25.17. The attractor diagram of the driven pendulum exhibits the sequence of regular and chaotic domains as a function of the parameter f. The phase-space trajectories corresponding to the frequency values marked at the upper margin are shown in Figure 25.15.

not really reflected about $\theta = 0$, since in the stroboscopic scanning the phase of the scanning moment is kept fixed and not shifted according to (25.81).)

However, for periodic trajectories of period length $n\, 2\pi/\Omega$ it may happen that $\theta(t + n\pi/\Omega) = -\theta(t)$ (mod 2π) holds. Such a symmetric trajectory is identical with its reflected partner, and there exists only one solution. This occurs, e.g., for the $n = 3$-vibration shown in Figure 25.16 (left), and in particular also for the $(n = 1)$-librations for small f. At about $f = 1.01$, the interesting case of a *symmetry-breaking bifurcation* occurs, which may be recognized by the kink in the attractor diagram Figure 25.17. The attractor continues to be periodic with $n = 1$, but looses its symmetry and therefore splits into a pair of distinct solutions which are reflected relative to each other. In the figure only the upper branch of this fork is included.

The attractor diagram provides a good survey of the various kinds of motion of a dynamic system. A more far-reaching quantitative measure of the stability of trajectories are the Lyapunov exponents σ_i discussed in Chapter 24. The driven pendulum—a three-dimensional system—has three Lyapunov exponents. One of these, let us call it σ_3, has always the value $\sigma_3 = 0$. It belongs to the degree of freedom ϕ, which according to (25.79) has the trivial linear time dependence $\phi(t) = \Omega t$. Thus, any perturbations along this direction neither increase nor shrink exponentially.

The *maximum Lyapunov exponent* σ_1 determines the stability of the system. Attractors with $\sigma_1 < 0$ are periodic, those with $\sigma_1 > 0$ are chaotic. Figure 25.18 shows the result of a numerical calculation of the maximum Lyapunov exponent, plotted over the same parameter range as in the attractor diagram Figure 25.17. One may clearly trace the sequence of regular and chaotic domains. At the bifurcation points the exponent σ_1 touches the zero line from below.

A conspicuous feature is that σ_1 never falls below the value -0.25. This is related to the fact that in the dissipative system under consideration the sum of all three Lyapunov exponents is determined by the negative of the friction coefficient β (here, $\beta = 0.5$):

$$\sum_{i=1}^{3} \sigma_i = -\beta. \tag{25.82}$$

Because $\sigma_3 = 0$ in the case of maximum stability $\sigma_1 = \sigma_2 = -\beta/2$.

The Poincaré cut introduced in Chapter 22 may further characterize the attractors. When choosing $\phi = \phi_0 \pmod{2\pi}$ as the cut condition, one just has a stroboscopic mapping at equidistant time points $t_n = t_0 + 2\pi/\Omega$. The three-dimensional phase space reduces to two dimensions (θ, ω), and the continuous trajectory turns into a cloud of points. The Poincaré cuts of *periodic attractors* simply consist of one or several fixed points, the number of which corresponds to the period length of the vibration. The situation is different, however, for the nonperiodic *strange attractors* which are characteristic for the occurrence of chaotic motion. Here, the cloud of points of the Poincaré cut covers extended partial domains of phase space more or less uniformly.

Figure 25.19(a) illustrates the situation for the chaotic attractor of the pendulum with a driving strength $f = 1.2$. The points were obtained by stroboscopic scanning of the trajectory shown in the fourth partial figure of Figure 25.15 over 2000 vibrational periods. The detailed shape of the Poincaré cut depends on the selected phase angle ϕ_0.

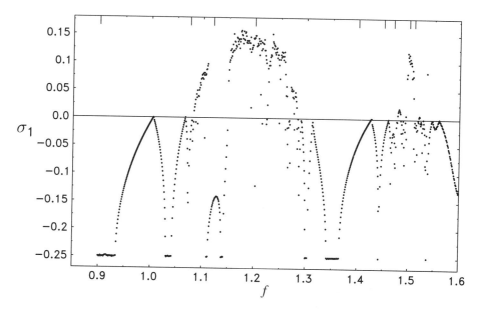

Figure 25.18. The largest Lyapunov exponent σ_1 of the driven pendulum as a function of the parameter f. Values $\sigma_1 > 0$ occur in the domains of chaotic motion which were perceptible already in the attractor diagram Figure 25.17. Compare also the analogous representation of the Lyapunov exponent of the logistic mapping in Figure 25.8.

The long curved object in Figure 25.19 at first glance appears as a strange bent one-dimensional curve. A closer look, however, shows the chaotic attractor to be a much more complex geometric object. In the sector magnification of a small partial range of the attractor (see Figure 25.19(b)), the seemingly single line dissolves into several closely spaced curves. But this is only the beginning, since a repetition of this operation would show that in each sector magnification the new lines again decay into several fractions, and the procedure may be repeated infinitely many times. (This process is limited in practice only by problems in the numerical integration of the equation of motion. By the way, in order to plot Figure 25.19(b), 100,000 periods had to be calculated.)

Thus, the attractor of the chaotic driven pendulum displays an infinite filigree ("puff-pastry") structure. Mathematically it is a *fractal* with broken dimension (see Chapter 24) since the points of the Poincaré cut occupy, roughly speaking, a larger volume in phase space than a one-dimensional

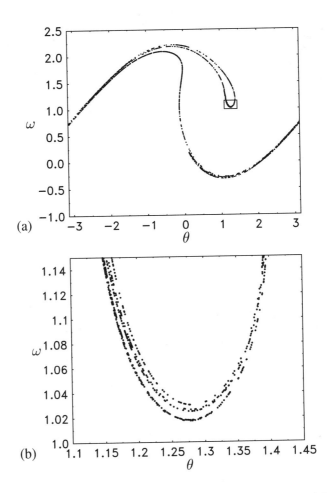

Figure 25.19. The Poincaré cut of a chaotic trajectory ($f = 1.2$) of the driven pendulum. The range in the upper figure marked by a box is represented below as a sector magnification, revealing the fractal structure of the strange attractor.

curve, but on the other hand they are too rare to cover a two-dimensional area. An analogous statement holds also for the full attractor in the three-dimensional phase space.

In Chapter 24, we dealt with the determination of the fractal dimension. A remarkable point is that the Lyapunov exponents σ_i may also be used to determine the dimension of a strange attractor. The existence of such a relation is not implausible, because the σ_i decide how "fast" a region of phase space spreads under the dynamic flow. Building on this consideration, Kaplan and Yorke[10] derived a formula for calculating a *Lyapunov dimension* D_L. For our special case (one positive and one negative Lyapunov exponent), the Kaplan–Yorke relation reads

$$D_L = 1 + \frac{\sigma_1}{|\sigma_2|} \quad \text{for } \sigma_1 > 0, \, \sigma_2 < 0. \tag{25.83}$$

The relation between D_L and the other dimension measures has not yet been cleared up in full. The originally assumed identification of Lyapunov dimension and capacity dimension D_f cannot be maintained, since counter-examples have been found. More recent speculations rather concern a possible relation with the information dimension $D_L = D_I$.

For the Poincaré cut 25.19, one gets from (25.83) with $\sigma_1 \simeq 0.14$, $\sigma_2 \simeq -0.64$, the Lyapunov dimension $D_L \simeq 1.2$. This value depends sensitively on the friction constant β. For a weaker damping, the strange attractor "blows up," and its dimension increases.

Example 25.5:   Chaos in celestial mechanics: The staggering of Hyperion

Hyperion is one of the more remote moons of Saturn. It revolves about Saturn with a revolution period of 21 days, on an ellipse with eccentricity $\varepsilon = 0.1$ and a large semiaxis $a = 1.5 \cdot 10^6$ km.

The motion of Hyperion is a particularly impressive example of chaotic staggering within our solar system. In this section we shall describe this behavior using a simplified model. The satellite *Voyager 2* among others has supplied pictures of the moon Hyperion. Hyperion is an asymmetric top that may be roughly described by a three-axial ellipsoid with the dimensions

$$190 \, \text{km} \times 145 \, \text{km} \times 114 \, \text{km} (\pm 15 \, \text{km}). \tag{25.84}$$

Hence, one obtains for the principal moments of inertia $\Theta_1 < \Theta_2 < \Theta_3$:

$$\frac{\Theta_2 - \Theta_1}{\Theta_3} \approx 0.3. \tag{25.85}$$

The striking prediction is that Hyperion performs a chaotic staggering motion in the sense that its rotational velocity and the orientation of its rotational axis vary significantly within a few revolution periods. This chaotic dancing, which must have happened also for other planetary satellites during their history (e.g., Phobos and Deimos with the planet Mars have been calculated), is implied by the asymmetry of Hyperion and by the eccentricity of the orbit.

To describe the change of the rotational velocity, we adopt the following model (see Figure 25.20): Hyperion H orbits Saturn S on a fixed ellipse with semimajor a and eccentricity ε. r represents the distance between Saturn and Hyperion, φ the polar angle of motion. Thus, the trajectory of Hyperion is given by

$$r(\varphi) = \frac{k}{1 + \varepsilon \cos \varphi}. \tag{25.86}$$

---

[10] J.L. Kaplan and J.A. Yorke, *Functional differential equations and approximation of fixed points*, H.-O. Peitgen and H.O. Walter (eds.), *Lecture Notes in Mathematics* **730**, Springer, Berlin (1979).

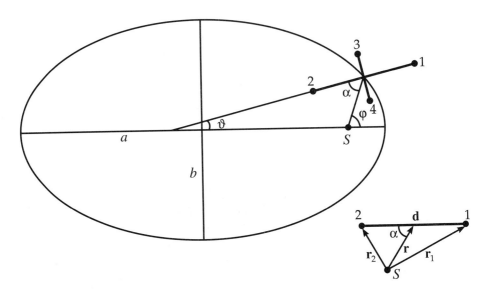

Figure 25.20. A simple two-dumbbell model for the asymmetric Saturn moon Hyperion.

Its asymmetric shape is simulated by four mass points 1 to 4 with equal mass m which are arranged in the orbital plane. Let the line 2–1 (distance d) be the (body-fixed) $\mathbf{e}_1$-axis, the line 4–3 (distance $e < d$) the (body-fixed) $\mathbf{e}_2$-axis. The $\mathbf{e}_3$-axis points perpendicular out of the image plane: $\mathbf{e}_3 = \mathbf{e}_1 \times \mathbf{e}_2$. The angle ϑ specifies the rotation of Hyperion about the $\mathbf{e}_3$-axis. It is defined as the angle between the semimajor a and the $\mathbf{e}_1$-axis. The moments of inertia obey

$$\Theta_1 = \frac{1}{2}me^2 < \frac{1}{2}md^2 = \Theta_2 < \frac{1}{2}m\left(d^2 + e^2\right) = \Theta_3. \tag{25.87}$$

In this model, the satellite shall rotate only about the $\mathbf{e}_3$-axis, i.e., the axis perpendicular to the orbital plane with the largest moment of inertia. This restriction is motivated because the tidal friction over very long times causes (1) the rotational axis of a moon to align along the direction of the largest moment of inertia, and (2) causes this direction to adjust perpendicular to the orbital plane. Moreover, the orbital angular momentum of Hyperion is assumed to be constant. This is a very good approximation, since the intrinsic angular momentum $\mathbf{L}_E$ of Hyperion is always very small relative to the orbital angular momentum $\mathbf{L}_B$, $|\mathbf{L}_E|/|\mathbf{L}_B| \approx (d^2 + e^2)/a^2 \approx 10^{-8}$. The gravitational field at the position of Hyperion is not homogeneous, and since Θ_1 and Θ_2 are distinct, the satellite experiences a torque that depends on its orbital point and its orientation ϑ, which will be calculated now. The tidal friction shall be neglected, however. The torque acting on the pair of masses (1,2) is

$$\mathbf{D}^{(1,2)} = \frac{d\mathbf{e}_1}{2} \times (\mathbf{F}_1 - \mathbf{F}_2), \tag{25.88}$$

where

$$\mathbf{F}_i = -\frac{\gamma m M \mathbf{r}_i}{r_i^3} \tag{25.89}$$

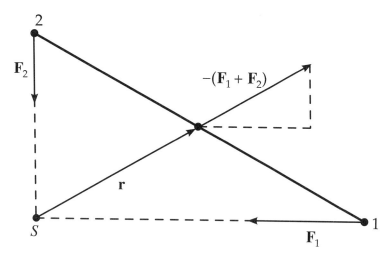

Figure 25.21. The forces causing the torque $\mathbf{D}^{(1,2)}$.

is the force acting on the mass point i, M being the Saturn mass. Figure 25.21 once more illustrates the torque that is caused by the gravitational forces $\mathbf{F}_1$ and $\mathbf{F}_2$ at the positions $\mathbf{r}_1$ and $\mathbf{r}_2$, and by the centrifugal force $\mathbf{F} = -(\mathbf{F}_1 + \mathbf{F}_2)$ at the position $\mathbf{r}$.

Since the length $d \approx 200$ km is small compared to the distance $r \approx 10^6$ km, the cosine law yields

$$r_i = r\sqrt{1 \pm \frac{d}{r}\cos\alpha + \left(\frac{d}{2r}\right)^2} \approx r\sqrt{1 \pm \frac{d}{r}\cos\alpha}. \tag{25.90}$$

The positive sign holds for r_1, the minus sign for r_2. α is the angle between $\mathbf{r}_1 - \mathbf{r}_2$ and $\mathbf{r}$. From (25.90), we obtain

$$\frac{1}{r_i^3} \approx \frac{1}{r^3}\left(1 \mp \frac{3}{2}\frac{d}{r}\cos\alpha\right). \tag{25.91}$$

Hence, for $\mathbf{D}^{(1,2)}$ we find

$$\begin{aligned}
\mathbf{D}^{(1,2)} &= \frac{d\mathbf{e}_1}{2} \times (\mathbf{F}_1 - \mathbf{F}_2) \\
&= \frac{d\mathbf{e}_1}{2} \times \frac{-\gamma m M}{r^3}\left[\left(1 - \frac{3}{2}\frac{d}{r}\cos\alpha\right)\left(\mathbf{r} + \frac{d}{2}\mathbf{e}_1\right) - \left(1 + \frac{3}{2}\frac{d}{r}\cos\alpha\right)\left(\mathbf{r} - \frac{d}{2}\mathbf{e}_1\right)\right] \\
&= \frac{3\gamma m M d^2 \cos\alpha}{2r^4}\mathbf{e}_1 \times \mathbf{r} \\
&= \frac{3\gamma m M d^2}{2r^3}\sin\alpha \cos\alpha\, \mathbf{e}_3 \\
&= \frac{3\gamma m M d^2}{4r^3}\sin 2\alpha\, \mathbf{e}_3 \\
&= \frac{3\gamma M \Theta_2}{2r^3}\sin 2\alpha\, \mathbf{e}_3 \\
&= \frac{6\pi^2 a^3 \Theta_2}{r^3 T^2}\sin 2\alpha\, \mathbf{e}_3.
\end{aligned} \tag{25.92}$$

In the last step of rewriting, γM has been expressed by the orbital period T and the semimajor a, using the third Kepler law:

$$\gamma M = \left(\frac{2\pi}{T}\right)^2 a^3. \tag{25.93}$$

Thereby, the reduced mass was replaced by the mass of Saturn.

The torque for the pair of masses (3,4) is obtained in the analogous way as

$$\begin{aligned}
\mathbf{D}^{(3,4)} &= \frac{e\mathbf{e}_2}{2} \times (\mathbf{F}_3 - \mathbf{F}_4) \\
&= \frac{e\mathbf{e}_2}{2} \times \mathbf{r}\, \frac{\gamma m M}{r^3} \left[\left(1 + \frac{3e}{2r}\sin\alpha\right) - \left(1 - \frac{3e}{2r}\sin\alpha\right)\right] \\
&= \frac{-3\gamma m M e^2}{2r^3} \sin\alpha \cos\alpha\, \mathbf{e}_3 \\
&= \frac{-3\gamma M \Theta_1}{2r^3} \sin 2\alpha\, \mathbf{e}_3 \\
&= \frac{-6\pi^2 a^3 \Theta_1 \sin 2\alpha}{r^3 T^2} \mathbf{e}_3.
\end{aligned} \tag{25.94}$$

The total torque $\mathbf{D} = \mathbf{D}^{(1,2)} + \mathbf{D}^{(3,4)}$ is therefore

$$\mathbf{D} = \frac{3}{2}\left(\frac{2\pi}{T}\right)^2 (\Theta_2 - \Theta_1) \left(\frac{a}{r}\right)^3 \sin 2\alpha\, \mathbf{e}_3. \tag{25.95}$$

Thus, the torque vanishes if $\Theta_1 = \Theta_2$. Besides, a configuration with α experiences the same torque as a configuration with $180^0 + \alpha$. The torque tries to rotate Hyperion in such a way that at any moment the $\mathbf{e}_1$-axis points toward Saturn. The expression (25.95) for the torque remains correct even if a more realistic mass distribution is assumed. With

$$\mathbf{D} = \frac{d\mathbf{L}}{dt} = \Theta_3 \frac{d^2\vartheta}{dt^2}, \tag{25.96}$$

the equation of motion for the eigenrotation of the satellite reads

$$\Theta_3 \ddot{\vartheta} = -\frac{3}{2}\left(\frac{2\pi}{T}\right)^2 (\Theta_2 - \Theta_1) \left(\frac{a}{r(t)}\right)^3 \sin(2(\vartheta - \varphi(t))). \tag{25.97}$$

Here, we set $\alpha = \varphi - \vartheta$. The equation (25.97) involves only one degree of freedom, ϑ, but the right side depends via the orbital radius $r(t)$ and the polar angle $\varphi(t)$ on the time and is therefore not integrable. An exception is the case of a *circular* orbit. Then the mean angular frequency

$$n = \frac{2\pi}{T} \tag{25.98}$$

equals the angular velocity ω, and $r = a$, $\varphi(t) = nt$. With $\vartheta' = 2(\vartheta - nt)$, the differential equation (25.97) therefore simplifies for $\varepsilon = 0$ to

$$\Theta_3 \ddot{\vartheta}' = -3n^2(\Theta_2 - \Theta_1) \sin\vartheta'. \tag{25.99}$$

This is the differential equation for the *pendulum*. The integral of motion is the energy E. To determine this quantity, we multiply (25.99) by $\dot{\vartheta}'$:

$$\Theta_3 \dot{\vartheta}' \ddot{\vartheta}' + 3n^2(\Theta_2 - \Theta_1)\dot{\vartheta}' \sin\vartheta' = 0, \tag{25.100}$$

which implies

$$\frac{d}{dt}\left(\frac{1}{2}\Theta_3 \dot{\vartheta}'^2 - 3n^2(\Theta_2 - \Theta_1)\cos\vartheta'\right) = 0. \tag{25.101}$$

Hence,

$$E = \frac{1}{2}\Theta_3\left(\frac{d\vartheta'}{dt}\right)^2 - 3n^2(\Theta_2 - \Theta_1)\cos\vartheta' \tag{25.102}$$

is an integral of the motion. Just as for the pendulum (see, e.g., Example 18.8) we have that for energies E larger than $E_0 = 3n^2(\Theta_2 - \Theta_1)$ the satellite rotates, while for $E < E_0$ it vibrates. Due to the (so far neglected) tidal friction the energy E will decrease more and more until it reaches the minimum value $E_{min} = -E_0$. From this follows $\dot{\vartheta}'_{min} = 0$ and $\vartheta'_{min} = 0$. Therefore, in the final state of satellites on a circular orbit the $\mathbf{e}_1$-axis always points toward the planet (bound rotation), as is known from the earth's moon.

As was stated already, the differential equation (25.97) for $\varepsilon \neq 0$ cannot be solved analytically. One may try, however, to get approximate solutions for $\varepsilon \ll 1$. This we shall do now. First, we introduce dimensionless quantities, the time $t' = nt = 2\pi t/T$ and $\omega_0^2 = 3(\Theta_2 - \Theta_1)/\Theta_3$. Then (25.97) turns into

$$\frac{d^2\vartheta}{dt'^2} = -\frac{\omega_0^2 a^3}{2r^3(t')}\sin\left(2(\vartheta - \varphi(t'))\right). \tag{25.103}$$

Since $r(t')$ and $\varphi(t')$ are periodic in 2π, the right side can be expanded into a Fourier-like Poisson series. One obtains

$$\frac{d^2\vartheta}{dt'^2} = -\frac{\omega_0^2}{2}\sum_{m=-\infty}^{\infty} H\left(\frac{m}{2},\varepsilon\right)\sin\left(2\vartheta - mt'\right). \tag{25.104}$$

In order to determine the coefficients $H(m/2, \varepsilon)$, $r(t')$ and $\varphi(t')$ must be known. $\varphi(t)$ is obtained, e.g., via the second Kepler law as the solution of the following differential equation:

$$\frac{d\varphi}{dt} = \frac{(1 + \varepsilon\cos\varphi)^2}{hk^2}. \tag{25.105}$$

Solving this differential equation and hence the determination of $H(m/2, \varepsilon)$ are beyond of the scope of this example. We only quote that the coefficients H are proportional to $\varepsilon^{2|m/2-1|}$ and were tabulated by Cayley[11] in 1859. For small ε, we have $H(m/2, \varepsilon) \approx -\varepsilon/2, 1, 7\varepsilon/2$ for $p = m/2 = 1/2, 1, 3/2$. Here the half-integer variable p has been introduced. If the argument of one of the sine functions varies only weakly with time, i.e., if

$$\left|2\frac{d\vartheta}{dt'} - m\right| \ll 1, \tag{25.106}$$

---

[11]The original paper is by A. Cayley, Tables of the developments of functions in the theory of elliptic motion, *Mem. Roy. Astron. Soc.* **29**, 191 (1859).

there occur resonances. It is then advantageous to rewrite the equation of motion (25.104) in such a way that it depends on the only slowly varying variable $\gamma_p = \vartheta - pt'$:

$$\frac{d^2\gamma_p}{dt'^2} = -\frac{\omega_0^2}{2} H(p,\varepsilon) \sin 2\gamma_p - \frac{\omega_0^2}{2} \sum_{n \neq 0} H\left(p + \frac{n}{2}, \varepsilon\right) \sin\left(2\gamma_p - nt'\right). \tag{25.107}$$

It turns out that the terms in the sum are oscillating so rapidly compared to the variation of γ_p that their total contribution to the equation of motion largely averages out, if ω_0^2 and ε are sufficiently small. In first approximation for small ω_0^2 and ε the high-frequency terms can be eliminated from the equation of motion by keeping γ_p fixed and averaging equation (25.107) over a period. One then obtains

$$\frac{d^2\gamma_p}{dt'^2} = -\frac{\omega_0^2}{2} H(p,\varepsilon) \sin 2\gamma_p. \tag{25.108}$$

This is again the pendulum equation. Integration of equation (25.108) analogous to the equations (25.100) to (25.102) again yields the energy

$$E_p = \frac{1}{2}\left(\frac{d\gamma_p}{dt'}\right)^2 - \frac{\omega_0^2}{4} H(p,\varepsilon) \cos 2\gamma_p. \tag{25.109}$$

Again, γ_p vibrates if E_p is smaller than $E_p^s = |H(p,\varepsilon)| \omega_0^2/4$, and rotates if E_p is larger. For $H(p,\varepsilon) > 0$, γ_p vibrates about 0; for $H(p,\varepsilon) < 0$, γ_p vibrates about $\pi/2$. The essential difference compared with the pendulum equation with $\varepsilon = 0$ is that here exists not only the synchronous ($p = 1$) solution but also resonances, depending on the initial condition. So, it may happen, e.g., that a satellite is captured by the tidal friction into a $p = 3/2$-state. This happens in the solar system for Mercury: During two circulations about the Sun it rotates exactly three times about its axis.

The question of to what extent the averaging over the high-frequency contributions is justified and whether these are indeed only weak perturbations is complicated and shall not be traced further here. But it is vividly clear that the high-frequency terms must not be neglected if the energy is close to the limit $E_p = E_p^s$. Then these terms will actually decide whether the satellite performs a full turn (in γ_p) or whether it vibrates back. There exists a band of energies $w_p \cdot E_p^s$ that are very close to E_p^s, with w_p defined by $w_p = (E_p - E_p^s)/E_p^s$. For energies within this band the high-frequency perturbations cannot be neglected. We shall only state without derivation that Chirikov found an analytical criterion for the width w_p of this band.[12] Chirikov's criterion predicts that for the parameters of Hyperion $\omega_0^2 = 0.89$ and $\varepsilon = 0.1$ the averaging over the high-frequency components is not possible, since the width of the band belonging to $p = 1$ and $p = 3/2$ is so large that the two bands overlap. In contrast, for $\omega_0^2 = 0.2$ and $\varepsilon = 0.1$ the averaging over the high-frequency components should be a good approximation. The following figures represent Poincaré cuts for these two cases. They show points in phase space taken at $\varphi = 0$.

Thereby, the differential equations have been solved numerically.[13] If the motion is quasiperiodic, then successive points form a smooth curve; chaotic trajectories seem to cover areas in a random manner.

Because of the symmetry of the inertial ellipsoid, the orientation ϑ is identical to that with $\vartheta + \pi$. Therefore, the graphs were plotted only for the range of ϑ between 0 and π. By averaging over the

---

[12] B.V. Chirikov, A universal instability of many-dimensional oscillators systems, *Phys. Rep.* **52**, 262 (1979).

[13] J. Wisdom, S.J. Peale, F. Mignard, The chaotic rotation of Hyperion. *Internat. Journal of Solar System Studies* **58**, 137 (1984).

high-frequency components, solutions were obtained in which $\gamma_p = \vartheta - pt'$ vibrates (for $E_p < E_p^s$). For each of these solutions, $d\vartheta/dt'$ has a mean value of exactly p, and ϑ takes all values between 0 and 2π. If one considers, however, only the points with $\varphi = 0$, i.e., times $t' = 2\pi n$, then γ_p exactly corresponds to ϑ modulo π. Therefore, a vibration in γ_p appears as a vibration in ϑ. The successive points of quasiperiodic vibrations therefore yield a simple curve in the vicinity of $d\vartheta/dt' = p$ that contains only a part of the angles between 0 and π. For nonresonant quasiperiodic trajectories ($E_p > E_p^s$), all γ_p's rotate, and successive points form a single curve that contains all angles ϑ.

As is seen from Figure 25.22, for small values of ω_0 and ε the resonant states and the nonresonant ones are separated by a narrow chaotic zone. The figure shows ten distinct trajectories. Three of them correspond to the quasiperiodic vibrations in the states $p = 1/2$, 1, and $3/2$. As predicted by the approximation, $\gamma_{1/2}$ vibrates about $\vartheta = \pi/2$; γ_1 and $\gamma_{3/2}$ vibrate about $\vartheta = 0$. Three further trajectories, always enclosing the resonant states, are chaotic. They fill narrow bands with points in a seemingly random manner. The last four trajectories show that each chaotic band is separated from the other bands by nonresonant quasiperiodic trajectories.

Figure 25.23 displays the situation for Hyperion. At least the chaotic zones of the states $p = 1$ and $p = 3/2$ are no longer separated from each other. There is a small remainder of the quasiperiodic $p = 1/2$ state; the quasiperiodic $p = 3/2$ state disappeared completely. Instead, there is a quasiperiodic state at $p = 9/4$, $\vartheta = \pi/2$, which is not given by the approximation (25.108). In total one sees 17 trajectories in the figure: eight quasiperiodic vibrations of the states $p = 1/2$, 1, 2, $9/4$, $5/2$, 3, and $7/2$, five nonresonant quasiperiodic rotations, and four chaotic trajectories.

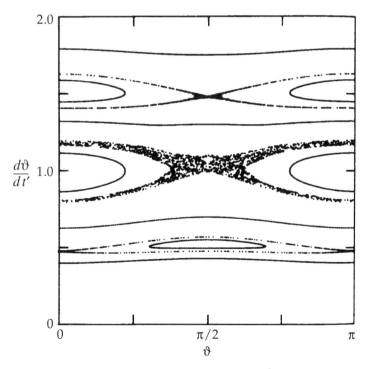

Figure 25.22. Transverse cut of the phase space for $\omega_0^2 = 0.2$ and $\varepsilon = 0.1$.

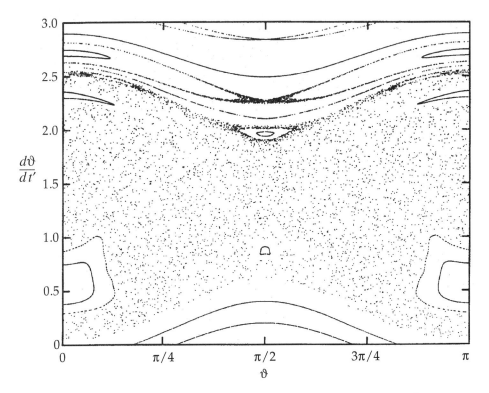

Figure 25.23. Transverse cut of the phase space for the parameters of Hyperion: $\omega_0^2 = 0.89$ and $\varepsilon = 0.1$.

A deeper study shows that the alignment of the rotational axis perpendicular to the orbital plane is not stable, both in the chaotic as well as in the synchronous state. This means that a small deviation of the rotational axis from the vertical increases exponentially. The time scale for the resulting staggering motion is of the order of magnitude of several orbital periods. The final stage of a "normal" moon is for Hyperion completely unstable. But if it tilts out of the perpendicular to the orbital plane, the equation (25.97) is no longer sufficient, and one has to solve the full nonlinear Euler equations. One then finds that the full three-dimensional course of motion is completely chaotic. All three characteristic Lyapunov exponents are positive (of the order of magnitude 0.1), which implies a strongly chaotic staggering. Even if one could have measured the spatial orientation of the rotation axis at the moment of *Voyager 1* passing Hyperion (in November 1980) with a precision of up to ten figures, it nevertheless would not have been possible to predict the orientation of Hyperion's axis at the moment when *Voyager 2* passed it (in August 1981).

Up to this point, the tidal friction, which causes a relatively very slow change of the initially Hamiltonian system, has been neglected. But one can roughly describe the history of Hyperion. Presumably, the period of the eigenrotation was initially much shorter than the orbital period, and Hyperion began its evolution in the range far above that which is shown in the figure. Over a time of the order of magnitude of the age of the solar system (circa 10^{10} years) the eigenrotation was decelerated, and the rotation axis straightened up perpendicular to the orbital plane. Thereby, the premises of our simplified model (25.97) are approximately justified. But when the evolution once had reached the chaotic domain, "the work of the tides lasting aeons was destroyed in a few days,"[14] since once it arrived in the chaotic domain, Hyperion began to stagger in a fully erratic manner (which continues to the present day).[15] Sometimes, its path will end up in one of the few stable islands of the figure. But this cannot be the synchronous state, since the latter state is unstable.

---

[14] J. Wisdom, Chaotic behavior in the solar system. *Nucl. Phys. B (Proc. Suppl.)* **2**, 391 (1987).

[15] The observations of *Voyager 2* are consistent with this statement. The staggering of Hyperion has also been observed directly from the earth (see J. Klavetter, *Science* **246**, 998 (1988), *Astron. J.* **98**, 1855 (1989)).

PART VIII

ON THE HISTORY OF MECHANICS*

---

*Here, we follow Friedrich Hund, *Einführung in die Theoretische Physik Bd. 1*, Mechanik, Bibliographisches Institut Leipzig (1951), and also I. Szabo, *Einführung in die Technische Mechanik*, Springer-Verlag, Berlin, Göttingen, Heidelberg (1956). For more on what is presented here, in particular on the history of statics, we refer to the works of P. Duhem:

Les origines de la statique, Paris (1905–1906),
Etudes sur Lionard de Vinci, Paris (1906),
Le systéme du monde, Paris (1913–1917),

and to P. Sternagel:

Die artes mechanical im Mittelalter (1966),

and to F. Krafft:

Dynamische und statische Beobachtungsweise in der antiken Mechanik (1970).

26 Emergence of Occidental Physics in the Seventeenth Century

Physics describes nature by theoretical-physical, mathematical theories. It uses general but precise concepts, builds upon certain laws ("axioms") and makes precise mathematical-logical statements on natural phenomena. An important role is played by the idealization of reality, i.e., the representation of reality by describing ideal situations. Precise statements can be made only about these ideal cases. In doing so, physics utilizes systematically worked-out procedures for proof (reproducibility) of its statements. The application of the laws of physics has led to a far-reaching control of nature which seems to grow continually.

Physics originated in the occidental culture. As a modern natural science it has characterized this society for about 300 years, i.e., since about the seventeenth century. One should note that the consideration of nature in Greek antiquity differs essentially from the present natural science in our society. In ancient Greece, say at the time of *Plato*[1] and *Aristotle*,[2] there existed little systematic knowledge of nature. What existed appears from a present-day view to be primitive and of low precision. Attempts at ultimate sharpening physical concepts were rarely known, although the clear line of thought in the then already highly developed mathematics might have served as a model. For example, *Eudoxus*[3] (400–347 BC) was one of the most outstanding mathematicians of that age who was close to Plato. The philosopher *Plato* in turn admired the inner logical consistency of mathematics. Partial disciplines of physics, such as astronomy, statics, and music theory, only later were quantified. But no genuine mechanics emerged. The analysis of motion and the precise formulation of the concept of force were not achieved in antiquity, although mathematics (e.g., by *Archimedes*,[4] 287–212 BC) was actually just as far developed as in the seventeenth century when the theory of motion originated. It seems to be a peculiar feature of the ancient culture that mechanics in its full meaning, and hence the beginning of comprehensive physics, could not develop in the ancient world. It is also remarkable that the natural scientists of the antiquity hardly influenced the general consciousness of that age. The specialized sciences were not of public concern. Scientists in Alexandria, Pergamon, or Rhodos belonged to a closed circle.

The cultures in East Asia and India, still existing nowadays, produced outstanding achievements in other disciplines and certain techniques (ceramics, dyeworks), but no physics. The conceptional-mathematical thinking about nature and the questioning of nature by determined experiments were not developed or promoted there.

Physical science emerged on one hand from the philosophic thinking about nature and on the other hand from technical problems (e.g., military resources, traffic, civil engineering, mining). In particular the first developed mechanics, which at the very beginning was merely statics, goes back to these two sources. A proper unification of these two approaches to nature was, however, not reached in antiquity. Archimedes (287–212 BC) represented the climax in ancient statics: the lever principle, the concept of the center of gravity of a body, and the well-known hydrostatic law named after him, were known to him in full clarity. Yet, Archimedes's discoveries fell into oblivion. The reasons are not known. Possibly his ideas were simply too hard to be understood. Whatever the reasons were, in the Middle Ages only the works of Aristotle were known, and these determined the further development of mechanics.

In the fourteenth century, statics enjoyed a time of prosperity, mainly due to distinguished men of the artist faculty of the university in Paris. The methods for decomposing and combining forces were developed there and utilized for solving statical problems. In addition, the concept "work of a force" was introduced, and the "virtual work" in virtual displacements was correctly used in simple cases. *Leonardo da Vinci*[5] (1452–1519) was a leading researcher in mechanics of his time. He performed the decomposition of forces for investigation of moments (lever law). For individual examples he traced the law of the parallelogram of forces back to the lever law. These relations were clearly and precisely formulated by Varignon (1654–1722) and *Newton*[6] (1643–1727) not before the seventeenth century.

The dynamics of mass points was created in the seventeenth century. Since this is a very important period in the history of mechanics, we will outline it in more detail. It should be mentioned that the formal completion and mathematical treatment of mechanics happened in the eighteenth century and culminated in the mechanics of *Lagrange*[7] (1736–1813).

The seventeenth century was presumably the most decisive period in the history of physics, probably the birth of physics as an exact science per se. In that time mechanics was created and completed in outline, monumental in its scientific clarity and beauty, and convincing in its predictive power and mathematical formulation. The processes of motion in the heavens and on earth were described in a consistent way. Methodically in mechanics one succeeded for the first time in sharply conceptualizing experiences (experiments). This was achieved mainly by using mathematical language, combined with the invention of very fruitful abstractions and idealized cases about which one was now able to make precise and final statements. The implications of the new physical knowledge were quickly realized in public consciousness, as is demonstrated by the science methodics of *Bacon*[8] (1561–1626), *Jungius*[9] (1587–1657) and *Descartes*[10] (1596–1650).

The new scientific spirit did not concern mechanics only. Actually the first book on physics in the meaning of the new science stems from the English physician *Gilbert*[11] (1540–1603) on the magnet. He started from well-planned experiments, generalized them, and in this way came to general statements on magnetism and geomagnetism. This book strongly influenced the intellectual-scientific evolution of that age. The great scientists of that era knew it (*Kepler*[12] praised it, *Galileo*[13] used it). However, it was not in magnetism that physics made a great breakthrough, but dynamics. The following stages were important:

(1) Kepler interpreted the processes in the sky as physical phenomena;

(2) Galileo succeeded in the correct conceptual understanding of simple motions. He invented the abstraction of the "ideal case";

(3) *Huygens*[14] and Newton cleared up and completed the new concepts.

In the following table, we list the most important researchers and thinkers of that era:

1473–1543	Copernicus
1530–1590	Benedetti
1540–1603	Gilbert (1600 *De Magnete*)
1548–1603	Stevin
1561–1626	Bacon of Verulam
1562–1642	Galileo (1638 *Discourses on Fall Laws*)
1571–1630	Kepler (1609 *Astronomia Nova*)
1587–1657	Jungius
1592–1655	Gassendi
1596–1650	Descartes (1644 *Principia Philosophiae*)
1608–1680	Borelli
1629–1695	Huygens (1673 *Pendulum Clock*)
1635–1703	Hooke
1643–1727	Newton (1686 *Principia*).

According to the opinion prevailing in antiquity—presumably going back to Aristotle—heaven and earth were greatly separated from each other. The heavens represented the perfect, unchanging, divine. The earthly, on the contrary, was changing and chaotic. This opinion was considered to be confirmed by the circular orbits of the celestial bodies and the ideally straight earthly motions. The stars were interpreted as being essentially different from the earth, which presumably prevented the progress of the heliocentric world system which had been invented already in antiquity. Only *Copernicus*[15] (1473–1543) made this breakthrough. As a proof for his doctrine, which states that the sun is in the center of the world, he had to offer only the simplicity and beauty. This was not yet a "physics of the heaven," but rather a kind of geometrical ordering of the world.

Only Kepler presumably realized the connection of the motions in the sky with physics. He asked about the forces when he put the orbiting velocity of planets that decreases outward in a causal connection with a force originating from the sun and decreasing with the distance from it (1596). He justified the validity of the heliocentric system with the sun as the origin of the force which causes the planetary motions. In his *Astronomia Nova Seu Physica Coelestis* he expressed (1609) the first planetary laws. But his conclusion, a $1/r$-dependence of the gravitational force is false. He concluded from the area law

$$r^2 \dot{\varphi} = \text{constant}$$

that the velocity $r\dot{\varphi}$ perpendicular to the vector sun-planet is inversely proportional to the radius

$$r\dot{\varphi} = \frac{\text{constant}}{r}$$

and, therefore, the force should also show such a dependence, which, as stated already, is wrong. He related this reasoning with the propagation of the force in the planetary plane. Thus, Kepler understood the problem set by the planetary motion rather clearly, but he still did not yet have the conceptual and mathematical tools for describing the curved motion.

Kepler on the one hand was a rational astronomer and physicist who searched for the laws of motion of planets and for the reasons behind them; on the other hand he was an esthete who considered the world as an ordered entity. In the *Harmonices Mundi* (1619), he ascribed a particular role in the structure of the world to the five regular bodies. He used them as mathematical "archetypes." The use of mathematics for describing connections, as soon became customary, was denied him, however.

Galileo does not match to Kepler in the physical interpretation of planetary motion. He considered the circular orbit to be natural and did not yet understand the general meaning of the laws of inertia. *Borelli*[16] (1608–1680), on the contrary, qualitatively described the motions of celestial bodies as an interplay of the attraction by the sun and of the centrifugal force. He understood the planetary motion as a problem of theoretical mechanics. But to solve this problem essential tools still had to be developed; first of all free fall and the throw had to be understood.

The idea of the inertia of bodies, their persistence in the state of uniform motion in the absence of forces, was hard to understand. The abstraction was accepted only laboriously. On the contrary, it was easier, more natural, to ask for the reason of changes of position and to look for relations between force and velocity, as was already done by Aristotle in nebulous form. He took the view that the thrown body had a certain immaterial ability which, however, gradually decreased (*vis impressa natura—liter deficiens*). In the fourteenth century this ability was called "impetus" (Buridan, 1295–1336?). The impetus normally remains constant, but gravity and air resistance can modify it. A falling body is accelerated because more and more new impetus is added by gravity (*Benedetti,*[17] 1530–1590). Galileo also shared this view, but he at first tried to combine this understanding with the doctrine of the Aristotelian school that force and velocity were proportional to each other, in a complicated nontransparent way.

Only on the third day of the *Discourses on Fall Laws*, clarity about the course of the fall motions is achieved. The ideal cases of uniform motion and of uniformly accelerated motion are outlined and mathematically grasped. The uniformly accelerated motion is compared with the experience of free fall and the fall on the tilted plane. Finally, on the fourth day of the discourses the tilted throw is analyzed: it is correctly understood as a composition of an (ideal) propagating motion and an (ideal) fall motion.

Although Galileo described the law of inertia (persistence of bodies in uniform motion if no forces are acting on them) and the proportionality between force and acceleration, he did not express the general law of motion. He also did not apply his knowledge of the laws of falling bodies on planetary motion. From his entire work one realizes how reluctantly he gradually gave up the old ideas.

But the new ideas gained acceptance. In 1644, Descartes made the first attempt to formulate general laws of motion. He talked about the persistence of bodies in the state of rest or of motion, about linear motion as most natural motion (meaning force-free motion) and about the "conservation of motion" in the impact of bodies. The latter is obviously the conservation of momentum (called motion) $\sum_i m_i \mathbf{v}_i$. Descartes however did not realize the vector character of the momentum (the "motion"). Presumably, therefore, his applications of this law are wrong.

The conceptual completion of mechanics was achieved by Huygens when treating curved motions. He studied the motion of a body on a given path in the gravitational field of the earth (tautochrone problem) and demonstrated clear understanding of the centripetal and centrifugal force in the treatment of the pendulum clock (1673). He explained these topics by infinitesimal considerations as uniformly accelerated deviation from linear motion. He realized the proportionality between centrifugal force and (centrifugal) acceleration, where the acceleration is already defined infinitesimally. Huygens realized the momentum was a vector quantity, and thus he correctly interpreted the momentum conservation law, as can be seen from his applications of this law to impacts.

Then Newton came up with his brilliant work *Philosophiae Naturalis Principia Mathematica* (1686/87), and systematically showed the connection between mass, velocity, momentum, and force. He demonstrated by the example of the gravitation law how a force (measured by the change of the momentum) is determined by the locations of the involved bodies. Finally he applied the laws of mechanics to the treatment of planetary motion, and he showed how the gravitation law follows by induction from Kepler's laws, and how the Kepler laws follow deductively from the gravitation law. This represented the final breakthrough and the proof of the new mechanics; so to speak the completion of Kepler's quest.

After the preceding considerations, the question might arise why Kepler failed to discover the acceleration law—and, hence, gravity—although it follows seemingly simply from his own law. But it is not for us to accuse Kepler for this reason of a lack of brilliance and imagination. It is beyond any doubt that he had both—the genius in empirical research and the imagination in far-reaching speculations.[†] The explanation is as follows: Kepler was a contemporary of Galileo, who survived him by twelve years. Although Kepler knew the Galilean mechanics, in particular the central concept of acceleration, the laws of inertia

---

[†]For example, in his thoughts concerning the possible number of planets, since he was convinced—like the Pythagoreans—that God had taken the choice on number and proportions according to a certain number rule.

and of throwing by correspondence and by hearsay, he nevertheless could not work it out to a consistent structure. (Note that Kepler died in 1630, while Galileo's *Discorsi* outlining his mechanics was published only in 1638.) But even more decisive is the fact that the theory of curved motion—invented by Huygens for the circle and completed by Newton for general orbits—was not available to Kepler. But without the concept of acceleration for curved motions it is impossible to derive the form of the radial acceleration from Kepler's laws by simple mathematical manipulations.

Newton's mechanics of gravitation, which emerges from the dynamical fundamental law and from the reaction principle, is by its nature a further development of the throw motion discovered by Galileo. Newton writes about this point: "That the planets are being kept in their orbits by central forces is seen from the motion of thrown objects. A (horizontally) thrown stone under the action of gravity will be deflected from the straight path and falls, following a bent line, finally down to the earth. If it is thrown with greater velocity, it flies farther away, and so it might happen that it finally would fly beyond the borders of the earth and would not fall back. Hence, the missiles thrown away from the top of a mountain with increasing velocity would describe more and more extended parabolic curves and finally—at a certain velocity‡—return to the top of the mountain, and in this way move around the earth." An overwhelming argument—by conception and compelling logic!

The English physicist *Hooke*[18] (1635–1703), who is known as the founder of theory of elasticity, also came close to the gravitation law. This is shown by the following statements made by him:* "I shall develop a world system which in every way agrees with the known rules of mechanics. This system rests on three assumptions: (1) All celestial bodies have an attraction directed toward their center (gravity); (2) all bodies that are put in straight and uniform motion move in a straight line until they are deflected by any force and forced into a curved path; and (3) attractive forces are the stronger the closer they are to the body they are acting upon. Which are the various degrees of attractions I could not yet determine by

---

‡The value for horizontal throw is correctly given by Newton from $mv^2/R = mg$ as $v = \sqrt{gR} = 7900 \, \text{ms}^{-1}$; for the vertical shot into space the necessary velocity is obtained from the energy law

$$\frac{1}{2}mv^2 = \gamma \int_R^\infty \frac{mM}{r^2} dr = \gamma \frac{mM}{R}$$

with $g = \gamma M/R^2$ as

$$v = \sqrt{2gR} = 11200 \, \text{ms}^{-1}.$$

Both results do not involve the friction losses in the air.

*There are still two other hints on the many-sided active genius of Hooke: In 1665, he writes the prophetic words: "I have often thought that it should be possible to find an artificial, glue-like mass which is equal or superior to that excretion from which the silkworms produce their cocoon and which can be spun to threads by jets." This is the basic idea of the man-made fiber that—although two-and-a-half centuries later—revolutionized the textile industry! In the same year he writes, anticipating the mechanical theory of heat (hence, also kinetic gas theory): "That the particles of all bodies, as hard as they may be, nevertheless vibrate, one needs in my opinion no other proof than that, that all bodies involve a certain degree of heat and that never before has an absolutely cold body been found."

experiments. But it is an idea that must enable the astronomers to determine all motions of celestial bodies according to one law."

These remarks show that Newton did not create the monument of his *Principia* out of nothing. But it took an immense mental strength and bold ideas to concentrate all that created by Galilei, Kepler, Huygens, and Hooke in physics, astronomy, and mathematics into one focus, and in particular to announce that the force that lets the planets circulate on their orbits around the sun is identical with the force that drives the bodies on the earth to the floor.

For this knowledge, mankind needed one-and-a-half millennia if one considers that in the *Moralia* (*De facie quae in orbe lunae apparet*) Plutarch[19] (46–120) states that the moon by the momentum of its orbiting is just in the same way kept from falling to earth as a body which is "rotated around" in a sling. It took the genius of Newton to realize what the "sling" for the planets is!

The new mechanics had a tremendous impact on the spirit of that age. Now there existed a second incontestable science besides mathematics. Furthermore, the exact natural sciences were born: Mechanics advanced to their model.

Let us summarize once again the most important stages in the evolution of mechanics from a present-day view: The essential part of mechanics and of its fundamental concepts is expressed in the basic dynamic law

$$\frac{d\mathbf{p}}{dt} = \frac{d}{dt}(m\mathbf{v}) = \mathbf{F}.$$

Here, basically the acceleration appears as the signature of an acting force $\mathbf{F}$; the law of inertia, i.e., the conservation of momentum

$$\mathbf{p} = m\mathbf{v}$$

and, hence, of the velocity $\mathbf{v} = \dot{\mathbf{r}}$ if no external forces are acting, is also contained therein. This law of inertia had already been realized in the ancient and scholastic mechanics (Philoponos, Buridan) by experience. Uniform motion as the ideal case of motion was described by Galileo. Descartes clearly formulated the law of inertia, and Huygens utilized it correctly. As already stated, in the basic dynamic equation the acceleration appears as a differential quotient of the velocity. Huygens clearly realized that. He also correctly understood the acceleration as a measure of the force, as well as the role of the mass in the momentum. Newton summarized everything in a sovereign manner and applied the fundamental law to celestial mechanics. In this sense Newton is the endpoint of the way to mechanics to which besides him also Galileo and Huygens contributed essentially. The general concept, however, is due to Kepler.

For the history of mechanics, we also refer to the outlines of the history of physics. For the sections treated here, see particularly

E.J. Dijksterhuis, *Val en worp*, Groningen 1924.
E. Wohlwill, *Galilei*, Hamburg and Leipzig 1909 and 1926.

The most important original papers were translated to German in the collection: Ostwald's classics of exact sciences. The main works of Kepler and Newton are available in German as

J. Kepler, *Neue Astronomie oder Physik des Himmels* (1609). German translation. Munich 1929.

I. Newton, *Mathematische Principien der Naturlehre* (1686/87). German Translation. Leipzig 1872.

E. Mach, *Die Mechanik in ihrer Entwicklung historisch-kritisch dargestellt* (1933).

R. Dugas, *A history of mechanics*, Neuenburg/Switzerland, (1955).

P. Sternagel, *Die artes mechanical im Mittelalter* (1966).

F. Krafft, *Dynamische und statische Betrachtungweise in der antiken Mechanik* (1970).

For modern english editions and translations of historical texts, we refer to

J. Kepler, W.H. Donahue (Translator), *New Astronomy* (1992), Cambridge Univ. Press.

I. Newton, I.B. Cohen, A. Whitman (Translators), *The Principia: Mathematical Principles of Natural Philosophy* (1999), Univ. California Press.

J.L. Lagrange, A.C. Boissonnade, V.N. Vagliente (Translators), *Analytical Mechanics* (2001), Kluwer Academic Publishers.

Finally, we mention some texts about the history of mechanics

E.J. Dijksterhuis, *The Mechanization of the World Picture: Pythagoras to Newton* (1986), Princeton Univ. Press.

E. Mach, T.J. McCormack (Translator), *The Science of Mechanics: A Critical and Historical Account of its Development*, 6th edition (1988), Open Court Publishing Company.

R. Dugas, *A History of Mechanics* (1988), Dover Pub.

Notes

[1] *Plato*, Greek philosopher, b. 427 BC, Athens–d. 347, Athens, was the son of Ariston and Periktione, from one of the most noble families of Athens. According to legend, he wrote tragedies in his youth. The meeting with Socrates, whose scholar he was for 8 years, became decisive for his turn to philosophy. After Socrates' death (399), he first went with other scholars of Socrates to the town of Megara to study with Euclid. He then broadened his horizons by extended travel (first to Cyrene and Egypt). He soon returned home and opened the war against the educational ideal of the Sophists by his first works. He quickly won enthusiastic followers while dealing with science, secluded from public life. Presumably, scientific intentions led him about 390 to Italy, where he became familiar with the Pythagorean doctrine and school organization. He was introduced to the court of the tyrant Dionysius of Syracuse. Dionysius was at first much interested in him, but according to legend handed Plato over as a prisoner to the envoy of Sparta, who sold him as a slave. After payment for release and return to Athens, Plato founded the Academy in 387. Despite his bad experience, he set his hopes for a full effectiveness to Syracuse. In 368, he followed an invitation by Dion, uncle of the younger Dionysius, who hoped to win the young ruler over to Plato's political principles. Dionysius, however, only tolerated Plato's ideas for a short time. A third trip (361–360) also failed, since Dionysius distrusted him and turned

against him. Plato spent the last years in Athens in continuing scientific activity within a circle of well-known scholars; according to legend, he died during a wedding meal.

Plato's works are all preserved, except for the lecture *On the good*, which can be reconstructed only in broad terms. But not all work recorded under his name is authentic. The authenticity of the 7th Letter and of *Laws* is controversial.

The most important and surely authentic papers from the early period are: *Apology, Protagoras, State I, Gorgias, Menon, Kratylos*; from the middle artistic period: *Phaidon, Banquet, State II–X*; from the last years: *Phaidros, Parmenides, Theaithetos, Sophistes, Timaios, Philebos.*

Almost all of his writings are dialogues that by language and structure are of great artistic beauty. In most of them, Socrates appears as the main host of conversation.

Plato's philosophy turns dialectics, which for his teacher Socrates had only the negative function of destroying the false knowledge on the good and the virtue, into an approach of realizing the good and the virtue—into a path to the "ideas." The ideas are not acts of imagination but are the content of that being represented by them, which in itself is independent of us. By this distinction of the sensual (which is with us) from the hypersensual (the later "transcendent"), Plato became a promoter of the later so-called metaphysics.

Since the innermost nature of love is the will for perpetuation, it comes to fulfillment only as love of the eternal ideas. All other love is a preliminary stage to that. To kill off the transitory sensual and to turn toward the everlasting ideas is the aspiration of the really philosophical man. The way toward this goal is the dialectics. Also, the nature of this method of perception of ideas is logic.

Plato's understanding of the role of the idea varies between the general idea and the idea a priori. Provided it is the latter one, it is not brought into man from the outside, but he remembers it as something he already knows but has forgotten. The understanding of the idea is remembrance (*anámnesis*). The method of remembering is that of the hypothesis. By this, Plato means the proof in the form of the statement-logical conclusion: If the first, then the second. But now the first, therefore the second. Or: But now not the second, therefore not the first: e.g., in *Menon*: If virtue is knowledge, then it is teachable. But now it is knowledge, therefore it is teachable. But now it is not teachable, since there are practically no teachers of virtue. Therefore, it is not knowledge, but only correct opinion inspired by the Gods; to transform it into knowledge that is able to self-satisfy is according to Plato the essential duty of philosophy.

The later form of dialectics, the method of division (*diaeresis*) of the species into sorts, is the draft of a logical method of proof. Aristotle rightly interprets it as a prelude of the class-logical conclusion discovered by him. All proving is proving on conditions. These can be proved themselves. In the *State*, Plato sketches the idea of a completion of this proving up to the omission of all assumptions (*anyipódeton*), i.e., the idea of a proof by and from the purely logical. The absolute is defined here as the good. In his later works, Plato interprets the ideas as numbers, i.e., as units that include a manifold in themselves, and sees their absolute principle in the One and the "Great-and-small" (interpretation controversial).

Plato remains aware of the limits of all human proofs. Where the dialectics ends, there remains the speculative speech that uses the language of myth. All knowledge of the sensual, the nature, does not go beyond well-founded presumption. Therefore, all natural-scientific speech is necessarily myth. Plato develops this idea in his dialogue *Timaios*, which was of particular influence in the Middle Ages.

The question of what is for man the good and the virtue as the way toward this goal is answered by Plato by means of his dialectics of ideas, at first in the *State* by his doctrine of the four cardinal virtues: wisdom, bravery, prudence, and justice. The sketch of an ideal state serves only for proving this doctrine and does not represent a plan that should be realized. This "ideal state" with its subdivision into the three orders of scholastic profession, military profession, and peasantry, and his doctrine on the community of possessions and women, and on the necessity that the kings should become philosophers and the philosophers should become kings, later on was interpreted as a political program and became efficient. In the late work *Philebos*, Plato sees the good of human life in the composition of knowledge (*epistéme*) and joy (*hedoné*), where all knowledge is admitted, but among the joys only those that are not mixed with grief, those that cannot impair the knowledge. Men must be educated in the spirit of such a life ideal if a real and stable state shall be possible. This restoration program demands a radical restriction of the influence of the traditional literature on the individual and on the community. Plato's philosophy and critics of art were also of extraordinary historical influence.

Plato's doctrine, Platonism, was first developed further in Plato's school, the Academy. One distinguishes the older, intermediate, and younger Academy. In the older one, whose first and most important leaders were Speusippos and Xenocrates, the Pythagorean attitudes of the late philosophy of Plato were emphasized. The ratio of ideas and numbers became the focus of interest; soon mythological elements joined. On the contrary, the leading men of the intermediate Academy, Arcesilaos (315–241 BC) and Carneades (214–129), intended to revive the critical-scientific attitude of Plato. In this way, there emerged an—although moderate—scepticism that believes that only probable insight is possible. The younger Academy considers the power of reason again as more positive and combines thoughts of various systems in an eclectic manner, in particular Platonic and Stoic thoughts. To the younger Academy belong Philo of Larissa (160–79) and Antiochos of Ascalon († 68 BC), heard by Cicero in Athens. The Platonism of the three academies is called older Platonism. The transition from this one to the new Platonism is mediated by the "intermediate" Platonism, with the main representative Plutarch (AD 50–125), who taught a religious Platonism with a strong emphasis on the absolute transcendence of God and assumed a series of steps of intermediate beings between God and the world.

In the Middle Ages until the twelfth century, only *Timaios* was known and had a strong impact. In the twelfth century, Henricus Aristippus translated *Menon* and *Phaidon*, and in the thirteenth century, W. von Moerbeke translated *Parmenides*. The new Platonism had more influence than Plato's original ideas. The historical evolution of the philosophy of the Middle Ages was largely determined by the confrontation between Platonism and Aristotelian philosophy. In the early scholastic, Platonism was dominated mainly by Augustinus; in particular, the school of Chartres had a Platonic orientation. In the high scholastic, Platonism formed a strong undercurrent in the doctrines of the Aristotelians (Albert, Thomas). It emerged as an independent movement among the mathematical-scientific thinkers (Robert Grosseteste, Roger Bacon, Witelo, Dietrich of Freiberg) and among the German mystics. The latter established the link to the Platonism of the early Renaissance (Nicholas of Cues).

Modern Platonism began during the Italian Renaissance. In 1428, Aurispa brought the complete Greek text of Plato's works from Constantinople to Venice. Latin translations soon emerged, the most important one by Marsiglio Ficino, who completed it 1453 and 1483. Followers of Platonism included Lionardo Bruni and the older Pico Della Mirandola, as well as Byzantines who had fled to Italy, among them the two Chrysoloras, Gemistos Plethon and Bessarion. Central to this movement was the Platonic academy in Florence, founded in 1459 by Cosimo de Medici and guided by Ficino. From there, Platonism spread all over Europe. However, only in England did a truly Platonic school emerge (Cambridge). But Plato's thoughts had lasting effects in the rationalistic systems of Cartesius, Spinoza, and Leibniz. Malbranche was even called the "Christian Plato." In the nineteenth century, German idealism brought a revival of Plato's system of thought. Hegel resorted not only to Platonism but even more to Plotinus and the New Platonism. Plato's influence is seen in the recent past in the phenomenology of Husserl and in world philosophy. A.N. Whitehead explicitly confessed to Platonism. Although his statement that all of European philosophy is only a footnote to Plato is exaggerated, it nevertheless rightly points out the immense influence of Plato's philosophy. Platonism is even more dominant in the philosophy and theology of the Christian East, where the Platonic tradition of Origenes and of the Greek church fathers survives; for example, W. Solowjew and N. Berdjajew are Christian Platonics. [BR].

[2]*Aristotle,* Greek philosopher, b. 384 BC, Stagira in Macedonia–d. 322, Chalcis on Euboea. He came at the age of 18 to Athens and became a student of Plato, where he remained at the Academy for almost two decades, first as a scholar and then as a teacher, finally opposing Plato with his own philosophy. After Plato's death (347), he lived for three years in Asia Minor with Hermias, with the ruler of Atarneus. In 343, he was called to the court of Phillip of Macedonia to be the tutor of Philip's son Alexander. When Alexander ascended the throne, Aristotle returned to his hometown; however, in 334 he returned to Athens. In Athens, he founded the Peripataetic school, so called because of the covered walks (*peripatoi*) surrounding the lyceum. He taught there for twelve years among an ever-increasing circle of scholars, until the revolt of Athens after Alexander's death became dangerous to him, a friend of the royal dynasty. He went to his estate at Chalcis on Euboea, where he soon died. [BR].

[3]*Eudoxus of Cnidus* (400–347 BC), Greek scientist. He was equally active as mathematician, astronomer, and geographer. His biography is not recorded in detail, but it is considered certain that he was a member of the Platonic school. Later, he conducted his own school in Cyzicus. Eudoxus gave a new definition for proportion. He developed the exhaustive method and applied it to many geometric and stereometric problems and theorems, which he could prove exactly for the first time. Possibly, the major part of Euclid's twelfth book is the work of Eudoxus. Eudoxus made a map of the stars that remained top-ranking for centuries. He subdivided the sky into degrees of longitude and latitude, gave an improved value for the solar year, and improved the calendar. He estimated the earth's circumference to be 400,000 stadia, edited a new map of the known continents, and wrote a geography of seven volumes.

[4]*Archimedes,* outstanding mathematician and mechanic of the Alexandrian era, b. about 285, Syracuse, killed by a Roman soldier during the capture of Syracuse. Archimedes was

close to the Syracusean dynasty. He wrote important papers on mathematics and mathematical physics, fourteen of which are preserved. He calculated the area and circumference of a circle, the area and volume of segments of the parabola, the ellipse, spiral, the rotation paraboloid, the one-shell hyperboloid, and others, and determined the center of gravity of these figures. For π, he gave a value between 3 1/7 and 3 10/71; he developed in his "sand calculation" a method of exponential notation of arbitrarily large numbers, and in the *Ephodos,* a kind of integration calculus. Even more important than his treatment of the equilibrium conditions of the lever is the treatise on swimming bodies, where the principle of Archimedes is given. Archimedes determined the ratio of the volumes of the straight circular cone, the half-sphere, and the straight circular cylinder as $1:2:3$. Uncertain are the inventions of the water screw named after him and of the composed tackle; legendary is the burning of the Roman fleet by concave mirrors. [BR]

[5]*Leonardo da Vinci,* Italian painter, sculptor, architect, scientist, technician, b. April 15, 1452, Vinci near Empoli–d. May 2, 1519, in the castle Cloux near Amboise; illegitimate son of Ser Piero, notary in Florence, and of a peasant girl. He was educated in his father's house, and at the age of 15, he went to Florence as an apprentice of A. Verrocchio, who taught him not only painting and sculpting but also gave him an extensive education in the technical arts. In 1472, he was admitted to the Florentine guild of painters, but remained in Verrocchio's studio. In this time, of common work the earliest of his preserved works emerged: an angel and the landscape in Verrocchio's painting of the baptism of Christ (Florence, Uffizi), two preachings (Uffizi and Louvre), and the madonna with the vase (Munich, Pinakothek). About 1478, he became a freelancer and worked at Florence for about 5 more years. From this era stems the portrait of Ginevra Benci (Vaduz, gallery Liechtenstein), the unfinished painting of St. Jerome (Vatican), and the also unfinished great panel painting of the worship of the kings (Uffizi), which he got as an order for the high altar of a monastic church, but which he gave up half-finished when he left Florence at the end of 1481, to start work with Duke Lodovico of Milan.

The end of the Sforza dynasty forced Leonardo to leave Milan (1499). Through Mantua, where he drew a portrait of margravine Isabella d'Este (Louvre), and Venice, where he drew up a defense plan against the threatening invasion of the Turks, he returned in April 1500 to Florence, where he began the painting of St. Anna Selbdritt (Louvre). In May 1502, Leonardo started work as the first inspector of fortress buildings with Cesare Borgia, the papal military leader, throughout whose territory Leonardo traveled for about 10 months: the Romagna, Umbria, and parts of the Toscana. From this activity emerged a large fraction of his maps and city maps that—masterpieces of surveying and representation—belong to the earliest records of modern cartography. Florence also asked for his advice as a war engineer; he worked on a plan to divert the Arno river, in order to cut off the main access road to Pisa with whom Florence was at war, and he designed the project for a channel to make the Arno navigable from the sea to Florence. Both of these plans were not realized, just like the draft proposed at the same time by sultan Bajasid II to built a bridge 300-m long across the Bosporus. In 1503, Leonardo got the order to paint a monumental wall painting for the large senate hall of the Palazzo della Signoria in Florence; there, the drafts for the battle of Anghiari were born, which became the classical model in many copies, e.g., by Rubens for the cavalry-fight painting of Renaissance and Baroque, even to Delacroix.

At the same time, Leonardo painted the *Mona Lisa*, presumably the world's most famous painting, and the standing *Leda* (preserved only as copies). At this time, Leonardo reached the peak of his artistic fame. The arising geniuses of the young generation either admired him without envy (Raphael) or accepted him reluctantly with jealousy (Michelangelo). Later, he painted only hesitantly, and more and more turned to scientific problems. Besides mathematical studies, he studied anatomy comprehensively. He dissected corpses and began an extensive treatise on the structure of the human body, where he promoted the anatomical drawing accompanying the text as a tool for teaching. He also extended his biological and physical studies; the experiments on the flight of man—already begun in Milan—led him to investigate the flight of birds, which he also summarized like a treatise. Besides the laws of air flow, he also tried to investigate those of water flow. These studies contain approaches for theoretical and practical hydrology; he recorded them as materials for a treatise on water. In these years, he tried to arrange his notes by the main topics of his planned "books," which as a whole comprise a theory of the mechanical primordial forces of nature, i.e., an entire cosmology.

In 1506, Leonardo, at the request of the king of France released by the Florentine Signoria under the pressure of the political situation, stopped the work on the Anghiari battle and returned to Milan. There he served until 1513 mainly as an adviser to the French governor Charles d'Amboise, for whom he designed a large domicile and the plans for a chapel (S. Maria alla Fontana). From this era also date the drawings for the tomb of General Giangiacomo Trivulzio that—like the Sforza monument—was planned as an equestrian statue but was not realized. There is no clue to two almost finished madonna paintings for His Most Christian Majesty. Also, in Milan Leonardo mainly dealt with scientific studies. He continued his great "anatomy," in connection with the anatomist Marc Antonio Della Torre of Pavia, and he extended his hydrological and geophysical investigations both theoretically and practically, as is testified by his project for an Adda channel between Milan and the lake of Como, and by his amazing geological observations on the origin of fossils. He also returned to his botanical studies; also in this field, he created exact demonstration drawings according to exactly defined principles of graphical representation, as in all of his research activities. He thereby founded the scientific illustration.

When at the end of 1513 Leo X had risen to the papal throne, the now sixty-year-old Leonardo went to Rome, presumably with the hope of acquiring orders through his patron, Cardinal Giuliano de Medici. But he did not get big orders such as Raffael, Bramante, and Michelangelo got. His years in Rome were occupied by research, in particular on mechanics and anatomy. Only one painting, his last one, the mysterious *John the Baptist* (Paris, Louvre), may have originated in this period.

In January 1517, Leonardo left Rome, following the invitation of Franz I. The country castle Cloux near Amboise was allocated to him as a residence, and he got the title *Premier peintre, architecte et mechanicien du Roi*. He did not paint anymore, however, because of paralysis of his hand, but mainly arranged his scientific materials; in particular, he worked on completing his "anatomy." Among the few artistic creations of this last era, the project of a large castle and park for the residence of the Queen Mother in Romorantin is known. The building could not be built. His ideas nevertheless had a lasting effect on the tremendous castle that was begun by Franz I when Leonardo was still alive, the building of Chambord.

The most stirring documents of his late work are the drawings of the end of the world (Windsor), where Leonardo exhibited his experiences of life devoted to studying nature in a unique synthesis of scientific and artistic imagination; they are the symbol of the primordial forces penetrating the world, which once had created and finally shall destroy it, but even in self-destruction shall still obey the laws of harmony.

Leonardo was buried in Amboise in the church of St. Florentin, which was destroyed during the French Revolution. His pupil and friend Francesco Melzi became heir to his enormous written work, which is almost completely done in mirror writing, familiar to him as a left-hander.

The greatness of Leonardo and his significance in the history of occidental culture rests on the fact that he, like nobody else, understood art and science as a unity of human will of perception and power of mental comprehension. As a painter, he was the first master of the classic style; his few artistic creations remained models of perfection for all following eras and styles. As a researcher and philosopher, he is at the borderline between the Middle Ages and modern thinking. Altogether an empirist, he tried to acquire an encyclopedic knowledge by means of experience and experiment. Guided by his imagination and less capable of abstract logical thinking, Leonardo must not—as was often tried—be considered as the founder of modern science as such. His achievements in the field of physics and pure mechanics are mediocre, often even questionable. But, since he performed his all-embracing observations on natural phenomena with ultimate objectivity and by virtue of his artistic talent was able to represent them in drawings, he became the pioneer of a systematic descriptive approach in the natural sciences. Also, in the field of applied mechanics he can be considered as the founder of elementary engineering, for which he developed the graphical principles of demonstration. [BR]

[6]*Isaac Newton*, b. Jan. 4, 1643, Woolsthorpe (Lincolnshire)–d. March 31, 1727, London. Beginning in 1660, Newton studied at Trinity College in Cambridge, particularly with the eminent mathematician and theologian I. Barrow. After getting various academic degrees and making a series of essential discoveries, in 1669 Newton became the successor of his teacher in Cambridge. From 1672, he was a member and, from 1703, president of the Royal Society. From 1688 to 1705, he was also a member of Parliament, from 1696, attendant, and from 1701, mint-master of the Royal mint. Newton's life's work, besides theological, alchemistic, and chronological-historical writings, mainly comprises works on optics and on pure and applied mathematics. In his investigations on optics, he described light as a flow of corpuscles and in this way interpreted the spectrum and the composition of light, as well as the Newton color rings, diffraction phenomena, and double-refraction. His main opus *Philosophiae Naturalis Principia Mathematica* (printed in 1687) is fundamental for the evolution of exact sciences. It includes the definition of the most important basic concepts of physics, the three axioms of mechanics of macroscopic bodies, e.g., the principle of *actio et reactio*, the gravitation law, the derivation of Kepler's laws, and the first publication on fluxion calculus. Newton also dealt with potential theory and with the equilibrium figures of rotating liquids. The ideas for his big work mainly emerged in 1665 to 1666, when Newton left Cambridge because of the plague.

In mathematics, Newton worked on the theory of series, e.g., in 1669, the binomial series, on interpolation theory, approximation methods, and the classification of cubic curves and conic sections. But Newton could not remove logical problems even with his fluxion calculus that was represented in 1704 in detail. His influence on the further development of mathematical sciences can hardly be judged, since Newton disliked publishing. For example, when Newton made his fluxion calculus public, his method was already obsolete compared with the calculus of Leibniz. The quarrel about whether he or Leibniz deserved priority for developing the infinitesimal calculus continued until the twentieth century. Detailed studies have shown that both of them obtained their results independently of each other. [BR]

[7] *Joseph Louis Lagrange*, mathematician, b. Jan. 25, 1736, Torino–d. April 10, 1812, Paris, at the age of 19 years, professor of mathematics in Torino. In 1766, he followed the call of Friedrich the Great to the Berlin academy of sciences. After Friedrich's death, Lagrange moved to Paris as a professor at the École Normale. He invented the principle named after him. Important for function theory is his *Théorie des Fonctions Analytiques, Contenant les Principes du Calcul Différentiel* (1789), and for algebra and number theory his *Traite de la Résolution des Équations Numériques des Tous Degrés* (1798). In the *Mécanique analytique* (1788), he generalized and condensed the principles of mechanics to the systems of equations named after him. [BR]

[8] *Francis Bacon*, English philosopher and politician, b. Jan. 22, 1561, London–d. April 9, 1626, London, son of Nicholas Bacon, nephew of Lord Burleigh; advocate and deputy. In the notorious trial of his patron Essex, Bacon convicted Essex of high treason. In 1607, he became Solicitor General; in 1613, Attorney General; in 1617, Keeper of the Great Seal; and, in 1618, Lord Chancellor. Ennobled as Baron Verulam and Viscount of St. Albans, in 1621 Bacon was thrown out by parliament because of passive corruption and was sentenced to high penalties and imprisonment, which, however, was remitted by the king's influence. He was a curiously split character: outstandingly talented, vastly well read, vain, excessively ambitious, and of frightening emotional frigidity. The reasons for his downfall were not only the proven and confessed failures, but equally the anger of the parliament about the egocentric and unauthorized policy of the king who utilized Bacon as a submissive tool.

Bacon left a large number of philosophical, literary, and legal writings. His philosophical life's work, the *Instauratio Magna* (i.e., great revival of philosophy), remained a fragment, an attempt (based on insufficient means) of a complete reconstruction of sciences on the basis of "unfalsified experience." His main piece, *Novum Organum* (the title indicates the contraposition to Aristotle, whose logical writings traditionally were summarized under the title *Organon*), is a method of scientific research, worked out down to the last detail, which shall serve to snatch the secrets of nature and to govern it (Bacon considered knowledge as the means for a purpose, "knowledge is power"). The starting point for any knowledge is experience. Experience and mind should be tightly linked in a "legitimate marriage," instead of the separation so far. Bacon constructed a complicated system of scientific induction, but he failed to appreciate the role of mathematics. His main piece is preceded by an inventory of all sciences (*De diguitate et augmentis*), where—according to the three mental abilities memory, imaginative power, and mind—three main sciences are distinguished: history, poetry, and philosophy. Bacon listed what had been achieved by each science and what still remained to be done.

Among Bacon's literary works, the essays suggested by Montaigne are timeless: 10 in the first edition (1597), and 58 in the last edition (1625). In these "dispersed meditations," Bacon presented practical life's wisdom in the various fields, general guiding principles of the conduct of life, beyond good and evil, in an antithetical style of epigrammatic brevity, realistic and plain. "Nova Atlantis" is the perfect description of a philosophical ideal state.

Bacon's legal writings testify to his absolute mastery of the subject. His plan of codifying the English law of his age was not completed.

In the second half of the nineteenth century, Bacon was also considered to be the author of Shakespeare's dramas (Bacon theory). [BR]

[9] *Joachim Jungius*, philosopher and scientist, b. Oct. 22, 1587, Lübeck–d. Sept. 17, 1657, Hamburg, in 1609, professor in Giessen. In 1622, he founded in Rostock the first scientific society of Germany for the cultivation of mathematics and natural sciences. In 1624, he became a professor in Rostock; in 1625, in Helmstedt; and in 1628, headmaster of the Johanneum and of the academic high school in Hamburg. He defended the principle "improvement of philosophy has to originate from physics (= natural sciences)." Jungius decisively contributed to the breakthrough of scientific chemistry and the renewal of atomism. He was also important as a botanist. [BR]

[10] *René Descartes*, b. March 31, 1596, La Haye–d. Feb. 11, 1650, Stockholm. Descartes was the son of a councilor of the parliament of Bretagne and was educated in a Jesuit college. He then began the study of law, and beginning in 1618, he participated in various campaigns. Beginning in 1622, Descartes traveled in many countries of Europe, then settled down 1628 in the Netherlands, and lived from 1649 in Sweden as a teacher of philosophy. The mathematical main achievement of Descartes is the foundation of analytical geometry in his *Géometrie* (1637), which also essentially influenced the further development of infinitesimal calculus. [BR]

[11] *William Gilbert*, English scientist and physician, b. May 24, 1544, Colchester–d. Nov. 30, 1603, London. Gilbert was from 1573 a practicing physician in London; from 1601, the private physician of Elizabeth I; and after her death, of King James I of England. In his fundamental work *De magnete, magneticisque corporibus et de magnode magnets Tellure physiologia nova* (London, 1600; facsimile edition, Berlin, 1892; English translation and comment by S.P. Thompson, in *The collectors series in science*, 1958), Gilbert summarized the knowledge of older authors to an impressing doctrine of magnetism and geomagnetism, and added a number of new observations and findings. The work, which in the second book also involves a special chapter on *corpora electrica*, on substances that—like amber (electrum)—after rubbing are capable of attracting light bodies, impressed several of his contemporaries, among others Kepler and Galileo. His treatise *De monde nostro sublunari philosophia nova* appeared posthumous (Amsterdam, 1561). [BR]

[12] *Johannes Kepler*, b. Dec. 27, 1571, Weil der Stadt–d. Nov. 15, 1630, Regensburg. Kepler was the son of a trader who also often served in the military. He first went to school in Leonberg, and later to the monastic school in Adelberg and Maulbronn. From 1589, Kepler studied in Tübingen to become a theologian, but in 1599, he took the position of professor of mathematics in Graz that was offered to him. In 1600, because of the Counter-Reformation Kepler had to leave Graz and went to Prague. After the death of Tycho Brahe (Oct. 24, 1601), as his successor Kepler became the imperial mathematician. After the death

of his patron, Emperor Rudolf II, Kepler left Prague and in 1613 went to Linz as a land surveyor. From 1628, Kepler lived as an employee of the powerful Wallenstein, mostly in Sagan. Kepler died unexpectedly during a visit to the meeting of electors in Regensburg.

Kepler's main fields were astronomy and optics. After extraordinarily lengthy calculations, he found the fundamental laws of planetary motion: the first and second of Kepler's laws were published in 1609 in *Astronomia Nova*, and the third one in 1619 in *Harmonices Mundi*. In 1611, he invented the astronomical telescope. His Rudolphian tables (1627) continued to be one of the most important tools of astronomy until the modern age. In the field of mathematics, he developed the heuristic infinitesimal considerations. His best-known mathematical writing is the *Stereometria Doliorum* (1615), where, e.g., Kepler's barrel rule is given.

[13] *Galileo Galilei*, Italian mathematician, b. Feb. 15, 1564, Pisa–d. Jan. 8, 1642, Arcetri near Florence, studied in Pisa. At the Florentine Accademia del Dissegno, he got access to the writings of Archimedes. On the recommendation of his patron Guidobaldo del Monte, in 1589 he received a professorship for mathematics in Pisa. Whether or not he performed fall experiments at the leaning tower is not proven incontestably; in any case, the experiments had to prove his false theory. In 1592, Galileo took a professorship of mathematics in Padua, not because of disagreements with colleagues but for a better salary. He invented a proportional pair of compasses, furnished a precision mechanic workshop in his home, found the laws for the string pendulum, and derived the laws of falling bodies first in 1604 from false assumptions and then in 1609 from correct assumptions. Galileo copied the telescope invented one year earlier in the Netherlands, used it for astronomical observations, and published the first results in 1610 in his *Nuncius Siderus*, the "star message." Galileo discovered the mountainous nature of the Moon, the abundance of stars of the Milky Way, the phases of Venus, the moons of Jupiter (Jan. 7, 1610), and in 1611 the sunspots, although for these Johannes Fabricius preceded him.

Only beginning in 1610 did Galileo, who returned to Florence as Court's mathematician and philosopher to the grand duke, publicly support the Copernican system. By his overeagerness in the following years, he provoked in 1614 the ban of this doctrine by the pope. He was urged not to advocate it further by speech or in writing. During a dispute on the nature of the comets of 1618, where Galileo was completely right, he wrote as one of his most profound treatises the *Saggiatore* (inspector with the gold balance, 1623), a paper dedicated to Pope Urban VIII. Since the former cardinal Maffeo Barberini had been well disposed toward him, Galileo hoped to win him as pope for accepting the Copernican doctrine. He wrote his *Dialogo*, the "Talk on the Two Main World Systems," the Ptolemyan and the Copernican, gave the manuscript in Rome for examination, and published it 1632 in Florence. Since he obviously had not included the agreed-upon changes of the text thoroughly enough and had shown his sympathy with Copernicus too clearly, a trial set up against Galileo ended with his renunciation and condemnation on June 22, 1633. Galileo was imprisoned in the building of inquisition for a few days. The statement "It (the earth) still moves" (*Eppur si mouve*) is legendary. Galileo was sentenced to unrestricted arrest, which he spent with short breaks in his country house at Arcetri near Florence. There, he also wrote a work important for the further development of physics: the *Discorsi e Dimonstrazioni mathematiche*, the "conversations and proofs" on two new branches of science: mechanics (i.e., the strength of materials) and the science branches concerning local motions (falling and throwing) (Leiden 1638).

In older representations of Galileo's life, there are many exaggerations and mistakes. Galileo is not the creator of the experimental method, which he utilizes no more than many other of his contemporaries, although sometimes more critically than the competent Athanasius Kircher. Galileo was not an astronomer in the true sense, but a good observer; and as an excellent speaker and writer, he won friends and patrons for a growing new science and its methods among the educated of his age, and he stimulated further research. Riccioli and Grimaldi in Bologna confirmed Galileo's laws of free fall by experiment. His scholars Torricelli and Viviani developed one of Galileo's experiments—for disproving the "horror vacui"—to the barometric experiment. Christian Huygens developed his pendulum clock based on Galileo's ideas, and he transformed Galileo's kinematics to a real dynamics.

Galileo was one of the first Italians who used their native language for presentation of scientific problems. He defended this point of view in his correspondence. His prose takes a special position within the Italian literature, since it is distinguished by its masterly clarity and simplicity from the prevailing bombast that Galileo had reproved in his literary-critical essays on Taso et al. In his works *Il Dialogo sopra i due massimi sistemi* (Florence 1632) and *I Dialoghi delle nouve scienze* (Leiden 1638), he utilized the form of dialogue that came down from the Italian humanists, to be understood by a broad audience. [BR]

[14] *Christian Huygens*, Dutch physicist and mathematician, b. April 14, 1629, Den Haag–d. July 8, 1695, Den Haag. After initially studying law, he turned to mathematical research and published among other things in 1657 a treatise on probability calculus. At the same time, he invented the pendulum clock. In March 1655, he discovered the first moon of Saturn, and in 1656, the Orion nebula and the shape of Saturn's ring. By then, he was already familiar with the laws of collision and of central motion, but published them—without proof—only in 1669. In 1663, Huygens was elected a member of the Royal Society. In 1665, he settled in Paris as a member of the newly founded French academy of sciences, from where he returned in 1681 to the Netherlands. After publishing in 1657 the small treatise *Horologium* and in 1659 his *Systema Saturnium, sive de causis mirandorum Saturni phaenomeno*, in 1673 emerged his main work: *Horologium oscillatorium* (the pendulum clock), which besides the description of an improved watch construction contains a theory of the physical pendulum. Further one finds treatises on the cycloid as an isochrone, and important theorems on central motion and centrifugal force. From 1675 dates Huygens's invention of the spring watch with a balance spring, from 1690 the *Tractatus de lumine* (treatise on light), which contained a first version of the wave theory (collision theory) of light, and based on that, the theory of double refraction of Iceland spar is developed. The spherical propagation of action around the light source is explained there by means of Huygens' principle. [BR]

[15] *Copernicus*, Coppernicus, German Koppernigk, Polish Kopernik, Nikolaus, astronomer, and founder of the heliocentric world system, b. Feb. 19, 1473, Thorn–d. May 24, 1543, Frauenburg (East Prussia). Beginning in 1491, he engaged in humanistic, mathematical, and astronomical studies at the university in Cracow. From 1496 to 1500, he studied civil and clerical law in Bologna. At the instigations of his uncle, bishop Lukas Watzelrode, in 1497 he was admitted to the chapter of Ermland at Frauenburg, but he took only the lower holy orders. In Bologna, he continued his astronomical work together with the professor of

astronomy Dominico Maria Novarra, made a short stay in Rome, and in 1501 temporarily returned to Ermland. Beginning in the autumn of 1501, he studied in Padua and Ferrara, graduating on May 31, 1503, as a doctor of canonical law, and then studied medicine. After returning home in 1506, he lived in Heilsberg as secretary to his uncle from 1506 until his death in 1512. He was involved in administrating the diocese of Ermland, and he accompanied his uncle to the Prussian state parliaments and the Polish imperial parliament. As chancellor of the chapter, Copernicus after 1512 lived mostly in Frauenburg, resided as governor of the chapter (1512–1521) in Mehlsack and Allenstein, and in 1523 was administrator of the diocese of Ermland. As a deputy, he represented the order chapter (1522–1529) at the Prussian state parliaments and there particularly supported monetary reform.

Contrary to Polish claims, Copernicus's German origin is established (the paternal family stems from the diocesan country Neiss in Silesia). Like his elder brother Andreas, he defended the concerns of Ermland against the Crown of Poland; both in writing and in oral speech, he utilized only German and Latin. Besides his administrative work, he also practiced as a physician. Toward the end of his life, in 1537, Copernicus had differences with Johannes Dantiscus, the newly elected bishop of Ermland.

As an astronomer, Copernicus completed what Regiomontan had imagined: a revision of the doctrine of planetary motion, taking into account a series of critically evaluated observations. Only on such a basis could one then think of reforming the calendar. The urgency of that reform was generally recognized at the beginning of the sixteenth century. Copernicus presumably was influenced by these considerations. In the course of his work, he then decided to accept a heliocentric world system, inspired by ancient writings. A brief, preliminary report on this topic is the *Commentariolus*, presumably written before 1514. Here, already the decisive assumptions are expressed: The sun is in the center of the planetary orbit—still considered as circular—and the earth circulates about the sun; the earth daily rotates about its axis and in turn is orbited by the moon. The wider public got the information on the Copernican doctrine only by the *Narratio prima* of Georg Joachim Rheticus (first report on the six books of Copernicus on the circular motions of celestial paths, 1540, German, by K. Zeller, 1943).

The main work of Copernicus, the "Six Books on the Orbits of Celestial Bodies" (*Die revolutionibus orbium coelestium libri VI*, 1543; German, 1879; new edition, 1939), was published in the year of death of the author. It was dedicated to Pope Paul III, but instead of the original foreword of Copernicus, it was introduced with a foreword by the Protestant theologian Andreas Oslanden that inverted the meaning of the whole subject. The doctrines of Copernicus remained uncontested by the church until the edict of the index congregation of 1616. The imperfections of the Copernican theory of planets were removed by Johannes Kepler. [BR]

[16] *Giovanni Alfonso Borelli*, b. 1608, Naples–d. Dec. 31, 1679, Rome, physicist and physiologist. In 1649, Borelli became a professor of mathematics in Messina and in 1656 moved to Pisa. In 1667, he returned to Messina. In 1674, he was forced to leave to Rome. There he lived until his death under the patronage of Christina, Queen of Sweden. His best-known work, *De motu animalium*, deals with the motions of the body of animals, which he traced back to mechanical principles. In a letter published in 1665 under the pseudonym Pier

Maria Mutoli, Borelli first expresses the idea of a parabolic trajectory of comets. Among his numerous astronomical works, there is also *Theoretica mediceorum planetarum ex causis physicis deducta* (Florence, 1666), treating the influence of the attracting force of Jupiter's moons on the orbital motion of Jupiter.

[17] *Giovanni Battista Benedetti*, b. Aug. 14, 1530–d. Jan. 20, 1590, Torino, first to recognize the buoyancy action of the surrounding medium in free fall. Taking up the ideas of Archimedes, he writes in his work *De resolutione omnium Euclidis problematum* (Venice 1553) that the fall velocity shall be determined by the difference of the specific weights of the falling body and the medium.

[18] *Robert Hooke*, English researcher, b. July 18, 1635, Freshwater (Isle of Wight)–d. March 3, 1703, London. Hooke was at first an assistant to R. Boyle; then from 1665 a professor of geometry at Gresham College in London; and from 1677 to 1682, secretary of the Royal Society. Hooke improved already-known methods and devices, e.g., the pneumatic pump and the composite microscope (described in his *Mikographia*, 1664). Hook was often involved in questions on priority, e.g., with Huygens, Hevelius, and Newton. He proposed, among others, the melting point of ice as the zero point of the thermometric scale (1664), recognized the constancy of the melting and boiling point of substances (1668), and for the first time observed the black spots on soap bubbles. He gave a conceptually good definition of elasticity and in 1679 established Hooke's law. [BR]

[19] *Plutarch* (Greek Plutarchos), Greek philosopher and historian, b. about AD 50, Chäronea, from an old bourgeois family–d. about 125. He was educated in 66 in Athens by the academician Ammonios to be a follower of Plato's philosophy. He visited, among other cities, Alexandria and Rome, where he had contact with prominent Romans, but lived permanently in his small hometown. There, he participated in communal politics; in Delphi he became a priest about 95. Plutarch was honored by the emperors Trajan and Hadrian. However, his life center remained in the vicinity of his homeland, where he closely associated with family and a circle of friends. This milieu feeds the ethos of education and the national pathos of his numerous writings. Despite the wealth of topics and the abundance of material, they nevertheless display an inner unity from the pleasant integrity of a personality formed by philosophy and religion.

Plutarch's works consist of two groups. The first group contains the biographies (*Vitae parallelae*), 46 comparative life descriptions of famous Greeks and Romans (e.g., Pyrrhos-Marius, Agesilaos-Pompeius, Alexander-Cesar). Great Romans were presented to the Greeks, in parallel and in contrast to their own history. Thereby an important step toward inner balance of the double-culture of the Greek-Roman era of emperors was made. The literary appeal of the biographies rests on the vivid representation and the description of characters, supported by memorable anecdotal features.

The second group, *Moralia*, contains popular, ethical-educational writings, but also strictly philosophical, metaphysical, religious-philosophical investigations, learned antiquarian studies, political treatises, etc. This writing was borne by the tradition of the Platonic school, without schoolmasterly narrowness, with vivid religious emphasis.

While Plutarch was read in the Byzantine empire, he was unknown to the West in the Middle Ages. In the early fifteenth century (Guarino and his scholars, then Pier Decembrio, L. Bruni), Plutarch's works were translated to Latin; from 1559 to 1572, by J. Amyot in classical form into French (fertilizing influence on the dramatic art in France in the seventeenth century); and in 1579, by North into English (influence on Shakespeare). In Germany, for a long time Plutarch was appreciated only by learned circles. Only toward the end of the eighteenth century was the interest turned again to him (Schiller, Goethe, Jean Paul, Beethoven, and Nietzsche). [BR]

Recommendations for further reading on theoretical mechanics

The textbooks on theoretical mechanics listed below represent only part of the wealth of excellent literature on this topic.

Classical textbooks on mechanics:

H. Goldstein: *Classical Mechanics*, 3rd edition (2001), Addison-Wesley Pub. Co.

A. Sommerfeld: *Mechanics (Lectures On Theoretical Physics Vol. 1)*, 4th edition (1964), Academic Press.

L.D. Landau and E.M. Lifschitz: *Mechanics*, 3rd edition (1982), Butterworth-Heinemann.

Problems and exercises for classical mechanics:

M.R. Spiegel: *Theory and Problems of Theoretical Mechanics* (Schaum's Outline Series), SI-edition (1980), McGraw-Hill.

More mathematical presentations of mechanics:

F. Scheck: *Mechanics: From Newton's Laws to Deterministic Chaos*, 3rd edition (1999), Springer.

J.B. Marion and S.T. Thornton: *Classical Dynamics of Particles and Systems*, 4th edition (1995), Saunders College Publishing.

We consider the work of H. Goldstein to be particularly suited as an addendum. Starting from the elementary principles, he outlines the formal Hamilton–Jacobi theory in a didactically brilliant manner. All typical applications (central force problem, rigid body, vibrations etc.) are discussed and expanded on in exercises and by special recommendations for further reading. The lectures by A. Sommerfeld, planned in a similar way, represent a gold mine because of the treatment of many special problems and the imaginative power demonstrated in the mathematical solution techniques.

Readers may gain an appreciation of the formal esthetics of the volume *Mechanics* from the textbook by Landau and Lifschitz.

Index

acceleration
    centripetal, 8, 11
    Coriolis, 8, 35
    gravitational, 11
    linear, 8
action function, 400
    Hamilton, 386
action quantum, Planck, 410
action variable, 391
action waves, 401
air resistance, 330
amplitude modulation, 89
angle variable, 391
angular frequency, 110
angular momentum, 67
angular velocity, 5, 7
anticyclone, 34
approximation, successive, 14
Arnold tongues, 493
asymptotic stability, 427
attractor, 423
    chaotic, 425, 464
    strange, 425, 467, 501
attractor diagram, 481, 484, 499
Atwoods fall machine, 77
autonomous system, 419, 421, 449

baker transformation, 466, 473
bascule bridge, 284
basin of attraction, 425

beat vibrations, 86
Bernoulli shift, 474, 486
Bessel functions, 152
Bessel's differential equation, 149
bifurcation, 452
    static, 452
    subharmonic, 459
    time-dependent, 457
bifurcation cascade, 480, 482
billiard ball, 184
body, rigid, 39, 165
boundary condition
    Dirichlet, 126
    periodic, 149
brachistochrone, 357
branching, 452
    Hopf, 456
    pitchfork, 454, 477
    saddle-node, 453
    transcritical, 455
butterfly effect, 462

canonical transformation, 380
Cantor set, 467, 471
capacity dimension, 470
Cardano formula, 118
casus irreducibilis, 122
catenary, 355
center of gravity, 43
    circular cone, 47

center of gravity (*cont.*)
  pyramid, 45
  semicircular disk, 46
center-of-mass coordinates, 72
central field, scattering in, 51
central forces, 68
central motion, 346
centrifugal force governor, 335
centripetal acceleration, 8, 11
chain, vibrating, 90
Chandler period, 226
chaos, 417
chaotic attractor, 425, 464
characteristic equation, 196
charged particle, 327
Chasles' theorem, 41
Chirikov criterion, 508
circular mapping
  dissipative, 491
  one-dimensional, 491
cluster property, 44
collision parameter, 53
conditionally periodic, 159
cone pendulum, 30
configuration space, 364
constraints, 271, 274
  generalized, 316
  holonomic, 271
  nonholonomic, 271, 314
  rheonomic, 271
  scleronomic, 271
continuity equation, 367
continuum, 39
contracting flow, 423
control parameter, 419
cooling
  of particle beam, 374
  stochastic, 369
coordinates
  generalized, 271, 274, 275, 420
  ignorable (cyclic), 291
coordinate system
  body-fixed, 192
  center-of-mass, 54, 72
  rotating, 3, 7
  for rotating earth, 10
Coriolis acceleration, 8, 35
Coulomb scattering, 57

coupled mass points, vibrations, 83
coupled pendulums, 87
couple of forces, 165
critical point, 425
cross section
  differential, 52
  Rutherford, 56
cycloid, 135, 306
cyclone, 34

D'Alembert principle, 279
deformable medium, 39
degeneracy, 141
degrees of freedom, 41
determinant of coefficients, 93
deterministic, 419, 462
deviation moments, 168, 192
devil's staircase, 493
differential cross section, 52
differential principles, 351
dimension
  broken, 465, 467
  capacity, 470
  fractal, 469
  Hausdorff, 472
  information-, 473, 503
  Lyapunov, 503
Dirichlet conditions, 126
discrete systems, 475
discretization, 444, 489
dispersion law, 110
displacement, virtual, 279
dissipation, 420, 421, 423
dissipation function, 328
dissipative circular mapping, 491
double pendulum, 303
dynamical system, 419

earth, nutation of, 225
eastward deflection, 13, 18
eclipse, 241
ecliptic, 236
eigenfrequency, 84, 141
eigenmodes, orthonormal, 158
eigenvalue, 195
eigenvector, 195
eigenvibrations, 84
electromagnetic field, 327

ellipsoid, rotating, 227
ellipsoid of inertia, 203
    quadratic disk, 210
    regular polyhedron, 226
energy law, 69
equation, characteristic, 196
equilibrium solution, 425
ergodic system, 463, 488
Euler angles, 249
Euler equations, 222
Euler–Lagrange equation, 355
externally excited system, 443

fall machine, Atwoods, 77
Feigenbaum constant, 481
figure axis, 216
fixed point, 425, 464, 475
    unstable, 427
Floquet multiplier, 448, 450
Floquet's theory of stability, 447
flow of a vector field, 420
force
    generalized, 277
    Lorentz, 324
    nonconservative, 328
Foucault's pendulum, 23
Fourier coefficients, 126
Fourier series, 125
fractal, 466
fractal dimension, 469
free fall, 10
frequency spectrum, (an)harmonic, 141
friction force, 214
friction parameter, 496, 501, 503
friction tensor, 329
function
    even (odd), 127
    homogeneous, 344
fundamental harmonic, 141

galaxy, flattening of, 68
generating function, 381
gravitational acceleration, 11
gyrocompass, 240
gyroscope, 238

Hamiltonian, 342
Hamiltonian mechanics, 420

Hamilton–Jacobi theory, 386
Hamilton principle, 351
Hamilton's equations, 341, 343, 363
harmonic oscillator, 384
Hausdorff dimension, 472
herpolhodie, 219
hockey puck, 182
holonomic, 271, 293, 330
Hopf branching, 456
hydrogen atom, 411
Hyperion, 503

inertial system, 3
information dimension, 473, 503
integrability condition, 272
integral principles, 351
iteration method, 14
iterative mapping, 475

Jacobi matrix, 426, 449

Kaplan–Yorke relation, 503
Kepler problem, 393
Kepler's law, 506
kinetic energy, 73
    of a rotating rigid body, 193
Koch curve, 467, 471

laboratory system, 3
Lagrange equations, 289
    nonholonomic constraints, 314
Lagrange multipliers, 314
Lagrangian, 289
Legendre transformation, 341
libration, 497
limit cycle, 432, 434, 440, 449, 458, 464
linearization, 425
Liouville theorem, 365, 421
Lissajous figure, 160
logistic mapping, 477, 481, 486
Lorentz force, 324
Lorenz, 462
Lyapunov dimension, 503

Lyapunov exponent, 461, 466, 510
   logistic mapping, 484, 488
   maximum, 463, 500

Mandelbrot, 465
many-body system, 66
mapping
   iterative, 475
   logistic, 477, 481, 486
   one-dimensional, 460, 476
   Poincaré, 446, 460, 475, 496
   stroboscopic, 444
mass, reduced, 74
mass density, 45
membrane
   circular, 146
   rectangular, 138
   vibrating, 136
Mercury, 508
mode locking, 492, 495
molecule
   asymmetric linear, 302
   three-atom, 297
   triangular, 300
momentary center, 49
moment of inertia, 166, 504
   circular cylinder, 168
   cube, 173
   rectangular disk, 170
   rigid bodies, 179
   sphere, 172
momentum
   generalized, 289
   linear, 66
momentum space, 364
monodromy matrix, 447
motion, central, 346

neutron star, vibrating, 229
Newton's equations
   arbitrary relative motion, 8
   rotating coordinate system, 3, 7
nodal line, 142
node, 100, 428, 454
nonconservative forces, 328
nonholonomic, 314, 331
nonlinear dynamics, 417
normal frequencies, 86, 302

normal vibrations, 84, 109
North Pole
   geometrical, 226
   kinematical, 226
nutation, 220
nutation cone, 219
nutation of earth, 225

one-dimensional mapping, 460
orbit, 420
orthogonality relation
   for trigonometric functions, 126
oscillator
   harmonic, 384
   nonlinear, 430
   Rayleigh, 433
   van der Pol, 433, 438
overtone, 111, 141, 154

parabola, Neil, 402
paraboloid, 319
path stability, 442
Peano curve, 471
pendulum
   cone, 30
   coupled, 87
   double, 303
   Foucault, 23
   periodically driven, 496
   physical, 171
   plane, 364
   rolling, 175
   string, 308
   theories, 346
   upright, 294, 295
pendulum length, reduced, 175
pendulum watch, 492
period, Chandler, 226
period-doubling, 459, 477, 478, 497
periodic attractor, 492
periodic solution, 443
perturbation calculation, 12
phase diagram, 364
phase flow, 420, 423
phase integral, 391
phase space, 364
phase-space density, 368
phase trajectory, 364

phase velocity, 110
physical pendulum, 171
pirouette, 69
pitchfork branching, 454, 477
Pivot forces, 230
Planck's constant, 411
plane, invariable, 218
Poincaré, 417, 462
Poincaré–Bendixson, theorem, 437
Poincaré cut, 445, 501, 508
Poincaré mapping, 446, 460, 475, 496
Poincaré recurrence time, 155
Poinsot ellipsoid, 217
point attractor, 477
Poisson bracket, 413
pole cone, 220
pole curve, 49
pole path, 49
pole trajectory, 219
polhodie, 219
potential
    generalized, 325
    scalar, 327
    vector, 327
    velocity-dependent, 324
precession, 235
    stationary, 259
precession velocity, 259
principal axes of inertia, 194
principal axis, 211
projectile, 52, 330

quantum hypothesis, 410
quantum number, 109
quasiperiodic, 491, 494, 508

Rayleigh oscillator, 433
recurrence time, Poincaré, 155
reduced mass, 74
relaxation vibration, 437, 439
resonances, 508
rheonomic, 271, 293
river, superelevation of bank, 19
rolling pendulum, 175
rosette path, 28
rotating coordinate system, 3, 7
rotating ellipsoid, 227
rotating tube, 35

rotation, 41
rotation about a fixed axis, 165
rotation about a point, 190
rotational velocity, 7
rotation energy, 193
rotation matrix, 201, 251
rotator, periodically kicked, 489
rotor, 428
Rutherford scattering, 56

saddle-node branching, 453
saddle point, 428, 454
Saros cycle, 241
Saturn, 503
scattering, in a central field, 51
scattering cross section
    Rutherford, 56
    square well potential, 60
scattering experiment, 52
scattering of two atoms, 64
scleronomic, 271
sea level, 20
secular equation, 426
self-exciting vibration, 437
self-similarity, 468, 484, 485, 494
Sierpinski gasket, 469
similarity transformation, 202
slant throw, 398
sleeping top, 259
solar system, 78
sound velocity, 110
spin, 255
spiral, 428, 430
stability, 425
    asymptotic, 427, 442
    orbital, 443
    time-dependent paths, 442, 446
stability of paths, 446
staggering motion, 503, 510
standard mapping, 491
Steiner's theorem, 169
stochastic cooling, 369
straight line, invariable, 218
strange attractor, 425, 467, 501
stretching and folding, 466, 486
string, vibrating, 105, 112
string pendulum, 308
string tension, 24, 105, 274, 280

stroboscopic mapping, 444
subcritical branching, 455
subharmonic bifurcation, 459
subharmonic cascade, 499
supercritical branching, 455
superposition principle, 109
superstability, 483, 484, 494
surface density, 47
symmetry axis, 211
symmetry-breaking bifurcation, 500
synchronization, 492
system of mass points, 66
system of principal axes, 194

target, 52
tautochrone problem, 133
tensor, 201
tensor of inertia, 192
    square, 198
    three mass points, 212
theorem of Chasles, 41
theorem of Liouville, 364
theorem of Steiner, 169
theory of chaos, 462
theory of stability, Floquet, 447
theory of top
    analytical, 220
    geometrical, 217
tidal forces, 241
tidal friction, 507, 508
top
    asymmetric, 267, 311
    free, 216
    heavy, 233, 261
    oblate, 216
    prolate, 216
    rolling circular, 207
    sleeping, 260
    spherical, 197, 216
    symmetric, 197
top moment, 236
torque, 165
    elliptic disk, 232
    of rotating plate, 228

torus, 464
total time derivative, 415
trace cone, 219
trace trajectory, 219
trajectory, 420, 497
transcritical branching, 455
transformation
    canonical, 380
    to center-of-mass coordinates, 72
    of kinetic energy, 73
transients, 423, 484
translation, 41
transverse cut, 445
tube, rotating, 293
turbulence, 417

van der Pol oscillator, 433, 438
variational problem, 351, 354
vector field, 419
vector product, 5
velocity
    angular, 7
    generalized, 277
    true, 7
    virtual, 7
velocity-dependent potential, 324
velocity field, 419, 422
vibrating chain, 90
vibrating membrane, 136
vibrating string, 105, 112
vibration, self-exciting, 437
vibration antinode, 100
vibrations of coupled mass points, 83
virtual displacement, 279, 516
virtual forces, 8
virtual work, 280, 281, 332, 516
volume density, 44
Voyager, 503, 510

wave equation, 107
wavelength, 110
wave number, 110
winding number, 492
work, virtual, 279, 281, 332, 516